Plant and Soil Science

Fundamentals and Applications

Join us on the web at

agriculture.delmar.com

Plant and Soil Science

Fundamentals and Applications

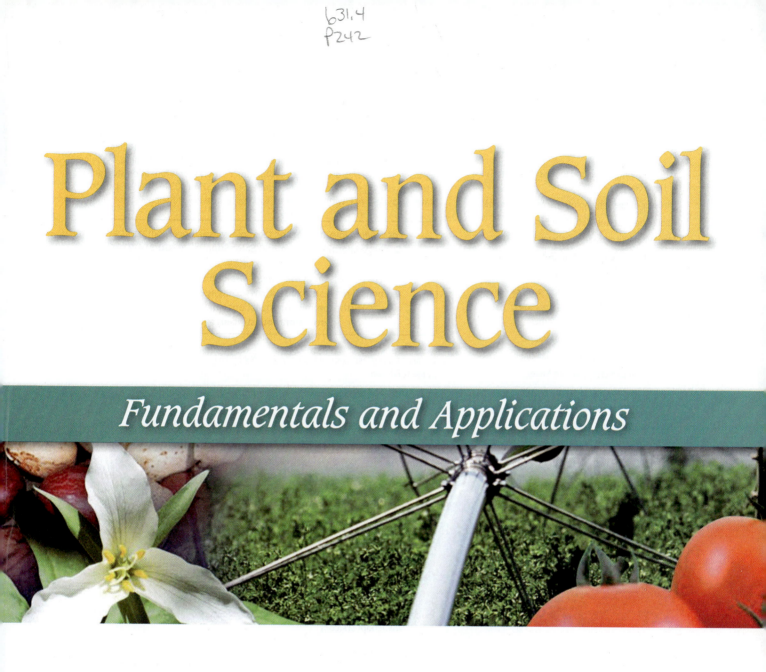

By Rick Parker

DELMAR
CENGAGE Learning

Australia • Brazil • Japan • Korea • Mexico • Singapore • Spain • United Kingdom • United States

Plant & Soil Science: Fundamentals and Applications
Author: Rick Parker

Vice President, Career and Professional Editorial:
  Dave Garza

Director of Learning Solutions: Matthew Kane

Acquisitions Editor: Benjamin Penner

Managing Editor: Marah Bellegarde

Product Manager: Christina Gifford

Editorial Assistant: Scott Royael

Vice President, Career and Professional Marketing:
  Jennifer McAvey

Marketing Director: Debbie Yarnell

Marketing Manager: Erin Brennan

Marketing Coordinator: Jonathan Sheehan

Production Director: Carolyn Miller

Production Manager: Andrew Crouth

Content Project Manager: Andrea Majot

Art Director: David Arsenault

Technology Project Manager: Tom Smith

Production Technology Analyst: Thomas Stover

For product information and technology assistance, contact us at
Cengage Learning Customer & Sales Support, 1-800-354-9706

For permission to use material from this text or product,
submit all requests online at **www.cengage.com/permissions**
Further permissions questions can be emailed to
permissionrequest@cengage.com

Library of Congress Control Number: 2008939361
Student Edition

ISBN 13: 978-1-4283-3480-9
ISBN 10: 1-4283-3480-7

Delmar
5 Maxwell Drive
Clifton Park, NY 12065-2919
USA

Cengage Learning is a leading provider of customized learning solutions with office locations around the globe, including Singapore, the United Kingdom, Australia, Mexico, Brazil, and Japan. Locate your local office at:
international.cengage.com/region

Cengage Learning products are represented in Canada by
Nelson Education, Ltd.

To learn more about Delmar, visit **www.cengage.com/delmar**
Purchase any of our products at your local college store or at our preferred online store **www.ichapters.com**

Notice to the Reader
Publisher does not warrant or guarantee any of the products described herein or perform any independent analysis in connection with any of the product information contained herein. Publisher does not assume, and expressly disclaims, any obligation to obtain and include information other than that provided to it by the manufacturer. The reader is expressly warned to consider and adopt all safety precautions that might be indicated by the activities described herein and to avoid all potential hazards. By following the instructions contained herein, the reader willingly assumes all risks in connection with such instructions. The publisher makes no representations or warranties of any kind, including but not limited to, the warranties of fitness for particular purpose or merchantability, nor are any such representations implied with respect to the material set forth herein, and the publisher takes no responsibility with respect to such material. The publisher shall not be liable for any special, consequential, or exemplary damages resulting, in whole or part, from the readers' use of, or reliance upon, this material.

Printed in the United States of America
1 2 3 4 5 6 7 12 11 10 09

Table of Contents

Part 2: Fundamentals of Soil Science 85

Chapter 5: Soil Materials and Formation 86

Chapter 6: Soil Characteristics, Classifications, and Use 109

Chapter 7: Soil Fertility and Management 131

Preface

Photosynthesis is the "reaction of life." Through photosynthesis, plants convert the sun's energy into a form that humans and other living creatures can use. To do this, they use carbon dioxide and produce oxygen. Photosynthesis is the original source of all important fuels, including oil, coal, wood, and natural gas. It is also the source of all foods, because all animals depend on plants at some point in the food chain. To work the magic that is photosynthesis, plants need a medium for their roots. Most often this is the soil, so to teach the science of using plants for human needs, an understanding of the soil is necessary.

As an author, my goal was to create a book that is useful and easy to use for both teaching and learning. To achieve this, a detailed outline guided the development of the book and the order of the chapters in the book. For each chapter, the outline became the chapter headings. Because these headings are in bold type and because the outline for each chapter is detailed, the book is easy to read and information can be found quickly.

Each chapter of the book features learning objectives, key terms (defined in the glossary), tables, charts, illustrations, and color photographs to aid the learning and teaching process. Learning objectives at the beginning of each chapter highlight the important concepts. The review questions at the end of each chapter are tied to the learning objectives to assess if the objectives were covered.

To further guide the learner and organize the 28 chapters, the book is divided into four major parts: (1) Introduction to Plants, (2) Fundamentals of Soil Science, (3) Fundamentals of Plant Growth and Propagation, and (4) Crops—Applied Plant and Soil Science.

The four chapters in Part 1 set the stage for learning plant science by covering the organization of the plant kingdom, the origins of cultivated plants, and the general structure and anatomy of cultivated plants. The anatomy of plants is discussed in the hierarchy of biochemicals, cells, tissue, and organs.

Six chapters in Part 2 provide a solid understanding of soils and other media for growing plants. These chapters cover soil materials and formation, land use and classification, soil fertility and management, soilless plant production, irrigation, and soils as related to organic farming.

Nine chapters in Part 3 include temperature, light, photosynthesis, and respiration and their role in plant growth and development. Chapters on plant growth at the cellular level and vegetative growth logically follow. Next, cultivated plants would be of little use to humans if they could not be propagated on a large scale; thus, the chapter on propagation details sexual and asexual propagation methods. Growing plants for human use also requires that plant pests be recognized and controlled. Finally, no text today would be complete without a chapter on genetic engineering and biotechnology.

Chapters 20 through 28 in Part 4 can be considered the practical application of Chapters 1 through 19 of the book. These eight chapters describe the production of major agronomic crops, grains, forage grasses, oil seeds, legumes, vegetables, small fruits, fruits, nuts, flowers, and sod.

Lastly, those who find learning about and using plants and soils enjoyable will want to be employed in the industry. Chapter 28 of the book provides ideas on necessary skills, education, and career paths in plant and soil science.

Every book on effective teaching and learning can be captured in an old Chinese proverb that says, "I hear, I forget; I see, I remember; I do, I understand." At the end of each chapter in this book, additional resources are listed and a "Knowledge Applied" section contains student activities. These sections are designed to get the students "doing" so that they begin "understanding" plant and soil science.

Acknowledgments

After nearly 40 years of teamwork and support no words can be found to sufficiently acknowledge my appreciation of the efforts of my wife, Marilyn. She is truly remarkable. Without her help, I would be left to do my own proofreading, prepare the art manuscript, and double-check final details. She catches many of the little oversights or "dumb mistakes" because she is willing to take that "one last look" before submission. In addition to her direct help with the details of writing a book, she is a source of constant encouragement in facing the trials and tribulations of everyday living.

Any photos in the text without a courtesy line were take by Marilyn or by me.

I also appreciate my association with Delmar Cengage Learning throughout the creation of this book and the many opportunities resulting from this relationship. David Rosenbaum, Christine Gifford and Andrea Majot at Cengage deserve a special thank-you.

About the Author

Rick Parker completed his undergraduate degree at Brigham Young University–Utah and received his Ph.D. in Reproductive Physiology from Iowa State University. After two post-doctorate positions and a stint as a coauthor with M.E. Ensminger, the noted author of numerous agriscience textbooks, he served as a division director and instructor at the College of Southern Idaho (CSI) in Twin Falls for 19 years. In this role, he worked with faculty in agriculture, information technology, drafting, marketing and management, and electronics. Dr. Parker also taught computer classes, biology, and agriculture classes at CSI. Currently, he serves as the director for AgrowKnowledge, the National Center for Agriscience and Technology Education, a project funded by the National Science Foundation (DUE #0434405). Additionally, he is the editor for the peer-reviewed *NACTA Journal*, which focuses on the scholarship of teaching and learning. He also teaches and conducts workshops for AgrowKnowledge.

Dr. Parker is also the author of the following Delmar Cengage Learning texts: *Aquaculture Science* (second edition), *Introduction to Plant Science*, *Introduction to Food Science*, and *Equine Science* (third edition). He is also the co-author of *Fundamentals of Plant Science*.

He and Marilyn, his wife of 40 years, are the parents of 8 children and grandparents to 20.

Dedication

To a world able to produce and distribute sufficient healthy foods to all humans, and to Justus Moroni Parker (1976-1997) who lives in our memory.

PART 1

Introduction to Plants

Chapter 1
The Plant Kingdom

THE ENERGY OBTAINED *from food is first converted from sunlight to usable, transferable energy by green plants. The oxygen supply in the earth's atmosphere is a result of photosynthesis by green plants. Fossil fuels come from plant material. Plants also create and modify local environmental conditions on which many species of animals and other plants depend.*

With more than 500,000 different kinds of plants, people need a way to communicate about them. This method cannot be limited by language and location differences. The method must also help group and determine the similarities and differences among plants.

After completing this chapter, you should be able to—

- Discuss the importance of taxonomic systems

- Describe how agriculture uses various formal and informal, scientific and nonscientific classification systems to describe crop plants

- Identify the basis of current systematic taxonomic systems and how to properly employ nomenclature of plants

- Give examples of crops used for food, beverages, fiber, industry, and oil

- Describe the binomial system for naming plants

- Name the two classes of the Spermatophyta division of plants

- List the differences between monocots and dicots and give examples of each

- Describe gymnosperms and angiosperms and give two examples of each

- Explain how geography influences the type of plants

KEY TERMS

Adaptation
Biome
Catkins
Classification
Climatic Zone
Cotyledon
Cultivar
Division
Genus
Kingdom
Mesic
Morphology
Nomenclature
Phyla
Species
Taxonomy
Variety
Xeric

Historically, plants have been considered one of the two kingdoms of living things, the other being animals. Plants, kingdom Plantae, are broadly distinguished from animals, kingdom Animalia, by being stationary, by manufacturing their own food, by having a continuous type of growth that is readily modified by the environment, and in possessing a less definite form when mature. Possibly 500,000 kinds of plants exist, and many do not fit well into either kingdom. Recent trends in classification recognize five different kingdoms based on evolutionary origins and relationships:

1. Monera, the bacteria and blue-green algae
2. Protista, including all other algae and the protozoans
3. Mycota, or fungi such as mushrooms and molds
4. Plantae, or plants, containing the mosses, ferns, seed plants, and several minor groups
5. Animalia, or animals.

EVOLUTION OF PLANTS

The earliest known forms of life date back about 3.5 billion years. These organisms were apparently bacteria and blue-green algae. The green algae are believed to have appeared about 1 billion years ago, as did the fungi. Photosynthetic algae began putting oxygen into the atmosphere, and by about 600 million years ago the atmospheric oxygen was sufficient to support life on land.

The earliest land plants occurred a little more than 400 million years ago, during the late Silurian period, and were similar to the whisk fern (*Psilotum*) of today. The first seed plants appeared about 350 million years ago, during the late Devonian period, and are called seed ferns because of the large fernlike leaves. The flowering plants date back to about 120 million years ago, or earlier, in the early Cretaceous period. Today, flowering plants total about two-thirds of all plants (Figure 1-1).

GEOGRAPHICAL DISTRIBUTION

Few places on the earth can be found where at least some of the 500,000 species of plants are not adapted to live. Only the polar zones, the highest mountains, the deepest oceans, and the driest deserts are devoid of plants—other than bacteria. Each plant species has a limited distribution, depending on its own particular requirements. Some species are broadly distributed, being tolerant of a wide range of conditions. Narrowly restricted species have limited tolerance to a specific factor, such as soil type.

Climate is the major factor affecting the distribution of plants and determining their structural adaptations. The greatest numbers of species are found near the equatorial regions in tropical climates, where moisture and temperature are seldom limiting factors to support plant life. The number of species in an area decreases toward the poles.

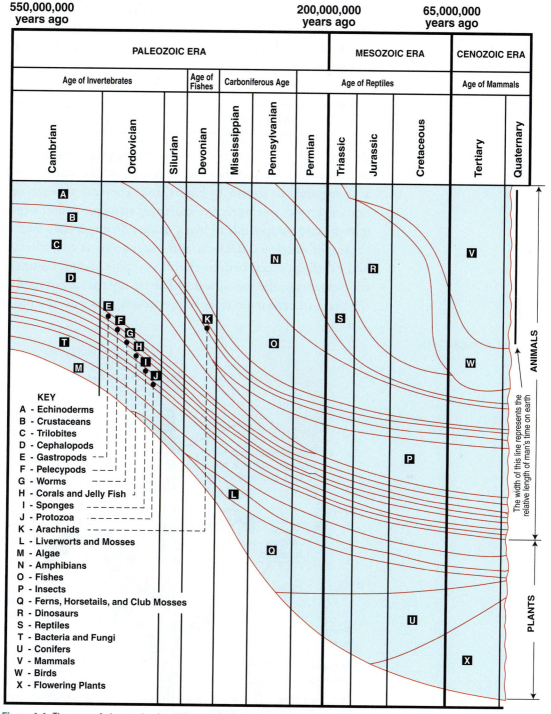

Figure 1-1 The story of plant and animal life on earth for the past 550 million years.

Plants adapt as they occupy drier or colder areas away from the tropics. In drier climates plants develop features known as xeromorphic (dry form) characteristics, such as smaller, thicker leaves, spines, dense hairiness, and water-storage organs. In colder zones plants grow closer to the ground, with the growing points protected at or just beneath the ground. Because of similar adaptations, plants of a specific **climatic zone** form a characteristic vegetation type. A large area occupying a given climatic zone and with a characteristic vegetation and associated groups of animal species is called a **biome**. The major biomes of the world include the tropical rain forest, desert, and tundra.

ROLES OF PLANTS

Because of their photosynthetic processes, green plants form the base of the food chain and thus the beginning of the energy flow through an ecosystem, as shown in Figure 1-2. They are also the only important organisms able to receive inorganic elements and incorporate them into organic compounds in living tissues. Plants form a vital link in the cycling of nutrients. Bacteria and fungi serve as the other major link in the cycling process because they decompose organic tissues and release the elements to the soil or water. The energy that is not used directly by the plant in carrying out its life processes goes into the production of

Figure 1-2 An overview of photosynthesis.

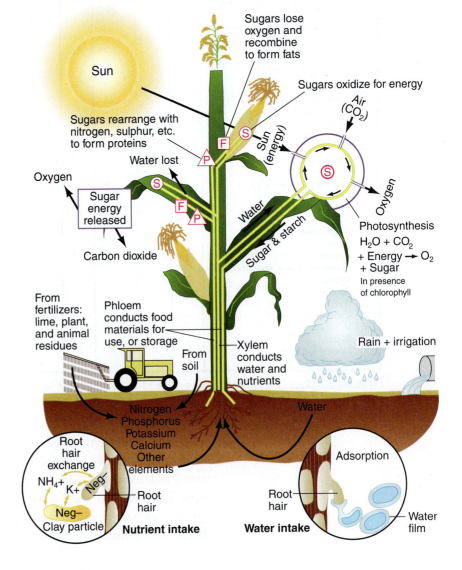

new tissues or biomass. Other organisms, including humans, use this as a food source.

Edible Plants

Edible, concentrated portions of various plants—such as seeds, fruits, and tubers—are used as a food source not only for the human populations of the world but also for livestock. Since prehistoric times, desirable types of plants have been selected and bred to obtain types that produce greater yields or larger seeds, fruits, or roots.

The most important food plants are the grains of the grass family, particularly wheat, rice, corn, sorghum, and barley (Figure 1-3). In many tropical countries the higher-protein cereal grains do not grow well, and the basic foods are the more starchy root crops, such as yams, sweet potatoes, potatoes (Figure 1-4), and manioc or cassava (Figure 1-5). Plant foods contribute about 88 percent of the world's calories and about 80 percent of the proteins. As a general rule, the more developed the country, the less of its diet is from plant foods.

Certain plants also provide the major beverages of the worlds, including coffee, tea, mate, and fruit juice. Beer is usually brewed from fermented

Figure 1-3 World production of grains. (Source: Food and Agricultural Organization statistics; http://faostat.fao.org)

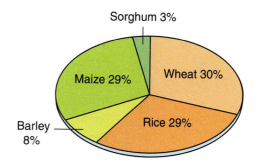

Figure 1-4 Harvesting potatoes, one of the world's starchy roots. (Photo courtesy of USDA ARS)

Figure 1-5 World production of starchy roots. (Source: Food and Agricultural Organization statistics; http://faostat.fao.org)

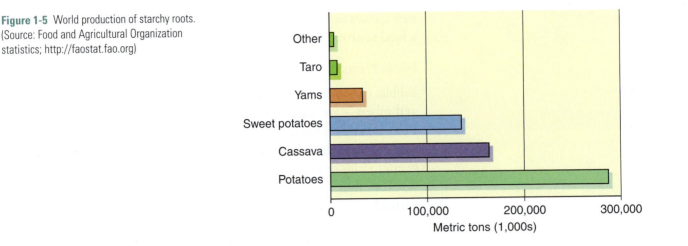

barley and hops; wines are made from grapes; and such spirits as whiskey and vodka come from grains or potatoes.

Industrial Uses

Scarcely a product exists in which plants have not played some important role, either as a component, an implement in its construction, or at least as the energy source (fossil fuels) of its production. Some of the more important plant products include wood, fibers, oils, and rubber. Many textile fibers are derived from plants, including cotton, flax, and hemp. The wood of trees is used to make tools, furniture, and houses. Such chemicals as acetic acid, methanol, and turpentine are obtained from trees. Fibers are elongated cells abundant in the bark and leaves of some plants. Cotton (Figure 1-6), which comes from the long hairs on the cotton seed, is the world's most important fiber (Figure 1-7). Other fibers are flax, jute, and kapok.

Figure 1-6 Cotton—the world's most important plant fiber—ready for harvest. (Photo courtesy of USDA ARS)

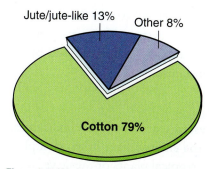

Figure 1-7 World cotton production. (Source: Food and Agricultural Organization statistics; http://faostat.fao.org)

Medicinal Uses

Certain plants are useful in medicine because they contain chemicals that have a desired physiological effect on the human system. Some of these chemicals and their plant sources are the antimalarial quinine from cinchona bark, the heart-stimulant digitalis from foxglove leaves, the antispasmodic atropine from belladonna (nightshade), and rauwolfia tranquilizers from the plants of the genus *Rauvolfia*. Most of the chemicals used are extremely poisonous under different concentrations and uncontrolled use. Many of these chemicals probably have been developed in the plant as a defense against predators.

Oils

Oils are stored as food reserves in the seeds and fruits of many plants. Most are used as human food, but some are used in industry. The most significant oil plant is the soybean. Also important are the coconut, sunflower, peanut, cottonseed, and rapeseed (Figure 1-8).

Figure 1-8 World oilseed production. (Source: Food and Agricultural Organization statistics; http://faostat.fao.org)

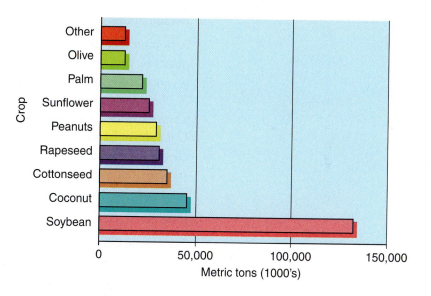

TAXONOMY

Theophrastus (370-285 BC) started to record names and arrange or group plants based upon growing habit and visible features. The problem with his system was that it used mostly local names and others outside his area did not understand his system of classification.

How to Classify Plants

Plants can actually be classified many ways, including—
+ Phylogenetic—how they look
+ Environmental—where they grow
+ Agricultural—what use they are grown for
+ Morphological—how their structure compares to each other

Systematics, or taxonomy, is an organizational system for descriptive classification of plants. Nomenclature is a system of assigning names to plants. The rules for assigning names to plants are established by the International Rules of Botanical Nomenclature.

Artificial Classification Systems

Climatic systems and agricultural systems are artificial classification systems. Climatic systems depend on the environment in which a plant grows; for example, temperate zone vs. tropical plants or mesic vs. xeric. Climatic systems can also depend on the season in which plants grow; for example, spring vs. summer vs. autumn crops, warm season vs. cool season, spring annuals vs. winter annuals, or annual or biennial.

Agricultural systems depend on the system used for production; for example, agronomic crops vs. horticultural crops. Or agricultural classification can depend on the part of the crop consumed; for example, seeds, vegetables, fruits, or stems. Finally, this system of classification could also depend on the use; for example, ornamentals or crop plants.

Botanical Systems

A botanical system was first developed by Theophrastus. He classified plants by gross morphology; specifically, form, size, texture, and use. He is considered the Father of Botany for these contributions.

Carl Linnaeus, also known as Carl von Linné (1707–1778), devised a system of categorizing plants, which was published in 1732 as *Hortus Uplandicus*. Later this was enlarged to a study of 935 genera and called *Genera Plantarum* (1737). The system is mostly based on flowering structure of the plants. Much of his taxonomy continues today and his work led to the modern taxonomic system and nomenclature of plants.

Taxonomic Classifications

The plant world is extremely diverse, ranging from the one-celled algae to the huge oaks and sequoias (redwoods). The plant world contains plantlike organisms like the mushroom, which contain no green coloring. Lichens and mosses are green plants but have no true roots and leaves and no flowers. Plants with true leaves and roots but no flowers reproduce by means of spores on the backs of the leaves or on separate stems. Finally, the flowering or seed-bearing plants make up most of the plant kingdom. This group of plants is the concern of this book.

Divisions

The plant kingdom traditionally was divided into two large groups, or subkingdoms, based mainly on reproductive structure: the thallophytes (subkingdom Thallobionta) and the embryophytes (subkingdom Embryobionta). The plant form of the thallophytes is an undifferentiated thallus lacking true roots, stems, and leaves. The subkingdom Thallobionta is composed of more than 10 divisions of algae and fungi

USDA Started in Patent Office

Henry Leavitt Ellsworth (1791–1858) could be called the founder of the U.S. Department of Agriculture (USDA). Mr. Ellsworth, a lawyer, farmer, leader of an agricultural society, and head of an insurance company in Hartford, Connecticut, became the head of the newly established Patent Office in 1839. He often received seeds and plants from naval and consular officers overseas, and he distributed these to farmers without any government authority or help.

For government aid to agriculture, he received $1,000 to collect and distribute new and valuable seeds and plants, to carry on agricultural investigations, and to collect agriculture statistics. Even with this small budget, Mr. Ellsworth had a great insight to the importance of agricultural improvement. He stated:

> If the application of the sciences be yet further made to husbandry, what vast improvements may be anticipated! Mowing and reaping will, it is believed, soon be chiefly performed on smooth land by horse power. Some have regretted that modern improvement make so important changes of employment—but the march of the arts and sciences is onward, and the greatest happiness of the greatest number is the motto of the patriot.

In 1849, Daniel Lee, MD, and former editor of the *Genesee Farmer* and professor of agriculture at the University of Georgia, was hired as a "practical and scientific agriculturist" to supervise agricultural matters in the Patent Office. Dr. Lee also prepared separate annual reports on agriculture. The first of these was dated 1849, and it eventually became the *Yearbook of Agriculture*, which was published every year until President Clinton ended the practice in 1993. Dr. Lee preached soil conservation, and he issued reports on tillage, runoff, drainage, insects, fertilizers, and the improvement of farm animals, rural science, and the statistics of weather. Dr. Lee considered agricultural progress to come from scientific inquiry, practical wisdom, intellectual honesty, and forthright advocacy.

Eventually the staff expanded in the Division of Agriculture of the Patent Office. Townend Glover (1813–1888), a British entomologist, was hired to collect statistics and other information on seeds, fruits, and insects in the United States. He thoroughly studied the insects of the southern United States and gathered an extensive collection of insects, birds, models of fruit, and herbarium plants. In 1865 a botanist and a chemist also joined the staff.

Isaac Newton grew up in Pennsylvania, where he managed a model dairy farm. As a delegate of the United States Agricultural Society, he repeatedly introduced a resolution calling on Congress to establish a Department of Agriculture. In 1861 he was appointed Superintendent of the Agricultural Division of the Patent Office. When the Department of Agriculture was organized on May 15, 1862, his friend President Lincoln made him the first Commissioner of Agriculture. Mr. Newton established an agricultural library and a museum. He also selected the grounds for the Department of Agriculture in Washington, DC. One year after Mr. Newton took office, the new department had a horticulturist, a chemist, an entomologist, a statistician, an editor, and 24 other staff members. Congress appropriated $100,000 for an office building in 1867.

Even before the Department was established, its advocates urged that it be made an executive department, headed by a secretary who would be a member of the Cabinet. Advocates argued that agriculture was the single most important economic activity of the nation. Finally, in 1889 the Congress elevated the U.S. Department of Agriculture (USDA) to Cabinet status.

(once considered plants). The subkingdom Embryobionta is composed of two groups: the bryophytes (liverwort and moss), division Bryophyta, which have no vascular tissues, and a group consisting of seven divisions of plants that do have vascular tissues. The Bryophyta, like other non-vascular plants, are simple in structure and lack true roots, stems, and leaves; therefore, they usually live in moist places or in water.

The vascular plants have true roots, stems, and leaves and a well-developed vascular system composed of xylem and phloem for transporting water and food throughout the plant; thus, they are able to inhabit land. Three of the divisions of the vascular plants are currently represented by only a very few species:

+ Psilotophyta, with only three living species
+ Lycopodiophyta (club mosses)
+ Equisetophyta (horsetails)

All the plants of a fourth subdivision, the Rhyniophyta, are known only as fossils (extinct).

The remaining divisions include the dominant vegetation of the earth today:

+ Polypodiophyta (the ferns)
+ Pinophyta (the cone-bearing gymnosperms)
+ Magnoliophyta (the angiosperms, or true flowering plants)

Pinophyta and Magnoliophyta are often collectively called spermatophytes, or seed plants (Figure 1-9).

Spermatophytes: Seed Plants

The seed plants are those which produce true seeds, each containing an embryo (a dormant plant) that germinates under favorable conditions. They have true leaves, stems, roots, and vascular tissue. The seed-bearing plants make up the greater part of the vegetation of the earth. They consist of two classes—

+ Gymnosperms
+ Angiosperms

Gymnosperms. The gymnosperms are the "naked-seeded" plants. They are all woody, perennial, and with few exceptions, evergreen plants. The reproductive organs are borne in structures called catkins or in cones with the seeds usually uncovered. Their leaves may be fernlike, scalelike, strap-shaped, or needle-shaped. This group is represented today principally by the cone-bearing trees (conifers). Examples of plant families that are gymnosperms include:

+ Cycadaceae—the Cycad family, cycads (*Cycas* spp.)
+ Ginkgoaceae—the Ginkgo family, maidenhair tree (*Ginkgo biloba*)
+ Pinaceae—the Pine family, pines (*Pinus* spp.), cedars (*Cedrus* spp.)
+ Podocarpaceae—the Yew family, podocarpus (*Podocarpus* spp.)
+ Taxaceae—the Florida yew (*Taxus floridana*); Florida torreya (*Torreya taxifolia*)
+ Zamiaceae—the zamia (*Zamia* spp.)

Figure 1-9 Seeds from spermatophytes—samples of seeds maintained at the Agricultural Research Service National Seed Storage Laboratory in Fort Collins, Colorado. (Photo courtesy of USDA ARS)

Angiosperms. The angiosperms include those species that have flowers and seeds always protected by a fruit. They form a large, complex group of flowering plants under two main subdivisions:

+ Monocotyledoneae (monocots)
+ Dicotyledoneae (dicots)

These divisions are based on the number of **cotyledons**, or seed leaves, found in the seed.

The Monocotyledoneae (monocots) have one cotyledon (embryonic leaf) in the seed. Generally they have parallel-veined leaves and flower parts in threes or sixes or in multiples of threes, but never fours or fives.

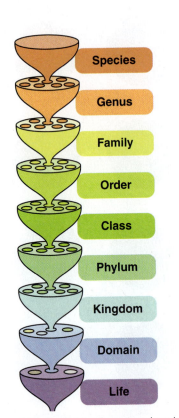

Figure 1-10 The eight major taxonomic ranks of the hierarchy of biological classification. Intermediate minor rankings are not shown.

Their stems are made up of fibrovascular bundles without pith or bark. A few of the monocots such as bamboos and palms are treelike. The majority are herbaceous plants such as grasses, cattails, lilies, irises, orchids, bananas, and bromeliads.

Dicotyledoneae (dicots) have two cotyledons in the seed. They can be recognized by their net-veined leaves and by their flower parts (sepals, petals, stamens, and pistils), which are generally in fours or fives or multiples of fours or fives.

Some dicots have many small flowers in catkins, as in willows and poplars. Others have flowers of separate petals and sepals, as in the rose and the southern magnolia (*Magnolia grandiflora*). Still others have flowers in which the petals or sepals, or both, are fused to form a trumpet-shaped flower, as in honeysuckle (*Lonicera* spp.) and allamanda (*Allamanda cathartica*).

Lower subdivisions. Starting at the level of classes, the lower subdivisions include:

+ Order
+ Family
+ Genus
+ Species
+ Form
+ Variety/cultivar/clone

Family is a group of closely related genera. The relations are based on structures or characters. When writing a family name, the first letter of family names is capitalized. Figure 1-10 provides a visual image of the hierarchy of biological classification.

Plant Identification and Nomenclature

Linnaeus latinized names and gave protocols for the plants. Most names were derived from Latin or Greek. The names are usually phonetic and give a clue to the plant's characteristics or habitat; for example:

+ Folius, phyllon, phyla refer to leaves
+ Flora, refers to the flowers

Table 1-1 shows how the Latin names for plants give a clue to the plant's characteristics or habitat.

Binomial nomenclature. Each plant has a two-word (binomial) name—genus and species. A third name is often given to the authority who named the plant. More names are given in agriculture for the cultivar name.

Table 1-1 Examples of Common Name, Scientific Name, and Meaning of Scientific Name

Common Name	Scientific Name	Meaning
Ivy grape	*Cissus rhombifolia*	ivy, diamond-shaped leaves
Red maple	*Acer rubrum*	maple, red
Sweet gum	*Liquidambar styraciflua*	liquid amber, flowing with gum
Pink pinxter azalea	*Rhododendron canescens*	rose tree, off-white hairs
Spider plant	*Chlorophytum comosum*	green plant, tufted

Binomial nomenclature is the scientific system of giving a double name to each plant or animal. The first, or genus, name is followed by a descriptive or species name. Modern plant classification, or taxonomy, is based on a system of binomial nomenclature developed by the Swedish physician, Carl Linnaeus. Prior to Linnaeus, people had tried to base classification on the leaf shape, plant size, flower color, and other features. None of these systems proved workable. The revolutionary approach of Linnaeus was to base classification on the flowers and/or reproductive parts of a plant and to give plants a genus and species name. This has proven to be the best system because flowers are the plant part least influenced by environmental changes. For this reason, knowledge of the flower and its parts is essential for anyone who is interested in plant identification.

Other subgroups. Other subgroup names can be the botanical variety or the cultivar. Botanical variety is a plant sub-grouping with inheritable differences from the general species. Cultivar came from the combination of the words "cultivated" plus "variety." Cultivars are genetically true through propagation, presenting a strong set of characters or attributes. If propagated asexually, by cuttings and not seeds, they are considered "clones" genetically identical or extensions of the original plant. Cultivar names are always capitalized but not italicized. Cultivars may be purebred hybrid lines, or made from a cross that yields predictably the same progeny (offspring).

Some plants hybridize (cross) naturally; others must be crossed intentionally. Hybrids within a genus and among species are denoted Genus X species.

RESEARCH

Much of the emphasis in research is on the development and use of new techniques and equipment. In the field of classification especially, the use of new chemical methods and computers has become important in discovering relationships and in handling data. The electron microscope, and in recent years the scanning electron microscope, are important tools of the plant anatomist in determining the structure and function of plants at the subcellular level. The scanning electron microscope is used particularly in studying the detail of leaf surfaces and pollen grains.

Studies of the structure and function of membranes, where much of the plant's activities take place, are widely pursued. Recent research in molecular botany includes the synthesis of chlorophyll and the interrelationships of nucleic acids and hormonal functions. The role of hormones and their interactions with phytochromes in affecting flowering continues to be an intensively studied area in plant physiology. Nearly all photosynthetic plants use the carbon from carbon dioxide to manufacture sugar molecules by employing one specific set of chemical reactions to fix, or transfer, the carbon atoms. This series of chemical reactions is called the C-3, or Calvin-Benson-Bassham, cycle because the three-carbon compound, phosphoglyceric acid, is formed during its operation. An area of recent interest has been the discovery of an alternative carbon-fixing pathway in a number of other plants. This is called the C-4, or Hatch-Slack, cycle because four-carbon compounds are produced during this process.

New Taxonomic Tools

In the future, three new taxonomic tools will be used to classify and group plants—
1. Chemical analysis—composition of the plant
2. Protein analysis
3. DNA analysis—genetic fingerprinting

SUMMARY

The plant kingdom contains mosses, ferns, and seed plants. For agriculture, the seed plants are the greatest concern. These belong to the class of plants called spermatophytes, which is divided into the gymnosperms and the angiosperms. Further, the angiosperms are divided into the monocots and the dicots. These plants are important to humans as sources of food, beverage, fiber, oil, and industry supplies.

Plants can be classified a variety of ways; for example, by use or by growing season. But this type of classification changes from area to area. To avoid confusion, specific plants are given latinized names in a binomial system. The binomial name can have a variety and cultivar name to help identify a plant. Research continues on understanding and identifying the relationships between plants. Tools such as chemical analysis and genetic fingerprinting may be used in the future to classify plants.

Review

True or False

1. As a general rule, the more developed the country, the less of its diet is from plant foods.

2. Green plants are at the top of the food chain.

3. Binomial nomenclature is the scientific system of giving a double name to each plant or animal.

4. New ways to classify plants in the future will include DNA analysis.

Short Answer

5. What are the most important food plants?

6. _____ is the major factor affecting the distribution of plants and determining their structural adaptation.

7. Name the two classes of the spermatophyte division of plants.

8. List the six lower subdivisions of plants.

Critical Thinking/Discussion

9. How do plants adapt to areas away from the equatorial regions?

10. Discuss the difference between monocots and dicots.

11. What is taxonomy?

12. Give examples of crops used for food, beverages, fiber, industry, and oil.

Knowledge Applied

1. Visit a museum, personally or on the Internet, and look at specimens of fossilized plants. Report on your findings.

2. Use a plant identification key to classify some unknown plants.

3. Make a pressed plant collection. Label the plants with their common and scientific names.

4. List all of the economically important plants grown in your area. Use their common name and their scientific name, and collect photographs of as many as possible. Describe how each plant is used by humans. (This could be done using some form of presentation software.)

5. Make a seed collection of monocots and dicots.

6. Visit a garden supply store or use a garden catalog to list and describe all of the varieties or cultivars of some of the common garden plants such as tomatoes, potatoes, carrots, cucumbers, and corn.

7. Develop an electronic collection of a group of plants by visiting one of the online databases listed in the Web sites section of Resources.

Resources

Goodman, P. (2004). *Plant classification*. London, UK: Hodder Wayland.

Judd, W. S., Campbell, C. S., Kellogg, E. A., Stevens, P. F., & Donoghue, M. J. (2007). *Plant systematics: A phylogenetic approach* (3rd ed.). Sunderland, MA: Sinauer Associates, Inc.

Mabberley, D. J. (2008). *Mabberley's plant-book: A portable dictionary of plants, their classifications, and uses* (3rd ed.). New York: Cambridge University Press.

Zomlefer, W. B. (1995). *Guide to flowering plant families*. Raleigh, NC: The University of North Carolina Press.

Web Sites

University of Minnesota, Plant Information Online: https://plantinfo.umn.edu/arboretum/default.asp.

U.S. Department of Agriculture Natural Resources Conservation Service Plants Database: http://plants.usda.gov/index.html.

Internet

Internet sites represent a vast resource of information. The URLs (uniform resource locators) for World Wide Web sites can change. Using one of the search engines on the Internet such as Google or Yahoo!, find more information by searching for these words or phrases: historical botany, plant taxonomy, plant classification, evolution of plants, plant kingdom, plant distribution, and plant uses.

Chapter 2
Origins of Cultivated Plants

MANY OF THE *cultivated plants in our fields, gardens, and landscapes originated from wild ancestors. The process of passing from a weedy ancestor to a cultivated plant is lost in the past for many species. Nevertheless, reconstructing such histories is a fascinating problem for the historical botanist. Exact points of origin can be an important source of germplasm (genetics) to the plant breeder and are critical to the improvement or maintenance of present plant characteristics.*

As plants were introduced and improved in the United States, researchers found new ways to maintain markets. The search for new uses and new plants continues today.

After completing this chapter, you should be able to—

- Give the origin of corn, potatoes, wheat, alfalfa, barley, tomatoes, and peaches

- Describe how early American farmers tried new crops and adapted to the markets

- List crops grown by North American Indians

- Describe how the USDA first helped the American farmer

- Explain the difference between utilization research and market research

- Give two examples where demand for a crop increased because new uses for the crop were discovered

- Explain why continuing research is necessary for the success of cultivated plants

- Discuss the importance of all plant germplasm

- Name two examples of new crops that could be used for industrial purposes

KEY TERMS

Agricultural Adjustment Act
Fertile Crescent
Hatch Act
Organization of Petroleum Exporting Countries Research and Marketing Act
Farm Bill

ORIGIN AND DEVELOPMENT OF CULTIVATED PLANTS

Most botanists agree that two main regions of the world are the homes of most cultivated food items:

+ Asia and Asia Minor in the Old World
+ Central Mexico to Chile in the New World

As explorations increased, other food items were added from other areas of the globe. Figure 2-1 shows the origins of many food plants from around the world.

In 1492, European countries had never heard of the potato, sweet potato, corn, peanut, tobacco, rubber, tomato, dahlia, marigold, or vanilla.

Figure 2-1 Centers of origin for (A) major vegetables; (B) fruits; and (C) grains and oil crops. (Image courtesy of USDA ARS)

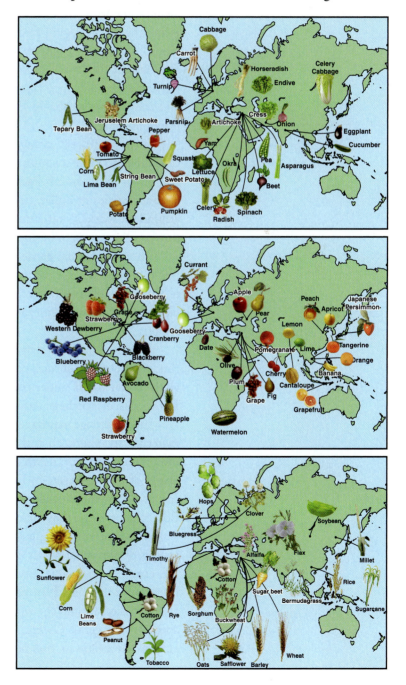

All of these plants and fruits were well known and widely used by the Aztecs and Incas of Central and South America. Besides these food products, whole families of plants are totally American and were unknown before the Spanish conquest. The two most notable examples are the pineapple family, with about 1,000 species, and the cactus family, with over 1,300 species.

The Old World had existed for centuries without American plants. Most commonly grown European vegetables have been in cultivation for over 2,000 years, and some much longer than that. The wild ancestors of many commonly cultivated vegetables are European plants, but the origin of the vegetable as a cultivated crop is most likely Asiatic or Near Eastern.

Much of the cultivated wheat, barley, common millet, buckwheat, and alfalfa came from the eastern end of the Mediterranean Sea and Central Asia. This is the region frequently known as the **Fertile Crescent**. Seed crops like pea, lentil, and chickpea are associated with the Near East. This same region also lays claim to the fig, date, pear, pomegranate, and possibly the apple. Peaches may have developed in China. Oats, sugar beets, rye, and cabbage are probably European. Eggplant and cucumber have been traced to India.

Citrus plants (e.g., oranges, lemons, grapefruit), banana, mango, several types of sugarcane, coconut, and tea were introduced from Southeast Asia and places in the South Pacific. Coffee is listed as an African plant, as are several varieties of millet, cowpea, and rice. Similar plants have developed around the world along common lines of latitude where light and temperature conditions might be equal.

Often the formal gardens of northern Europe are considered as the origin of many of our common landscape plants. Early settlers who came to America with plants used in their homeland support this theory. More likely, however, most common European plants originated in the Asiatic regions.

Eastern Asia is known for its great concentration of barberry, oak, rhododendron, fir, spruce, pine, azalea, and flowering cherry species. Also, all the cotoneasters under cultivation have been imported from China, the Himalayas, or northern Asia.

The last ice age may have set the stage for our present collection of plant materials. In areas such as the eastern United States and China, where the mountain ranges run north and south, certain plants retreated southward ahead of the ice sheet. In Europe, however, most of the mountains run east and west and blocked the migration of plants species ahead of the ice. Botanists presume that many species were lost during this period.

Woody ornamental plants do not have a single seat of origin. Individual species of any given genus have been found growing wild in many areas. For example, the genus *Taxus* (yew) has three primary species that have been found in Japan, Europe, and North America. Viburnum species can also be traced around the world. The ginkgo or maidenhair tree is the only species in the genus *Ginkgo*, which in turn is the only genus in the Ginkgo family. The origin of this plant has been traced to eastern China. Fossil records suggest that the plant has been growing on earth for 150 million years and was native in America at one time, but not as a well-known landscape specimen.

NEW CROPS AND NEW USES

The search for new crops and new ways to use old crops has interested Americans inside and outside the agricultural sector since America's colonial days. Over the past century that interest has intensified. The major recent theme in this quest has been how to better use the surpluses produced by American farmers, in industrial as well as agricultural ways.

A second important theme includes the need to find substitutes for commodities currently in use. Substitutes cut our dependence on foreign imports, replace critical materials in short supply during wartime, or exchange products that harm the environment for environmentally sound ones.

Historically, these themes have tended to alternate, with times of crisis also being times when research in these areas has received the most support. The quests to use agricultural surpluses and to find substitute commodities have come together to reinforce each other. Both areas enjoy many achievements, and conditions today promise even more success in the future.

Early American Adaptations

From the beginning, American farmers were responsive to market signals and looked for new crops and markets that would increase their incomes. Many of the crops new to the colonists had long been grown by North American Indians; for example, corn, potatoes, squash, and tobacco.

Corn (maize) was first introduced into eastern North America around 200 AD. When the first European settlers arrived they learned how to grow corn from the Indians and depended on corn for their survival. Corn also remained a staple of Native Americans.

Some of these crops were taken back to Europe by the first Spanish explorers, where their adoption depended on how well they fit the needs and expectations of European farmers and consumers.

Before Columbus discovered the New World, Europe's major food crops offered only meager yields and left Europeans vulnerable to famine. Corn succeeded in Europe because it offered a high-producing alternative to other grains and required relatively little work. Potatoes had such great yields that they became a staple in the diets of the poor, from Ireland to Germany. On the other hand, tomatoes, suspected of being poisonous and not offering a ready substitute for anything, required several centuries to become an important food crop except in a few areas. Corn was restricted mainly to animal food because Europeans found it too coarse to eat.

European adoption of American crops revealed a pattern that would repeat itself in the United States: Farmers replaced low-yielding or high-labor crops that they had grown for centuries with ones that were cheaper to produce—and which consumers would buy.

Early American farmers soon faced a surplus problem that threatened to hold back the development of commercial agriculture. In a country where over 90 percent of the people were farmers, most markets had to come from overseas. To simply duplicate the crops of Europe would have made it difficult for the American farmers to compete with European farmers. Many American farmers, therefore, turned to newer crops that they could export to their advantage, such as tobacco, rice, and indigo. In the nineteenth century, farm exports, especially cotton, provided the income that the young America needed for industrialization.

Pioneer farmers also found an industrial use for something otherwise considered a nuisance. As they cleared the fields of trees, they were able to sell tree ashes as potash for use in making lye. This became a profitable source of income even before crops were planted.

American farmers had a continuing interest in trying out new varieties and new species, and this interest was one of the main reasons for establishing the U.S. Department of Agriculture (USDA). The first specifically agricultural activity of the federal government was to encourage United States embassies to collect seeds from different parts of the world for distribution to farmers in the United States. When the Department was founded in 1862, seed distribution and research were an integral part of its efforts.

Beginnings of Agricultural Research

The **Hatch Act** of 1887 greatly expanded research by setting up experimental stations at each of the land-grant colleges. Initial agricultural research focused on improving productivity by using better agricultural practices and varieties. Ironically, the very success of that research created the potential for production to exceed demand, depressing prices below the level of profitability. By the turn of the century people recognized that new uses for farm products would have to be found if surpluses were to be avoided and by-products efficiently used.

Within the USDA, utilization research increased. In 1920—at a time when farm prices were collapsing because of a contraction in exports—an Office of Development Work was set up in the Bureau of Chemistry to find ways to chemically break down farm products into substances that industry could use. World War I also provided an incentive for research into industrial uses for agricultural products. When imports of medicinal plants were cut off, USDA' Bureau of Plant Industry helped establish such plants in Florida. Similarly, the disruption of dye imports from Germany prompted USDA to begin research on dye materials. This led to the establishment of the vat dye industry in the United States.

One of the most successful utilization researchers early in this century was George Washington Carver at Tuskegee Institute in Tuskegee,

Soybeans—A Relatively Recent Successful Import

Soybeans were first processed in the United States with hydraulic presses about 1911. Those beans were importations from Manchuria. The earliest record of crushing American-grown soybeans was in a cottonseed oil mill at Elizabeth City, North Carolina, in December 1915. The Chicago Heights Oil Manufacturing Company processed small lots at intervals between 1917 and 1923, using both hydraulic and screw presses.

In September 1922, the A. E. Staley Manufacturing Company began operations in Decatur, Illinois, in a mill that was equipped with screw presses designed for crushing soybeans.

Eventually four modern processing plants operated in Decatur. That town became known as the soybean capital of the world. The combined capacity of the mills exceeded 100,000 bushels of soybeans a day, a large part of which was handled in solvent-extraction equipment.

Since their introduction, soybeans have found their way into a diverse array of foods such as soy milk, tofu, soybean curd, soy yogurt, soy burgers, soy loaf, and soy sausage. Soy oil is the most widely used edible oil in the United States; it is found in mayonnaise, salad dressing, process cheese products, dessert frostings, and much more. Soy components such as protein and oil are ingredients in dozens of everyday foods from granola to potato chips to a sandwich. Soy lecithin is used in almost all chocolate treats.

Soybeans have been incorporated into a host of nonfood products. These range from soy oil-based ink to lipstick, plastics, flooring, paints, and stain-removing cleaners, and now biodiesel (Figure 2-2).

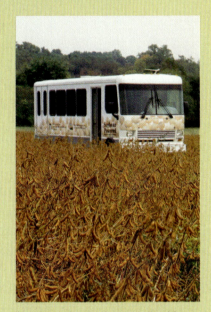

Figure 2-2 Bus running on biodiesel from soybean oil, shown traveling through a soybean field that is ready for harvesting. (Photo courtesy of USDA ARS)

Most soybean varieties used in the United States were developed between 1980 and 1994, during which time scientists released 66 varieties and 280 breeding lines.

Alabama. Carver saw that the South was suffering from the overproduction of cotton, almost to the exclusion of other crops. Not only were cotton prices low, but the boll weevil was also beginning to devastate cotton farms.

Carver believed farmers would turn to other crops if enough new uses could be found for them to create a sufficient market. He concentrated on peanuts and sweet potatoes, two crops Southerners already knew how to grow but that were not widely planted for commercial purposes. Over the years, Carver developed hundreds of new products from peanuts and sweet potatoes. The best known product to consumers is probably peanut oil. The USDA Extension Service helped distribute information throughout the South about converting from cotton to other crops. Largely because of Carver's research, peanut acreage quadrupled between 1910 and 1940. Carver also found new uses for cotton and soybeans. In fact, soybeans can be called the most successful "new" crop in the United States this century. Figure 2-3 illustrates some continuing research with soybeans.

Agricultural Adjustment Act. Surplus production became an even greater problem during the Depression of the 1930s, when domestic markets fell and exports nearly disappeared. Congress took a major step toward the expansion of utilization research when it passed the **Agricultural Adjustment Act (AAA)** of 1938. This act created four regional research laboratories as part of what is now the Agricultural Research Service. Each laboratory specialized in the crops grown in its region—for example, cotton in the South; wheat, fruits, vegetables, and alfalfa in the West; animal products, milk, and tobacco in the East; and grain crops, soybeans, and other oilseeds in the North.

The basic institutional structure of today's national, regional, and state experimental station laboratories was established with the AAA of 1938. Advisory committees of the AAA and research on utilization and marketing were strengthened by the **Research and Marketing Act** of 1946, which sought to change the imbalance between production and postproduction research. That act also set up the mechanism for contracting with private research facilities, permitting the government to draw on the expertise of private sector scientists as well as those of its own. This institutional framework supported research into the problems of particular areas and sought to supply quick answers to questions of national importance.

World War II. World War II showed how rapidly the research establishment could respond to a national crisis. The war redirected the regional laboratories toward finding substitutes for critical materials and other war needs. Some notable successes came out of this effort, including synthetic rubber, replacements for chemical cellulose, dehydrated foods, and the extraction of wheat starch, which was used to supplement corn in feeding livestock. Most important, in 1943 USDA scientists discovered a way to mass-produce penicillin, making this miracle drug widely available for the first time.

Figure 2-3 Soybean research in which USDA Agricultural Research Service scientists use gene silencing (biotechnology) to shut down the genes in soybeans that make the protein that is thought to cause most soybean allergies in humans. (Photo courtesy of USDA ARS)

Postwar Research

After the war, attention again focused on crops in surplus. In 1957, with farm productivity soaring from the greatly increased use of chemicals and machinery, Congress created a Commission on Increased Industrial Use of Agricultural Products to recommend new research in this area. The Commission noted that the greatest advances came "when utilization research in agriculture has had the benefits of adequate expenditures and large teams of workers," as it had during the war. It urged a broadly expanded program of research, including more basic research. Funds were increased.

The first two decades following the war saw some striking accomplishments from the regional laboratories, whose efforts were directed toward particular agricultural problems. The development of frozen food technology in the 1940s and 1950s enabled consumers to enjoy fresher-tasting foods all year round and helped even out seasonal swings in vegetable and fruit prices. Development of frozen orange juice concentrate was a particular achievement. Concentrate produced from Florida oranges climbed from 226,000 gallons in 1945 to more than 84 million gallons in 1960. Frozen foods paralleled and accelerated a consumer trend toward convenience foods (Figure 2-4).

That same trend helped ensure the success of another USDA product, instant potato flakes. Potato flakes were developed at a time when potato consumption was falling. Their introduction gave potatoes a new market. By 1969, dehydrated potatoes represented over 9 percent of the total potato crop, and per capita consumption had turned upward.

Another crop in trouble by the 1950s was cotton. During and after World War II, cotton faced increasing competition from synthetic fibers, which required less ironing than cotton fabrics. Research on cross-links between cotton fibers led to chemical finishes that imparted to cotton and cotton blend fabrics the wrinkle resistance they needed to satisfy

Figure 2-4 Most modern food products are prepared for the consumer, as seen in ready-to-eat frozen grocery items. (Photo courtesy of USDA ARS)

consumers. These new fabrics helped slow the trend toward polyester and nylon.

New Sources of Demand

Since the 1960s, rising productivity has kept interest high in using surplus crops. In addition, the past several decades have also been a time of renewed interest in finding agricultural substitutes for industrial materials. New demands for research have appeared. One such incentive came from the dramatic oil price hikes by the Organization of Petroleum Exporting Countries in 1973 and 1979. Not only did energy costs soar, but Americans became painfully aware of their dependence on foreign oil. This kindled interest in organic substitutes for petroleum. Research to replace part of the petrochemicals used to manufacture plastics with biodegradable cornstarch derivatives promises to increase the demand for corn. Soy ink, made from soybeans, is beginning to replace petroleum-based ink products for some uses. New research to replace gasoline with ethanol or other fuels from renewable resources could sharply cut our reliance on oil imports and provide a large market for corn.

Another attraction of organic fuels is that they likely will generate less pollution than petroleum products. Interest in the environment has brought a call to substitute biological pesticides for chemical ones and to find biodegradable and recyclable materials to conserve natural resources and reduce waste disposal problems.

Another source of demand for utilization research has been the expanded interest in nutrition by increasingly vocal and discriminating consumers. The widespread desire to reduce consumption of fat, for example, led animal scientists to breed leaner beef and pork. As people increase their intake of grain products, higher-protein wheat and oats have been developed. Research on brans, such as rice bran, can add fiber to the diet and possibly reduce cholesterol.

In the 1970s, strong exports of grain, soybeans, and other crops made the United States even more a part of the global market than it had been in the past. Trade deficits in the 1980s emphasized both the vulnerability of farmers to shifts in world markets and the urgent need to find alternatives to traditional imported products. This, and the desire to replace nonrenewable substances with renewable ones, revived the effort to find new plants. Plant explorers from the Agricultural Research Services each year bring back thousands of new plants from abroad for testing.

Historically, introducing new plants has been harder than finding new uses for old ones because new crops must be able to be grown in a new location and must be accepted by farmers and the marketplace. Nevertheless, a number of potentially useful plants have been experimented with, including guayule for rubber, kenaf as a substitute for wood pulp in making paper, crambe for industrial oil, and switchgrass for ethanol production. Using biotechnology, plant breeders should be able to develop new and better varieties faster than in the past and be able to better predict their chemical properties.

The Future

Today, opportunity and necessity combine to create a more favorable climate for the introduction of new crops and new uses than at any other period except that of wartime. The utilization research establishment has been solidly in place for over fifty years, and it has the facilities and experience to start a major expansion if needed. Plant scientists looking for new varieties or species to introduce have an even longer history to draw on, plus the advantage of using the tools of biotechnology (Figure 2-5).

The demand for research is great and has never come from so many different directions. New crops and new uses for old ones promise to restore the balance of trade, reduce our dependence on imports such as oil, and make us more competitive in agricultural exports. They may also make it possible for us to replace our nonrenewable resources with renewable ones and in many cases benefit the environment as well.

Another potential advantage would be reduced spending on farm programs because of stronger demand for program commodities or lower supply of surplus crops caused by a shift to new crops.

Finally, utilization research could benefit some rural economies by bringing new rural factories to process raw materials. Research could lead to greater farm sales from land that in the past has been diverted by government programs to reduce production.

Farm Bill. Proposals in the 2007 Farm Bill include funding for a bioenergy and biobased product research initiative because advances in technology play an important role in the future of renewable energy. The purpose of this funding is for scientists, farmers, and entrepreneurs to coordinate efforts to continue improvements in crop yields and work to reduce the cost of producing alternative fuels. Provisions in the 2007 Farm Bill also include funding for renewable energy systems and efficiency to support small alternative energy and energy efficiency projects that directly help farmers, ranchers, and rural small businesses. Funding is also included for the development of cellulosic ethanol production.

Figure 2-5 Pollinating blackberries. Many new plant varieties have been produced by controlling pollination. (Photo courtesy of USDA ARS)

Other proposals in the 2007 Farm Bill are targeted toward the support of fruit and vegetable producers by providing funding for research in specialty crops. This initiative will include fundamental work in plant breeding, genetics, and genomics to improve crop characteristics such as product appearance, environmental responses and tolerances, nutrient management, and pest management.

For more details on the current Farm Bill as well as changes and progress of the bill, visit the USDA Web site [http://www.rurdev.usda.gov/rd/farmbill.html].

Biotechnology. Using the tools of genetic engineering allows the incorporation and expression of genes from other species into plants. Some examples of application of gene technology in agriculture include:

+ Herbicide-resistant plants; for example, crop plants resistant to the commercial herbicide Roundup, notably soybean, corn, tomato, potato, wheat, cotton, and alfalfa
+ Insect-resistant plants using Bt toxin gene from bacteria and expressed in plants to provide resistance to insects without the need for insecticides used in crops like corn and cotton
+ Controlled ripening of tomato and other crops like two varieties of tomato on the market engineered for delayed ripening
+ Crops bioengineered for disease resistance, as employed in papaya and squash
+ Enhanced nutritional benefits like "golden rice," a transgenic rice with in the genes for the production of vitamin A

Benefits through the use of biotechnology or genetic engineering of plants seem endless; for example:

+ Nutritionally enhanced foods that contain more starch or protein, more vitamins, more antioxidants, and fewer trans-fatty acids
+ Foods with improved taste, increased shelf life, and better ripening characteristics
+ Trees that make it possible to produce paper with less environmental damage
+ Ornamental flowers with new colors, fragrances, and increased longevity
+ Plants that may be used as "manufacturing facilities" to inexpensively produce large quantities of materials, including therapeutic proteins for disease treatment and vaccination, textile fibers, biodegradable plastics, or oils for use in paints, detergents, and lubricants
+ Plants produced with entirely new functions, enabling them to detect and/or dispose of environmental contaminants like mercury, lead, and petroleum products
+ Plants that grow in high concentrations of salt

PROTECTION OF GERMPLASM

Less than one-tenth of one percent of about 350,000 (about 350) available plant species are used for agriculture. Plant breeders, geneticists, and biotechnologists are concerned about preserving, and in some cases cataloging, this invaluable resource for the future. Genes to incorporate traits like disease resistance and salt tolerance into tomorrow's crop plants will come from this vast germplasm pool. Some plants will be selected from this germplasm pool for cropping in the future, for new uses or products, and to meet the need for crops that are adapted to adverse environments. Plants and plant products for tomorrow will come from the yet untapped germplasm pool that exists today only if its importance is recognized and preserved.

In agriculture, the main emphasis is still on increased food production, with breeding programs to develop high-yield strains, especially those yielding more protein. Computers are also used to simulate the growth of several food crops and study the factors involved.

SUMMARY

As world exploration and civilization spread, so did the cultivated plants. The New World and the Old World eventually shared the cultivated plants their civilizations developed. Farmers in the New World began producing crops that could be exported to the Old World. As production increased or as markets diminished, farmers and researchers found new uses for old crops, or new crops to replace supplies of products from other sources. The development of technology brought forth many new products and increased the demand for specific crops. In the future, the quest will continue to improve old crops through biotechnology, to find new crops that can produce products in short supply, to create a demand for old crops through new products, and to tap into the great potential that exists in the worldwide plant germplasm.

Review

True or False

1. Many of the cultivated plants originated from wild ancestors.

2. George Washington Carver was one of the most successful utilization researchers early in this century.

3. In 1943 USDA scientists discovered penicillin.

4. Introducing new plants is easier than finding new uses for old plants.

Short Answer

5. What common beverage is listed as an African plant?

6. What new crop exported in the nineteenth century provided the income that our young country needed to industrialize?

7. Frozen orange juice _____, _____ rubber, extraction of _____ starch for livestock, instant potato _____, and cotton that needed less _____ were all successes from USDA research before, during, and after World War II.

8. Name the three plants that have been used to substitute experimentally for rubber, wood, and industrial oil.

9. Early farmers sold tree ashes as _____ for use in making lye.

Critical Thinking/Discussion

10. What are the two themes for new ways to use old crops?

11. Why did corn and potatoes succeed as crops in Europe?

12. What is the AAA and what did it create?

13. The first, and continuing, crude oil price increases sparked research into what areas?

14. What has caused an increase in nutrition research?

15. What was the first specifically agricultural activity of the federal government for United States embassies?

Knowledge Applied

1. Go to a grocery store and make a list of all the products from one crop that are the result of technology being applied to the processing of the crop; for example, potatoes, peanuts, oranges, and wheat.

2. Make a list of all the crops that supply starch to the diets of humans and indicate which civilizations or ethnic groups use each type of starch.

3. Conduct a taste test of the different crops used for starches in human diets (see number 2).

continues

Knowledge Applied, *continued*

4. Make a list of five plants that could become a crop of the future. Choose one of these plants and develop a computer presentation on this plant. Include in the presentation the following: common name, scientific name, value, growing conditions, yield, and additional processing.

5. Visit a facility where a crop is processed for human use; for example, a sugar factory, cotton mill, oilseed processing plant, potato processing plant, or a flour mill.

6. Develop a presentation describing a plant that you would like to genetically engineer. Describe what new characteristics you would like to give the plant and the source (plant, animal, bacteria) of the genes to engineer your new plant.

Resources

Brown, J., & Caligari, P. (2007). *An introduction to plant breeding*. Ames, IA: Blackwell Publishing.

Collinson, A. S. (1988). *Introduction to world vegetation*. London: Unwin Hyman Ltd.

Food and Agricultural Organization. (1996). *Food for all*. Rome, Italy: Author.

Simpson, B. B., & Ogorzaly, M. C. (2000). *Economic botany: Plants in our world* (3rd ed.). New York: McGraw-Hill.

Sleper, D. A., & Poehlman, J. M. (2006). *Breeding field crops* (5th ed.). Ames, IA: Blackwell Publishing.

United States Department of Agriculture. (1950-51). *Crops in peace and war: The yearbook of agriculture*. Washington, DC: Author.

United States Department of Agriculture. (1961). *Seeds: The yearbook of agriculture*. Washington, DC: Author.

United States Department of Agriculture. (1986). *Research for tomorrow: The yearbook of agriculture*. Washington, DC: Author.

United States Department of Agriculture. (1992). *New crops, new uses, new markets: The yearbook of agriculture*. Washington, DC: Author.

Internet

Internet sites represent a vast resource of information. The URLs (uniform resource locators) for World Wide Web sites can change. Using one of the search engines on the internet such as Google or Yahoo!, find more information by searching for these words or phrases: plant biotechnology, new foods, plant genetic engineering, agricultural research, food origins, germplasm, utilization research, export, or Fertile Crescent.

Chapter 3
Structure of Cultivated Plants

TO UNDERSTAND PLANT GROWTH, *food production, and the relation of plants to human welfare, the basic parts of a plant and their functions need to be known. A green plant's physiological processes are functions adapted to fulfilling its needs for energy, nutrients, water, reproduction, and dispersal. A plant's structural features, although diverse in the various groups, are each especially adapted for carrying out the functions of photosynthesis, assimilation, respiration, and growth. Structural features are also used to identify and classify plants.*

After completing this chapter, you should be able to—

- Use the terms that describe vegetative parts of the plant
- Identify and use terms that describe the reproductive parts of plants
- Name three types of roots
- List three functions for roots and for stems
- Describe a stem
- Name four types of stems
- List all of the parts of a typical leaf
- Give three types of venation in leaves
- Explain the difference between an incomplete and a complete flower
- Identify all of the parts on a complete flower
- Describe the functions of the essential reproductive organs of the flower
- Name three types of flowering characteristics found in plants
- Discuss four ways flowers vary
- Name the two large categories of fruit
- List four terms used to describe fleshy fruits and four terms used to describe dry fruits
- Discuss how plant structure is used to classify plants
- Identify all of the parts of a typical seed

KEY TERMS

Aerial Roots
Anther
Auricles
Axillary
Berry
Blade
Branched
Calyx
Cambium
Carpel
Climbing
Complete Flower
Compound Leaf
Corolla
Cotyledons
Creeping
Crowns
Dehiscent
Dichotomous
Dicotyledoneae (Dicots)
Dioecious
Drupe
Dry Fruits
Embryo
Epicotyl
Fibrous Roots
Filament
Fleshy Fruits
Fleshy Roots
Floral Bracts
Germination
Gymnosperm
Hypocotyl
Incomplete Flower
Indehiscent
Knees
Ligule
Monocotyledoneae (Monocots)
Monoecious
Nodes
Opposite
Ovary
Palmate
Parallel

Key terms continue next page

PLAN OF THE ENTIRE PLANT

The major type of plant that people deal with each day is the angiosperm. This type of plant produces seeds that are enclosed in a fruit. Angiosperms are divided into **monocot** (one seed leaf) and **dicots** (two seed leaves). The three basic parts of a plant include roots, stems, and leaves—for both a monocot and dicot. The flowers, fruits, and seeds, as well as the growing points of roots and stems, can be identified when a mature plant is examined (Figure 3-1).

Roots

Roots are often the overlooked part of the plant because they are located in the soil. Root functions include—

- Anchoring the plant
- Absorbing water and minerals from the soil and transporting them to the stem
- Storing food produced by the above-ground portion of the plant

Root systems consist of a main or primary root, rootlets or secondary roots, and root hairs. Figure 3-2 shows the three basic types of root systems:

1. Taproot
2. Storage taproot
3. Fibrous root

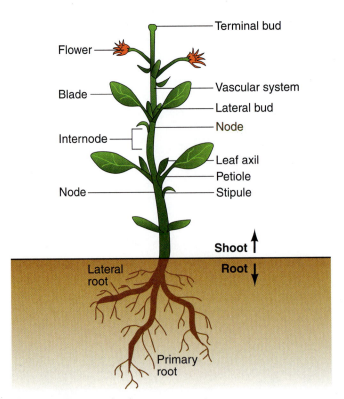

Figure 3-1 The parts of a plant.

Figure 3-2 Types of root systems. *A,* taproot; *B,* storage taproot; and *C,* fibrous root.

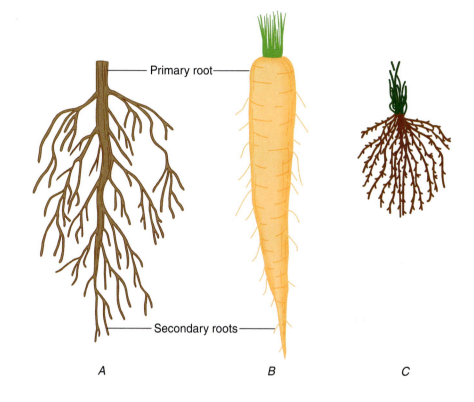

The **primary root** arises from the **embryo**. Branches of the primary (tap) root are often fibrous and are called **secondary roots**. The primary root system of many plants is short-lived and is replaced by a secondary root system. These secondary root systems become the permanent roots of many monocotyledons (monocots) such as the grasses. Roots coming from bulbs, corms, rhizomes, or tubers are adventitious roots, as are roots developed from aerial stems (stolons and runners) and cuttings from stems or leaves. Red mangrove (*Rhizophora mangle*), Screw pine (*Pandanus* spp.), and banyan (*Ficus* spp.) send down adventitious roots, which become supports for heavy, horizontal branches.

Root hairs are specialized cell extensions that penetrate into the openings between soil particles (Figure 3-3). The outside wall of the cell

Figure 3-3 Root hairs in soil particles.

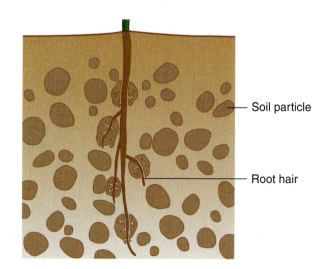

becomes distended to form a tubular outgrowth, which makes contact with the soil and absorbs water and dissolved minerals from it. Water and soluble nutrients enter the root hairs, pass into the rootlets, and travel through the main root into the stems and leaves. Root hairs are formed in great numbers near the tips of roots. In most plants they are short-lived. If a plant is transplanted carelessly, it is the loss of many of these small root hairs with their water-absorbing cells that will cause the plant to wilt.

Taproots are prominent primary roots from which all other lateral rootlets or secondary roots grow. They may divide, become fleshy, and often penetrate deeply into the soil. Many herbaceous perennials, such as the carrot (*Daucus carota*) and the dandelion (*Taraxacum officinale*), and some trees such as oaks (*Quercus* spp.) and citrus (*Citrus* spp.) have taproots.

Fibrous roots have no distinguishable primary root and are composed of a number of fine, threadlike roots of the same kind and size originating at the base of the stem. Fibrous roots often spread out near the surface of the soil, rather than penetrating straight down or deep. Grasses, most annuals, and some shrubs have fibrous roots.

Fleshy roots become food reservoirs that retain surplus food during the winter or adverse periods to be used by the plant when it is able to renew its growth. Carrots, turnips (*Brassica rapa*), and beets (*Beta vulgaris*) have main or taproots containing food. Sweet potatoes (*Ipomoea batatas*) and dahlia (*Dahlia* spp.) tubers are secondary roots transformed into tuberous roots packed with food.

Aerial roots form freely on many land and water plants in a favorable, moist atmosphere. These roots enable climbers such as philodendrons (*Philodendron* spp.) to attach themselves to a host. The aerial roots of air plants or epiphytes such as some orchids not only attach the plant to its host but also absorb water from the air. Many aerial roots are fleshy or semifleshy, functioning as reservoirs for water storage.

Knees (pneumatophores) are developed by bald cypress (*Taxodium distichum*) and pond cypress (*Taxodium ascendens*), when grown in swampy ground. As the water or ground cuts off the air from the roots, these trees develop woody knees, which protrude above the surface to enable the plant to obtain air.

Stems

The **stems** of plants are important in forming the major above-ground structural part of the plant (Figure 3-4). They act as the attachment point for leaves, flowers, and fruit. Stems also contain the water and food distribution system (vascular system) of the plant.

The **vascular system** is made of xylem and phloem. Water and plant nutrients are taken in by the roots and go through xylem vessels to leaves and growing points. Food and other materials made by photosynthesis are distributed throughout the plant by the phloem vessels. The vascular system is arranged differently in monocots (corn, grass, lilies) than in dicots (bean, squash, poplar trees).

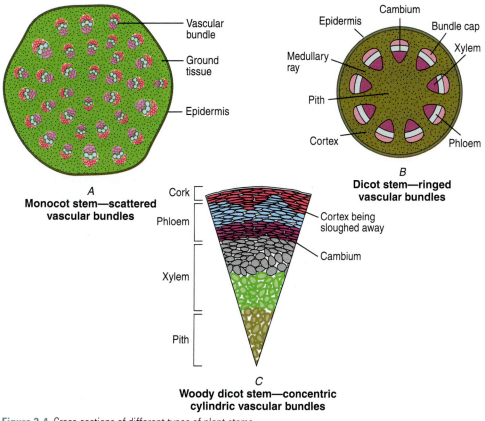

Figure 3-4 Cross-sections of different types of plant stems.

The xylem and phloem are in bundles scattered throughout the stem of monocots, but the xylem and phloem are segregated into circular zones separated by the cambium layer in dicots. The cambium layer produces new xylem and phloem cells. These are discussed in Chapter 4.

A stem is the structure that develops from a bud to bear leaves—either full-sized or rudimentary—and buds. Stems have swellings at certain points called nodes. A node is the point on a stem where a leaf is or was attached. The area between nodes is termed the internode. Stems usually grow upward to the light, but may be below the ground (subterranean).

The functions of stems are to support and display leaves, fruit, and flowers; to carry water and nutrients from the roots to the leaves; and to carry food produced by the green parts back to the roots. In plants that lack leaves, such as cacti, all food is produced in the green stem. Stems may be annual, biennial, or perennial, although some plants with perennial roots have annual stems.

Stems may be adapted for food storage, such as tubers, corms, bulbs, or rhizomes. They may also be specialized as runners, tendrils, or thorns. The tendrils of grapes (*Vitis* spp.) are modified stems. Other types of stems include:

+ **Crowns**—short and inconspicuous stems. Such plants are said to be stemless or acaulescent, such as the gerbera daisy (*Gerbera* spp). Crown is also the name for the base of the stems where roots arise.

+ **Simple**—stems without branches (sidegrowths), such as in papaya (*Carica papaya*) and corn (*Zea mays*).
+ **Branched**—stems with more than one terminal bud, sidegrowth, or branch.
+ **Climbing**—stems too weak to support themselves, which lean or twist about other plants or posts for support, such as in bougainvillea (*Bougainvillea* spp.) and the flame vine (*Pyrostegia venusta*).
+ **Creeping**—stems that rest on the surface of the ground, sending down roots at the nodes or joints, such as in wedelia (*Wedelia trilobata*) and trailing lantana (*Lantana montevidensis*).
+ **Rhizomes**—prostrate, usually thickened, subterranean stems, with leaves coming from one side and roots from the other, such as canna (*Canna x generalis*) and star-leaf begonia (*Begonia heracleifolia*).
+ **Stolons**—slender, modified stems growing along the surface of the ground and rooting at the nodes, as in the strawberry (*Fragaria x ananassa*) and St. Augustine grass (*Stenotaphrum secundatum*).

Leaves

The main function of leaves is to make food and other chemicals (proteins, fats, and oils) that are used by the plant for growth and reproduction. However, leaves are not the only place these materials are made. Any portion of the plant that is green (fruits and stems) may also produce these materials.

Parts. The main parts of a complete leaf are the blade, petiole, and stipule (Figure 3-5). The blade is the broad, thin part of the leaf. The narrow cylindrical part is the petiole.

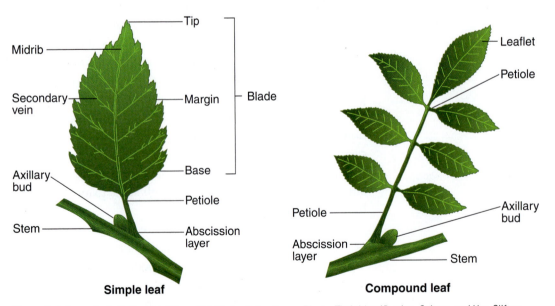

Figure 3-5 Parts of a leaf. (Reprinted from F.M. Birdwell, *Landscape Plants: Their Identification, Culture, and Use.* Clifton Park, NY: Delmar Cengage Learning, 1994.)

Figure 3-6 The different parts of a grass leaf.

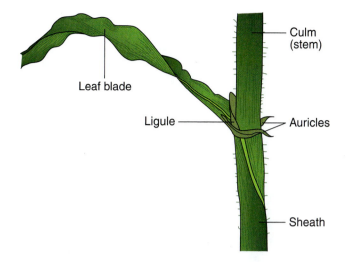

The stipules are the pair of small leaflike appendages at the base of the petiole. Stipules are usually small structures as in the apple leaf, but some of them, such as in the pansy and garden pea, are large and much like another leaf. All leaves do not have stipules.

Some leaves have no apparent petiole. The blade is attached directly to the stem. In this case, the leaves are described as sessile.

The parts and terms used to describe a grass leaf are shown in Figure 3-6. The sheath is that portion of the leaf base of the blade that surrounds the stem. Many grasses have a collarlike extension at the top of the sheath called the ligule. The ligule varies in size and shape and may be completely absent in some grasses. Examples of long and membranous ligules are found on red top and annual bluegrass. In Bermuda grass, the ligule is only a ring of short white hairs.

The auricles are appendages that surround the stem at the junction of the blade and sheaths on some grasses. For example, auricles are not present on oats but are large and clasping on barley.

Simple or compound leaves. A leaf consisting of only one blade is known as a simple leaf. Examples are the grasses, oak, and apple leaves. A compound leaf consists of several leaflets (Figure 3-5). The leaflets may be joined to the end of the petiole or along the central axis of the leaf (the rachis) like sumac or locust leaves.

Shape of the leaf. The shape and margin of the leaf are usually characteristic of the species. Many plants can be identified by the shape of the leaf alone. Several different shapes and margins of leaves are shown in the sketches in Figures 3-7 and 3-8. Terms or words used to describe the leaves are also given in the figures.

Arrangement of veins in the leaf. The arrangement of the veins in the leaf is termed venation. There are four general arrangements of the principal veins in leaves that are easily recognized. These are illustrated in the drawings in Figure 3-9. The grasses have parallel veins. The veins extend parallel from the base to the apex or tip of the leaf. Some of the ferns and the ginkgo tree have forked or dichotomous venation.

Needle-like Awl-like Scale-like Oblong Linear

Lanceolate Oblanceolate Spatulate Ovate Obovate

Elliptic Oval Cordate Peltate

Reniform Pinnately lobed Palmately lobed

Figure 3-7 Names given to various leaf shapes. (Reprinted from F.M. Birdwell, *Landscape Plants: Their Identification, Culture, and Use.* Clifton Park, NY: Delmar Cengage Learning, 1994.)

When the secondary veins extend from the midrib, like the divisions of a feather, the venation is **pinnate**. In **palmate** venation, the principal veins extend from the petiole near the base of the blade similar to the bones in the hand.

Margins

Entire · Serrate · Double serrate · Serrulate · Crenate

Dentate · Incised · Undulate · Lobed · Spinase

Tips

Acute · Obtuse · Acuminate · Mucronate · Cuspidate

Truncate · Emarginate · Round · Retuse

Bases

Acute · Cuneate · Cordate · Oblique · Truncate

Rounded · Hastate · Sagittate · Obtuse

Figure 3-8 Leaf margins are helpful in identifying plants. (Reprinted from F.M. Birdwell, *Landscape Plants: Their Identification, Culture, and Use.* Clifton Park, NY: Delmar Cengage Learning, 1994.)

Arrangement of the leaves on the stem. Leaves usually have definite arrangement on the stem. The leaves may be alternately arranged with a single leaf at each node. On some plants the leaves occur two at a node on opposite sides of the stem. In this case, the leaves are said to be **opposite.**

Figure 3-9 Principle veins in leaves: parallel, dichotomous, pinnate, and palmate venation.

Grass, an example of parallel venation

Ginkgo, an example of dichotomous venation

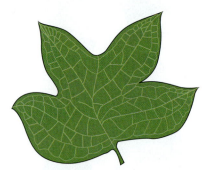

Poplar, an example of pinnate venation

Sweergum, an example of palmate venation

The leaf arrangement is termed **whorled** if there are three or more leaves at each node (Figure 3-10).

Flowers

Flowers are the sexual reproductive part of angiosperm plants. Many plants have **complete flowers** that contain both male and female structures.

Figure 3-10 Examples of various leaf arrangements. (Reprinted from F.M. Birdwell, *Landscape Plants: Their Identification, Culture, and Use.* Clifton Park, NY: Delmar Cengage Learning, 1994.)

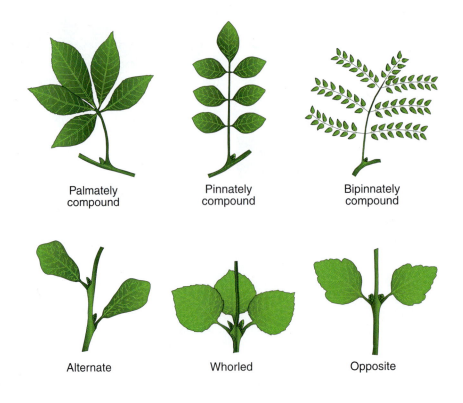

Palmately compound

Pinnately compound

Bipinnately compound

Alternate

Whorled

Opposite

Figure 3-11 A typical complete flower.

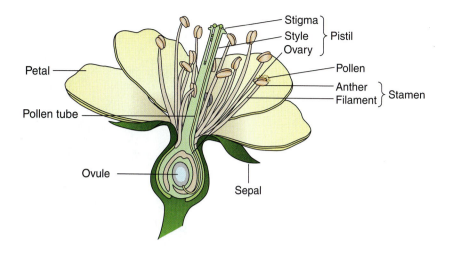

The male part of the flower (**stamen**) produces pollen, and the female part (**pistil**) receives the pollen and eventually forms seeds. Examples of plants having complete flowers are tomatoes, apples, beans, and roses. Figure 3-11 shows a typical complete flower.

Some plants have separate flowers containing only male or female parts. Often these male or female flowers are both on the same plant. Examples are pecan, oak, watermelon, squash, and corn (Figure 3-12). Other flower types and terms used to describe flowers include—

◆ Single Flower—one flower borne at the end of an elongated stalk or branch of the main axis of the plant, as in tulip and the magnolia (*Magnolia grandiflora*). The **peduncle** is the stalk that bears the single flower at the top (and is also the main stem or axis of a flower cluster). The **pedicel** is the stalk of an individual flower in a cluster.

◆ Cluster—three or more flowers gathered closely together in simple or branched groups to increase their conspicuousness, such as in penta (*Pentas* spp.), ligustrum (*Ligustrum japonicum*), firethorn (*Pyracantha coccinea*), and mango (*Mangifera indica*).

◆ Inflorescence—general term for the arrangement of flowers or groups of flowers on a plant. There is great diversity in this arrangement among different types of plants, but they generally remain characteristic for a particular type and may be useful in identifying species. There are two main types of inflorescences, each of which is further subdivided; they are the racemose type and the cymose type.

Racemose inflorescences. In this type of flower, the axis of the inflorescence continues to grow (it is an indeterminate inflorescence) and the flowers are borne in the axes of the reduced leaves or bracts, with the oldest flowers at the base and the newest flowers near the growing tip (Figure 3-13).

◆ Raceme—flowers on short pedicels of about equal length along the main axis, as in snapdragon (*Antirrhinum majus*) and allamanda (*Allamanda cathartica*).

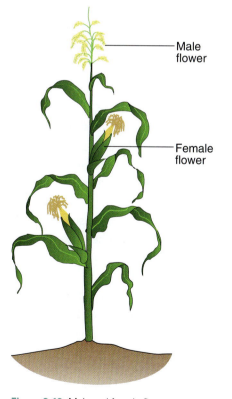

Figure 3-12 Male and female flowers on a corn plant.

Figure 3-13 A flowering plant, the petunia.

- Panicle—compound raceme (the branches have branches), with individual flowers replaced by simple racemes, as in crape myrtle (*Lagerstroemia indica*) and orange jessamine (*Murraya paniculata*).
- Spike—like a raceme, but flowers lack pedicels and are sessile or almost sessile. Examples are copperleaf (*Acalypha wilkesiana*) and bottlebrush (*Caliestemon* spp.).
- Spadix—type of spike; a fleshy axis bearing the sessile, generally fleshy flowers close together, commonly surrounded and partially enclosed by a spathe. Examples are garden calla (*Zantedeschia aethiopica*) and anthurium (*Anthurium* spp.).
- Catkin—spike that normally produces only staminate or pistillate flowers, and at maturity falls away as a unit. Examples are river birch (*Betula nigra*) and oaks.
- Corymb—pedicels of older flowers longer than those of younger flowers, which brings all of them to nearly the same level. Pedicels come from different points on the main peduncle, giving the inflorescence a rather flat-topped or convex look with the outside flowers opening first. Examples are candytuft (*Iberis* spp.) and ixora (*Ixora coccinea*).
- Umbel—short axis that causes pedicels to appear to arise from a common point. This gives the inflorescence a knoblike look. The outer flowers open first. Examples are dill (*Anethum graveolens*) and crinum lily (*Crinum* spp.).
- Head—similar to umbels, but sessile flowers are very close together. What popularly passes for a "flower" in the Compositae family is really an inflorescence with many small, true flowers, such as in the sunflower (*Helianthus annuus*) shown in Figure 3-14. Heads may be globular or almost spherical as in the buttonbush (*Cephalanthus occidentalis*) and alsike clover (*Trifolium hybridum*), or they may be composed of rays and disks. Ray flowers form a fringe of radiating

Figure 3-14 A flower head, such as the sunflower. (Photo courtesy of USDA ARS)

irregular, unsymmetrical flowers on the edge of the head. Disk flowers cover the remainder of the head and are regular, symmetrical, and usually less showy. Heads may also have flowers that are all irregular with no differentiation into rays and disks.

+ Involucre—cluster or whorl of bracts or leaves directly under a flower or a cluster of flowers, which is often conspicuous. They are often found under umbels and heads. Examples are the sunflower and cups of acorns.

Cymose inflorescences. In flowers of this type, the upward growth of the floral axis is stopped early by the development of a terminal flower. The first flower to open (the oldest) is at the tip; with younger flowers appearing lower down on the axis. The floral axis ceases to elongate after the first flower opens and is a determinate inflorescence.

+ Cyme—one terminal flower and two or more side flowers coming from the end of the axis. Examples are frangipani (*Plumeria* spp.) and Brazilian nightshade (*Solanum seaforthianum*).
+ Scorpioid cyme—floral axis curves over, carrying the flowers along the top of the curve, as in forget-me-nots (*Myosotis* spp.).
+ Fascicle—flowers are very closely crowded on almost the same plane, as in the Sweet William (*Dianthus barbatus*).

Flower positions. The two flower positions include **terminal** and **axillary**.

1. Terminal—flowers or clusters of flowers are carried on the ends of the axis or branches, as in the Southern magnolia and oleander (*Nerium oleander*).
2. Axillary—flowers or clusters of flowers arise at the junction of the stem or axis and the leaf, as in the periwinkle (*Vinca major*), beautyberry (*Callicarpa americana*), and hibiscus (*Hibiscus rosa-sinensis*).

Some plants, such as ixora, have both terminal and axillary flowers.

Flower parts. A flower is a highly differentiated and specialized branch of the stem bearing modified leaves or flower parts. It is the site of sexual reproduction in these plants and is their most distinctive structure. The great variety of forms acts as a guide in separating flowering plants into the major groups. Before classifying flowers, the terms for the individual structures in the flower need to be known.

Accessory organs. Accessory organs of the flower include the **perianth, calyx, corolla, receptacle,** and **floral bracts.**

+ Perianth—the outer floral parts, composed of the calyx and the corolla.
+ Calyx—the ring of sepals making up the outermost, leaflike part of the flower. Sepals are commonly green, but may be almost any color and serve primarily as protection for the other floral parts. (Figure 3-15)
+ Corolla—the inner set of leaflike parts lying just within the calyx and composed of petals. Petals are generally white or brightly colored to attract pollinating insects to their nectar. They also serve as protection for the innermost organs.

Figure 3-15 Drawing of the parts of a flower.

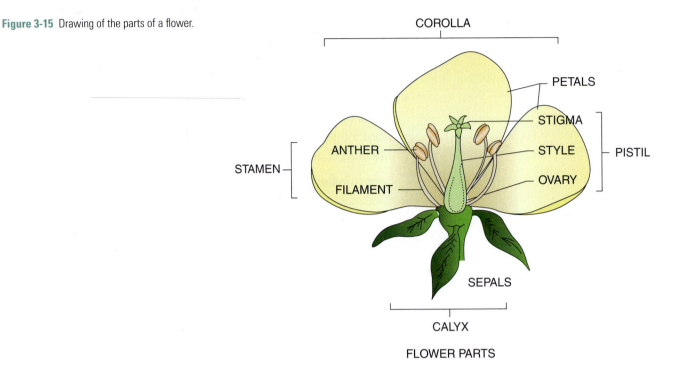

FLOWER PARTS

♦ Receptacle or torus—the apex of the pedicel upon which the organs of a flower are developed.

♦ Floral bracts—modified leaves that can simulate petals and add the conspicuous part to otherwise inconspicuous flowers.

Essential organs. Essential organs are the reproductive structures whose presence usually determines the survival of the species. These include the stamens and pistil and their associated parts.

♦ **Stamens**—male reproductive organs attached to the receptacle, in some species inside the corolla or in other species to the corolla itself. Each stamen is composed of the filament, anther, and pollen grains.

1. **Filament**—thin stalk that attaches the anther to the rest of the flower.

2. **Anther**—lobed, oblong, baglike appendage at the top of the filament that produces the pollen grains, which develop the male germ cells. Anthers are generally yellow and when young have from one to four cavities (cells) in which pollen grains arise. When mature, the anther usually contains two cavities from which the pollen grains are released by the formation of apical pores or longitudinal slits in the cavity wall.

3. **Pollen** grains—usually appear as tiny specks barely visible to the unaided eye, but are produced in such quantity that they often form a layer of powder. Each grain is usually one-celled, spherical, ovoid, or disklike in appearance, whose surface is marked with ridges, spines, and germ spores. Pollen grains, collectively known as pollen, are so characteristic of the different species that they are used for identification purposes.

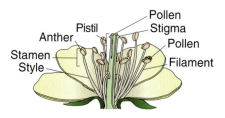

Figure 3-16 Pollen sticking to the stamen.

Stamens are called opposite when they are opposite to the petals. They are alternate when they alternate with the petals (Figure 3-16). Stamens are usually free or separate from each other, but in a few families they are united either by their filaments as in tomatoes (*Lycopersicon esculentum*), or by their anthers as in the sunflower.

✦ **Pistils**—the female reproductive organs usually occur in the very center of the flower and are often surrounded by the stamens, petals, and sepals. Flowers may have just one simple pistil as in the sweet pea (*Lathyrus odoratus*), or two, three, four, five, or more separate pistils as in larkspur (*Consolida* spp.). A **carpel** refers to either a simple pistil or one of the segments of a compound pistil. Pistils are usually flask- or bottle-shaped. They are composed of three parts—

1. **Style**—the elongated stalk or neck connecting the ovary with the stigma.
2. **Ovary**—enlarged, bulbous, basal part of the pistil that bears the ovules (the egg-containing units that become the seeds after fertilization) attached either to its central axis or to its inner wall. The place on the wall of the ovary to which ovules are attached is called the placenta. Each ovule usually contains one egg, the female gamete or sex cell. The ovule normally develops into a seed. Generally there are two or more ovules per carpel. In some plants only one may mature into a seed. The ovary normally develops into a fruit containing seeds.
3. **Stigma**—the tip of the style or pistil especially adapted to receive the pollen grains, which is expanded into a bulb or disk or divided into two or more slender parts.

A pistil is said to be compound when several or many carpels become united. Carpels are united so that a compound ovary often contains as many cavities as there are carpels. In some flowers, a compound ovary becomes "one-celled" by the disappearance of the partitions between the different carpels as in primrose (*Primula* spp.). Union of carpels may be so complete that it includes the styles and stigmas as well as the ovaries.

Flower Classification

Flowers are classified according to the presence or absence of their parts. Complete flowers are made up of a calyx, a corolla, stamens, and a pistil or pistils (the four "regular parts"). **Incomplete flowers** lack one or more of the four regular parts of a complete flower as in peperomia (*Peperomia* spp.). Perfect flowers have both stamens and pistils, but not necessarily sepals or petals. Imperfect flowers lack either stamens or pistils, and may or may not have sepals or petals. Naked flowers are without petals or sepals as in the calla lily (*Zantedeschia* spp.). Apetalous flowers lack petals as in silverthorn (*Elaeagnus pungens*). Staminate (male) flowers have a stamen or stamens, but no pistils. Pistillate (female) flowers have a pistil or pistils, but no stamens.

Three terms are applied to plants based on their flowering characteristics—

1. **Monoecious** plants bear both staminate and pistillate flowers on the same plant as in oak and corn.
2. **Dioecious** plants bear staminate flowers on one plant and pistillate flowers on a different plant. These plants are called male and female plants. Holly (*Ilex* spp.) and the Brazilian pepper (*Schinus terebinthifolius*) are examples.
3. **Polygamous** plants bear staminate, pistillate, and hermaphroditic (bisexual—both sexes present and functional in the same flower) flowers on the same plant. An example is the red maple (*Acer rubrum*).

Flower Forms

The diversity of flower form is a very important factor in plant identification. Flowers vary in the number of their floral parts, but they also vary in four other major ways relative to their parts—

1. Flowers vary in the degree to which floral parts are united.
   - Gamopetalous—petals united to form a tubular or rotate corolla with the united part known as the tube and the spreading or flat part known as the limb. These flowers take different forms as—
     - Funnelform—the tube gradually widens upward and flares into the limb without any particular point of demarcation. Examples include oleander and the bush morning-glory (*Ipomoea carnea*).
     - Rotated—the tube is short and the limb is flat and circular. The tomato and elderberry (*Sambucus canadensis*) are examples.
     - Urn-shaped (urceolate)—broad tube and slightly recurved, short limb, as in blueberries (*Vaccinium* spp.).
     - Salverform—slender tube and an abruptly widened, flat limb. Annual phlox (*Phlox drummondii*) and periwinkle are examples.
   - Gamosepalous—flower with united sepals. Periwinkle and hibiscus are examples.
   - Polypetalous—petals of corolla composed of separate parts. Southern magnolia, camellia (*Camellia japonica*), and rose (*Rosa* spp.) are examples.
   - Polysepalous—sepals of calyx composed of separate parts. Southern magnolia (*Magnolia grandiflora*) and rose are examples.
2. Flowers vary in the placement of floral parts on the receptacle.
   - Hypogynous—sepals, petals, and stamens are attached to a convex or conical receptacle at the base of the ovary. Sepals are arranged in the outermost or lowest layer, followed by petals and stamens, with carpels or ovary innermost. The ovary is called superior and the perianth is inferior or hypogynous. The tomato is an example (Figure 3-17).
   - Perigynous—sepals, petals, and sometimes stamens borne on the edge or margin of the receptacle so that they appear to

Focus on Modified Crops to Improve World's Food Supply

A new agriculture system driven by consumer demands will revolutionize what crops are grown in the future, and biotechnology will be at the heart of this change. Value-enhanced food that contains added health and nutritional benefits, which could potentially reduce risks for such afflictions as osteoporosis and heart disease, are possible through biotechnology. Agricultural scientists will be able to develop industrial polymers and fiber-based products from renewable resources such as corn and soybeans. This new agricultural system has the potential to add significant value to current grain and oilseed-based feed and food ingredients markets well into the new century.

Among key trends in the new agriculture are the enormous influence being brought to the creation of products and services by consumers who demand better food, the rapid expansion of biotechnology and information technology, new business partnerships, and expanding global potential.

Agriculture of the future will focus on customer-driven innovation to improve the quality and quantity of the world's food supply. This is a significant departure from previous developments in agricultural biotechnology that primarily focused on the delivery of pest control through genetics.

For instance, the high-oleic soybeans will meet consumers' demands for lower fat and better-tasting oils. This newly developed soybean also will provide economic opportunities for farmers. Even now, with new high-oil corn there are numerous nutritional benefits for enhancing livestock feed. In fact, the future is bright for such value-added crops. Value-enhanced corn could occupy 20–25 percent of corn production areas in the United States within 10 years. Similar growth could be expected in soybean production. Some 20–25 percent of U.S. soybean acres could be planted with value-enhanced soybeans in 10–15 years.

Internationally, this new agriculture offers enormous potential. Domestic U.S. agriculture will remain competitive, and firms focused on markets here will enjoy strong demand for their products. But, for globally focused firms, the increased global feed and food demand will mean opportunities no matter what happens in the United States.

Key growth areas include China, Europe, and Argentina because value-enhanced crops are consumer-driven everywhere. The changes driving the new agriculture mean that this is a defining moment in the history of agriculture that provides opportunities to feed the world, maybe even eliminate hunger, and certainly make life better for generations to follow.

Figure 3-17 A flowering tomato.

form a cup around the pistil. The peach (*Prunus persica*) is an example.

+ Epigynous—sepals, petals, and stamens appear to arise from the top of the ovary. The concave receptacle not only surrounds the ovary, but is fused with it. In this case, the ovary is called inferior and the perianth is called superior or epigynous. An example is the apple (*Malus pumila*).

3. Flowers vary in the number of subdivisions of each of the four regular parts. The number of sepals and petals is established in all but a few of the families regarded as the most primitive—three or multiples of three of each in the monocotyledonous plants, four or five of each in the dicotyledonous plants, and reduced to none in some plants.

   There is a large and indefinite number of stamens in many flower types, but in some there is a definite number, often the same as or twice the number of petals, or even further reduced to one or two. Some flowers have a large number of separate pistils, but in others they are more or less united to form a compound pistil, and in many there is just one simple pistil.

4. Flowers vary in the symmetry of flower forms.
   + Regular or actinomorphic—floral parts, especially the corolla, are arranged symmetrically so that when quartered all sections are equal. The rose, camellia, and bush morning-glory are examples.
   + Irregular or zygomorphic—floral parts are not arranged symmetrically, and when divided horizontally the two parts are unequal and dissimilar. There are three types of irregular flowers—
     + Papilionaceous (peas or beans)—five petals of three distinct types. The sweet pea (*Lathyrus odoratus*) and lupine (*Lupinus* spp.) are examples. The three types of petals are—
       (1) Standard or banner—large petal in the uppermost part or back of the flower.

(2) Keel—two usually narrow and elongated petals in front of and usually below the standard.

(3) Wings—two petals placed to the right and left of the keel and more or less clasping the keel.

+ Labiate (mints)—tube of corolla usually deeply slit into two irregular lobes, the upper lobe erect and made up of two petals, and the lower lobe spreading or open and composed of three petals. Bilabiate means two-lipped or double-lipped corolla. Scarlet salvia (*Salvia splendens*) and the snapdragon (*Antirrhinum majus*) are examples.

+ Orchidaceous (orchids)—three sepals and three petals, with one petal, usually the lower one, modified to form the very different and variable lip. Stamens are reduced and united with the pistil to form the column. The cattleya orchid (*Cattleya* spp.) and lady slipper orchid (*Cypripedium* spp.) are examples.

Fruit

One of the characteristics of angiosperms is that their seeds are enclosed in a fruit. The fruit serves as protection for the seed and often as a means of dispersal; for example, cocklebur and beggarweed. The botanical concept of a fruit is often not what the average person would call a fruit. Examples of this are bean pods, cucumbers, eggplants, and peanut hulls.

Fruits are the ripened and seed-bearing ovaries of flowers. Fruits are nearly as varied in color, form, size, texture, and number as are flowers, making them valuable tools in plant identification. Botanists use the term fruit in a much broader way than does the layperson. Fruits are divided into two large categories—

1. Dry fruits
2. Fleshy fruits

Dry Fruits. Dry fruits are generally gray, brown, or another dull color, with a very thin and dry ovary wall, so that the food is largely confined to the seeds (Figure 3-18). These may be further subdivided based on the number of seeds and whether the fruit remains closed at maturity (indehiscent) or opens naturally (dehiscent).

+ Achene are small, hard, indehiscent, one-cavitied, one-seeded fruit with a thin, almost inseparable wall, as in the sunflower (*Helianthus annuus*).

+ Samara are indehiscent, one- or two-seeded winged fruit, as in the red maple (*Acer rubrum*).

+ Nuts are hard-shelled, usually one-seeded, indehiscent fruits, such as the walnut (*Juglans* spp.) or pecan (*Carya illinoinensis*).

+ Grain or caryopsis are one-seeded, indehiscent fruit of most grasses, including the cereals. The enclosed seed is almost inseparable from the enveloping ovary wall. This fruit is little more than a seed for all practical purposes.

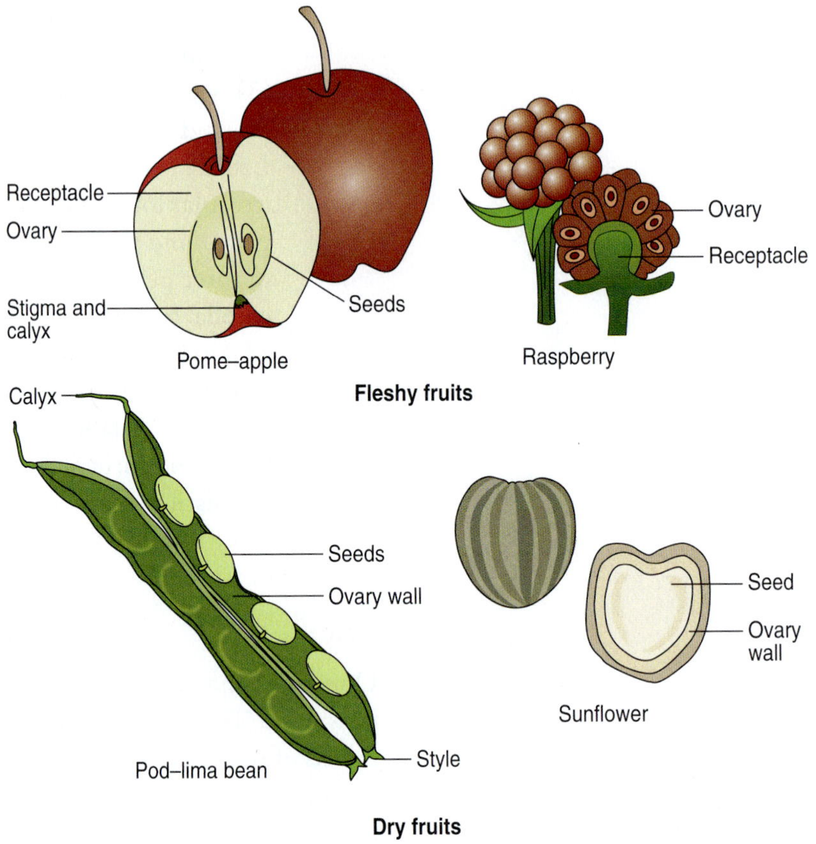

Figure 3-18 Two examples of fleshy fruits—apple and raspberry—and two examples of dry fruits—lima bean and sunflower.

Receptacle
Ovary
Stigma and calyx
Seeds
Pome–apple

Ovary
Receptacle
Raspberry

Fleshy fruits

Calyx
Seeds
Ovary wall
Pod–lima bean
Style

Seed
Ovary wall
Sunflower

Dry fruits

- Capsule is a dehiscent fruit composed of two or more carpels, generally with several or many seeds in each carpel, as in the poppy (*Papaver* spp.).
- Silique describes several-seeded fruit with two carpels that pull away from the central partition at maturity, as in the candytuft (*Iberis odorata*) and the mustard (*Brassica* spp.).
- Legume is a pod formed from a simple pistil, dehiscent along both sides, as seen in the pea (*Pisum sativum*) and candle bush (*Cassia alata*).
- Follicle is a several-seeded fruit formed from a single carpel and splitting open along one side only. Two or more follicles may be produced by each flower, as in the periwinkle (*Vinca major*) and the milk weed (*Asclepias* spp.).

Fleshy Fruits. Fleshy fruits are usually juicy and brightly colored, contrasting with their background to make them more noticeable to animals, who are responsible for their dispersal (Figure 3-18). All fleshy fruits are indehiscent, and considerable fleshy tissue is developed as the ovary changes into the fruit.

- **Drupe,** or **stone fruit,** is a simple fruit produced from a single carpel, usually one-seeded, with an outer fleshy layer of tissue called the pericarp and an inner, heavy stony layer called the endocarp. Examples include the peach (*Prunus persica*), coconut (*Cocos nucifera*), and the mango (*Mangifera indica*).

+ A **berry** is described as one or more carpels developed within a thin covering, very fleshy within, with the seeds embedded in the common flesh of a single ovary, such as the guava (*Psidium guajava*), the tomato (*Lycopersicon esculentum*), and the blueberry (*Vaccinium* spp.).

+ A pepo is a berrylike fruit of large size, with a tough or very firm and hard outer wall that is developed from the receptacle, such as the watermelon (*Citrullus lanatus*), the cucumber(*Cucumis sativus*), and the squash (*Cucurbita* spp.).

+ Hesperidium is the berrylike fruit of all the citrus group. These fruit have a thick rind with numerous oil glands, and an interior fleshy part composed of wedge-shaped compartments, with or without seeds.

+ A **pome** is fruit developed largely from the receptacle that surrounds the carpels or inedible core parts, as in the apple (*Malus pumila*) and the sand pear (*Pyrus pyrifolia*).

+ Aggregate describes a fleshy fruit developed from the receptacle of a single flower, which becomes enlarged and bears many simple, true fruits resembling achenes or drupes, as in the strawberry (*Fragaria x annassa*).

+ Multiple describes fleshy fruits derived from many closely clustered flowers, such as in the mulberry (*Morus* spp.) and the pineapple (*Ananas comosus*).

SEEDS

As the fertilized egg within the ovule develops into an embryo, the ovule walls develop into a **seed coat**, forming the ovule into a seed. The **seed** serves as the unit of dispersal for the new plant. It also provides some protection from injury and drying and some nourishment for the young plant until it can make its own food.

Seeds consist of an outer coat or wall, which is usually very tough, hard, or woody, within which is the embryo. Seeds normally have just one embryo, but sometimes have more than one, as the citrus and the mango. This results in two or more new plants growing from one seed. Seeds are developed as a result of the fertilization of the egg in the ovule of the ovary of a flower.

Typically, seeds are oval or globular and range in size from dustlike orchid seeds to the huge seed of the avocado (*Persea americana*), with some plants bearing seeds of even greater size. Seeds vary greatly in color, texture, longevity, and methods of dispersal. Some of the modifications of seeds that aid in dispersal are coverings of spines, hooks, bristles, cotton, or plumes, or

Figure 3-19 Structure of a seed.

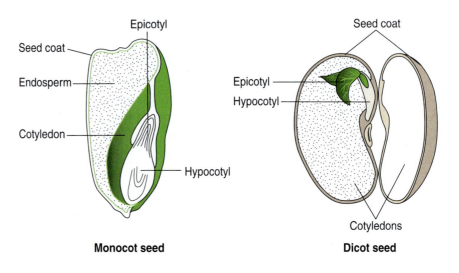

Monocot seed

Dicot seed

having wings and arils. They also vary in the types and abundance of food they contain.

Seeds serve as a means of sexual reproduction in angiosperms. The seed contains a miniature plant (embryo), which is divided into the **epicotyl** and **hypocotyl** (Figure 3-19). The epicotyl forms all plant parts above the first node of the stem and the hypocotyl forms the lower stem and roots. The embryo is surrounded by stored food (cotyledon in beans and endosperm in corn) and a final protective covering called the seed coat. The fact that seeds represent a concentrated food supply explains why they are valued by people and other animals as a food. The hard seed coat protects the tiny plant from physical injury and loss of moisture.

Germination of the seed occurs when the seed coat is softened by water and there are favorable temperatures. The embryo receives nutrients for growth from the stored food within the seed. The epicotyl grows upward out of the soil to form stems and leaves, while the lower portion of the hypocotyl grows deeper into the soil, forming roots. The epicotyl responds negatively to gravity and grows away from it, while the hypocotyl grows toward a gravitational attraction.

Cotyledon or seed leaves are the primary leaves in the embryo. In monocotyledons, the embryo contains only one seed leaf, and in dicotyledons, there are two. In monocots, such as grasses, the single cotyledon remains inside the seed and acts as a digestive organ for the embryo. Cotyledons of dicots are often pushed up out of the seed at germination and reach above ground where they develop green color and act as true leaves. In other dicots, such as the pea, the cotyledons remain below the surface and simply provide stored food for the young plant.

Gymnosperms produce seeds, but not true fruits, because they have no carpel. Such seeds are said to be naked and are borne on the inside of the scales or cones, but in plants such as the junipers and podocarpus they are imbedded in a fleshy fruitlike organ.

USING PLANT STRUCTURE

More than a quarter of a million different species of plants have been identified and named. Many of these plants can be identified by—

+ Characteristics of the leaf and its parts
+ Characteristics of the stem
+ Arrangement of the leaves on the stem
+ Flower and fruit types

In many cases, identification keys are needed to correctly identify a plant. Some of the keys are based on vegetative parts, such as the leaf and stem of the plant, in combination with the characteristics of the flower and fruit. Knowledge of these different vegetative parts of the plants is needed before identifying plants from memory or from keys.

SUMMARY

For monocots and dicots, the three basic parts of the plant are the roots, the stems, and the leaves. The three main types of roots include the taproot, the storage taproot, and the fibrous root. Root hairs are specialized cell extensions that actually make contact with the soil to absorb water and nutrients. Stems act as attachment points for leaves, flowers, and fruit. They also contain the vascular system of the plant. Like roots, stems exhibit many adaptations.

The main function of the leaves is to manufacture food and other chemicals. The main parts of a complete leaf are the blade, petiole and the stipule. Leaf shape, venation, and arrangement of the stem demonstrate considerable variation. Flowers are the reproductive parts of plants. Throughout the plant kingdom they exhibit extreme diversity in their design and location on the plant. A flower is a highly differentiated and specialize branch of the stem. The stamens and pistils and their associated parts are essential reproductive organs of the flower.

In angiosperms, the seeds are enclosed in the fruit. Fruits can be divided into two large categories: dry fruits and fleshy fruits. Seeds are the unit of dispersal for a new plant. Each seed contains an embryo.

All the variations in root, stem, flower, fruit, and seed structure are used to classify and identify plants. Identification keys are based on differences in these structures.

Review

True or False

1. Primary roots arise from the embryo.

2. Dicots do not have stems.

3. Nodes are found on the leaves.

4. Flowers can be classified as complete and incomplete.

5. Fruits are either dry or wet.

Short Answer

6. Name three types of roots.

7. List four types of stems.

8. Identify the three types of venation found in leaves.

9. List all the main parts of a typical leaf.

10. Identify all of the parts of a typical seed.

Critical Thinking/Discussion

11. Describe the difference between a complete flower and an incomplete flower.

12. Discuss four ways in which flowers vary.

13. Explain how plant structure is used to classify plants.

14. Use four terms to describe fleshy fruits.

15. Discuss the functions of stems.

16. Describe the types of dry fruits.

17. List and describe the function of the essential reproductive organs of the flower.

18. Describe a stem.

Knowledge Applied

1. Collect three plants with each of the characteristics listed below—

   A - Plants that have leaves with blade, petiole, and stipules.

   B - Plants that have leaves with blade and sheath.

   C - Plants with simple leaves.

   D - Plants with compound leaves.

   E - Plants with alternate leaves.

   F - Plants with opposite leaves.

   Place the plants between newspapers for drying. A weight, such as a heavy book, should be placed on the paper to hold the plants flat. If large plants are collected, a portion of the leaves and stems can be removed. After drying the plants, use cellophane tape or glue to fasten each plant to a single sheet of paper.

   Complete the table below for each of the plants by using the appropriate term under leaf margin and leaf shape.

Collection	Plant Number	Leaf Shape	Leaf Margin
Leaves with blade, petiole and stipules	1 2 3		
Leaves with blade and sheath	1 2 3		
Plants with simple leaves	1 2 3		
Plants with compound leaves	1 2 3		
Plants with alternate leaves	1 2 3		
Plants with opposite leaves	1 2 3		

2. Saw off a pine tree limb or tree trunk. Draw and label the growth rings. How old is the limb or tree?

3. Collect the flowers of 10 different plants. Which of the flowers are complete flowers? Which of the flowers have more than one stigma? Select one of the flowers, draw it, and then label its parts.

4. Collect and list 10 fruits with seeds contained within. Which fruit is most like the common concept of what a fruit is? Which fruit is most unlike this concept? Explain. Do any of the fruits aid in the dispersal of the seed away from the mother plant? How? Examine the seeds within the fruit. Which of these seeds appear to be those of a dicot?

5. Select a plant from a field, yard, or garden. Draw the plant and label as many parts of a mature plant as you can.

Resources

Beck, C.B. (2005). *An introduction to plant structure and development.* New York: Cambridge University Press.

McMahon, M., Kofranek, A. M., & Rubatzky, V. E. (2006). *Hartmann's plant science: Growth, development, and utilization of cultivated plants* (4th ed.). Englewood Cliffs, NJ: Prentice Hall.

Raven, P. H., Evert, R. F., & Eichhorn, S. E. (2004). *Biology of plants* (7th ed.). New York: W. H. Freeman.

Reiley, H. E. (2006). *Introductory horticulture* (7th ed.). Albany, NY: Delmar Cengage Learning.

Stern, K. R. (2005). *Introductory plant biology* (10th ed.). New York: McGraw-Hill.

United States Department of Agriculture. (1961). *Seeds: The yearbook of agriculture.* Washington, DC: Author.

Internet

Internet sites represent a vast resource of information. The URLs (uniform resource locators) for World Wide Web sites can change. Using one of the search engines on the Internet such as Google or Yahoo!, find more information by searching for these words or phrases: complete flower, monocot, dicot, germination, root systems, vegetative plants, reproductive organs of flowers, fleshy fruits, dry fruits, plant classifications, plant structure, and seeds.

Chapter 4
Anatomy of Plants

WITH AN UNDERSTANDING *of crop plant growth, agriculturists can maximize or optimize crop plant growth and/or cropping potential. An understanding of plant growth begins with an understanding of plant architecture. This architecture or structure has a hierarchy: biochemicals, cells, tissues, and organs.*

After completing this chapter, you should be able to—

- Compare eukaryotes with prokaryotes
- Describe the basic chemical composition of cells
- Draw and label all of the parts of a plant cell
- Describe the function of the parts of a cell
- List the two generalized types of tissues in plants
- Name the four categories of meristems
- Identify four types of permanent tissue
- Describe the function of xylem and phloem
- Name the types of cells found in xylem and phloem
- Identify plant tissues and describe how they are organized
- Describe the anatomy of the primary root, stems, and leaves
- Explain the difference between primary and secondary growth
- List six steps in normal cell division
- Describe what happens in each of the six steps of cell division

KEY TERMS

Cell
Cell Wall
Chloroplasts
Cytoplasm
DNA
Endoplasmic Reticulum
Epidermis
Eukaryotes
Golgi Apparatus
Leucoplasts
Meristems
Microtubules
Mitochondria
Nucleus
Palisade Cells
Parenchyma
Peroxisomes
Phloem
Plasmolemma
Prokaryotes
Protoplast
Ribosomes
Stele
Stomata
Tissue
Vacuoles
Vesicles
Xylem

THE CELL AND ITS STRUCTURE

The cell is the basic structural and physiological unit of crop plants, within which chemical reactions of life occur providing metabolites for plant life and for human use. Two types of cells exist—prokaryotes and eukaryotes.

Prokaryotes

Prokaryotes are cells without a nucleus. They have genetic materials but are not enclosed within a membrane. These cell types include bacteria and cyanophytes. The genetic material is a single circular DNA and is contained in the cytoplasm, as there is no nucleus. Recombination happens through transfers of plasmids (short circles of DNA that pass from one bacterium to another). Prokaryotes do not have internal organelles. They do not engulf solids, nor do they have centrioles or asters.

Eukaryotes

Eukaryotes are cells with a nucleus, where the genetic material is surrounded by a membrane much like the cell's membrane. Eucaryotic cells are found in humans and other multicellular organisms including plants and animals, as well as algae and protozoa. They have both a cellular membrane and a nuclear membrane. Also, the genetic material forms multiple chromosomes that are linear and complexed with proteins that help it "pack" and are involved in regulation.

Plants differ from animal cells in that they have large vacuoles, cell wall, chloroplasts, and a lack of lysosomes, centrioles, pseudopods, and flagella or cilia. Animal cells do not have the chloroplasts, and may or may not have cilia, pseudopods, or flagella, depending on the type of cell.

Eukaryotic Composition

Cells are 90 percent fluid (cytoplasm) and consist of free amino acids, proteins, glucose, and numerous other molecules. The cell environment—the contents of the cytoplasm and the nucleus as well as, the way the DNA is packed—affect the gene expression/regulations, and thus are very important parts of inheritance. The approximations of the elements of a cell include:

+ 59 percent hydrogen (H)
+ 24 percent oxygen (O)
+ 11 percent carbon (C)
+ 4 percent nitrogen (N)
+ 2 percent others—phosphorus (P), sulphur (S), and so forth

Molecules that make up the cell include:

+ 50 percent protein
+ 15 percent nucleic acid
+ 15 percent carbohydrates
+ 10 percent lipids
+ 10 percent other

In eukaryotic plant cells, the cellular contents are bounded by membranes. The first membrane from the outside in is the middle lamella. It is made of pectic substances.

Structures of a Cell

Structures of a cell include the cell wall, plasma membrane, the protoplasts, and the organelles. The organelles include plastids, mitochondria, microfilaments, endoplasmic reticulum, the nucleus, vesicles, vacuole, and Golgi apparatus (Figure 4-1).

Cell Walls. The primary cell walls are made of hemicellulose. Some cells have a thickening of the primary cell wall to add stability to the cell. This is specific to some tissues. Secondary cell walls are made of cellulose, lignin, suberin, and cutin.

Plasma Membrane. The plasma membrane is also known as the plasmolemma, or cytoplasmic membrane. It is made of a phospholipid bilayer membrane.

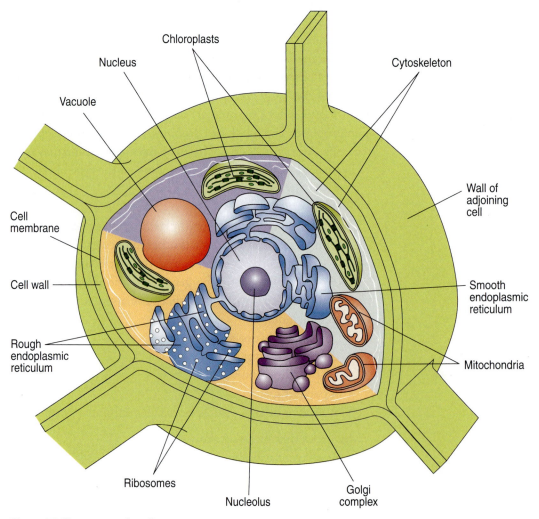

Figure 4-1 The structure of a cell.

What is a Cell and What is Living?

Cells are structural units that make up plants and animals. There are many single-cell organisms. What all cells have in common is that they are small sacks composed mostly of water. The sacks are made from a phospholipid bilayer. The membrane is a semipermeable membrane. It allows some things to pass in or out of the cell and blocks others. It also controls the transport of substances through active methods, selecting what should enter a cell.

Cells are a complex system in and of themselves. When an individual cell is added to its environment, whether it is a single- or multiple-celled organism, a complex web of reactions occur. One organism, like a plant, can have the same genetic material in every cell, yet there are many types of cells in the plant. These cells are of different shapes and sizes and carry out very different functions, but all of them developed from just one cell at the time of fertilization!

Cells are 90 percent fluid (cytoplasm), which consists of free amino acids, proteins, glucose, and numerous other molecules. The cell environment—the contents of the cytoplasm and the nucleus—affect the gene expression or regulations and are very important parts of inheritance. Cells are living.

So, what is living? This topic can start many long discussions and depends on an individual's definition. Some definitions include:

 + The quality that distinguishes a vital and functional being from a dead body or purely chemical matter
 + The state of a material, complex or individual, characterized by the capacity to perform certain functional activities, including metabolism, growth, and reproduction
 + The sequence of physical and mental experiences that make up the existence of an individual

Under these varying definitions, life may or may not include a virus that is only "alive" if it can insert its genetic material into a living cell. Perhaps the best definition of living is that the substance can react to its environment, grow, improve, and reproduce.

Protoplast. Protoplast refers to the inside of the cell or the cellular contents. The cytoplasm is the liquid matrix of the protoplast. Cytoplasm includes the water solutes, proteins, and so forth that stream through the protoplast.

Organelles. Organelles are the internal structures within the protoplast (cell). They include the plastids—

+ Leucoplasts are for the storage of oil, starch, and proteins.
+ Chloroplasts (Figure 4-2) are double-membrane plastids with chlorophyll, used in photosynthesis, storing starch, and contain genetic information (DNA).

Other organelles include the mitochondria, the nucleus, vacuoles, the endoplasmic reticulum, ribosomes, the Golgi apparatus, the peroxisomes, and the microtubules.

Mitochondria (Figure 4-3) are double-membrane bound and they are the site of respiration and the production of respiratory energy. Mitochondria convert foods into usable energy—production of adenosine triphosphate (ATP)—through aerobic respiration. The inner membrane shapes differ between different types of cells and it forms projections called cristae.

The nucleus is contained in the double-membrane nuclear envelope. It contains the chromosomes, which are long strands of deoxyribonucleic acid (DNA). DNA provides the genetic code. Note: DNA is also found in the chloroplast and mitochondria.

Figure 4-2 Chloroplast.

Figure 4-3 Mitochondrion.

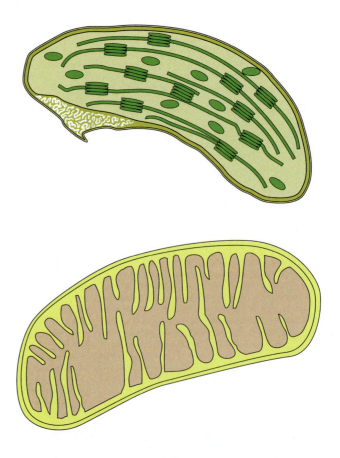

Vacuoles are surrounded by the tonoplast. They occupy the major volume of the cell and contain water solution and dissolved substances—sugars, organic acids, and pigments. The vacuole is the storage reservoir for water, sugars, salts, and other biochemicals.

Endoplasmic reticulum (ER) is important for protein synthesis. It is a transport network for molecules destined for specific modifications and locations. There are two types: rough ER has ribosomes and tends to have more of a sheetlike structure; smooth ER does not have ribosomes and tends to be more of a tubular network.

Half of the ribosomes are on the ER and the other half are free in the cytosol. Ribosomes translate the RNA into proteins.

Golgi apparatus is important for glycosylation, and secretion.

The peroxisomes use oxygen to carry out catabolic reactions.

Microtubules are made from tubulin, and they compose centrioles and cilia.

TISSUES

Tissues are large groups of organized cells of similar structure that perform specific functions in the plant. The two generalized types of tissues are meristematic and permanent.

Meristematic Tissues or Meristems

Meristems contain actively dividing cells that form new tissues. They are found in root and shoot tips, at nodes, and in the cambium of plants. Four categories of meristems are—

- ✦ Apical
- ✦ Subapical
- ✦ Intercalary
- ✦ Lateral/cambial

Apical Meristems. These are the shoots or root (at the apex or tip). They produce new buds, leaves, or modified leaf parts such as flower structures in shoots, and produce new root extension in roots. A number of permanent tissues form at meristems, including the epidermis, xylem, phloem, leaves, and shoots.

Intercalary Meristems. These meristems are separated by zones of mature tissues just above the node or at the base of leaves in many monocot species such as grass. They are not found in the dicots.

Lateral Meristems. These meristematic regions are found laterally along shoots. They are cylinders of actively dividing cells forming the conductive tissue of plants and the protective bark (cork) covering. They form the vascular cambium, capable of producing new conductive tissue of xylem and phloem. Most of the thickening and girth expansion of trees is due to xylem tissue that has been laid down by the vascular cambium. Cork cambium produces the bark of plants. It is a lateral ring of meristematic tissue found in woody plants, producing cork on the outside of the ring and parenchyma on the inside of the ring.

Permanent Tissues

Permanent tissues may be simple or complex. Simple permanent tissues are uniform. They have only one type of cell structure. These include the epidermis cells, parenchyma cells, and schlerenchyma cells.

Epidermis Cells. Epidermis cells are a single layer of cells on the exterior of stems, leaves, flowers, and fruits, and depending on the origin, roots (Figure 4-4). Sometimes the epidermis cell is two to three layers thick. These cells usually contain no pigment. The only exceptions are the stomata guard cells with chlorophyll and the apple epidermis with anthocyanin. Sometimes the cells elongate to form pubescence or hairlike structures, such as root hairs on roots or trichomes on leaves.

Parenchyma. The parenchyma is made of cells that have thin cell walls and large vacuoles. The cortex of shoots or fruits is parenchyma tissue, as is the area between phloem. In leaves, parenchyma cells contain chloroplasts for photosynthesis.

Collenchyma cells have a thickening of primary cell wall. They have a support function in some tissues.

Sclerenchyma cells have thick cell walls, which make plant fibers, or gritty cells, as in pears.

Complex Permanent Tissues. The complex permanent tissues include the conductive tissues, the xylem, and phloem (Figure 4-5). These tissues move water and solutes around the plant.

Figure 4-4 Cross-section of a root showing the stages of root hair development.

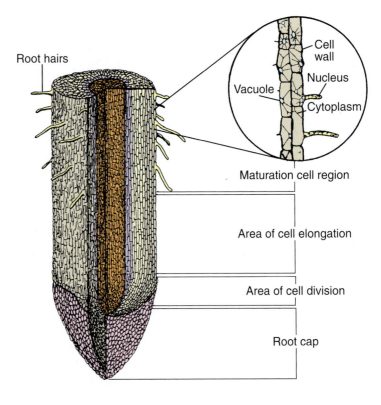

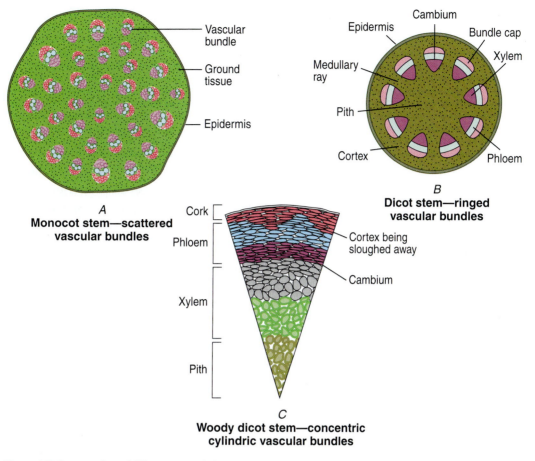

A
**Monocot stem—scattered
vascular bundles**

B
**Dicot stem—ringed
vascular bundles**

C
**Woody dicot stem—concentric
cylindric vascular bundles**

Figure 4-5 Cross-sections of different types of plant stems.

Xylem. Xylem conducts water and dissolved nutrients, amino acids, proteins, and remobilized sugars from roots to aerial portions of the plants. Water and solution moves with the water lost from transpiration or from differences in concentrations in various parts of the plants. Types of cells in xylem include vessels, tracheids, fibers, and parenchyma cells. Vessels are joined end-to-end, and the end cell walls dissolve for conduction of water. Tracheids are elongated, conductive cells, the contents of which are non-living. Fibers are thick support cells.

Phloem. Phloem conducts soluble sugars and metabolites such as proteins, hormones, dissolved minerals, and salts from leaves to other portions of the plant. Cell types in phloem include sieve tube cells, companion cells, phloem parenchyma, and phloem fibers. Sieve tube cells are long, slender tubes with porous ends (these occur only in angiosperms). Companion cells are associated with sieve tube elements. They provide energy to sieve tube cells and aid in the conduction and movement of solutes into and out of the sieve tube cells. Phloem parenchyma provides short- and long-term storage for solutes moving through the phloem. Phloem fibers provide support.

ANATOMY OF PRIMARY ORGANS

Tissues are found in the roots, stems, and leaves.

Structure of Primary Roots

Root systems may consist of one major root (taproot) or of a profuse mass of similar-sized branches. Penetration into the soil is accomplished by cell division, largely by the elongation of cells just behind the tip. A protective cap covers the tip (Figure 4-6). Just behind the region of elongation are the root hairs, which are small projections of the epidermal cells. The tremendous combined surface area of the myriad root hairs is responsible for absorption. Once absorbed, water and minerals pass through the cortex, or root wall, into the center of the root, called the stele or vascular cylinder. Here these substances are conducted upward through the tracheids and vessels of the xylem.

Structure of Primary Stems

Support is provided by various thick-walled cells found in the xylem or in strands outside the xylem. In herbaceous stems, turgor, or internal water pressure, is also important, as evidenced by the limp shape of a wilted plant.

Water and minerals are transported in the xylem, and manufactured food is transported in the phloem. In monocot stems the conducting tissues occur in separated, usually scattered, bundles, whereas in dicot stems

Figure 4-6 Cross-section of a root tip. (Image courtesy of Stephen DiCerbo)

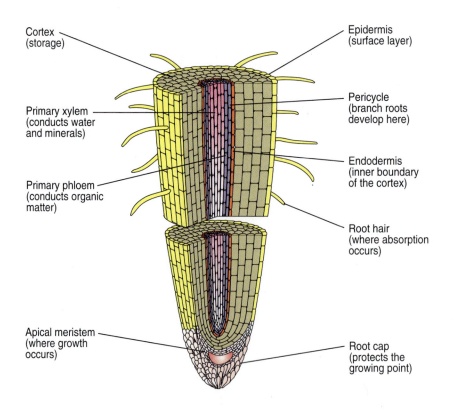

Cortex (storage)

Epidermis (surface layer)

Primary xylem (conducts water and minerals)

Pericycle (branch roots develop here)

Primary phloem (conducts organic matter)

Endodermis (inner boundary of the cortex)

Root hair (where absorption occurs)

Apical meristem (where growth occurs)

Root cap (protects the growing point)

the vascular tissues are arranged in a ring, with the primary xylem on the inside, the primary phloem on the outside, and a layer of dividing cells, called the vascular cambium, between them (Figure 4-5). The term wood in its commercial sense refers to secondary xylem. Secondary xylem is produced by the vascular cambium inward toward the center of the stem between itself and the primary xylem, increasing the thickness of the stem. The yearly production of secondary xylem usually forms a ring around that of the previous year, and these rings can be used to determine the age of a tree. In a similar manner, secondary phloem is produced by the vascular cambium outward toward the surface of the stem between itself and the primary phloem. This, too, contributes to the thickness of the stem.

Structure of Leaves

A flat, broad, thin structure gives more surface area for light interception and penetration. Where high light intensities are harmful, leaves may reduce the effects of the light by orientating themselves vertically, by becoming thickened or covered with hairs, or by having a highly reflective surface.

Intake of carbon dioxide and release of oxygen occurs through small pores (**stomata**) in the leaf surface. The stomata are mostly on the lower surface and are able to close at midday. The cells within the leaf may be formed into two layers, the upper, tightly packed with elongated **palisade cells**, and the lower, loosely packed with spongy tissue. Photosynthesis occurs mostly in the palisade cells. Figure 4-7 shows a cross-section of a leaf.

Figure 4-7 Cross-section of a leaf. (Image courtesy of Stephen DiCerbo)

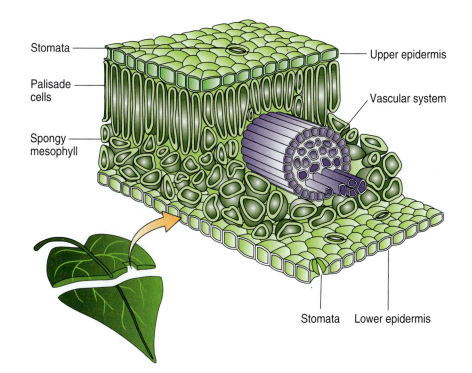

PRIMARY AND SECONDARY GROWTH

One of the general characteristics of plants, compared to animals, is that they tend to grow continuously throughout their lives. Growth serves not only to increase a plant's size but also to provide the plant with a limited means of movement and orientation for placing itself in a more favorable position with regard to light, nutrients, reproduction, and dispersal. The growth of plants involves both the production of new cells and their subsequent enlargement. Following enlargement, a cell undergoes differentiation to become a part of a specific tissue.

The two aspects of plant growth are primary and secondary. Primary growth takes place in young, herbaceous organs, resulting in an increase in length of shoots and roots. Secondary growth follows primary growth in some plants and results in an increased girth as layers of woody tissue are laid down. Monocots and herbaceous dicots typically exhibit only primary growth.

The formation of new cells takes place in regions known as meristems. At the tip, or apex, of each stem and root is an apical meristem, where cells are actively dividing (Figure 4-6). Each apical meristem produces three other meristems (protoderm, ground meristem, and procambium) called primary meristems. Tissues derived from the primary meristems are called primary tissues and include epidermis, cortex, pith, and primary xylem and phloem (vascular tissues). The elongation of cells produced by the primary meristems accounts for most of the increased length of stems and roots.

Cell Division

Mitosis is the reproduction of cells in which the genetic material of the cell is duplicated exactly. The cells simply divide and produce new cells like themselves. Daughter cells have the same genetic makeup as parent cells. The steps of mitosis are shown in Figure 4-8 and described in the next paragraphs.

Step 1: The Resting Stage, or Interphase. This is the period between one division and the next. Individual chromosomes are not visible and the nuclear membrane is visible.

Although chromosomes cannot be seen, they are present inside the nucleus. Remember that chromosomes are the parts in the nucleus that control inherited traits.

During this period, a most important event takes place; each chromosome makes an exact copy of itself. For example, if there are four chromosomes (or "two pair") in a resting cell, there are eight after copying.

Step 2: Preparing to Divide, or Prophase. The two identical chromosomes are joined together. There are then two complete sets of chromosomes in the nucleus. The chromosomes are now coiled tightly. The coiled chromosomes can be seen through a microscope. The nuclear membrane begins to disappear at this time. Given four chromosomes (two pair)—a; A; B; b—duplication in prophase creates aa; AA; BB; bb.

Figure 4-8 Mitosis is the process of division or duplication of a typical cell.

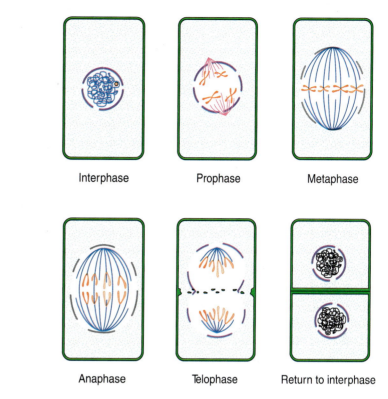

Interphase   Prophase   Metaphase

Anaphase   Telophase   Return to interphase

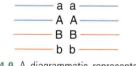

Figure 4-9 A diagrammatic representation of the lining up of chromosomes at metaphase, Step 3 of mitosis.

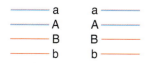

Figure 4-10 A diagrammatic representation of four chromosome pairs at the end of anaphase, Step 4 of mitosis.

Step 3: Beginning of Mitosis, or Metaphase. There is no longer a division between the nucleus and the cytoplasm. Chromosomes become thicker and shorter, and are now easier to see. The nuclear membrane has faded away. If the cell in Step 1 had four chromosomes, four double chromosomes could now be seen.

After making a copy, each chromosome appears as a doubled chromosome. The original and copy appear side by side and are attached to each other. The pairs of identical chromosomes line up in the center of the cell. Thin fibers, called spindle fibers, attach to the chromosomes. These fibers define the direction in which the chromosomes will later move into the two daughter cells.

At the end of Step 3, the chromosome pairs are lined up as in Figure 4-9.

Step 4: Mitosis Continuing, or Anaphase. The pairs of identical chromosomes separate from one another. The chromosomes move toward the center of each new cell. As one member of each pair moves to one end of the cell, the other member of the pair moves to the other end of the cell. There are now two groups of identical chromosomes at opposite ends of the cell. These two groups are composed of four identical chromosomes each. The chromosomes are once again single-stranded.

Diagrammatically, the four chromosome pairs will look like the illustration in Figure 4-10 after Step 4.

Step 5: Mitosis, or Telophase. A nuclear membrane begins to form around both sets of chromosomes. The chromosomes uncoil and no longer appear as distinct structures. The cytoplasm separates as new cell membrane (or membranelike material) forms in the middle of the old

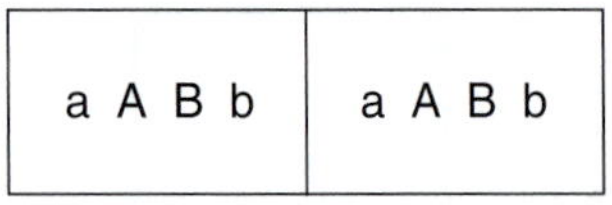

| a A B b | a A B b |

Figure 4-11 A diagrammatic representation of two cells after telophase, Step 5 of mitosis.

parent cell. This is called furrowing in animal cells, in which the cell pinches on all sides until two daughter cells are formed. Because of the structure of plant cells, they do not furrow; instead a "cell plate" is formed between the two daughter cells.

The original parent cell has become two new daughter cells with identical chromosomes. These two new cells are smaller than the original cell from which they came. Each one may grow and may divide again. There are now two cells and each again has four chromosomes, the same number as the original parent cell. The two cells after telophase would look like the illustration in Figure 4-11.

Step 6: Interphase. The new cells return to the Interphase.

SUMMARY

Plant cells are eukaryotic cells. Their cellular contents are bounded by the cell wall and the plasma membrane. Within the protoplast, the organelles of the cell exist. These organelles include the mitochondria, nucleus, vacuoles, endoplasmic reticulum, ribosomes, Golgi apparatus, peroxisomes, and the microtubules. Organelles serve specific functions in plant cells. Cells organized in groups that perform a specific function and are of similar structure form the tissues of plants. Meristematic and permanent are the two generalized types of plant tissues.

Meristems are actively dividing into new tissues. Root tips, stem tips, nodes, and the cambium contain meristematic tissues. Permanent tissues can be simple or complex. The simple permanent tissues include the epidermis cells, the parenchyma, collenchyma, and sclerenchyma cells. Xylem and phloem are complex permanent tissues. These tissues are found in the primary roots, the primary stems, and the leaves.

Growth increases the plant size and to some extent provides a means of movement. Growth occurs through an increase in cell number (mitosis) and cell size. Plants exhibit primary and secondary growth.

Review

True or False

1. Xylem is the site of secondary growth.

2. Most plants used for crops are composed of prokaryotic cells.

3. In the cell, the endoplasmic reticulum contains the DNA.

4. Parenchyma is a meristematic tissue.

5. Metaphase is the first step in mitosis.

Short Answer

6. List four of the chemical elements and four of the biochemical molecules found in plant cells.

7. Name four categories of meristems.

8. List the six steps of mitosis.

9. Name six parts of a plant cell.

10. Identify types of permanent tissue in plants.

Critical Thinking/Discussion

11. Define growth and explain the difference between primary and secondary growth.

12. Describe the function of xylem and phloem.

13. Explain the function of the cell wall, the endoplasmic reticulum, the nucleus, the chloroplast, the mitochondria, and ribosomes.

14. What is the difference between meristematic and permanent tissues in plants?

15. Diagram the structure of xylem and phloem.

Knowledge Applied

1. From a biological supply company, obtain prepared microscope slides of roots, stems, and leaves. Identify the epidermis cells, parenchyma cells, xylem, phloem, stomata, guard cells, and palisade cells. Make sketches of each type of cell or tissue. As an alternative, search the Internet for sites with photographs of tissue cross-sections.

2. Draw and label your own diagram of a typical plant cell. Label all of the components of the cell.

3. Draw and label a microscopic cross-section of a stem showing the xylem and phloem.

4. Draw and label a microscopic cross-section of a growing root.

5. Conduct an experiment to find out what would happen to a leaf if its lower surface were covered with a film of petroleum jelly. Explain the results.

6. Make a microscopic cross-sectional drawing that compares the monocot stem to the dicot stem.

Resources

Asimov, I. (1954). *The chemicals of life.* New York: New American Library.

Connors, J. J., & Cordell, S., eds. (2003). *Soil fertility manual.* Tucson, AZ: Potash & Phosphate Institute.

Janick, J., ed. (1989). *Classic papers in horticultural science.* Englewood Cliffs, NY: Prentice Hall.

McMahon, M., Kofranek, A. M., & Rubatzky, V. E. (2006). *Hartmann's plant science: Growth, development, and utilization of cultivated plants* (4th ed.). Englewood Cliffs, NJ: Prentice Hall.

Schmidt, D., Allison, M. M., Clark, K. A., Jacobs, P. F., & Porta, M. A. (2005). *Guide to reference and information sources in plant biology* (3rd ed.). Portsmouth, NH: Libraries Unlimited.

Equipment and Supplies

Carolina Biological Supply Company, Carolina Science and Math Catalog 66, 2700 York Road, Burlington, NC 27215-3398.

Fisher-EMD, 4901 W. LeMoyne Street, Chicago, IL 60651.

Nebraska Scientific, 3823 Leavenworth Street, Omaha, NE 68105-1180.

Internet sites represent a vast resource of information. The URLs (uniform resource locators) for World Wide Web sites can change. Using one of the search engines on the Internet such as Google or Yahoo!, find more information by searching for these words or phrases: cell structure, plant growth, anatomy of plants, plant tissues, plant cells, and cell division.

PART 2

Fundamentals
of Soil Science

Chapter 5
Soil Materials and Formation

SOIL IS THE SOFT outer covering of the earth. It is one of our most important natural resources. Soil is necessary for plants to grow. Soil provides food for plants, which in turn furnish food for humans and animals. Five factors influence soil formation: parent material, time, climate, organisms, and topography. Soil profiles are used to study soils.

After completing this chapter you should be able to—

- Discuss what soil is and where it comes from
- Describe soil layers and how they differ
- Discuss how plants depend on soils for growth
- Define a soil body
- List examples of the five soil-forming factors
- Describe how soils develop
- Describe the horizons of the soil profile

KEY TERMS

Alluvial Fan
Alluvial Soil
Chemical Weathering
Colluvium
Delta
Ecosystem
Eluviation
Eolian Deposit
Floodplains
Frost Wedging
Glacial Drift
Glacial Outwash
Glacial Till
Hydrolysis
Igneous Rock
Illuviation
Lacustrine
Leaching
Levee
Loess
Marine Sediment
Master Horizon
Metamorphic Rock
Mineral Soil
Organic Soil
Oxidation-Reduction
Parent Material
Pedology
Pedon
Physical Weathering
Plow Layer
Polypedon
River Terrace
Root Wedging
Sedimentary Rock
Soil Genesis
Soil Horizon
Soil Profile
Soil Series
Solum
Subsoil
Talus
Topography
Topsoil
Weathering

INTRODUCTION TO SOILS

Soil is created through a variety of processes and over long periods of time. Heating and cooling or **weathering**, the forces of water and wind, all help to create soil. Decomposed plant and animal material also add to the makeup of soil. Healthy soil (Figure 5-1) is critical to the success of agriculture. Care and conservation of the soil is vital to the continued success of agriculture.

Definition of Soils

Soil is the soft material that covers the surface of the earth and provides a place for the growth of plant roots. Soil is a vital part of the **ecosystem**. An ecosystem is a natural community in which different kinds of plants and animals live closely together, getting the things they need to sustain life from within the community and its natural resources.

Soil provides plants with food and water. It provides a home for many small animals as well. Without soil, most plants would not survive. Along with air and water, soil provides the basis for natural ecosystems. Soils are just as important in agriculture as in natural ecosystems. The long-term health of agriculture, and of human society, depends on the health of the soil. Soil is a very slowly renewable resource.

Study of Soils

Pedology is the study of soil formation, also known as **soil genesis**, and soil classification and mapping, covered in Chapter 6. Modern pedology dates to the eighteenth and nineteenth centuries in Germany, the United States, and especially Russia. Early researchers developed concepts of the soil as an evolving body arising from weathered rocks of the crust under a variety of influences.

Figure 5-1 Students and instructor closely inspect good soil.

The Soil Body

The soil is a collection of natural bodies of the earth's surface containing living matter that is able to support the growth of plants. It ends at the top where the atmosphere or shallow water begins. It ends at the bottom at the farthest reach of the deepest rooted plants. The soil varies across the landscape: in one area it may be mostly made of decayed plant parts, in another place it may be mostly sand.

Not everything can be learned about a soil just by standing on the surface. One must dig a hole to see what it looks like below the surface. Because a soil scientist cannot dig up acres of ground to study a whole body of soil, the soil is broken up into small parts that can be easily studied. This small body is called the **pedon** (Figure 5-2). A pedon is a section of soil, extending from the surface to the depth of root penetration of the deepest rooted plants, but generally examined to a depth of 5 feet.

Generally, a pedon has dimensions of about 1 meter by 1 meter, and about 1.5 meters deep (about 3 feet × 3 feet × 5 feet). Soil scientists use the pedon as a unit of soil easily studied by digging a pit in the ground (Figure 5-3).

The traits of a pedon are set by the combination of factors that formed it. In the landscape near the pedon being studied are other pedons that are probably very similar. As one moves across the landscape, however, one will reach a pedon that is different because the combination of factors that formed it were different. A collection of pedons that are much the same is called a **polypedon**. These polypedons are mapped into units called a **soil series**.

Formation of Soils

All soil begins with solid rock. Since the earth's beginning, wind, running water, rain, earthquakes, landslides, and other forces of nature have

Figure 5-2 Two soil pedons show how each relates to the polypedons and the total landscape.

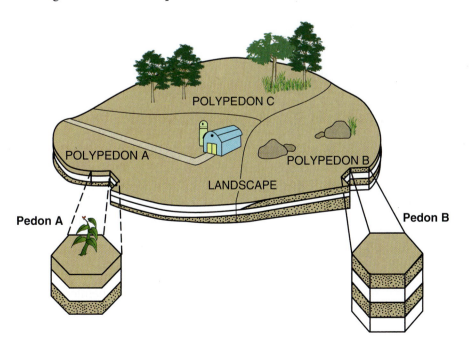

Figure 5-3 Soil pit dug in a grain field so the instructor can show and describe the soil to students.

changed rocks into soil. These changes started millions of years ago and take place very slowly. Soil is still being formed today.

The natural process by which rock is broken into smaller pieces is called weathering.

Physical weathering is the disintegration of rock by temperature, water, wind, and other factors. For instance, in cold climates **frost wedging** occurs when water freezes and expands in rocks or in cracks in the rock, causing it to break apart. The alternate expansion and contraction of rock caused by heating and cooling cycles also stresses the fabric of rock. Both cause rock to fracture, or outer layers to peel away. Rain, running water, and windblown dust also wear away at rock surfaces.

Chemical weathering changes the chemical makeup of rock and breaks it down. The simplest process is a solution. Rainwater is mildly acidic and can slowly dissolve many soil minerals, as in the solution of lime, calcium carbonate:

$$CaCO_3 + 2H^+ \rightarrow H_2CO_3 + Ca^{+2}$$

In this reaction, lime dissolves in acidic water by reacting with hydrogen ions to form carbonic acid (H_2CO_3) and calcium ions. In **hydrolysis**, water reacts with minerals to produce new, softer compounds, as in the hydrolysis of the feldspar mineral orthoclase to a softer feldspar mineral ($HAlSi_3O_8$):

$$NaAlSi_3O_8 + H_2O^- \rightarrow HAlSi_3O_8 + Na^+ + OH^-$$

Hydration also involves water here but the molecule itself joins the crystalline structure of the mineral (Fe_2O_3), creating a softer, more easily weathered mineral:

$$Fe_2O_3 + 3H_2O^- \rightarrow 2Fe(OH)_3$$

Much of the chemical weathering involves water interacting with crystalline minerals to create new materials by dissolution, hydrolysis, or hydration. Not surprisingly, moisture availability is an important variable in soil formation.

Figure 5-4 Lichens grow on rocks and aid in the formation of soil.

Oxidation-reduction and other reactions are also important in chemical weathering.

Plants also play an important role in rock crumbling. Roots can exert up to 150 pounds per square inch of pressure when growing into a crack in rock. Root wedging from the pressure pries apart stone.

Lichens growing on bare rock (Figure 5-4) form mild acids that slowly dissolve rock. When lichens die, its dry matter is added to the slowly growing mixture of mineral particles and organic matter. When a small bit of soil forms in a rock crevice, plants begin to grow from seed that has blown into the crevice, continuing the cycle.

Soil formation does not stop when a layer of young soil covers the surface. The new soil continues to slowly age and develop over thousands of years. Soil scientists identify five factors operating during the process of soil formation and development. A brief overview of the rocks of the earth's crust will be helpful before describing the five factors.

ROCKS AND MINERALS

The original source of most soils is rock—the solid, unweathered material of the earth's crust. Solid rock breaks into smaller particles, which are the parent materials of soil. Rock is a mixture of minerals that supply plant nutrients when broken down. Geologists classify rock into three broad types: igneous, sedimentary, and metamorphic. Refer to Figure 5-5 to aid in understanding the discussion of rocks in the next paragraphs.

Igneous Rock

The basic material of the earth's crust is igneous rock, created by the cooling and solidification of molten materials from deep in the earth. Igneous rocks, such as granite, contain minerals that supply 14 of the required plant nutrients (see Chapter 7). Granite, which is mined for

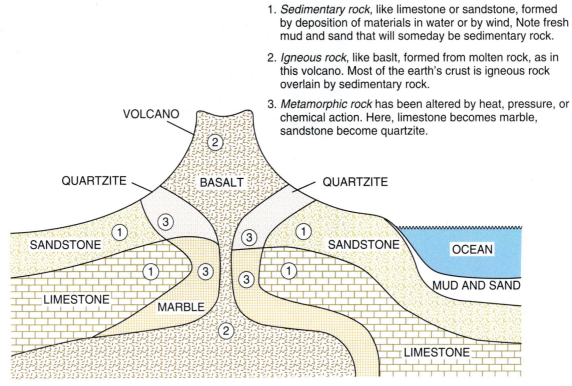

1. *Sedimentary rock*, like limestone or sandstone, formed by deposition of materials in water or by wind, Note fresh mud and sand that will someday be sedimentary rock.

2. *Igneous rock*, like baslt, formed from molten rock, as in this volcano. Most of the earth's crust is igneous rock overlain by sedimentary rock.

3. *Metamorphic rock* has been altered by heat, pressure, or chemical action. Here, limestone becomes marble, sandstone become quartzite.

Figure 5-5 Three types of rock: sedimentary, igneous, and metamorphic.

monuments and building material, is a hard, coarse-grained rock made of feldspar, quartz, and other minerals. Feldspar, a fairly soft mineral containing potassium and calcium, weathers easily to clay. Quartz, a very hard and resistant mineral, weathers slowly to sand. Table 5-1 lists the nutrient content of two sample igneous rocks: granite and basalt. Granite tends to weather slowly to create acidic parent materials high in sand, while basalt, a softer, finer-grained rock, weathers more quickly to less acidic materials low in sand.

Table 5-1 Composition of several igneous and sedimentary rocks of Minnesota, according to the Minnesota Geological Survey

Minerals (Percent)	Gray granite	Basalt	Hinckley sandstone	Platteville limestone
Quartz	64	49	94	7.5
Feldspars, other	20	20	2	—[a]
Calcite, dolomite	7	15	5	90
Elements (Pounds per Ton of Rock)				
Ca	69	150	—	704
K	66	17	—	—
Mg	36	66	—	18
Fe	23	35	15[b]	—
P	5	5	—	—
Mn	—	3	—	—

[a] Dashes mean only trace amounts.
[b] Estimated.

Sedimentary Rock

Igneous rock comprises only about one-quarter of the earth's actual surface, even if most of the crust is igneous. This is because **sedimentary rock** overlays about three-quarters of the igneous crust. Sedimentary rock forms when loose materials like mud or sand are deposited by water, wind, or other agents, slowly cemented by chemicals and/or pressure into rock. Much of the sedimentary rock covering North America was deposited in prehistoric seas.

The parent materials of many American soils derive from sandstone and limestone. Sandstone, which consists of cemented quartz grains, weathers to sandy soils. Generally, these soils are infertile and droughty. Limestone is high in calcium and weathers easily to soils high in pH, calcium, and magnesium. Table 5-1 lists contents of a typical sandstone and limestone.

Metamorphic Rock

If igneous and sedimentary rocks are subjected to great heat and pressure, they change to form **metamorphic rock**. For instance, limestone is a fairly soft, gritty rock. When subjected to heat and pressure, it changes to marble, which is harder and can be cut and polished. Soils arising from metamorphic parent materials resemble soils from the original sedimentary or igneous rock.

FIVE FACTORS OF SOIL FORMATION

The five factors influencing soil formation include:
1. Parent material
2. Time
3. Climate
4. Organisms
5. Topography

All five of these factors acting together result in the many different kinds of soils. Some have suggested that human activities might be named a sixth factor because most soils have been modified to some degree by humans.

Parent Material

Parent material refers to the rock or other material in which the soil formed. The parent material provides the original minerals from which the soil develops. The soil forms by weathering of these original minerals and through additions of new material or losses of other material. Weathering refers to the changing of the original material to new materials through the action of natural forces. It may be physical weathering, in which larger pieces of rock are simply broken into smaller ones by water, wind, plant roots, ice, or gravity, or it may be chemical weathering, in which minerals are changed to different minerals by reactions with other substances in the environment. Rain is very important in weathering. Most rain is slightly acidic. These acids dissolve some soil minerals and change others to new forms. Rainwater passing through the soil moves dissolved substances deeper in the soil. This is called **leaching**.

Some parent materials are very resistant to weathering. These materials result in shallow soils with little development. Other parent materials weather easily, resulting in deep soils. The most common parent material is limestone rock. Other rocks include sandstone and shale. Most soils are formed in beds of sand, silt, or clay put there by water or wind. Many of the best soils have formed in silty, wind-deposited material called **loess**.

The parent materials provide the starting point for soil formation. The final product depends on the effect the other four factors have on weathering and the additions and losses of material. Figure 5-6 shows the distribution of various parent materials across the United States.

Glacial Ice. Glacial ice carried parent materials over the northern part of North America (Figure 5-7) during numerous glacial periods over the past 2 million years. The last four periods left the most evidence, and the most recent glacier, that of the Wisconsin period, reached its peak expanse about 18,000 years ago and melted back out of the United States about 10,000 to 12,000 years ago. Most of the glacial deposits that cover the northern states come from the Wisconsin ice sheet. Glaciers expanded out

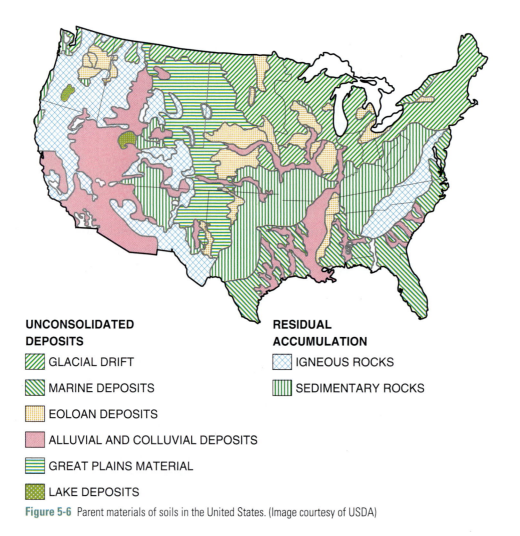

UNCONSOLIDATED DEPOSITS

- GLACIAL DRIFT
- MARINE DEPOSITS
- EOLOAN DEPOSITS
- ALLUVIAL AND COLLUVIAL DEPOSITS
- GREAT PLAINS MATERIAL
- LAKE DEPOSITS

RESIDUAL ACCUMULATION

- IGNEOUS ROCKS
- SEDIMENTARY ROCKS

Figure 5-6 Parent materials of soils in the United States. (Image courtesy of USDA)

Figure 5-7 Glaciers have covered much of North America several times.

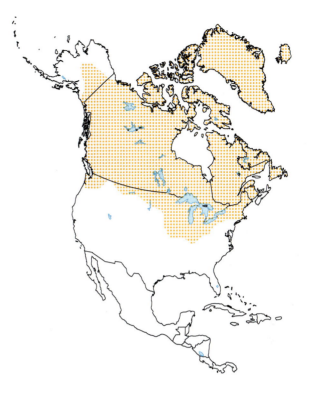

of several centers in Canada, carving and grinding the earth, picking up and transporting soil, gravel, rocks, and other debris. As the glaciers melted and shrunk between glacial periods, transported material remained in deposits called **glacial drift**. In the process they left behind a very distinctive landscape over much of the northern United States and Canada.

Glaciers deposited materials in many ways, so there are several kinds of glacial drift. During the melting process, some debris simply dropped in place to form deposits called **glacial till**. Because there was no sorting action in the deposition, glacial till is extremely variable, and so are the soils derived from it. Till soils often contain pebbles, stones, and even boulders. Other materials carried by the glacier washed away in meltwater to form sediments in streams and lakes. During the process, the materials were sorted by size. Coarser material, being larger and heavier, was deposited near the glacier and in streams and rivers to form **glacial outwash** Outwash soils tend to be sandy. Smaller particles reached glacial lakes to form **lacustrine** deposits on the lake bottoms.

Wind. Some parent materials were carried by wind, leaving **eolian deposits**. For example, some soils in Nebraska formed from sand dunes, deposits of sand carried by rolling in the wind. Most eolian soils in the United States are a result of the last glacial period.

After the last glaciers melted and meltwaters subsided, large expanses of land were exposed to a dry climate with strong westerly winds. Winds picked up silt-sized (medium) particles and deposited them in the Mississippi and Missouri River valleys and elsewhere. These loess soils—wind-deposited silt—are important agricultural soils in much of Iowa, Illinois, and neighboring states.

Figure 5-8 Water- and marine-deposited soils. (1) Flood plains form along rivers from materials deposited during flooding; (2) alluvial fans are deposited at the base of slopes by running water; (3) deltas form when smaller particles drop out as a river enters an ocean; (4) river terraces are old flood plains left above a new river level.

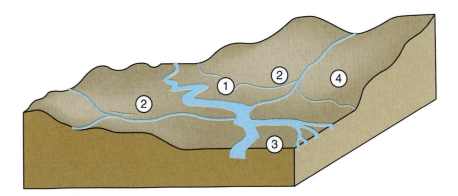

Water. **Alluvial soils** are soils whose parent materials were carried and deposited in moving fresh water to form sediments (Figure 5-8). Alluvial materials can be deposited in several ways. **Alluvial fans** form below hills and mountain ranges where streams flowing down the slope deposit material in a fan shape at the base. As water speed slows abruptly at the foot of the slope, large particles drop out first. As a result, alluvial fans are generally sandy or gravelly. Finer materials are carried away in rivers.

Flooding rivers also leave behind deposits. Coarser materials are often deposited in low ridges, or **levees**, along the river bank. Away from the river, floodwaters spread over large flat areas called **floodplains**. Here the water will be shallow and slow-moving; fine particles will settle out. Floodplains tend to be fertile because new soil is added at each flood, but the soils tend to stay wet. Levees, being coarser and elevated, dry more quickly. Floodplain soils are especially important along the Mississippi River and its tributaries, and along rivers that flow into the ocean on the East and Gulf coasts. Many important soils of California are from river alluvium.

Sometimes a river will cut deeply into its floodplain to flow at a lower elevation. This establishes a new riverbed and floodplain, while the old floodplain is left higher as a **river terrace**. An example of river terrace soils is some soil of the San Joaquin Valley of California.

Lacustrine deposits form under still, fresh water. Most of our lacustrine soils remain from giant glacial lakes that have since dried up. Examples include Glacial Lake Agassiz of northern Minnesota, North Dakota, and Canada, and Glacial Lake Bonneville of Utah. When glacial runoff water ran into the lake, the heaviest materials were left near the shore, while the smallest particles were carried to the center of the lake. Thus, lacustrine soils are sandy near the old shoreline and grade to soils with smaller particles toward the old lake center.

Marine sediments form in the ocean. Many scattered soils of the Great Plains and the Imperial Valley of California are beaches of prehistoric seas that once covered the United States. Other beach soils are common along the Atlantic coastline and the Gulf of Mexico.

These beach soils all tend to be sandy soils. Deltas, in contrast, have very small particles and tend to be wet. Deltas form when rivers flowing into an ocean deposit sediments at the mouth of the river. The Mississippi River Delta of Louisiana is a prime example, as is the Rio Grande Valley of Texas and Mexico.

Gravity. Some parent materials move simply by sliding or rolling down a slope. This material, called colluvium, is scattered in hilly or mountainous areas. An example of a colluvial material is a talus—sand and rocks that collect at the foot of a slope. Avalanches, mudslides, and landslides are other examples.

Volcanic Deposits. The ash blown out of a volcano and deposited nearby or carried some distance by wind forms a chemically distinct, dark, and lightweight parent material. The Pacific Northwest, Hawaii, and Alaska are areas of the United States where such deposits are common.

Organic Deposits. Characteristics of the soils formed from parent materials described so far are set by mineral particles in the soil. Mineral soils contain less than 20 percent organic matter, except for a surface layer of plant debris. Organic soils, containing 20 percent or more organic matter, form under water as aquatic plants die. Low oxygen conditions under water retard decay of these dead plants, so they tend to pile up on the lake bottom. Eventually the lake fills in and is replaced by an organic soil. Organic soils are extensive in Minnesota, Wisconsin, and Florida.

Time

Soils change over time, undergoing an aging process. Initially, a thin layer of soil forms on the parent material. Such a young, immature soil takes as little as a hundred years to form from well-weathered parent materials under warm, humid conditions. Under other conditions, it may take hundreds of years.

Weathering of the young soil continues, and many generations of plants live and die, so the young soil becomes deeper and higher in organic matter. If there is enough rainfall, leaching begins to carry some material deeper into the soil, creating the soil profile described later in this chapter.

As soils age, biological processes tend to increase the nitrogen content, while leaching tends to reduce phosphorus. Thus, young soils tend to be low in nitrogen but high in phosphorus, whereas older soils are the opposite. Mature soils are generally productive, but as soils continue to age, they become more severely weathered, more highly leached, and often less productive. In general, as soil ages it becomes deeper, develops distinct layers, and becomes more acidic and leached.

However, the aging process is not static. Time zero for a soil usually begins when some dramatic event such as a landslide or a glacier changes everything and resets the clock. Such events can happen at any time. A soil might age through the years until it reaches some steady state and remains unchanged thereafter, but this is rare. Soils can erode away, be buried, or even become the parent material for a new soil. If soil factors change, the direction of soil development can be deflected into a new path. For instance, if forest invades prairie, the soil embarks on a new path toward a forest-type soil.

Climate

Climate first affects soils by causing physical and chemical weathering of rock. However, climate continues to affect soil development long beyond this initial stage. The main effects are due to temperature and rainfall.

Temperature affects the speed of chemical reactions in the soil—the higher the temperature, the faster a reaction. Chemical weathering in soils occurs mostly when the soil is warmer than 60°F. Thus, in cold areas, like tundra, soils develop slowly. In warm areas, like the tropics, soils develop more rapidly.

Another result of temperature is its effect on organic matter. Warmth promotes greater vegetation, so more organic matter is added to the soil. However, warm temperatures also speed the decay and loss of organic matter. Thus, soils of warm climates tend to be low in organic matter.

Rainfall affects soil development mainly by leaching. Leaching moves materials deeper into the soil via water moving downward through the soil. Leached materials include lime, clay, plant nutrients, and other chemicals. These materials are then deposited in lower parts of the soil.

High rainfall areas also tend to grow more vegetation, so the soils of humid areas tend to have more organic matter than soils of drier regions. In summary, rainfall tends to cause leaching and the accumulation of organic matter.

The United States is a good example of the effects of climate on soil (Figure 5-9). The climate of the United States cools from south to north. This is reflected in an increase in average organic matter content from south to north. Also, the most weathered soils in the United States are in the South. The average rainfall of the United States increases from west to

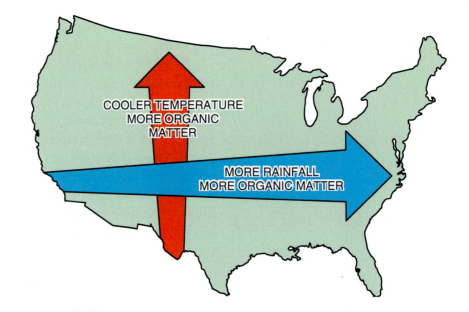

Figure 5-9 One effect of climate on United States soils. The average organic matter of soils increases to the east and north because of cooler temperatures and higher rainfall. Other factors also affect organic matter, like vegetation, which alters these general rules regionally.

Figure 5-10 Native vegetation of the United States. (Courtesy of USDA)

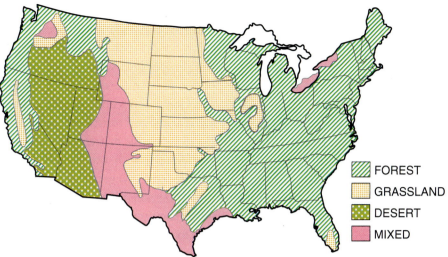

FOREST
GRASSLAND
DESERT
MIXED

east. As a result, the organic matter content of the United States' soils also tends to increase from west to east.

Soil color also follows north-south trends. Because organic matter is black, soils tend to appear darker as one moves from warmer to cooler climates. Because of changes in chemical reactions involving iron, soils tend to appear redder as one moves from cooler to warmer climates.

Organisms

Organisms that live in soil—like plants, insects, and microbes—actively affect soil formation. The actual properties of a developing soil are influenced especially by the type of plants growing in it. Figure 5-10 shows the parent vegetation of soils of the United States.

Mineral soils having the highest organic matter content form under grasslands. Grasses usually have a dense mat of fibrous roots, some of which die each year. This keeps the organic matter content high and the

Figure 5-11 Soils of the Sonoran Desert near Phoenix, Arizona, are influenced by the sparse, unique vegetation and arid climate.

soil color dark. In a forest, much of the organic material is above ground in the trees. When the leaves fall or the tree dies, the material falls to the soil, where it creates a surface layer of organic matter that does not mix with deeper layers. As a result, forest soils have less organic matter than prairie soils and are lighter in color. The type of trees also influences the soil. Compared with hardwoods (deciduous trees), softwood (conifer) foliage is acidic and resistant to decay; therefore, their soils tend to be thinner, lower in organic matter, and more acidic.

Deserts, with very sparse vegetation, have the least organic matter (Figure 5-11). Vegetation also affects the location of nutrients and other ions in the soil. Plants absorb ions in the roots and carry them to the tops, where they are returned to the soil surface when leaves drop. This recycles ions from deeper in the soil to the surface and helps reduce their loss from leaching. Deep tree roots, for example, extract ions from deep in the soil, leaving the surface horizon of forest soils enriched in ions.

We tend to stress vegetation as the main living factor in soil formation, but other life impacts soil as well, such as burrowing animals that bring subsoil to the surface, earthworms that create large, deep pores and speed organic matter decay, or nitrogen-fixing bacteria.

Topography

Topography, or the soil's position in the landscape, influences soil development mainly by affecting water movement. Water runs off slopes, making them drier, and collects in low areas, making them more moist. This, in turn, affects leaching, chemical reactions, and types of vegetation. Slope effects vary according to a number of characteristics: steep or south-facing slopes are drier than gentle or north-facing slopes, and the top portion of a slope is drier than its bottom portion.

If enough water runs off a slope, it may carry away soil as fast as it is formed. Thus, soil may be thin on a slope and thick at its base. The effect of topography is most obvious in rolling fields, where sloped areas are light brown from topsoil loss and lower areas are black from accumulating topsoil and organic matter.

Because running water tends to carry off smaller particles, soils in lower areas may be finer than those of higher areas. Depressions may also intersect the water table at least part of the year, keeping them wet for long periods.

Humans

Humans may be considered just another living entity that modifies soil, but their action can be so rapid, dramatic, and different from other life that they might be considered a separate, sixth factor in soil formation. Very few soils have been unaffected by human activities. Effects may be as subtle as the deposition of air pollutants distant from any human habitation to as massive as earthmoving during road construction. The latter resets the time clock for this new soil material to zero, and the earth moved by the machinery is the parent material for this new soil.

Dirt

To the soil scientist it is not dirt but soil. Still, the word dirt often means soil, and soil means dirt. For example, soil or dirt is the part of the earth's surface consisting of humus and disintegrated rock or it can be the state of being covered with unclean things, sort of like being dirty. Here are some examples of how dirt is used in expressions when soil might not work:

+ Pay dirt was most commonly used as a phrase in times of mining, particularly in the times of the California gold rush circa 1849.
+ A dirt dive is to rehearse a skydive on the ground. A soil dive just does not sound right.
+ Dirt bikes and dirt jumping are names given to the practice of riding special motor bikes over shaped mounds of dirt/soil.

+ A dirt ball is a baseball term meaning a pitched ball that lands in the dirt, usually just in front of or alongside the plate. Unrelated to baseball, a dirt ball can also mean a person who has a nasty or unethical character undeserving of respect. Using the term soil ball in these instances just would not be the same.
+ Dirt cheap means very cheap. Good soil is hardly dirt cheap.
+ Finally, The Nitty Gritty Dirt Band is an American country-folk-rock band existing in various forms since its founding in California in 1966. Can you imagine The Nitty Gritty Soil Band?

Dirt can be used to refer to soil unless a soil scientist is nearby, but soil cannot always replace dirt in our vocabulary.

SOIL PROFILE

Soils change over time in response to their environment, represented by the soil-forming factors. Soil scientists have classified the causes of those changes into four soil-forming processes:

1. Additions: materials may be added to the soil; some examples are fallen leaves, wind-blown dust, alluvium, and man-made materials like air pollutants and compost.

2. Losses: materials may be lost from the soil as a result of deep leaching, erosion from the surface, or as gases filtering out of the soil.

3. Translocations: materials may be moved within the soil by leaching deeper into (but not out of) the soil, being carried upward with evaporating water, or by being moved by animals like ants or earthworms.

4. Transformations: materials may be altered in the soil; for example, organic matter decay, weathering of minerals to smaller particles, or chemical reactions.

Each of these processes occurs differently at different depths. For instance, organic matter tends to be added at or near the surface, not deep in the soil. Some material moves from high in the soil to be deposited lower. As a consequence, different changes occur at different depths, and horizontal layers develop as a soil ages (Figure 5-12).

Figure 5-12 Idealized soil profile showing some of the soil horizon relationships. See text for detailed description. (Courtesy of USDA NRCS)

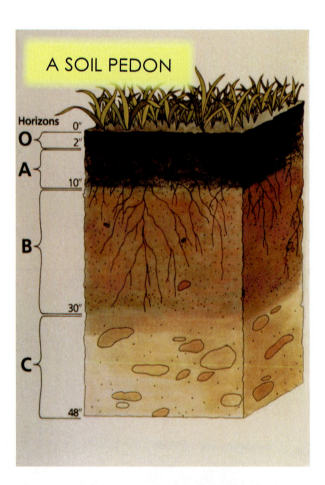

These layers are known as **soil horizons**, visible wherever the earth is dug deep enough to expose them. The **soil profile** is a vertical section through the soil extending into the unweathered parent material and exposing all the horizons. Each horizon in the profile differs in some physical or chemical way from the other horizons.

In a very young soil, weathering and plant growth produce a thin layer of mineral particles and organic matter atop parent material (the O horizon). The thin layer of soil labeled the A horizon is a surface mineral horizon enriched with organic matter. The parent material below the A horizon of this young soil is termed the C horizon. It is defined as a subsurface mineral layer only slightly affected by soil-forming processes. Thus, this young soil has an AC soil profile.

As the young soil ages, the soil increases in depth. In addition, clay-sized particles and certain chemicals leach out of the A horizon, moving downward in the profile to create a new layer, the B horizon.

Master Horizons

The A, B, and C horizons are known as **master horizons**. They are part of a system for naming soil horizons in which each layer is identified by a code: O, A, E, B, C, and R. These horizons are shown in Figure 5-13, and are described as follows.

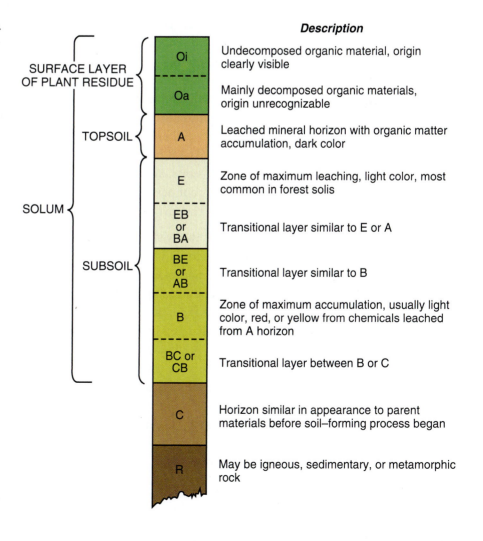

Figure 5-13 Main horizons of the soil profile. Bold lines divide O, A, E, B, C, and R horizons. Broken lines show the subhorizons.

SURFACE LAYER OF PLANT RESIDUE

TOPSOIL

SOLUM

SUBSOIL

Description

Oi	Undecomposed organic material, origin clearly visible
Oa	Mainly decomposed organic materials, origin unrecognizable
A	Leached mineral horizon with organic matter accumulation, dark color
E	Zone of maximum leaching, light color, most common in forest solis
EB or BA	Transitional layer similar to E or A
BE or AB	Transitional layer similar to B
B	Zone of maximum accumulation, usually light color, red, or yellow from chemicals leached from A horizon
BC or CB	Transitional layer between B or C
C	Horizon similar in appearance to parent materials before soil–forming process began
R	May be igneous, sedimentary, or metamorphic rock

+ The O horizon is an organic layer made of wholly or partially decayed plant and animal debris. The O horizon generally occurs in undisturbed soil because plowing mixes the organic material into the soil. In a forest, fallen leaves, branches, and other debris make up the O horizon.

+ The A horizon, called **topsoil** by most growers, is the surface mineral layer where organic matter accumulates. It is darker than the horizons below. Over time, this layer loses clay, iron, and other materials to leaching. This loss is called **eluviation**. Materials resistant to weathering, such as sand, tend to remain in the A horizon as other materials leach out. The A horizon provides the best environment for the growth of plant roots, microorganisms, and other life.

+ The E horizon, the zone of greatest eluviation, is very leached of clay, chemicals, and organic matter. Because the chemicals that color soil have been leached out, the E layer is very light. Many soils have no E horizon; it is most likely to occur under forest vegetation in sandy soils in high rainfall areas.

+ The B horizon, or **subsoil**, is often called the "zone of accumulation" where chemicals leached out of the A and E horizon accumulate. The term for this accumulation is **illuviation**. The B horizon has lower organic matter content than the topsoil and often has more clay. The A, E, and B horizons together are known as the **solum**, the portion of the soil profile most affected by soil-forming processes and that usually contains most plant roots.

+ The C horizon lacks the properties of the A and B horizons. It is the soil layer that is little touched by soil-forming processes and is usually the parent material of the soil. It may also include very soft, weathered bedrock that roots can penetrate.

+ The R horizon is underlying hard bedrock, such as limestone, sandstone, or granite. It may be cracked and fractured, allowing some root penetration. The R horizon is identified only if near enough the surface to intrude into soil.

Subdivisions of the Master Horizons. As soils age, they may develop more horizons than the basic master horizons. Some of these layers are between the master horizons both in position and properties. These transitional layers are identified by the two master letters, with the dominant one written first. Thus, an AB layer lies between the A and B horizons and resembles both, but is more like the A than the B. Figure 5-13 shows these layers.

A soil layer can be further identified by a lowercase letter suffix that tells some trait of the layer. An Ap layer is a surface layer disturbed by humankind, so that the old layers were mixed up. For instance, plowing would mix up an O, A, and AB horizon if they were all in the top 8 inches of soil. The Ap horizon is the same as the **plow layer**, the top 7 or 8 inches of soil in a plowed field. A Bt horizon is a B horizon in which clay has accumulated, usually by illuviation.

Further subdivisions are noted by a number following the letters. Thus, one could have a soil with both a Bt1 and a Bt2 horizon. Such coding means that the Bt horizon of the soil has two distinct layers in it.

SUMMARY

Soils form from minerals broken up by the action of weathering and plant roots and from the addition of decaying plant parts. Young soils continue to age—growing deeper, being leached by rainfall, developing layers, and changing over time. This soil-forming process involves the addition, loss, translocation, or transformation of soil materials, and is governed by the five factors of parent material, climate, organisms, topography, and time. Humans could be considered as a sixth factor in soil formation. Over time, soil deepens and develops recognizable horizons, and may finally become severely weathered and highly leached.

Residual soils develop directly from bedrock (igneous, sedimentary, or metamorphic). Most mineral soils come from parent materials moved from one area to another by ice, water, wind, or gravity. Organic soils are composed of decaying plants. Each type of parent material is responsible for a different soil.

Parent materials are acted on by climate and living organisms. Soils develop quickly in warm areas with high rainfall, then age into heavily weathered soils low in organic matter. In cooler regions, organic matter accumulates and weathering is less extreme. In arid climates, sparse plant growth inhibits the formation of organic matter. Grassland soils tend to be high in organic matter, forest soils lower, and dryland soils have the lowest amount of organic matter.

Topography affects soil formation by changing water movement and soil temperature. Low areas often have deep, rich soils that drain slowly. Erosion causes soils on slopes to be thin. Time is a factor because soil development is a continuing process. Young soils tend to be thin, with little horizon development. Mature soils are deeper and productive, with several recognizable horizons. Old soils are severely weathered, highly leached, and less productive.

Soil profiles, which develop over time, are divided into master horizons. These horizons, in turn, may also contain layers. Each layer is named by a code system that identifies its position in the profile and provides some information about it.

Review

True or False

1. All soils begin as clay.

2. Soil horizons are different colors.

3. A pedon is a section of soil, extending from the surface to the bedrock.

4. Water and wind are examples of physical weathering.

5. Rock is a mixture of minerals.

6. Rain is slightly acid.

7. The R horizon is also called topsoil.

Short Answer

8. List the five factors that control soil formation.

9. Identify four parent materials.

10. Define a soil body.

11. Define soil.

12. Where would you expect to find the least amount of organic matter in a soil?

13. List the four soil-forming processes.

14. Name the master horizons in a soil profile.

Critical Thinking/Discussion

15. Describe the difference between a soil profile and a layer.

16. Discuss what soil is and where it comes from.

17. Describe soil layers and how they differ.

18. List examples of the five soil-forming factors.

19. Describe how soils develop.

20. Running water removes soil from the surface. Which of the four soil-forming processes does this exemplify, and how might the soil-forming factors affect it?

21. Describe the horizons of the soil profile.

22. What do alluvial fans, floodplains, deltas, and terraces have in common? How are they different?

23. Organic matter tends to increase from west to east in the United States because of increasing rainfall. Yet, some of the highest organic matter soils are in the Plains states, which are relatively dry. Explain why.

Knowledge Applied

1. Mix one-half pint of topsoil and one-half pint of subsoil and place in a quart jar. Add water until jar is three-fourths full. Put the lid on the jar, tightly. Place one hand on the bottom of the jar and one on the lid. Shake the jar for 3 minutes. Let the soil settle overnight. Draw a diagram of the different layers. Label the layers sand, silt, and clay.

2. Examine a soil profile. Find a bare road bank or some other place where soil has been removed for some reason. (Do not use a hole or fresh bank more than 4 feet deep, as it could fall on you.) Study the soil profile on the bank or on the side of the hole. Describe the soil below. Include thickness of each horizon, texture, color, roots, rocks or anything else you see or feel in the soil.

3. Obtain samples of common soil-forming rocks and minerals. Find more information about each from a simple field guide to rocks and minerals. What plant nutrients does each contain?

4. Try to scratch feldspar with quartz, and vice versa. Which is harder? How does this help to understand weathering?

5. To observe the effects of freezing on physical weathering, pat a handful of clay soil into a ball. Inject water into the ball with a syringe; then freeze overnight. Observe the results.

6. Draw a soil profile containing seven distinct horizons in the correct order, and label them. Indicate topsoil, subsoil, and solum.

7. Develop a presentation to describe the major parent materials and vegetation as well as climate that contributed to the soils of your state/area, and discuss how these influenced your soils.

8. A case study: Eighteen thousand years ago, during the peak of the Wisconsin Glaciation, northern Kentucky was probably free of ice. However, the climate was cold, and the likely vegetation was taiga, which is relatively open conifer forest. How do you think today's soils of northern Kentucky would compare with those of 18,000 years ago?

Resources

Ashman, M. R., & Puri, G. (2002). *Essential soil science*. Ames, IA: Blackwell Publishing.

Brady, N. C., & Weil, R. (2007). *Nature and properties of soil* (14th ed.). Englewood Cliffs, NJ: Prentice Hall.

Gregorich, E. G., & Carter, M. R. (1997). *Soil quality for crop production and ecosystem health*. Burlington, MA: Elsevier Science.

Plaster, E. J. (2008). *Soil science and management* (5th ed.). Albany, NY: Delmar Cengage Learning.

Troeh, F. R., & Thompson, L. M. (2005). *Soils and soil fertility* (6th ed.). Ames, IA: Blackwell Publishing.

Internet

Internet sites represent a vast resource of information. The URLs (uniform resource locators) for World Wide Web sites can change. Using one of the search engines on the Internet such as Google or Yahoo!, find more information by searching for these words or phrases: soil formation, soil profile, topsoil, weathering, alluvial soil, rock formation, and subsoil.

Chapter 6
Soil Characteristics, Classifications, and Use

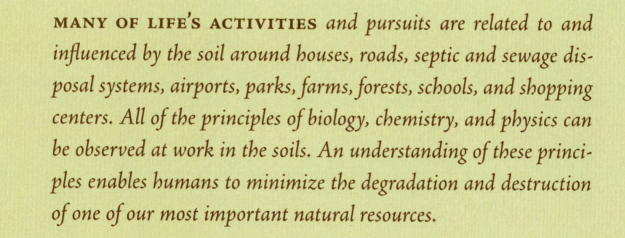

MANY OF LIFE'S ACTIVITIES and pursuits are related to and influenced by the soil around houses, roads, septic and sewage disposal systems, airports, parks, farms, forests, schools, and shopping centers. All of the principles of biology, chemistry, and physics can be observed at work in the soils. An understanding of these principles enables humans to minimize the degradation and destruction of one of our most important natural resources.

After completing this chapter, you should be able to—

- Discuss the different definitions of soil

- List soil components

- Discuss what creates soil texture

- Classify soils based on texture

- List the basic physical properties of soils

- Identify the chemical properties of soil

- Use the soil texture triangle to name soil

- Discuss how the physical and chemical properties of soil affect plant growth

- Describe soil particle sizes

- Calculate the bulk density of soil

- Discuss how iron and organic material (humus) influence the color of some soils

- Explain the relationship between field capacity water and the permanent wilting point

- Identify the pH of neutral, acidic, and alkaline soil

- Describe sodic and saline soils

- Discuss the effect of organic material on a soil

- Explain cation exchange capacity (CEC)

- Name three ways soils are classified

KEY TERMS

Acidic
Alkaline
Bulk Density
Cation Exchange Capacity (CEC)
Field Capacity
Hydroxyl Ions
Neutral
Organic Matter
Permanent Wilting Point
pH
Saline
Sodic
Soil Classification
Structure
Texture

SOIL DEFINITIONS

Depending on the discipline providing the definition, soil can be defined traditionally, geologically, by components, by soil taxonomy, or as a part of the landscape.

By traditional definition, soil is the material that nourishes and supports growing plants. It can include rocks, water, snow, and even air, all of which are capable of supporting plant life.

Geologically, soil is defined as the loose surface of the earth as distinguished from solid bedrock.

In terms of its components, soil is a mixture of mineral matter, **organic matter**, water, and air.

Soil taxonomy defines soil as a collection of natural bodies of the earth's surface, in places modified or even made by man or earthy materials, containing living matter and supporting or capable of supporting plants out-of-doors. Its upper limit is air or shallow water. Its extremes run from deep water to barren areas of rock or ice. Its lower limit is the lower limit of biologic activity, which generally coincides with the common rooting depth of native perennial plants.

As a portion of the landscape, soil is the collection of natural bodies occupying portions of the earth's surface that supports plants. The properties of soil are due to the integrated effect of climate and living matter, acting upon parent material, as conditioned by topography, over periods of time.

PHYSICAL PROPERTIES

The physical properties of soil are composition, **texture**, structure, bulk density, depth, color, and water-holding capacity.

Composition

Soils are made up of four substances, as shown in Figure 6-1. The two solid substances in soils are small mineral particles and organic matter. The small mineral particles come from breakdown of rocks. The organic matter is plant and animal material, both dead and living. The other two substances in soil are water and air.

In an average soil, about 45 percent of the volume is mineral particles and about 5 percent is organic matter. The other 50 percent is air and water. The amounts of air or water vary greatly, depending on how wet the soil is at any given time. Most of the organic matter, more than 99 percent, is made up of dead, decomposing plants and animal materials. A very small amount of the organic material consists of tiny living organisms called microbes. Although small in amount, these microbes are very large in importance. Among other things, they act as nature's recyclers. Microbes break down wastes and dead plant and animal material, and convert it into forms that can be used as food by living plants. Other soil microbes take nitrogen gas from the air and convert it into forms that

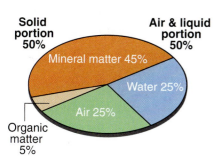

Figure 6-1 Composition of average soil.

plants can use as food. Although often overlooked because they cannot be seen, soil microbes are very important in many ways.

Textures

The small mineral particles in soils are called sand, silt, and clay. Some soils also have larger particles called stones, cobbles, or gravel. Soil texture refers to the amount of sand, silt, and clay in the soil. Soil texture can be best understood by rubbing samples of moist soil between the thumb and finger. Table 6-1 gives the particle sizes for the different components of soil (and see Figure 6-3, which shows the soils triangle for classifying soils according to their textures).

Table 6-1 Particle Size of Soils

Name of Separate	Diameter Limits (in mm)
Very coarse sand	2.00–1.00
Coarse sand	1.00–0.50
Medium sand	0.50–0.25
Fine sand	0.25–0.10
Very fine sand	0.10–0.05
Silt	0.05–0.002
Clay	less than 0.002

Sand particles are the largest in size. Sand can be seen with the naked eye and feels gritty between the thumb and finger.

Silt is intermediate in size, and in many other ways, between sand and clay. These particles are too small to be seen by the eye, but are visible under a microscope. Silt feels smooth, like talcum powder.

Clay is the smallest particle (Figure 6-2). Clay usually feels slick and sticky when wet, firm when moist, and hard when dry. Combinations of sand, silt, and clay are usually described as follows—

+ Fine texture—soils made up of mostly clay particles.
+ Medium texture—soils that are silty or loamy in nature. To the touch, they feel somewhere between fine and coarse texture.
+ Coarse texture—soils that have a high content of sand. They feel somewhat like table salt.

Soil texture determines the amount and size of spaces between soil particles. These spaces determine how easily water moves through soil and the amount of water soil will hold. The size of particles also affects the ease of plowing or working soil, as well as the crops grown. Texture affects the use of soil for foundations or waste disposal.

Soil Texture Triangle. The soil texture triangle (Figure 6-3) may be used to determine the textural name of a soil by mechanical analysis—actually measuring the percentage of sand, silt, and clay present in the soil. After the percentages of silt and clay are determined, these amounts are plotted on the soil triangle. This is done by projecting lines inward from the point on each side of the triangle, which represents the percentage of that particular type of soil. The line drawn from the silt side is placed parallel to the sand

Figure 6-2 Relative sizes of sand, silt, and clay. (Image courtesy of USDA NRCS)

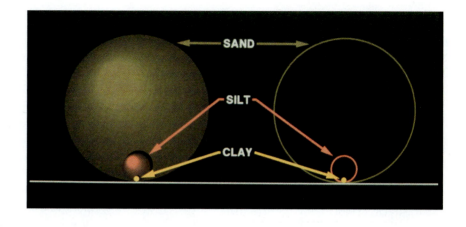

side of the triangle. The line projected from the clay line runs parallel to the silt line. The location of the point at which these two lines intersect indicates the name of the soil. The name of the section of the triangle in which the point is located is the name of the soil.

For example, a sample of soil contains 15 percent clay, 70 percent silt, and 15 percent sand. First, consider the clay. The base line at the bottom of the triangle is 0 percent clay. Read up the left side of the triangle to 15 percent clay. Draw a pencil line parallel to the base line for clay and through the 15 percent point for clay. Next, consider the silt. The zero line for silt is along the left edge of the triangle. Read down the right side of the triangle to just past 70 percent silt, and draw another line parallel to the zero line for silt. The lines cross in the area designated silt loam. (See dashed lines on Figure 6-3, which show this calculation.)

Figure 6-3 U.S. Department of Agriculture Texture Triangle. (Image courtesy of USDA NRCS)

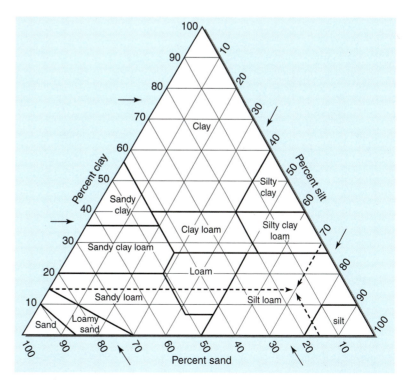

Uniqueness of Clay

Gravel, sand (very coarse to very fine), silt, and clay make up the soil. Clay is the smallest of these soil particles. A classification of these particles compares the diameter of clay to the others.

Name of particle	Diameter limits (mm)
Gravel	>2.00
Coarse sand	1.00–0.50
Fine sand	0.25–0.10
Silt	0.05–0.002
Clay	<0.002

Clay particles are plate-shaped and are often composed of complex compounds such as kaolinite, illite, and montmorillonite. Clay is sticky and capable of being molded (highly plastic) when wet, and very hard and cloddy when dry. Clay possesses some unique characteristics because of its small particle sizes and because all these particles have the same negative charge.

Clay in water forms a colloid. Because all clay particles have the same kind of electrical charge, they repel one another. Because the clay particles are so closely packed, they cannot move freely but maintain the same relative position. If a force is exerted upon the clay, then the particles slip by each other. This permits clay to take on a shape, as in pottery or brick making. When the force is removed, the particles retain their new position because the same electrical forces as before act upon them. Clays become permanently hard when baked or fired.

If sand or sandy is part of the name, the type of predominant sand must be stated; for example, very coarse sand, coarse sand, medium sand, fine sand, or very fine sand. The sand separate that occurs in an amount greater than any other separate is used to indicate the name; for example, fine sandy loam indicates a predominance of fine sand.

In case lines cross on a line between two class names, customarily the name that favors the finer fraction is used. For example, if the lines all cross at 40 percent clay, the name clay is used rather than clay loam.

By practicing on samples containing known percentages of the various fractions, an individual can become quite proficient at estimating the various soil fractions in the field.

Soil Structure

Structure refers to the arrangement of soil particles. Soils made up of practically all sand or all silt do not show any appreciable structural arrangement because of a lack of the binding properties provided by clay. A well-developed structure usually indicates the presence of clay. Soil structure is classified into three major classes and several subclasses. These classes are the structureless, which includes single grain and massive; the with-structure class, which includes granular, platy, wedge, blocky, prismatic, and columnar; and the structure-destroyed class, which includes puddle (Figure 6-4).

Soil structure is of particular importance in the absorption of water and the circulation of air. A desirable structure should have a high proportion of medium-sized aggregates and a significant number of large pores through which water and air can move. Structure of both the B horizon and the A horizon (discussed in Chapter 5) is very crucial to proper drainage, infiltration, and productivity. In soils with poor structure, root penetration

Figure 6-4 Examples of soil structure. (Image courtesy of USDA NRCS)

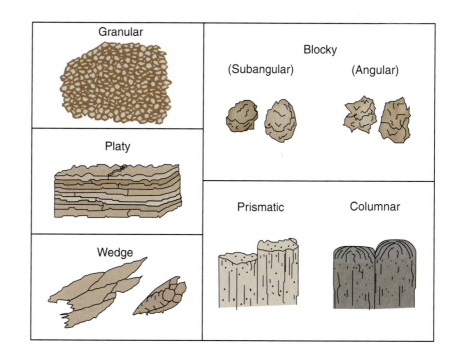

is limited, thus reducing the plant's access to water and nutrients. Structure of the A horizon receives a great deal of attention because of its relation to seedbed preparation, erosion potential, aeration, water infiltration, and overall soil health.

Three very important aspects of soil structure include:

1. Arrangement into aggregates of a desirable shape and size
2. Stability of the aggregate
3. Configuration of the pores—whether or not they are connected by channels or isolated

Aggregates stable in water permit a greater rate of absorption of water and greater resistance to erosion. Aggregates unstable in water tend to slake and disperse. These aggregates, when exposed to raindrops, are particularly subject to dispersion and the resultant crusting of soils. Crusting greatly affects seeding emergence and increases runoff and erosion.

The stability of aggregates is related to the kind of clay, the chemical elements associated with the clay, the nature of the products of decomposition or organic matter, and the nature of the microbial population. The expanding type of clay is more likely to produce unstable aggregates. An excess of sodium associated with clays tends to cause dispersion. A high proportion of hydrogen and/or calcium is associated with aggregation. Fungal growth also appears to have a binding effect on soils.

Although the kind of clay and amount of organic matter affect soil structure, there are other factors that also affect soil structure. The following are known to improve structure:

+ Freezing and thawing
+ Wetting and drying
+ Action of burrowing insects and animals
+ Growth of root systems of plants

All of these factors have a loosening effect on the soil, but no part in aggregate stability. The loosening of the soil is a necessary part of aggregate formation, not aggregate stability.

Bulk Density

Bulk density refers to the weight of the oven-dry soil with its natural structural arrangement. The pore space is a part of the volume of soil measured for bulk density. Bulk density is determined by dividing the weight of oven-dry soil in grams by its volume in cubic centimeters. Any variation in bulk density is largely because of the difference in total pore space. Because finer textured soils have higher percentages of total pore space, it follows that finer textured soils have smaller bulk density values. Obviously then, compacted soils have lower percentages of total pore space and, therefore, higher bulk densities. High- and low-bulk densities have great influences on engineering properties, water movement, rooting depth of plants, and many other physical limitations for soil interpretations.

Soil Depth

Soil depth refers to the total depth of topsoil, subsoil, and parent material, which will allow growth of plant roots. The depth of soil can cause the yield of a crop to be high or low. Deep-rooted plants, such as alfalfa, will not grow well when planted on a shallow soil. Such plants require deep soils so the roots can go down easily to depths of several feet.

Soil depth can be determined by measuring the distance between the soil surface and the layer that is unfavorable for root growth. Soil depth may be classified as indicated in Table 6-2.

Table 6-2 Soil Depth

Soils	Depth
Deep	35 inches or more
Moderately deep	20–35 inches
Shallow	10–20 inches
Very shallow	10 inches or less

Soil Colors

Color of soil is an important feature in recognizing different soil types, but color is also an indicator of certain physical and chemical characteristics. Color in soils is primarily due to two factors, humus content (organic matter) and the chemical nature of the iron compounds present in the soil. Humus has a dark brown, almost black, color. A very high content of humus may mask the color of the mineral matter to such an extent that the soil appears almost black, regardless of the color status of the iron compounds.

Iron is an important color material because iron appears as a stain on the surfaces of mineral particles. About 5 percent or more of mineral soils is iron. In unweathered soils, in which the iron exists as an unweathered mineral, iron has little or no influence on color. Iron that has the greatest effect on color is weathered from primary minerals and exists in the oxide or hydroxide form. Table 6-3 lists the forms of iron in weathered soils.

Table 6-3 Iron and Its Effect on Soil Color

Form	Chemical formula	Color
Ferrous oxide	FeO	Gray
Ferric oxide (hematite)	Fe_2O_3	Red
Hydrated ferric oxide	$2Fe_2O_3 \cdot 3H_2O$	Yellow

Water Relations

Every soil can be placed in a particular soil group using a soil textural triangle as presented in Figure 6-3. The size, shape, and arrangement of the soil particles and the associated voids (pores) determine the ability of a soil to retain water. Large pores in the soil can conduct more water more rapidly than fine pores. In addition, removing water from large pores is easier and requires less energy than removing water from smaller pores.

Sandy soils consist mainly of large mineral particles with very small percentages of clay, silt, and organic matter. In sandy soils there are many more large pores than in clayey soils. In addition the total volume of pores in sandy soils is significantly smaller than in clayey soils (30 to 40 percent for sandy soils as compared with 40 to 60 percent for clayey soils). As a result, much less water can be stored in sandy soil than in the clayey soil. It is also important to realize that a significant number of the pores in sandy soils are large enough to drain within the first 24 hours because of gravity, and this portion of water is lost from the system before plants can use it.

Plant Water. To study soil-water-plant relationships, it is convenient to subdivide soil water into water available to the plant and water unavailable to the plant. After the soil has been saturated with water one can observe a vertical, downward movement of water due to gravity. The exact time depends on the soil type; the drainage of the gravitational water generally takes a little longer for clayey soils. Most gravitational water moves out of the root zone too rapidly to be used by the plants. The remaining water is stored under tension in the pores of various sizes. The smaller the pore, the greater the tension and the more energy required to remove its water. As a result, plants have the ability to remove water only from certain sized pores. The removal of water from very small pores requires too much energy and, consequently, this water is not available to the plant. There is also some water that is very closely bound to soil particles. This water is called hygroscopic water. It is also very difficult to remove and is not available to the plants.

The range of water available to plants is between field capacity and the permanent wilting point. The soil is at field capacity when all the gravitational water has been drained and a vertical movement of water due to gravity is negligible. The permanent wilting point is defined as the point at which there is no more water available to the plant. The permanent wilting point depends on plant variety. Figure 6-5 compares the amount of available water in two different soils.

Soil Profile

The soil profile (Figure 6-6) refers to the arrangement and properties of the various soil layers. Layers (or horizons) are formed during the development of the soil. These layers are topsoil, subsoil, and parent material.

Topsoil is the surface or very top layer. It ranges from a depth of a few inches to several feet. This layer is darker than other layers because it contains organic matter. Organic matter is what is left of decayed plants and animals. The topsoil is usually softer and more easily worked than the underlying areas.

Subsoil is the layer just under the topsoil. It is usually lighter in color as there is little or no organic matter. It is usually firmer and more difficult to penetrate than the topsoil.

Figure 6-5 The amount of available water in two different soils.

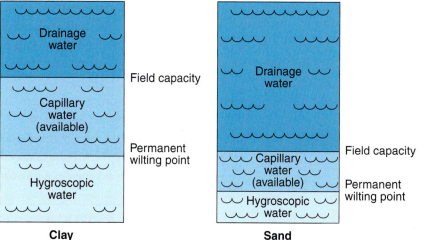

Figure 6-6 A soil profile.

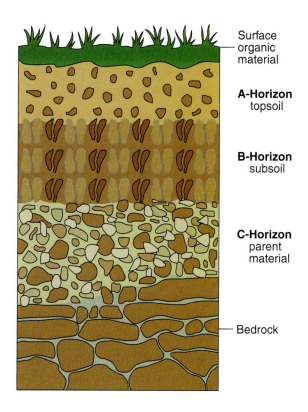

Parent material is the lower layer. It is the material from which the topsoil and subsoil have developed. It may be very firm and difficult for roots to penetrate, or it may be soft enough to allow root growth.

The surface layer or topsoil is usually called the A horizon. It is usually brown in color, and often darker than the layer under it. It is darker because it is near the surface, where there is a lot of organic matter from decaying plants.

In most soils, the subsoil layer under the A horizon is called the B horizon. It is usually lower in organic matter and higher in clay content than the A horizon. It may be red, brown, yellow, or gray in color.

Beneath the B horizon is the C horizon. The C horizon is the material from which the A and B horizon formed. The C horizon may form from rock, or it may form in loose material (sand, silt, clay, or gravel) put in place by water, gravity, glaciers, or wind. In some soils, there may be solid rock under the C horizon.

These horizons vary from soil to soil in thickness, texture, color, and other properties. In some soils, the B horizon may be very firm and compact, stopping movement of roots and water. In other soils, one or more horizons may be missing altogether.

Chapter 5 provides a more complete discussion of the soil profile.

CHEMICAL PROPERTIES

The chemical properties of **pH, cation exchange capacity, sodic, saline,** and organic matter relate to plant growth and the availability of nutrients.

Soil pH

The soil's water contains dissolved mineral salts. This liquid is known as the soil solution. The way the soil solution reacts determines the acidity, alkalinity, or neutrality of the soil. Some soil contains more hydrogen ions (H+) than **hydroxyl ions** (OH−). This makes them **acidic**. Other soils contain more hydroxyl ions than hydrogen ions. They are termed **alkaline**. When a soil contains equal concentrations of hydrogen and hydroxyl ions, it is termed **neutral** (Figure 6-7). The exact relationship between the hydrogen and hydroxyl ions is expressed as a pH number:

+ pH of 7.0 is neutral
+ pH of less than 7.0 is acidic
+ pH of more than 7.0 is alkaline

The pH of a soil can be measured precisely only using an instrument known as a pH meter. Commercial growers can either send soil samples to a soil testing laboratory to be tested for a fee or they can purchase their

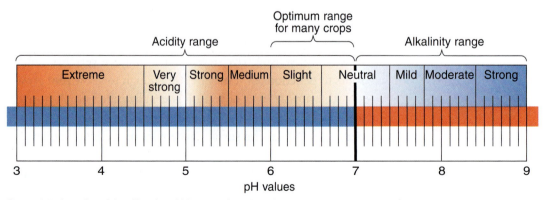

Figure 6-7 A portion of the pH scale, which ranges from 1 to 14.

own portable pH meter to obtain immediate results. There are also pH test kits on the market, but their results are imprecise compared with those obtained with a pH meter. In situations where the crop is very sensitive to soil pH, the pH meter test should be used to check pH level.

Additions to the soil that increase the number of H+ ions will lower the pH of the soil. Soil additions that increase the number of OH− ions will raise the soil pH. Many of the materials used to improve the structure and texture of the soil will also modify its pH. For example, peat moss is highly acidic, and its addition to the soil as a source of organic material will have a direct impact on the acidity of the soil solution. Limestone has the opposite effect, contributing alkalinity to the solution. These changes in the pH may or may not be desirable.

Additives should be used with caution and with knowledge of their total impact. It is easier to adjust the pH of a greenhouse bench or pot crop where the soil mix is totally within the control of the grower than it is to change the pH of a field, especially as many soils have a strong buffer resistance to pH change. Buffering occurs when hydrogen ions that are held in absorbed form dissociate from the clay particles and enter into the soil solution to replace those hydrogen ions neutralized by the addition of lime. No pH change will result until enough lime is added to deplete the supply of hydrogen ions that constitute a reserve of acidity in the soil. If the reserve of acidity is strong and the field is large, a significant change in pH may be impossible. Where strong buffer resistance is not a factor, the pH of nursery fields can be altered to improve crop production.

Within the pH range of 4.0 to 9.0, the availability of many mineral nutrients is determined by the acidity or alkalinity of the soil. For example, many plants will exhibit a distinctly patterned yellowing or chlorosis, when grown in soil having a high pH. The cause of the chlorosis is lack of iron in the plant tissue, and it results because iron compounds, which are needed by the plant, are precipitated out of the soil solution and rendered unavailable to the plants. At a lower pH, the iron will remain in the soil solution and be available for plant uptake.

Cation Exchange Capacity

As shown in Figure 6-8, cations held by soils can be replaced by other cations. This means they are exchangeable. Calcium can be exchanged for H+ and/or K+ (potassium ions) and vice versa. Generally, cations with higher numbers of charges (higher charge density) are more tightly held (absorbed) by colloids. Those cations can be exchanged or replaced by large numbers of cations with lower charge density (mass action). The total number of exchangeable cations a soil can hold (the amount of its negative charge) is called its cation exchange capacity or **CEC**. The higher a soil's CEC, the more cations it can retain. Soils differ in their capacities to hold exchangeable K+ and other cations. The CEC depends on amounts and kinds of clay and organic matter present. CEC increases as organic matter increases.

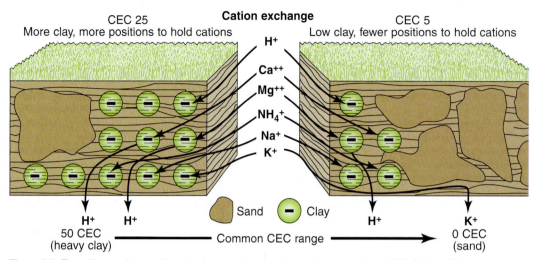

Figure 6-8 The soil's capacity to hold and exchange cations is cation exchange capacity, or CEC. Calcium (Ca), magnesium (Mg), potassium (K), sodium (Na), hydrogen (H), and ammonium (NH_4) are the positively charged nutrient ions and molecules. The negatively charged particles are clay, which attracts, holds, and releases positively charged nutrient particles. Organic matter particles are also negatively charged.

Expressing CEC. The CEC of a soil is expressed in terms of milligram equivalents per 100 grams of soil and is written as meq/100 g. Clay minerals usually range from 10 to 150 meq/100 g in CEC values. Organic matter ranges from 200 to 400 meq/100 g. So the kind and amount of clay and organic matter content greatly influence the CEC of soils.

Using the CEC. Where soils are highly weathered and organic matter levels are low, CEC values are low. Where less weathering has occurred and organic matter levels are usually higher, CEC values can be quite high. Clay soils with high CEC can retain large amounts of cations against potential loss by leaching. Sandy soils with low CEC retain smaller quantities of cations. This makes timing and application rates important in planning a fertilizer program. For example, it may not be wise to apply K on very sandy soils in the fall to serve next spring's crops, especially in areas in which fall and winter rainfall is high. But enough fall K for one or more future crops can be safely applied in a single application on soils with high CEC. Also, splitting N applications, using nitrification inhibitors, and timing applications to meet peak crop demands are important to reduce the potential for nitrate leaching on sands as well as finer textured soils.

Percent base saturation, the percent of total CEC occupied by the "basic" cations, has been used in the past to develop fertilizer programs. The idea is that certain nutrient ratios or "balances" are needed to ensure proper uptake by the crop for optimum yields. Research has shown, however, that cation saturation ranges have little or no utility in most agricultural soils. Under field conditions, ranges of nutrients can vary widely with no detrimental effects, so long as individual nutrients are present in sufficient levels in the soil to support optimum plant growth.

Sodic and Saline Soils

Sodic or alkali soils contain excessive amounts of sodium (Na) on the soil CEC sites. Soils are usually classified as sodic if Na saturation exceeds 15 percent of the CEC. They usually have pH values of 8.5 and above. Excess Na disperses the soil, limiting the movement of air and water because of poor physical properties. Water tends to pond on sodic soils.

Such soils can be reclaimed by replacing the Na on the CEC complex with calcium (Ca). Gypsum (calcium sulfate) is the most common treatment, but elemental sulfur (S) can be used if the soil is calcareous. Successful reclamation requires that the Na be leached out of the root zone. Poor water movement can make that task difficult. Deep ripping and/or manure application has been used to improve internal water movement.

Saline-Sodic Soils. Sometimes sodic soils will also be saline (salty). Saline-sodic soils are typically characterized by an Na saturation greater than 15 percent of the CEC but a pH of 8.4 or less. Their reclamation is the same as that for sodic soils.

ORGANIC MATTER

Soil organic matter consists of plant and animal residues in various stages of decay. Adequate levels benefit soil in four ways:

1. Improves physical condition and structure
2. Increases water infiltration
3. Decreases erosion losses
4 Supplies plant nutrients

Most benefits are derived from products released as organic residues decompose in the soil. Figure 6-9 shows a field with plenty of plant residue.

Figure 6-9 A field with lots of plant residue, requiring extra N. Ridge-till soybeans emerge from warmed ground in the row, which is surrounded by corn residues from the past year. (Photo courtesy of USDA NRCS)

Release of Nitrogen

Organic matter contains about 5 percent total nitrogen (N), so it serves as a storehouse for reserve N. But N in organic matter is in organic compounds and not immediately available for plant use because decomposition usually occurs slowly. Although a soil may contain much organic matter, fertilizer N is needed to assure nonlegume crops an adequate source of readily available N, especially those crops requiring high N. Other essential plant nutrients are also contained in soil organic matter. Plant and animal residues contain variable amounts of mineral nutrients such as Mg, Ca, S, and the micronutrients. As organic matter decomposes, these nutrients become available to growing plants.

Organic matter decomposition tends to release nutrients. However, some nutrients such as N and S can be temporarily tied up during the process. Microorganisms decomposing the organic matter require N to build protein in their bodies. If the organic matter being decomposed has a high C:N (C-to-N) ratio, meaning low N, these organisms will use available soil and fertilizer N. This process is called immobilization.

When residues of cotton and corn stalks or oat and wheat straw are incorporated into the soil, additional N should be applied if a crop is to be planted soon after that. If not, crops may suffer temporary N deficiency. Eventually, N immobilized in the bodies of soil organisms becomes available as the organisms die and decay. This is called mineralization. With conservation tillage and the buildup of residues as crop yields increase, N management requires extra attention until a new soil equilibrium is reached. Extra care should be taken to avoid a deficiency from too little N. At the same time, rates used should not exceed crop need so that potential NO_3^- leaching can be minimized. See Chapter 7 for more detail on N management.

Variability

Some soils contain very little organic matter. In tropical areas, most soils are inherently low in organic matter because warm temperatures and high rainfall speed up decomposition. In cooler areas, where decomposition takes place more slowly, native organic matter levels can be quite high. With adequate fertilization and good management practices, more crop residues are produced. In high-yield corn fields, as much as eight tons of residues are left after the grain has been harvested. These residues maintain or increase organic matter levels in soils and benefit physical, chemical, and microbial soil properties. Residues should be returned regularly to sustain crop production. The important point is to keep sufficient amounts of residues passing through the soil.

SOIL CLASSIFICATION

Soils have been classified in some manner or another for hundreds of years. They have been grouped according to their—

+ Agronomic use—such as good wheat soil, poor corn soil
+ Color—such as black soil, red soil

+ Organic matter content—such as mineral soil, muck soil, peat soil
+ Texture—such as sand, sandy loam, loam
+ Moisture condition—such as wet soil, dry soil

Scientific Soil Classification

Each of these groupings serves a particular purpose, but they are not very scientific according to our present knowledge of soil genesis, morphology, and classification. A scientific system of soil classification was adopted for use in the United States on January 1, 1965. The development of this classification began in 1951 and has now been revised ten times. Gradually, a workable system evolved that can be used to classify soils anywhere in the world.

Basically, this system treats soils as individual three-dimensional entities that can be grouped together according to their similar physical, chemical, and mineralogical properties. These divisions are based on broad differences in measurable and visible characteristics of certain kinds of soil horizons. The presence or absence of these definite kinds of layers indicates either the dominant active soil-forming process that has caused that soil to develop as it has, or the lack of such development.

Of the 12 soil orders, many fall into definite geographic ranges that are indicative of the importance of climate. The soil orders are listed and described in Table 6-4.

Table 6-4 The Twelve Soil Orders with Brief Descriptions

Order	Description
Gelisols	Soils with permafrost within 6 feet of the surface
Histosols	Organic soils
Spodosols	Acid forest soils with a subsurface accumulation of metal-humus complexes
Andisols	Soils formed in volcanic ash
Oxisols	Intensely weathered soils of tropical and subtropical environments
Vertisols	Clayey soils with high shrink/swell capacity
Aridisols	Calcium carbonate ($CaCO_3$)-containing soils of arid environments with subsurface horizon development
Ultisols	Strongly leached soils with a subsurface zone of clay accumulation and less than 35% base saturation
Mollisols	Grassland soils with high base status
Alfisols	Moderately leached soils with a subsurface zone of clay accumulation and greater than or equal to 35% base saturation
Inceptisols	Soils with weakly developed subsurface horizons
Entisols	Soils with little or no structural development

Each of the orders is divided into several suborders, each of which in turn consists of soils grouped together according to their somewhat similar developmental characteristics. For complete descriptions of the soil orders and their use check out the soils section of the NRCS website (http://soils.usda.gov/technical/soil_orders/)

Simple Classification of Land

Simple classifications are often useful to people who are buying or selling land. They are not as scientific as the present-day practical systems. An example of a simple classification of land is illustrated by the way a citrus farmer might classify his citrus land: good citrus land, fair citrus land, poor citrus land, and land unfit for citrus. Another grower might classify his land as: good corn land, fair corn land, and poor corn land or as good pasture land, fair pasture land, and poor pasture land. A similar classification is sometimes made for timberland as well as for land used in other ways.

Banks and other lending agencies often place a monetary classification on land, such as $5,000 per acre land, $2,000 per acre land, and $1,000 per acre land.

Practical Classifications of Land

Several practical land classification systems are presently in use. For example, the agronomist may classify land on the ability of soils to grow the common field crops. The forester may classify land in still another way—according to the rate of tree growth, using good, fair, poor, and so on, as land classes. One of the most widely used systems of land classification is one that was developed by the Soil Conservation Service as a basis for its work throughout the nation. This system divides all lands into eight capability classes, each being based on the capability of land, degree of erosion, and so on. The land capability classes are listed in Table 6-5.

Table 6-5 Land Classes and Safe Uses

Uses	I	II	III	IV	V	VI	VII	VIII
Recreation and Wildlife	X	X	X	X	X	X	X	X
Forestry	X	X	X	X	X	X	X	
Limited Grazing	X	X	X	X	X	X		
Intensive Grazing	X	X	X	X	X			
Limited Cultivation	X	X	X	X				
Moderate Cultivation	X	X	X					
Cultivation	X	X						
Very Intensive Grazing	X							

No Cultivation below Class IV

SUMMARY

Various definitions apply to soils. The physical and chemical properties of soils influence their ability to grow plants. Physical properties of soil include composition, texture, structure, bulk density, depth, color, and water-holding capacity. Chemical properties of soils include pH, cation exchange capacity, sodic, saline, and organic matter.

Soils are classified according to the percent of sand, silt, or clay they contain. Besides those three classifications there is sandy clay, silty clay, sandy clay loam, silty clay loam, clay loam, sandy loam, silt loam, and loamy sand.

Soils are very different in their ability to provide nutrients. For high yields of crops in agriculture, some nutrients almost always have to be added to the soil. The soil also provides plants with water. The amount of water plants can store and release depends on soil texture. Medium textures provide plants with the most water. Sandy soils provide the least. Clay soils are between the other two types. Soils with lots of stones or gravel provide less water to plants.

Soil pH, cation exchange capability, organic matter, and drainage are variable. These factors influence the ability of a soil to grow a crop. Use, value, and soil type have all been used to classify soil types.

Review

True or False

1. Silt is the smallest particle of soil.

2. Soil color depends on the amount of sand.

3. Soils with a pH of less than 7.0 are called alkaline soils.

4. Humus (organic material) causes soils to appear red.

5. Sodic soils contain a high content of iron.

Short Answer

6. What is the pH of neutral soil?

7. What are the four benefits of soil organic matter?

8. List four factors known to improve soil structure.

9. What is the new system of classifying soils based on?

10. A 100 cc sample of soil weighs 400 grams. What is its bulk density?

11. List three ways soils are classified.

12. Name three basic physical properties of soils and three chemical properties of soils.

Critical Thinking/Discussion

13. Define soil at least two different ways.

14. What is the cation exchange capacity (CEC) of a soil?

15. How do iron and organic material (humus) influence the color of some soils?

16. How does organic material affect a soil?

17. What are the differences between soil texture and soil structure?

18. Explain the relationship between field capacity water and the permanent wilting point.

19. How do the physical and chemical properties of soil affect plant growth?

20. Explain the scientific classification of soils that uses 12 orders.

Knowledge Applied

1. Mix one-half pint of topsoil and one-half pint of subsoil and place in a quart jar. Add water until jar is three-fourths full. Put the lid on the jar, tightly. Place one hand on the bottom of the jar and one on the lid. Shake the jar for 3 minutes. Let the soil settle overnight. Draw a diagram of the different layers. Label the layers sand, silt, and clay.

2. Collect samples of builder's sand, heavy clay soil, or modeling clay and medium-textured topsoil. Take about a tablespoon of each and put separately on a plate or baking pan. Add water to each until moist. Take each moist sample in turn and rub it between your thumb and finger. Describe how each one feels. Form a ball from each sample, then press on the ball with your thumb. Which ball breaks apart easiest? Which ball will press in the most without cracking? Which ball is best for molding into shapes? Try to press each sample into a long, thin ribbon between your thumb and finger. Which sample makes the longest ribbon? Can you make a ribbon with all three? If not, which one will not form a ribbon?

3. Use the soil texture triangle to name other soils in your area.

4. Collect soil samples and determine their bulk density.

5. Measure the pH of various soil samples and classify them as acidic, neutral, or alkaline (basic).

6. Based on the chapter feature on clay, develop a presentation that demonstrates some of the uniqueness of clay.

Resources

Ashman, M. R., & Puri, G. (2002). *Essential soil science.* Ames, IA: Blackwell Publishing.

Connors, J. J., & Cordell, S., eds. (2003). *Soil fertility manual.* Tucson, AZ: Potash & Phosphate Institute.

Gregorich, E. G., & Carter, M. R. (1997). *Soil quality for crop production and ecosystem health.* Burlington, MA: Elsevier Science.

Kirkham, M. B. (2004). *Principles of soil and plant water relations.* New York: Academic Press.

Scott, H. D. (2000). *Soil physics: Agriculture and environmental applications.* Ames, IA: Blackwell Publishing.

Troeh, F. R., & Thompson, M. L. (2005). *Soils and soil fertility* (6th ed.). Ames, IA: Blackwell Publishing.

United States Department of Agriculture. (1957). *Soil: The yearbook of agriculture.* Washington, DC: Author.

Wolf, B. (2000). *The fertile triangle: The interrelationship of air, water, and nutrients in maximizing soil productivity.* Binghamton, NY: Haworth Press.

Internet

Internet sites represent a vast resource of information. The URLs (uniform resource locators) for World Wide Web sites can change. Using one of the search engines on the Internet such as Google or Yahoo!, find more information by searching for these words or phrases: soil classification, soil orders, cation exchange capacity, soil chemistry, organic matter, soil structure, water holding capacity, humus, soil pH, and soil texture.

Chapter 7
Soil Fertility and Management

SIXTEEN ELEMENTS *are absolutely necessary for normal plant growth. Many of these elements are the same as those required by humans. In addition to carbon, hydrogen, and oxygen, which the plant gets from the air and water, another 13 elements are required by plants, which they obtain from the soil. These are usually divided into three classes: primary nutrients, secondary nutrients, and micronutrients. Functions of the elements in plant metabolism and symptoms are related to their deficiencies. Based on soil tests, fertilizers are applied to provide plants with some of these essential nutrients for optimal growth.*

After completing this chapter, you should be able to—

- Name the essential plant nutrients and describe their role in plant growth

- Explain how plant nutrients behave in soil

- List plant nutrients and fertilizer materials that are compatible with the environment

- Describe soil testing for determining plant nutrient needs and fertilizer sources

- Discuss how plant nutrients behave in the soil

- Describe what soil pH is and how it is managed

- Name the 16 elements essential for plant growth

- Categorize the 16 essential elements into those supplied by the air and water, primary nutrients, secondary nutrients, and micronutrients

- Draw the nitrogen cycle

- Describe the effect organic matter has on soil fertility

- Discuss how soil pH influences the availability of nutrients

- Describe the deficiency signs of five nutrients

- Explain how many pounds of nitrogen, phosphate, and potash are in a bag of fertilizer

- Use conversion factors when working with fertilizers

- List five types and sources of fertilizers

- Discuss the process of soil erosion, its importance and prevention

ESSENTIAL NUTRIENTS AND THEIR ROLES IN PLANTS

Chemical elements needed by plants for normal growth and development are called nutrients. Although as many as 20 chemical elements have been identified in certain plants, only 16 have been found to be needed for plants to grow and mature properly. These 16 elements are considered the essential plant nutrients and are shown in Table 7-1, along with their chemical symbols.

Table 7-1 Sixteen Essential Plant Nutrients

Nonmineral		Primary		Secondary Mineral		Micronutrients	
Name	Symbol	Name	Symbol	Name	Symbol	Name	Symbol
Carbon	C	Nitrogen	N	Calcium	Ca	Boron	B
Hydrogen	H	Phosphorus	P	Magnesium	Mg	Chlorine	Cl
Oxygen	O	Potassium	K	Sulfur	S	Copper	Cu
						Iron	Fe
						Manganese	Mn
						Molybdenum	Mo
						Zinc	Zn

The essential nutrients are divided into mineral and nonmineral elements. Carbon, hydrogen, oxygen, and nitrogen are the nonmineral nutrients, and the remaining 12 are the mineral nutrients. The roles of the various plant nutrients and their deficiency symptoms are indicated in Table 7-2.

Hydrogen and oxygen are supplied to plants from carbon dioxide and water through photosynthesis. The mineral nutrients are supplied by the soil through nutrient uptake processes. The sugars produced by the photosynthetic process account for most of the increase in plant growth.

The 12 mineral nutrients and nitrogen are divided into three groups, depending upon the amount of each used by plants. Those used in largest amounts are the primary or **macronutrients** and include nitrogen, phosphorus, and potassium. Nutrients used in intermediate amounts by plants are the secondary nutrients, which include calcium, magnesium, and sulfur. The seven **micronutrients** or minor elements are boron, copper, zinc, iron, manganese, molybdenum, and chlorine, and are required by plants in small or micro amounts.

To remember the elements from the air, the primary nutrients, and the secondary nutrients, remember: "C HOPKNS CaFe Mg." Group the micronutrients as "B Mn Cu Zn Mo Cl" to remember them.

FERTILIZERS

Fertilizers are added to the soil, plant, or other growing medium to supply the essential macronutrients and micronutrients to growing plants.

Table 7-2 The Roles of Nutrients in Plants and Their Deficiency Symptoms

Nutrient	Function	Deficiency Symptoms
Nitrogen	Promotes rapid growth; chlorophyll formation; synthesis of amino acids and proteins.	Stunted growth; yellow lower leaves; spindly stalks; pale green color.
Phosphorus	Stimulates root growth; aids seed formation; used in photosynthesis and respiration.	Purplish color in lower leaves and stems; dead spots on leaves and fruits.
Potassium	Increases vigor, disease resistance, stalk strength, and seed quality.	Scorching or browning of leaf margins on lower leaves; weak stalks.
Calcium	Constituent of cell walls; aids cell division.	Deformed or dead terminal leaves; pale green color.
Magnesium	Component of chlorophyll, enzymes, and vitamins; aids nutrient uptake.	Interveinal yellowing (chlorosis) of lower leaves.
Sulfur	Essential in amino acids, vitamins; gives green color.	Yellow upper leaves; stunted growth.
Boron	Important to flowering, fruiting, and cell division.	Terminal buds die; thick, brittle upper leaves with curling.
Copper	Component of enzymes; chlorophyll synthesis and respiration.	Terminal buds and leaves die; blue-green color.
Chlorine	Not well defined; aids in root and shoot growth.	Wilting; chlorotic leaves.
Iron	Catalyst in chlorophyll formation; component of enzymes.	Interveinal chlorosis of upper leaves.
Manganese	Chlorophyll synthesis.	Dark green leaf veins; interveinal chlorosis.
Molybdenum	Aids nitrogen fixation and protein synthesis.	Similar to nitrogen.
Zinc	Needed for auxin and starch formation.	Interveinal chlorosis of upper leaves.
Carbon	Component of most plant compounds.	
Hydrogen	Component of most plant compounds.	
Oxygen	Component of most plant compounds.	

Nitrogen

Nitrogen is one of the earth's most abundant and mobile nutrients. It is a part of every plant cell. Soils may contain as much as 5,000 pounds of nitrogen per acre. The air we breathe contains 37,000 tons (78 percent by volume) of elemental nitrogen over each acre of the earth's surface and is the origin for all other nitrogen sources.

Nitrogen is a part of chlorophyll, which gives plants their green color. When plants do not receive enough nitrogen, their leaves lose their normal green color and turn yellow. Tips of lower or bottom leaves turn yellow first (Figure 7-1). Nitrogen-deficient plants usually grow slowly with spindly stalks and stems. Preventing these symptoms from occurring through proper fertilization is very important to growing healthy plants.

The manufacturing of most nitrogen fertilizers begins with the production of ammonia (NH_3). Ammonia is a gas that is produced when nitrogen from the atmosphere is combined with hydrogen (H) from natural gas or

Figure 7-1 Patterns of deficiency symptoms of the primary nutrients in lower leaves of corn.

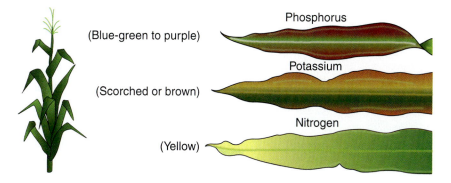

other fossil fuels. The ammonia produced can be applied directly or processed further into various solid and liquid nitrogen fertilizers (Figure 7-2).

The most common sources of nitrogen fertilizers used are urea, ammonium nitrate, and nitrogen solutions.

Nitrogen and the Environment

Nitrogen can be very mobile in the soil and is subject to many physical, chemical, and biological processes. As a result, significant losses of the nutrient can occur. Although some nitrogen is lost to surface and ground waters, most of the nitrogen lost from soils goes back into the atmosphere. The main pathways through which nitrogen not used by plants is lost from the soil are leaching, erosion, denitrification, and volatilization.

Nitrification. Regardless of the form of nitrogen applied to the soil, it ends up in the nitrate (NO_3^-) form. A major step in the process is nitrification. Through nitrification, ammonium (NH_4^+) from either decomposed organic materials or inorganic fertilizers is converted to the nitrate form. The only ammonium not going through this process is that trapped by soil clays. Some soils may contain large amounts of nitrogen in this form. The process of nitrification is carried out by bacteria in the soil and is illustrated by the following equation:

$$NH_3 \quad + \quad Nitrosomonas \quad \rightarrow \quad NO_2^-$$
$$(ammonia) \qquad Bacteria \qquad (nitrite)$$

$$NO_2^- \quad + \quad Nitrobacter \quad \rightarrow \quad NO_3^-$$
$$\qquad Bacteria \qquad (nitrate)$$

Figure 7-2 Fertilizers manufactured from ammonia (NH_3).

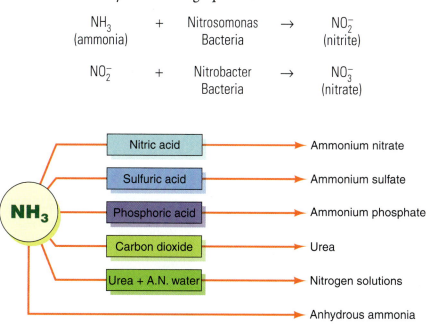

Once in the nitrate form, nitrogen becomes a part of the soil solution. As such, it is available to be used by plants. Nitrate can also be tied up by microorganisms, removed through leaching, or lost through denitrification. Because of these processes and others, as much as 50 percent of the nitrogen added to soils may never be used by crops.

Leaching. Nitrate nitrogen is the form most subject to leaching losses. Nitrates (NO_3^-) are held only slightly by soil clays and humus (colloids). As a result, they remain in the soil solution and are free to move with the soil water. On the other hand, ammonium (NH_4^+) is attached to soil particles or trapped by soil clays and is not free to move with soil water.

Leaching losses of nitrates during the growing season in the medium- and fine-textured soils are estimated to be less than 5 percent. Losses are greater from sandy or gravelly soils (coarse-textured soils).

Erosion. Nitrogen is lost through erosion in association with either water or sediment. Nitrate nitrogen moves primarily in runoff water, and ammonium and organic forms of nitrogen move with sediment. Sediment loss from fields can be greatly reduced by using approved soil conservation practices. Runoff can be reduced somewhat with soil conservation practices.

Denitrification. When soils become saturated with water, air is removed from soil pores. Under anaerobic conditions (absence of oxygen), some bacteria convert significant amounts of nitrate nitrogen to the elemental (N_2) form, which is a gas and is lost back to the atmosphere.

Losses through denitrification are estimated to be 15–30 percent of the total applied nitrogen from an area that has been flooded for three to five days. Fields flooded for longer periods likely lose even greater amounts. Denitrification losses from well-drained, medium-textured soils that are saturated but not flooded are generally small.

Volatilization. Volatilization occurs when urea is converted to ammonium carbonate—$(NH_4)_2CO_3$—when applied to warm moist soils. Ammonium carbonate, in turn, breaks down to form ammonia gas (NH_3). If the reaction takes place on the soil's surface, nitrogen is lost into the atmosphere (Figure 7-3). Although ammonia losses are usually less than 10 percent, they can be greater when urea is top-dressed on warm, moist soils followed by three or more days of good drying conditions. Incorporating with equipment or injecting urea directly into the soil

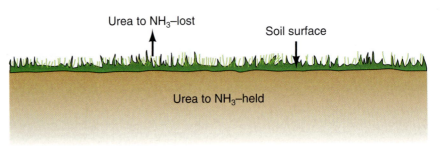

Urea to NH_3–lost

Soil surface

Urea to NH_3–held

Figure 7-3 Ammonia volatilization.

usually eliminates volatilization. Urea can be incorporated with enough water to wet the top 2 to 3 inches of soil.

Nitrogen Cycle

Before nitrogen can be used by plants, it must first be removed from the atmosphere, either naturally through nitrogen fixation, which occurs in green plants called legumes, or commercially through chemical processes. Once added to the soil, nitrogen usually goes through several changes or processes before returning back to the elemental form. The overall process through which atmospheric nitrogen is converted from elemental (N_2) to usable nitrogen back to elemental or atmospheric nitrogen is called the nitrogen cycle (Figure 7-4).

Nitrogen Fixation

Nitrogen fixation is the process whereby elemental nitrogen is removed from the atmosphere by soil bacteria called **rhizobia**. These bacteria live in the nodules (knots) on roots of legume plants such as alfalfa, clover, peas, beans, and vetch. Through the nitrogen-fixing process, legumes are able to provide their own nitrogen supply and do not normally need nitrogen applied from other sources. Average amounts of nitrogen fixed by various legumes are given in Table 7-3.

Organic Matter (Humus)

After plants mature and die, they decompose (break down) into soil organic matter. The rate of decomposition and the amount of nitrogen

Figure 7-4 The nitrogen cycle.

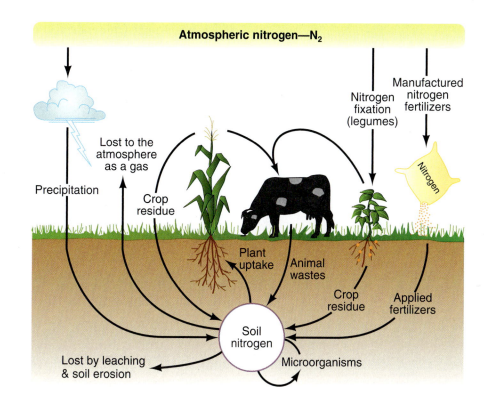

Table 7-3 Average Amounts of Nitrogen Fixed by Certain Legumes

Plant	Pounds N/Acre/Year
Alfalfa	194
Ladino Clover	179
Red Clover	114
Cowpeas	100
Soybeans	100
Vetch	80

released from decaying plants or other organic materials depend largely on the **carbon:nitrogen (C:N) ratio** of the decaying material.

When fresh supplies of nitrogen-rich plant residues or other organic materials with C:N ratios of about 25:1 or less are added to the soil, organic nitrogen is quickly converted to inorganic forms that can be used by growing plants. The conversion process is called mineralization and occurs in residues such as legumes, animal manures, and municipal sludges. However, when carbon-rich organic materials (C:N ratios greater than 25:1) are added to the soil, nitrogen is tied up or immobilized rather than released. Soil organisms that decompose organic materials have a narrow C:N ratio and require relatively large amounts of nitrogen. If sufficient nitrogen is not present in the fresh organic materials, organisms will take nitrogen from the soil during decomposition. Thus, when carbon-rich materials such as straw, sawdust, and bark are added to the soil, they decompose very slowly and may cause nitrogen deficiencies in growing plants unless fertilizer nitrogen is added. Organic matter that has completely decomposed and is more or less stable with the soil environment (mineralization = immobilization) is called humus and has a C:N ratio of about 12:1 (Table 7-4). Humus is responsible for controlling the release of much of the nitrogen contained in soils. The average amount of nitrogen in humus is about 5 percent. Thus, a soil with 1 percent organic matter will contain about 1,000 pounds of nitrogen per acre. However, only about 2 percent of that amount (20 pounds per acre) is released annually for crop use.

Table 7-4 Carbon:Nitrogen Ratios of Some Organic Materials

Material	Average Carbon:Nitrogen Ratio
Alfalfa	12:1
Sewage Sludge	12:1
Livestock Slurry	15:1
Swine Lagoon	17:1
Clovers	20:1
Cornstalks	40:1
Wheat Straw	80:1
Sawdust	400:1

Phosphorus

Unlike nitrogen, phosphorus is very immobile in the soil and moves primarily as soil particles are moved. It is lost from the soil through plant removal and soil erosion. Very little phosphorus moves through the soil.

Because of its immobility and the relatively high need for phosphorus by young plants, an adequate supply of phosphorus must be near the plant's root system early in growth. As a result, many plants respond to phosphorus placed near the seed, especially when soils are low in phosphorus and/or cool and wet.

Soil Reactions. The reaction of phosphorus in the soil is closely related to soil pH. Its maximum availability occurs under slightly acid soil conditions (Figure 7-5). At lower pH levels, phosphorus readily reacts with iron and aluminum to form insoluble compounds. If soils are alkaline or have been overlimed (high pH), phosphorus reacts with calcium to form insoluble calcium phosphates.

Phosphorus Sources. Diammonium phosphate and triple superphosphate are the main sources of fertilizer phosphorus used (see Table 7-5). Diammonium phosphate is the basic ingredient used in making dry bulk fertilizers, while triple superphosphate is available in bags or bulk. Other sources of phosphorus include animal manures, sludges, plant residues, and ground rock phosphate.

The most important source of phosphorus used in manufacturing phosphorus fertilizers is the mineral apatite. In the United States, important deposits of apatite are located in Florida, North Carolina, Wyoming, Montana, and Tennessee. In the high-phosphate region of middle Tennessee, apatite is sufficient in the soils to grow most plants without the use of fertilizer phosphorus. The chemical formula for apatite is $Ca_5(PO_4)_3$.

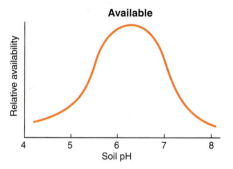

Figure 7-5 Soil pH and phosphorus availability.

Table 7-5 Properties of Certain Commercial Fertilizers

Fertilizer	Percent N-(P_3O_4)-(K_2O)	Major Components	Form	Percent Water Solubility
Urea	46-0-0	$(NH_2)_2CO_2$	Dry	100
Ammonium Nitrate	34-0-0	NH_4NO_3	Dry	100
Nitrogen Solutions	(28-32)-0-0	$(NH_2)_2CO + NH_4NO_3$	Liquid	100
Ammonium Sulfate	21-0-0	$(NH_4)_2SO_4$	Dry	100
Anhydrous Ammonia	82-0-0	NH_3	Gas	100
Sodium Nitrate	16-0-0	$NaNO_3$	Dry	100
Diammonium Phosphate	18-46-0	$(NH_4)_2HPO_4$	Dry	95–100
Monoammonium Phosphate	11-48-0	$NH_4H_3PO_4$	Dry	95–100
Triple Superphosphate	0-(42-50)-0	$Ca(H_2PO_4)_3$	Dry	95–100
Ammonium Polyphosphate	10-34-0	$(NH_4)_3HP_2O_7 + NH_4H_3PO_4$	Liquid	100
Muriate of Potash	0-0-60	KCl	Dry	100
Sulfate of Potash	0-0-50	K_2SO_4	Dry	100

Potassium

Potassium is second only to nitrogen in amounts used by plants. Although it is relatively immobile in the soil, some movement or leaching losses of potassium occurs in soils containing large amounts of sand. Potassium deficiencies are most likely to occur in plants growing in sandy soils. The most common potassium deficiency symptom in plants is scorching or browning along leaf margins (edges) of lower or bottom leaves (Figure 7-1).

Soil Reactions. Although the total quantity of potassium found in soils is generally greater than that of any other nutrient, amounts available to plants at any one time are relatively small. Much of the potassium found in soils is tied up by soil minerals. In some soils considerable amounts of added potassium become trapped (fixed) between soil clay particles. Like trapped ammonium, this fixed potassium is not available for plants to use but may become available later as the potassium is slowly released by the clay particles.

Potassium Sources. Most commercially produced potassium fertilizers are mined from deposits of water-soluble potassium minerals found in ancient seas. The largest known potassium deposit is in Saskatchewan, Canada, and is the source for most of the muriate of potash (potassium chloride) used in the United States. Muriate of potash is the primary source of fertilizer potassium. A second source of commercially produced potassium fertilizer is potassium sulfate.

Table 7-5 lists the properties of some of the commercial fertilizers.

SECONDARY AND MICRONUTRIENTS

Secondary and micronutrients are as important as the primary nutrients for good plant growth. The main difference is that they are used by plants in much smaller amounts than the primary nutrients. Information about commonly available secondary and micronutrient sources is given in Table 7-6.

Sulfur

In most soils, sulfur (S) is present primarily in the organic fraction that becomes available upon decomposition of organic matter and crop residues. The available sulfate (SO_4^-) ion remains in soil solution much like the nitrate (NO_3^-) ion until it is taken up by the plant. In this form it is subject to leaching as well as microbial immobilization. In water-logged soils, it may be reduced to elemental S or other unavailable forms. Sulfur may be supplied to the soil from the atmosphere in rainwater. It is also added in some fertilizers as an impurity, especially the lower grade fertilizers. The use of gypsum $(CaSO_4)$ also increases soil S levels.

Sulfur is taken up by plants primarily in the form of sulfate (SO_4^-) ions and reduced and assembled into organic compounds. It is a constituent of the amino acids cystine, cysteine, and methionine and, hence, proteins that contain these amino acids. It is found in vitamins, enzymes, and coenzymes.

Table 7-6 Some Common Sources of Secondary and Micronutrients

Nutrient Materials	Percent Nutrient	Pounds of Material for 1 Pound Nutrient
Boron		
Borax	11.3	8.8
Fert. Borate-granular	14.3	7.0
Solubor	20.5	4.9
Calcium		
Calcitic Limestone	35	2.8
Calcium Sulfate (Gypsum)	22.5	4.4
Iron		
Iron Sulfate	19	5.3
Iron Chelates	5–14	7–20
Magnesium		
Dolomitic Limestone	9	11.1
Magnesium Sulfate (Epsom Salts)	9	11.1
Potassium-Magnesium Sulfate	11	9.1
Manganese		
Manganese Sulfate	27	3.7
Manganese Chelates	12	8.3
Zinc		
Zinc Sulfate	36	2.8
Zinc Chelates	10–14	7–10

Sulfur is also present in glycosides, which give the characteristic odors and flavors in mustard, onion, and garlic plants. It is required for nodulation and N fixation of legumes. As the sulfate ion, it may be responsible for activating some enzymes.

Calcium

Calcium (Ca) occurs in the soil solution as a divalent cation (Ca^{2+}). It is supplied to plants by soil minerals, organic materials, fertilizers, and by **liming** materials. There is a strong preference for Ca^{2+} on the cation exchange sites of most soils and it is the predominant cation in most soils with a pH of 6.0 or higher.

Calcium, an essential part of plant cell wall structure, provides for normal transport and retention of other elements as well as strength in the plant. It is also thought to counteract the effect of alkali salts and organic acids within a plant. Calcium is absorbed as the cation Ca^{2+} and exists in a delicate balance with magnesium and potassium in the plant. Too much of any one of these elements may cause insufficiencies of the other two.

Magnesium

Soil minerals, organic material, fertilizers, and dolomitic limestone are sources of magnesium (Mg) for plants. Magnesium occurs as a divalent cation (Mg^{2+}) and is held on the exchange sites like calcium (Ca^{2+}) and potassium (K^+).

Magnesium is part of the chlorophyll in all green plants and is essential for photosynthesis. It also helps activate many plant enzymes needed for growth. Magnesium, a relatively mobile element in the plant, is absorbed as the cation Mg^{2+} and can be readily translocated from older to younger plant parts in the event of a deficiency.

Micronutrients

Of the 16 elements known to be essential for plant growth, six are required in such small quantities that they are referred to as micronutrients. These are Fe, Mn, Zn, Cu, B, Mo, and Cl. Micronutrients have become of more widespread concern than was the case earlier.

Micronutrients are most apt to limit crop growth under the following conditions:

- Highly leached acid sandy soil
- Muck soils
- Soils high in pH or lime content
- Soils that have been intensively cropped and heavily fertilized with macronutrients

Four of the micronutrients occur predominantly as cations in the soil solution. They are iron (Fe^{2++}), copper (Cu^{2+}), manganese (Mn^{2+}), and zinc (Zn^{2+}). Two occur predominantly as anions. These are molybdenum (MoO_4^-) and chlorine (Cl^-). Boron occurs as the neutral molecule, H_3BO_3.

Iron. Iron (Fe) is a constituent of many organic compounds in plants. It is essential for the synthesis of chlorophyll, which gives rise to the green pigment of plants. Iron deficiency can be induced by high levels of Mn. High Fe can also cause Mn deficiency.

Copper. Copper (Cu) is essential for growth and activates many enzymes. A deficiency interferes with protein synthesis and causes a buildup of soluble N compounds. Excess quantities of Cu may also induce Fe deficiency.

Manganese. Manganese (Mn) is mainly absorbed by plants in the ionic form Mn^{2+}. It activates many enzymes. Manganese may substitute for Mg by activating certain phosphate-transferring enzymes, which in turn affect many metabolic processes. High Mn concentration may induce Fe deficiency in plants.

Manganese availability is closely related to the degree of soil acidity. Deficient plants are usually found on slightly acid or alkaline soils.

Zinc. Zinc (Zn) is essential for plant growth because it controls the synthesis of indoleacetic acid, which dramatically regulates plant growth. Zinc is also active in many enzymatic reactions.

Molybdenum. Molybdenum (Mo) functions largely in the enzyme systems of N fixation and nitrate reduction. Plants that can neither fix N nor incorporate nitrate into their metabolic system because of inadequate Mo become N-deficient. Molybdenum is required in minute amounts.

Boron. Boron (B) primarily regulates the metabolism of carbohydrates in plants. The need varies greatly with different crops. Rates required for responsive crops may cause serious damage to B-sensitive crops. Boron deficiency may occur on both alkaline and acid soils but is more prevalent on the calcareous, alkaline soils.

NUTRIENT AVAILABILITY AND PLANT UPTAKE

Plant nutrients must occur in the soil in available forms if they are to be taken up and used by plants. Forms that are available to plants are shown in Table 7-7.

The available forms of nutrients are located at or near the surface of soil particles such as clay, silt, and humus. As growing plant roots come in contact with these soil particles and the soil solution that surrounds these particles, the available nutrients are taken up by the plant.

Although soils may contain large quantities of plant nutrients, amounts available for plant use at any given time are relatively small. For example, a soil may contain several hundred pounds per acre of total phosphorus, but contain only 15 to 20 pounds available for plant uptake. The remainder will be unavailable and cannot be used by plants unless it is converted to an available form. Nutrients unavailable for plants to use occur in soils in one or more of the following forms:

+ Insoluble chemical compounds—phosphorus and the micronutrients included in many of these
+ Unweathered or undecomposed soil minerals or rock fragments—most nutrients included
+ Organic matter or plant residues—the main ones are nitrogen and sulfur
+ Trapped-by-soil particles—lots of potassium and some ammonium may be in this form

Table 7-7 Available Forms of Essential Plant Nutrients[1]

Nutrient Element	Chemical State
Nitrogen	NO_3^-, NH_4^+
Phosphorus	PO_4^-, HPO_4^-, $H_2PO_4^-$
Potassium	K^+
Calcium	Ca^{++}
Magnesium	Mg^{++}
Sulfur	SO_4^-
Boron	BO_3^-, $B_4O_7^-$
Chlorine	CL^-
Copper	Cu^{++}
Iron	Fe^{++}
Manganese	Mn^{++}
Molybdenum	MoO_4^-
Zinc	Zn^{++}

[1] The plus (+) and minus (−) signs indicate the charge on each of the nutrients.

One of the main properties of soils that influence nutrient availability is pH. Soil pH is especially important for maintaining fertilizer nutrients in available forms. If the pH is not suitable, available fertilizer nutrients may rapidly become unavailable by forming insoluble chemical compounds (will not dissolve in water).

NUTRIENT INTERACTIONS

The term **nutrient interaction** refers to how one nutrient may help or hinder the uptake of another. In some instances, a certain nutrient may increase the uptake of one nutrient and decrease the uptake of another nutrient. Also, a given nutrient may increase the uptake of a nutrient under one set of conditions while reducing uptake of the same nutrient under other conditions.

Some examples of nutrient interactions include the following:

+ Ammonium-potassium: Ammonium nitrogen may interfere with the uptake of potassium. Excessive rates may create potassium deficiencies in some crops.
+ Potassium-magnesium: Potassium has been shown to reduce the uptake of magnesium when high rates were applied. Excessive rates may produce magnesium deficiencies.
+ Phosphorus-nitrogen: The uptake of phosphorus is often increased by the presence of nitrogen.
+ Phosphorus-zinc: High levels or rates of phosphorus may reduce zinc uptake.

SOIL pH

Soil test results indicate when a soil is too acid (needs lime) for good production. If not limed as needed, soils continue to become more acid. Some plants (acid-loving) such as blueberries, azaleas, and rhododendrons prefer soils high in acids or low in soil pH.

What Is Soil pH?

Soils contain both acids and bases. The relative amounts of each are expressed by pH. The pH scale ranges from 0 to 14. Soils with values below 7.0 contain more acids than bases and are referred to as acid or sour. Those with values above 7.0 contain more bases than acids and are referred to as alkaline or sweet. If the soil pH is 7.0, the soil is neither acid nor alkaline but neutral—contains equal amounts of acids and bases. How soil pH affects availability of plant nutrients is shown in Figure 7-6.

Causes of Acid Soils

Several factors are responsible for soils becoming acid. As nutrients such as calcium, magnesium, and potassium are removed from the soil

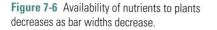

Figure 7-6 Availability of nutrients to plants decreases as bar widths decrease.

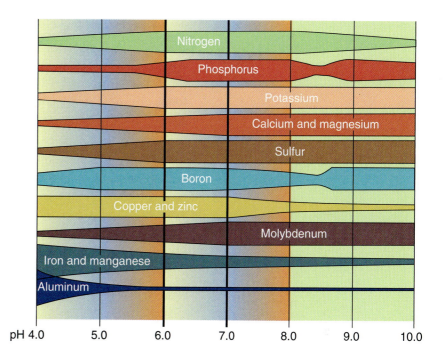

through soil erosion, leaching, and crop removal, they are replaced by soil acids. In addition, the use of acid-forming fertilizers will increase acid levels in the soil. In particular, the conversion of ammonium fertilizers to nitrates (the nitrification process) produces soil acids.

Adjusting Soil pH

Soil pH can be adjusted to any desired range. The important thing to remember is that adjustments should be made based on soil test results. Lime is used to adjust the pH upward. The lime neutralizes soil acids and increases soil pH.

Soil pH is lowered by applying elemental sulfur. Apply 2 pounds of elemental sulfur per 1,000 square feet for each one-tenth (0.1) unit the pH is to be lowered. Iron and aluminum sulfate may also be used to lower soil pH. The desired soil pH range for selected crops is shown in Table 7-8.

Lime Sources

Often the most common and economical liming material available is ground or agricultural limestone. Limestones containing both calcium and magnesium are dolomitic. Those containing only calcium are calcitic. When soils are acidic and magnesium levels are low, dolomitic limestone should be applied to increase both pH and magnesium levels.

If other materials are used, equivalent amounts should be applied (Table 7-9). For example, one ton of ground limestone is approximately equal to 1,500 pounds of calcium hydroxide or 1,100 pounds calcium oxide.

Table 7-8 Desired pH Ranges for Selected Plants

Crop	pH Range	Crop	pH Range
Alfalfa	6.6–7.0	Millet	6.1–6.5
Apples	6.6–7.0	Oats	6.1–6.5
Azaleas	5.0–5.5	Okra	6.1–6.5
Barley	6.1–6.5	Onions	6.1–6.5
Beans (snap)	5.6–6.5	Peaches and Pears	6.6–7.0
Bedding Plants	6.1–6.5	Peas	6.1–6.5
Beets and Carrots	6.1–6.5	Peppers	5.6–6.0
Blueberries	4.3–5.0	Potatoes (Irish)	5.1–6.0
Blackberries	5.6–6.0	Rhododendrons	5.0–5.5
Cabbage	6.1–6.5	Roses	6.1–6.5
Cantaloupes	6.1–6.5	Shade Trees (most)	6.1–6.5
Corn	6.1–6.5	Sorghum	6.1–6.5
Cotton	6.1–6.5	Soybeans	6.1–6.5
Clover (most)	6.1–6.5	Strawberries	5.6–6.0
Cucumbers	5.6–6.0	Tobacco (burley)	6.1–6.5
Dogwoods	6.1–6.5	Tomatoes	6.1–6.5
Grasses (most)	6.1–6.5	Watermelons	5.6–6.0
Hemlock	5.5–6.0	Wheat	6.1–6.5

Table 7-9 Equivalent Amounts of Liming Materials

Material	Equivalent Pounds	Chemical Formula
Calcitic limestone	2000	$CaCO_3$
Dolomitic limestone	2000	$CaCO_3 + MgCO_3$
Calcium hydroxide (hydrated or slaked lime)	1480	$Ca(OH)_2$
Calcium oxide (burnt or quick lime)	1120	CaO

Functions of Lime

Applying lime to the soil serves five functions:

1. Sweetens the soil—most plants do not grow well in acid or sour soils
2. Improves the availability of plant nutrients
3. Increases the effectiveness of applied nitrogen, phosphorus, and potassium
4. Increases the activity of microorganisms, including those responsible for nitrogen fixation in legumes and the decomposition of organic matter
5. Improves plant growth and crop yields

SOIL TESTING

The most accurate method for determining the amounts of lime and fertilizer to apply to soils for best plant growth is to test the soil. A soil test measures soil pH and the amounts of available nutrients the soil contains. By knowing the amounts of nutrients already present in the soil, it is much easier to determine the kinds and amounts of fertilizers to apply. Most plant fertilization problems are associated with the lack of and improper use of nitrogen, phosphorus, potassium, and lime.

Collecting the Sample

Each soil sample should represent the area for which lime and fertilizer are to be applied. A large composite sample should be collected consisting of small portions (cores or strips) of soil taken from several locations. For field crops, soil portions should be taken from an area not to exceed 10 acres. For lawns and gardens, soil cores should be collected at random from 8 to 10 locations.

Several types of tools may be used for collecting soil samples. One is the soil tube or probe. A uniform portion of soil is collected by pushing the tube into the ground to the desired depth and removing a soil core.

If a shovel or spade is used to collect soil samples, make a V-shaped cut into the soil to the proper depth. Remove a 1-inch thick vertical slice up to the same depth from the smoothest side of the cut. From this, remove a 1-inch strip of soil the length of the slice. Repeat the procedure at a sufficient number of sites within the sampling area.

Place cores or strips of soil for each composite sample into a clean container and thoroughly mix. The soil should be dry enough to allow easy mixing. From this, remove enough of the dry soil to fill a soil sample box for mailing to the laboratory.

FOLIAR SYMPTOMS

Foliar symptoms of nutrient deficiencies vary considerably between plant species. In addition, insect, disease, nematode, drought, and pesticide damage often produce foliar symptoms. General descriptions of symptoms produced by nutritional deficiencies should be used only in conjunction with other available information when trying to diagnose field problems.

Tissue Testing

Tissue testing is divided into separate phases that are quite different—rapid tissue testing and dry tissue testing.

Rapid Tissue Testing. Green tissue is usually taken for this type of analysis. Various chemicals included in these "quick tests" are used to test each element using green tissue or the extract from green tissue. These tests may give usable information, but accuracy is sacrificed for speed and field-use convenience. Confirmation using dry tissue testing of suspected nutritional problems is recommended.

Dry Tissue Testing. This type of testing usually involves the use of dried leaves or plants. The entire dried leaf or plant is ground up and the total content of selected nutrients is determined. Because tissue testing is standardized, results of analyses done on a split sample by different

EPCOT Center and the Land

When Walt Disney decided to branch out from his Disneyland theme park in southern California, he set out to build an Experimental Prototype Community of Tomorrow, nicknamed EPCOT. So what exactly is EPCOT? It is a kind of permanent world's fair and interactive science museum spread across a sprawling tract of land broken up into pavilions. One of these pavilions is The Land.

The Land, the largest pavilion in Future World, is a tribute to food and farming. It went through a renovation in 1994 and serves both as a theme park attraction and a science experiment in agriculture and **hydroponics**.

Living With The Land

This is the most popular attraction in The Land pavilion and features a gentle boat ride through a brief history of agriculture and through some of the experimental greenhouses in this pavilion. Guests are first taken through a series of biomes representing, amongst others, a tropical rain forest, a desert, the American prairie, and a small turn-of-the-century family farm. From there guests travel into the experimental greenhouses for a view of new methods of growing the major food crops and the growing of rare new crops that may help alleviate the earth's ever-growing dietary needs. Growing methods include hanging plants that are simply sprayed with nutrient solution instead of planted, plants growing in liquids, and a special drip irrigation system designed for arid climates. Other experiments include a fish farm and growing plants in space.

The guides on these tours offer an informative and educational talk on the plants being grown and the experimental techniques being used. It is worth noting that all of the plants grown in the greenhouses are genuine plants, no matter how unreal their growing conditions may seem, and supply the Garden Grill Restaurant and Sunshine Season Food Fair with vegetables and fish for guests to enjoy.

Research Activities of the Science and Technology Group at The Land

Research also takes place at The Land. Recent topics of research have included vegetables and nitrogen, endangered plant species, bacteria survival, and biological control.

Vegetables and Nitrogen

The effect of different concentrations of nitrogen in the hydroponic nutrient solution used to grow the pepper variety "Midal" was studied in the greenhouses at The Land. The study determined the following factors:

1. Yield of marketable fruit
2. Amount of hydroponic nutrient solution used to grow the peppers
3. Concentration of nitrate-nitrogen in hydroponic solution unused by the peppers

The results were used to reduce the amount of unused nutrient solution entering the environment.

Endangered Plant Species

Six species of endangered plants native to central Florida were successfully reproduced by tissue culture in the Biotechnology Laboratory at The Land. The project will lead to the replenishment of these six species, found naturally on Walt Disney World company property.

LDEF Bacteriology

Scientists at The Land studied seeds from the Long Duration Exposure Facility (LDEF) to determine the survival of bacteria after exposure to the harsh conditions of space. Amazingly, many different bacteria survived after six years in space.

Biological Control of Aphids

Populations of Green Peach Aphids, a major vegetable pest, can be controlled using natural enemies. Studies at The Land show that aphid population growth rates can be reduced by minimizing nutrient supply to the plants. Reductions in pest population growth rates via nutrient minimization greatly help ensure success with biological control.

Epcot is more than just entertainment!

laboratories should be comparable. This is not generally the case with soil tests because of differences in extractants.

There are several drawbacks to leaf analysis for estimating fertilizer requirements. Leaf analysis is more time-consuming and more expensive than either rapid tissue testing or soil analysis. The plant part sampled and stage of development is critical to interpretations of the results.

CHOOSING THE FERTILIZER SOURCE

There are many kinds of fertilizers available that can be used to supply recommended nutrients. Materials that provide one or more nutrients for plant use are called fertilizers. If a fertilizer contains only one primary nutrient, it is called a straight material; for example, urea or muriate of potash. Materials containing each of the three primary nutrients are referred to as complete or mixed fertilizers. Small quantities of fertilizer are usually purchased in bags and larger amounts are purchased in bulk.

The numbers on the fertilizer bag indicate the grade and guaranteed analysis of the material expressed in percent by weight as total nitrogen (N), available phosphate (P_2O_5), and soluble potash (K_2O). The pounds of each nutrient in 100 pounds of a 5-10-15 grade fertilizer are calculated as follows:

+ 5 pounds of N (100 pounds × .05 = 5)
+ 10 pounds of P_2O_5 (100 pounds × .10 = 10)
+ 15 pounds of K_2O (100 pounds × .15 = 15)

A 50-pound bag of fertilizer of the same analysis would contain:
+ 2.5 pounds of N (50 pounds × .05 = 2.5)
+ 5.0 pounds of P_2O_5 (50 pounds × .10 = 5.0)
+ 7.5 pounds of K_2O (50 pounds × .15 = 7.5)

Solids or Liquids?

Solid and liquid fertilizers perform equally well when equivalent amounts are properly applied. Although there are advantages and disadvantages to each, it makes little difference to plants which form is used. Nutrient availability of water-soluble dry fertilizers and liquid fertilizers are similar. Therefore, fertilizers should be selected based on economics, market availability, and other factors, not whether they are solids or liquids.

When calculating the nutrient content of liquid fertilizers, it may be necessary to determine the pounds of nutrients per gallon rather than per 100 pounds. For example, one gallon of a liquid 5-10-15 fertilizer weighing 11 pounds would contain:

+ 0.55 pounds of N (11 pounds × .05 = 0.55)
+ 1.10 pounds of P_2O_5 (11 pounds × .10 = 1.10)
+ 1.65 pounds of K_2O (11 pounds × .15 = 1.65)

The number of gallons of the aforementioned liquid necessary to provide the same amount of plant nutrients as 100 pounds of a solid or dry 5-10-15 fertilizer would be 9.1 gallons (100 pounds +11 pounds per gallon = 9.1 gallons).

Conversion Factors

Some of the common conversion factors for working with fertilizers include the following:

+ Phosphorus (P) = $P_2O_5 \times 0.44$
+ P_2O_5 = Phosphorus (P) $\times 2.29$
+ Potassium (K) = $K_2O \times 0.83$
+ K_2O = Potassium (K) $\times 1.20$
+ Parts per million $\times 2$ = pounds per acre
+ Pounds per acre $\div 2$ = parts per million
+ Percent $\times 10,000$ = parts per million
+ Parts per million $\div 10,000$ = percent

FERTILIZER PLACEMENT

The characteristics of the soil, the kind of crop, and the nature of the fertilizer materials should be considered when choosing methods of fertilizer application (Figure 7-7). Points about fertilizer placement are listed here:

+ Provide adequate quantities of plant nutrients within the root zone.
+ Irregular distribution can lower fertilizer efficiency.
+ Early stimulation of the seedling is usually advantageous. At least part of the fertilizer should be placed within reach of young seedling roots.
+ The rate and distance of fertilizer movement depend upon the character of the soil. Nutrient elements may move upward during dry periods, and may be carried downward by rain or irrigation water.
+ Soil-supplied nutrients are of little or no benefit to the plant when in dry soils. Excessive concentrations of fertilizer in contact with seed, roots, or legume inoculant may be injurious. Crops vary in their tolerance to soluble fertilizer salts.

Figure 7-7 Fertilizing a grain field following harvest.

- Water-soluble fertilizer of relatively low plant food content has a greater salt content per unit of plant food and has a greater tendency to produce salt injury than does a fertilizer containing more concentrated materials.
- Nitrogen and potassium carriers are more readily soluble than phosphatic fertilizer materials and cannot be safely concentrated in as large amounts near seeds or plant roots.
- Reduction of soil moisture content increases the concentration of salts in the soil solution. Soil drying increases the possibility of injury.
- Except in strongly acidic, sandy soils, phosphates move slowly from the point of placement. Therefore, phosphatic materials should be placed where they will be readily reached by plant roots.
- Placement of fertilizer in bands reduces contact with the soil, thereby delaying the conversion of phosphorus to forms not available to plants. Banded fertilizer rates may be safely reduced by as much as 50 percent compared with broadcast rates.

Use of GIS and GPS

Through **GPS (Global Positioning System)** and **GIS (Geographic Information System)**, farmers can optimize fertilizer applications for consistent yields. GPS and GIS allow the farmer to apply straight, accurate swaths every time, with minimal overlap and skip without foam markers, row markers, or flags, driving faster and more precisely, even in reduced light. Additionally, the use of this technology means the use of less fertilizer since it is applied only to the area that it is needed.

GPS and GIS are part of a suite of technologies that make precision agriculture possible. Application of precision agriculture has two spatial requirements: simultaneous knowledge of where the farm equipment is as it moves across a field and the value of one or more variables such as fertility as a function of position within the field. These two requirements each contain a "where" and a "what." Using GPS, the spatial precision needed for where varies from a few meters to a few centimeters. The second requirement, the what, is where remote sensing is used. By using satellite data to determine soil conditions and plant development, these technologies can lower the production cost by fine-tuning seeding, fertilizer, chemical and water use, all of which potentially increase production and lower cost. Additionally, precision agriculture may have a significant impact far beyond the farm by reducing the use of extra fertilizers and chemicals.

Foliar Fertilization

Foliar fertilization is the process of feeding nutrients to plants through their foliage (leaves, stems, blooms). It should be considered a supplementary method to feeding plants through the root system.

Foliar fertilization is not a practical method of applying large amounts of nutrients because only small amounts of fertilizers can be applied per application. Repeated applications are expensive, and if large amounts of nutrients are applied, severe burning of the crop is likely to occur, resulting in reduced plant growth. The most common use of foliar fertilization is when rapid uptake of a small amount of a particular nutrient (usually a micronutrient) is needed to correct a deficiency problem.

OTHER FERTILIZER SOURCES

Nonmanufactured materials can be used as fertilizers. These may be used by persons wanting to use only naturally occurring materials, as in the case of "organic" farmers or gardeners. The plant nutrient content of some nonmanufactured materials is presented in Table 7-10.

Animal Manures

Manures have been used as fertilizers for centuries. Although their plant nutrient content is generally low, manures contain some quantities of all of the essential elements. In many situations, the application of even a modest quantity of manure provides enough of a deficient nutrient (especially a micronutrient) to dramatically increase plant growth.

The nutrient content of manure is quite variable. Factors influencing the quality include the age and kind of animal, the feed it consumed, the

Table 7-10 Plant Nutrient Content of Some Nonmanufactured Materials Used as Fertilizer

| Material | Average Percentage of | | | | | |
	N	P_2O_5	K_2O	Ca	Mg	S
Sodium nitrate, Chilean	16	—	0.2	0.1	0.05	0.07
Blood, dried	13	1.5	0.6	0.24	0.10	0.17
Bone meal, raw	3.9	22	—	22	—	—
Bone meal, steamed	2.2	27	—	25	—	—
Castor pomace	5.2	1.8	1.1	0.41	0.32	—
Cocoa shell meal	2.4	1.0	2.7	0.94	4.03	0.09
Cotton seed meal	6.4	2.6	1.7	0.24	0.42	0.30
Dolomite	—	—	—	21	11	0.3
Limestone (calcite)	—	—	0.3	32	3	0.1
Gypsum	—	—	0.5	22	0.4	17
Tankage	2.8	3.1	1.1	3.0	0.3	0.61
Guano, Peruvian	12	11	2.4	8.8	0.6	1.10
Potassium chloride	—	—	61	0.09	0.11	0.11
Peat	1.9	0.2	0.2	1.1	0.36	0.26
Rock phosphate	—	32	0.2	33	0.16	—
Sewage sludge	5.6	5.1	0.4	1.3	0.57	0.98
Seaweed (kelp)	0.2	0.1	0.6	2.1	0.74	1.39
Soybean meal	6.8	1.6	2.4	0.26	0.31	0.21
Tobacco stems	2	0.7	0.6	3.6	0.36	0.38
Tung nut meal	4.3	1.7	1.3	0.46	0.52	—
Wood ashes	—	1.8	5.5	23	2.2	0.4

Table 7-11 Average Percentage Composition of Macronutrients in Fresh Manure

Type of Manure	Percentage of:					
	N	P_2O_5	K_2O	Ca	Mg	S
Cattle	0.6	0.3	0.5	0.3	0.1	0.04
Horse	0.6	0.3	0.6	0.3	0.1	0.04
Sheep	0.9	0.5	0.8	0.4	0.1	0.06
Swine	0.6	0.5	0.4	0.5	0.1	0.1
Poultry (layers)	1.5	1.3	0.5	3.0	0.3	0.4
Poultry (broilers)	3.1	3.0	2.0	2.0	0.4	0.7

amount and kind of litter or bedding used, and the manner in which the manure was handled. Representative values shown in Table 7-11 should be used only as general guidelines and should not be used for making important planning or management decisions.

Some interesting generalities about manures:

+ Approximately 16 tons of manure (at 75 percent moisture) are produced for each ton of livestock per year, regardless of species.
+ About 500 pounds of absorbent litter are needed to absorb 600 pounds of liquid in each ton of manure.
+ Addition of superphosphate to manure prevents loss of ammonia by volatilization. Use 40 pounds per ton for horse and sheep manure and 25–30 pounds per ton for poultry, swine, or cattle manure. However, intensive livestock and poultry operations already have an excess of phosphorus in the manure and would certainly not add more, even if it saves nitrogen.

COMPOST

Compost is recycled organic matter. It is a dark, easily crumbled, partially decomposed collection of plant products. Compost is created by a biological process in which numerous bacteria break down the plant tissue. As the process continues, the bacterial breakdown is increased by the actions of fungi, protozoans, centipedes, millipedes, sow bugs, earthworms, and other organisms. Because these organisms require oxygen and water, compost needs air and moisture. Finished compost has some characteristics of humus, the organic matter of the soil. Composting is an environmentally beneficial way to dispose of organic waste material such as plant residues and manure (Figure 7-8). These by-products are turned into useful compost.

Value

Good compost consists primarily of decomposed or partially decomposed plant and animal residues, but may also contain a small amount of soil. Compost improves both the physical condition and fertility of the soil when added to the landscape or garden. It is especially useful for improving soils low in organic matter.

Figure 7-8 Composted dairy manure from a large Idaho dairy being loaded for spreading on a field.

The organic matter in compost improves heavy clay soils by binding soil particles together, which makes them easier to work. Binding soil particles also helps improve aeration, root penetration, and water infiltration, and reduces crusting of the soil surface. In sandy soils, additional organic matter also helps with nutrient and water retention.

Although compost contains nutrients, its greatest benefit is in improving soil characteristics. Compost also is a valuable mulching material for garden and landscape plants. It may be used as top-dressing for lawns and, when it contains a small amount of soil, as a growing medium for houseplants or for starting seedlings.

Making Compost

Composting is a method of speeding natural decomposition under controlled conditions. Raw organic materials are converted to compost by the action of microorganisms. During the first stages of composting, microorganisms, primarily bacteria, increase rapidly. As organic materials decompose, certain kinds of microorganisms predominate. As they complete a certain function, some of these microorganisms decline while others build up and continue the decomposition process.

As microorganisms decompose the organic materials, temperatures in the pile rise. The center of a properly made heap should reach a temperature of 110–140°F in four to five days. At this time the pile will begin "settling," which is a good sign that the pile is working properly. The pH of the pile will be very acidic at first, at a level from 4.0 to 4.5. By the time the process is complete, the pH should rise to approximately 7.0–7.2.

The heating in the pile will kill some of the weed seeds and disease organisms. However, this only happens in areas where the most intense temperatures develop. In cooler sections toward the outside of the pile, this

Figure 7-9 Water erosion in a field. (Photo courtesy of USDA NRCS)

Figure 7-10 Wind erosion adjacent to a newly plowed field. (Photo courtesy of USDA NRCS)

may not happen. Proper turning is important to allow a more complete process to occur.

The organisms that break down the organic materials require large quantities of nitrogen. Therefore, adding nitrogen fertilizer or other materials that supply nitrogen is necessary for rapid and thorough decomposition. During the breakdown period, the nitrogen is tied up and is not available for plant use. This nitrogen is released when the decomposition is completed and the compost is returned to the ground.

SOIL MANAGEMENT

Soil erosion, conservation, compaction, and drainage are serious management concerns.

Erosion

Erosion is the removal of soil material by wind or water moving over the land (Figures 7-9 and 7-10).

Erosion is a natural process. Most hills and valleys are the result of very slow erosion by water. The Grand Canyon is a result of natural erosion. Erosion becomes a problem when it is accelerated by human activities.

Excessive erosion occurs when soil on slopes is left without a vegetative cover. The bare soil is exposed to rain and flowing runoff water, which loosens soil and moves it down the slope. This can happen when soil is tilled and left bare. It can also happen when road banks, ditches, or construction sites are left bare. The steeper the slope, the greater the erosion problem. Two types of erosion are of concern: sheet and rill erosion and gully erosion (Figure 7-11).

Gully erosion   Sheet and rill erosion

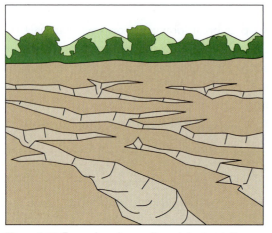

Figure 7-11 Types of erosion – gully, and sheet and rill.

+ Sheet and rill erosion is the more-or-less uniform removal of topsoil from a field. Soil washes from the field in thin layers, or sheets, and from small channels or rills. Sometimes immediately after a rain, fields can be seen with thousands of tiny rills running downhill.

+ Gully erosion is caused by rain water running over the soil surface. Gullies are deep ditches cut by flowing water. They are actually large rills. Gully erosion may destroy a field. How fast gullies grow depends on the amount of rain, the steepness of slope, the kind of soil, and how the land is used.

Importance of Erosion. Erosion is important in agriculture because it removes the A horizon, or topsoil. Soil from the A horizon usually has more nutrients, more organic matter, less clay, and is easier to till than the B horizon below it. So if the A horizon is lost to erosion, plants do not grow as well and consequently less food is produced.

Erosion is also an important environmental concern. The eroded soil leaves fields, road banks, and construction sites with runoff water. It is carried into streams and lakes. This eroded soil, called sediment, hurts fish and wildlife. It also makes the water less useful for swimming, boating, and drinking. The water quality in the streams and lakes is lowered.

Soil Conservation

Preventing or stopping erosion is called soil conservation. Conservation means to save or to use wisely.

The best way to control erosion is to keep the soil covered. This can be done with living plants, or with a mulch of dead plant material like leaves or crop residue. Crop residue is the plant material left in a field after the useful part of the crop is removed. Preparing the land for planting in a way that leaves crop residue on the soil surface is an excellent way to prevent erosion. This is called conservation tillage.

The soil is also protected by living plants like grass and trees. When crops have to be grown in bare soil, they can be grown in strips with strips of grass between them. This is called **strip cropping**. The grass strips stop the flow of water and filter out sediment.

Terraces also help reduce erosion from bare fields. Terraces are low dams or dikes built across slopes to catch runoff water and eroded soil before it leaves the field. Contour rows do the same thing when slopes are not so steep.

Contour Cropping. **Contour cropping** (Figures 7-12 and 7-13) reduces erosion; it is most effective on deep, permeable soils and on gentler slopes (2 to 6 percent) that are less than 300 feet long. The effectiveness of contouring is reduced greatly on steeper or longer slopes because of possible breakover of rows by runoff water. Contouring can reduce erosion losses up to 50 percent compared with up-and-down hill tillage on slopes of

| Contour farming | Terraces | Strip cropping |

Figure 7-12 Types of conservation farming: contour farming, terracing, and strip cropping.

Figure 7-13 Contour farming and terraces.
(Photo courtesy of USDA NRCS)

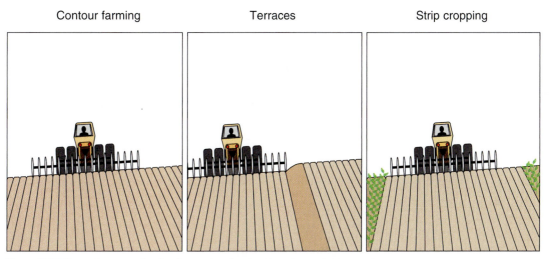

2 to 6 percent. On steeper slopes (18 to 24 percent), contour cropping without supplementary practices reduces erosion losses by only about 10 percent. Grass waterways are necessary to carry the runoff water safely from the contour rows.

Strip Cropping. The practice of alternating contour strips of sod and row crops is even more effective than contouring alone (Figure 7-12), reducing erosion to one-fourth of that resulting from up-and-down hill tillage. Strip widths should be governed by the percent slope and should vary from up to 100 feet on 2 to 6 percent slopes to 60 feet or less on slopes of 18 percent or greater.

Terraces. Terraces are channels and ridges built across slopes to intercept runoff water and shorten the effective length of a slope. They are generally more effective than either contouring or strip cropping and are designed especially for longer slopes. Many terraces (Figure 7-13) are designed with gradual slopes to lead water safely into grass waterways or other suitable outlets. The number and spacing of terraces depend on the soil type, slope, and cropping practices. Terraces should by designed by qualified soil conservation technicians. New, improved designs allow easier farming with modern machinery and reduce the number of point rows.

Diversion terraces are designed to divert larger flows of water away from buildings, gullies, ponds, or fields below long slopes.

Grass Waterways. Grass waterways are natural or constructed outlets or waterways protected by grass cover. They serve as safe outlets for runoff water from contour rows, terraces, and diversions. Natural drainage areas are good sites for waterways and often require a minimum of shaping to produce a good channel. They should be designed to be wide and flat to accommodate farm machinery, and be able to carry the runoff safely from the watershed above.

Conservation Tillage

Tillage practices that leave significant amounts of residue on the soil surface during the entire year are currently receiving wide attention as erosion-control practices. Such practices control erosion and save time, fuel, labor, and money. Many systems are available, and the system should be fitted to the situation. No one best tillage system works for all soils, and conservation tillage may need to be combined with other practices to achieve the desired level of erosion control.

Because all properly executed conservation tillage systems leave significant quantities of crop residue on the soil surface, evaporation is reduced and fields are moister than if had they had been plowed. Adequate drainage is essential in conservation tillage systems if yields are to be maintained. Surface and subsurface drainage should be adequate to remove excess water quickly. If adequate drainage cannot be achieved, use of elevated seedbeds, such as ridges or wide beds, may be beneficial.

Effective erosion control is often achieved on flatter slopes by using a chisel plow, disk, or field cultivator instead of the moldboard plow. One pass using these tools usually performs adequate tillage while leaving large

Figure 7-14 No-till farming is practiced by planting seed through the straw residue protecting the surface of the soil. This example shows cotton planted in rye residue. (Photo courtesy of USDA ARS)

quantities of residue on the surface. Specially designed planters are capable of planting precisely in the resulting seedbed. Not all reduced tillage, however, is conservation tillage. Multiple passes across the field, for example, bury all residue, in which case the conservation benefits are lost.

No-Till. This involves planting with no prior tillage (Figure 7-14). Seed is placed in a slot formed by the planter, and weed control is achieved entirely by surface applied and contact herbicides. When managed properly, **no-till** production works well on many soils.

Ridge Planting Systems. **Ridge planting** systems have achieved some popularity. They are quite similar to no-till in that the soil is not disturbed between harvest and planting. Ridges are usually reformed yearly by cultivation of the standing crop in June. This mandatory cultivation has allowed some farmers to reduce herbicide usage (and production costs) significantly. A major benefit of ridging over no-till seems to be the opportunity for earlier planting on more poorly drained soils.

Soil Compaction

Current concern about soil compaction in the plow layer and in the subsoil is closely related to the increased potential for damage to soil structure by soil compaction from long-term changes in soil management. These changes include the use of much larger and heavier farm equipment, increased specialization in crop production, the increased traffic and tillage necessary for application and incorporation of fertilizers, insecticides and herbicides, and earlier seedbed preparation and planting of many crops when soils are often wet and susceptible to compaction. Figure 7-15 illustrates how the tractor load affects soil compaction.

Definition. Soil compaction refers to the packing effect of a mechanical force on the soil. This packing effect decreases the volume occupied by pores and increases the density and strength of the soil mass. In addition, the number of large pores decreases, which slows both the rate of infiltration and drainage from the compacted layer. This occurs because large pores are the most effective in transporting water when the soil is saturated. Because

Figure 7-15 The curves under the tires represent pressure being applied at different depths. The pressure in the tire stays the same, but the load on the tire increases, compacting the soil at a deeper depth.

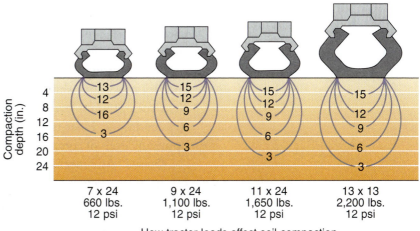

How tractor loads affect soil compaction

compacted soil has less air-filled pore space than before it was compacted, the exchange of gases slows down and the likelihood of aeration-related problems increases. The increased soil strength (ability of the soil to resist being moved by an applied force) of the compacted soil means that roots must exert greater force to penetrate the soil.

Consequences. Soil compaction can have both good and bad effects on plant growth. A moderate amount of compaction in the row promotes good seed–soil contact, fast germination, and prevents excessive drying out around the seed. Recently, corn planters have been designed specifically to provide the benefits of moderate compaction. Roots growing through a medium-textured soil with a bulk density of about 1.2 grams per cubic centimeter (comparable to a nontracked soil after a secondary tillage operation) will probably not have a high degree of branching.

The increased root branching and secondary root formation caused by moderate compaction means that a given volume of soil can be more thoroughly explored for nutrients. This could increase uptake of nonmobile nutrients such as phosphorus. Moderate compaction may also decrease water loss by evaporation.

When compaction exceeds the optimum level, root growth decreases. From the standpoint of crop production, the effect of soil compaction on water flow and storage may be more serious than the direct effect of increased soil strength on root growth.

Soil compaction has the potential to—
+ Decrease the infiltration rate
+ Increase runoff and decrease water storage
+ Increase the water content above a compacted subsurface layer by slowing the internal drainage of water
+ Decrease root growth and reduce the soil volume explored by roots, which can decrease nutrient and water uptake

Severe subsurface compaction due to heavy traffic can cause nitrogen and moisture stress by restricting the depth and extent of rooting, and by increasing denitrification.

Research from North America and Europe indicates a yield response to compaction. Starting with very low bulk density conditions, yields initially increase as compaction increases and they reach a maximum when the level of compaction is optimum for the given soil, crop, and climatic conditions. As compaction is increased beyond the optimum, yields decline. Wet soil conditions during the growing season shift the response curve to the left and dry conditions shift the curve to the right. Thus, under wet conditions the optimum level of compaction is considerably lower than under dry conditions. Finer-textured soils have higher clay contents and lower optimum levels of compaction than coarser-textured soils.

Causes. Any mechanical force exerts pressure on the soil. This force can be great, such as from a tractor, cow, combine, or tillage implement, or it can come from something as small and isolated in effect as a plant root, or it can be a major physical phenomenon, such as the last glacier.

The principal mechanical force considered here is vehicular wheel traffic. Tractors and combines are of special concern because the weight of large tractors increased from less than 3 tons in the 1940s to approximately 20 tons today for the big four-wheel-drive units. Large harvesting equipment often carries several tons of grain in addition to its own weight. These machines and others associated with many of our crop-management practices weigh enough to pack the soil, especially if the soil is wet during tillage, planting, or harvest (Figure 7-16).

Wheel traffic compaction resulting from weights of less than 5 tons per axle is generally restricted to the upper foot of the soil. Compaction in this zone can be removed largely by moldboard-plowing the compacted layer. Wheel tracks are often the most obvious cause of surface compaction, but they are by no means the only cause. Livestock and tillage equipment can also produce compaction.

Plant Responses. The effect of compaction on plant growth and yield depends on the crop grown and the environmental conditions that the plant encounters, which are determined by the soil and the weather. The compaction effect, which varies with years and crops grown, can be affected by management factors such as the nutrient status of the soil. Response of various crops—including soybeans, corn, wheat, potatoes, and sugar beets—to surface compaction has been studied.

Subsoil Compaction

Subsoil compaction is probable on many soils under wet soil conditions when there is traffic with loads greater than 10 tons per axle. The types of equipment most likely to have loads in this range are combines, four-wheel-drive tractors, and loaded farm trucks. Plowing with a tractor wheel in the furrow packs soil below the depth reached by normal tillage operations and can be another source of subsoil compaction.

Figure 7-16 Signs of soil compaction in a farm field—heavy equipment working in a field where conditions are too wet.

Plant Response. The plant response to subsoil compaction, as with surface compaction, depends on the crop, soil conditions, and climatic conditions in a particular year. Subsoil compaction can affect factors such as water availability, nitrogen uptake, and possibly potassium uptake in some instances. These factors in turn can affect crops growth and yield. Subsoil compaction can also reduce yields by delaying planting and other field operations.

Deep compaction decreases internal drainage, causing the compacted soil to be wetter and colder than the uncompacted plots. The wetter soil causes a nitrogen deficiency which, for instance, severely decreases corn yield on the compacted plots even though 60 percent more than the recommended rate of nitrogen was applied.

Control. Four common strategies are used in dealing with compaction—
1. Acceptance
2. Alleviation
3. Controlled traffic
4. Avoidance

Avoidance is the most desirable, when it is physically and economically possible. The old adage of "stay off the field until it's fit to work" still applies. However, the possibly severe economic repercussions of delaying planting, harvesting, or other operations may outweigh compaction damage or loss. The dilemma the farmer faces in a wet spring or fall is not easy to resolve.

Two methods can be used to alleviate and lessen the damage caused by compaction. Compaction can be prevented or the adverse effects reduced. To reduce the adverse effects the grower must know how compaction is affecting plant growth and yield to be effective. Moldboard tillage of the compacted depth has been effective in removing surface compaction in studies in some locations.

Subsoiling. Subsoiling has been very effective in removing or fracturing compacted layers in certain areas of the United States. In parts of the southeastern United States, such as South Carolina, certain sandy coastal plain soils are readily susceptible to pan formation. On these soils, annual in-row subsoiling at planting has proved effective in fracturing dense soil layers to allow crop roots to proliferate in the underlying soil, greatly increasing the soil water reservoir available to the crop. Both farmer experience and research have clearly documented the yield advantage of subsoiling under such conditions.

To increase the probability of obtaining beneficial effects from subsoiling, the following steps should be considered:
1. Determine that a compaction problem actually exists.
2. Determine that subsoiling can effectively disrupt the compacted layer. This requires favorable soil moisture conditions at the time of subsoiling. Also, the compacted layer must be within the depth disturbed by subsoiling.
3. Avoid recompacting the soil loosened by the subsoiling operations.

Drainage

Soil drainage affects the type of plants that can be grown and the rate at which they grow. Well-drained soils can grow a wide variety of plants. Poorly drained soils can grow only a few plants that are especially adapted to growing in saturated conditions, such as cypress trees or rice. This is because most plant roots need air to live and absorb water and nutrients. In poorly drained soils, water pushes out all the air. The range of plants that can be grown on wet soils can be improved by improving the drainage. This can be done by digging ditches or putting special tubes in the soil for drainage.

SUMMARY

Sixteen nutrients are essential for plant growth. Carbon, hydrogen, oxygen, and nitrogen are the nonmineral essential elements. The remaining 12 essential minerals are classified as macronutrients and micronutrients, depending on the amount needed by the plants. Fertilizers applied to fields, plants, water, and other growing media supply macronutrients and micronutrients to plants. Soil nitrogen cycles with atmospheric nitrogen by the action of plants, bacteria, precipitation, and animals. Besides the actual amount of minerals in the soil, pH, nutrient interactions, and soil type affect the availability of minerals. Soil pH can be altered by liming or by applying elemental sulfur. Soil testing determines the need for liming and fertilizers. Tissue testing can also be used to determine the need for fertilizer.

Fertilizer sources vary, and most are compared by the amount of nitrogen, phosphate, and potassium. Fertilizers may be dry or liquid, depending on use. The characteristics of the soil, the kind of crop, and the type of fertilizer help determine the method of application. Soil pH, organic matter, and drainage are variable. These factors influence the ability of a soil to grow a crop.

Soil erosion must be managed through conservation practices. Composts can be made and added to soil to increase the organic matter and improve the condition of the soil. Soil compaction can affect plant growth.

Review

True or False

1. The atmosphere contains 78 percent oxygen.

2. The decomposition rate of organic matter depends on the carbon:nitrogen ratio.

3. Potassium is a micronutrient.

4. Zinc is essential for plant growth.

5. Lime improves the availability of plant nutrients.

6. Soil drainage affects the type of plants that can be grown and the rate at which they grow.

Short Answer

7. List three primary nutrients.

8. Name three secondary nutrients.

9. _____ is a micronutrient affecting chlorophyll synthesis.

10. The uptake of _____ is often increased by the presence of nitrogen.

11. Name two compounds used to adjust soil pH.

12. Give the chemical name of four commercial fertilizers.

13. Indicate the conversion factors for the following:

    Phosphorus (P) = P_2O_5 × _____

    P_2O_5 = Phosphorus (P) × _____

    Potassium (K) = K_2O × _____

    K_2O = Potassium (K) × _____

14. A 100-pound bag of fertilizer is marked 0-15-15. How many pounds of nitrogen, available phosphate, and potassium are in the bag?

Critical Thinking/Discussion

15. Discuss how soil pH influences the availability of five nutrients.

16. Describe a nutrient deficiency of nitrogen, phosphorus, and potassium.

17. Explain why manure is an unreliable fertilizer.

18. Draw the nitrogen cycle and explain nitrogen fixation, nitrification, denitrification, and volatilization.

19. What is erosion?

20. What is the difference between no-till and conservation tillage?

21. Discuss the use of compost as a fertilizer source.

Knowledge Applied

1. Study nitrogen, phosphorus and potassium deficiencies. Fill five one-gallon containers with clean, dry sand. Punch holes in the bottom of each container. Number the containers 1, 2, 3, 4 and 5. Treat each container as follows:

   Container No. 1 - Add 1/3 teaspoon each of ammonium nitrate, triple superphosphate, and muriate of potash and 1 tablespoon of calcium hydroxide or hydrated lime. Mix the materials thoroughly with the sand in the container.

   Container No. 2 - Treat as No. 1 but leave out the ammonium nitrate.

   Container No. 3 - Treat as No. 1 but leave out the triple superphosphate.

   Container No. 4 - Treat as No. 1 but leave out the muriate of potash.

   Container No. 5 - Do not add any lime or fertilizer.

   Plant four corn seeds 1-inch deep in each container. Place containers in a warm area where they will receive eight or more hours of direct light daily and keep the sand moist but not wet. Observe the growth and color of the corn plants in each container. Complete the following table.

{PRIVATE} Days After Planting	Avg. Height of Plant (inches) Container Number					Color of Plants Container Number				
	1	2	3	4	5	1	2	3	4	5
7										
14										
21										
28										
35										

   Which container produced the healthiest plants? Why?

   Which container produced the least growth? Why?

   Which nutrient deficiency symptoms could you identify? Describe their appearance.

   Which plant nutrient would you most likely expect to be deficient? Why?

   Why did you not see any symptoms of micronutrient deficiencies?

2. Submit a soil sample for a soil test, or using a soil test kit conduct your own soil test on a sample that you collect. Determine the pH, phosphorus, potassium, and nitrogen. Discuss recommendations for fertilizer for different crops. Finally, describe the texture of the soil.

3. From a garden shop, farm chemicals business, or similar business collect and label samples of some of the fertilizers in Table 7-5.

4. Calculate the number of pounds each of nitrogen (N), phosphate (P_2O_5), and potash (K_2O) contained in the following: One ton of a dry 6-12-12 grade of fertilizer. Show your calculations.

5. In 1,000 gallons of a liquid 6-12-12 grade of fertilizer, if each gallon weighs 11 pounds, determine the number of pounds of N, P_2O_5, and K_2O. Show your calculations.

   continues

Knowledge Applied, *continued*

6. A soil test report recommends that 120 pounds N, 60 pounds P_2O_5, and 30 pounds K_2O be applied per acre to a soil to be used to grow corn. How many pounds of the following materials will be needed to supply the above nutrients to the corn? (Use information in Table 7-5 and show your calculations.)

____ pounds of ammonium nitrate to provide the 120 pounds of N.

____ pounds of triple superphosphate to provide the 60 pounds of P_2O_5.

____ pounds of muriate of potash to provide the 30 pounds of K_2O.

7. Create a pneumonic to help remember each of the macronutrients and micronutrients; for example, "C HOPKNS CaFe Mg."

8. Using digital photographs, find samples of erosion within the community and share with the class.

Resources

Ashman, M. R., & Puri, G. (2002). *Essential soil science*. Ames, IA: Blackwell Publishing.

Brady, N. C., & Weil, R. (2007). *Nature and properties of soil* (14th ed.). Englewood Cliffs, NJ: Prentice Hall.

Connors, J. J., & Cordell, S., eds. (2003). *Soil fertility manual*. Tucson, AZ: Potash & Phosphate Institute.

Coyne, M. S., & Thompson, J. A. (2005). *Fundamental soil science*. Albany, NY: Delmar Cengage Learning.

Krishna, K. R. (2003). *Agrosphere: Nutrient dynamics, ecology, and productivity*. Enfield, NH: Science Publisher.

Scott, H. D. (2000). *Soil physics: Agriculture and environmental applications*. Ames, IA: Blackwell Publishing.

Tan, K. H. (2000). *Environmental soil science* (2nd ed.). Boca Raton, FL: CRC Press.

United States Department of Agriculture. (1993). *Soil testing and plant analysis for fertilizer recommendation*. Beltsville, MD: Author.

White, R. E., & Bertola, G. (2005). *Principles and practice of soil science: The soil as a natural resource* (4th ed.). Ames, IA: Blackwell Publishing.

Internet

Internet sites represent a vast resource of information. The URLs (uniform resource locators) for World Wide Web sites can change. Using one of the search engines on the Internet such as Google or Yahoo!, find more information by searching for these words or phrases: plant nutrients, plant macronutrients, plant micronutrients, plant growth elements, nitrogen cycle, fertilizer production, organic fertilizer, value of manure, fertilizer sources, fertilizer application, commercial fertilizer production, soil test, and carbon:nitrogen ratio.

Chapter 8
Soilless Plant Production

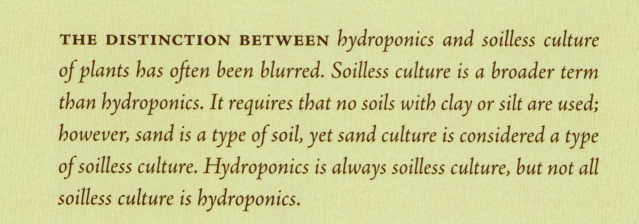

THE DISTINCTION BETWEEN *hydroponics and soilless culture of plants has often been blurred. Soilless culture is a broader term than hydroponics. It requires that no soils with clay or silt are used; however, sand is a type of soil, yet sand culture is considered a type of soilless culture. Hydroponics is always soilless culture, but not all soilless culture is hydroponics.*

After completing this chapter, you should be able to—

- Describe media used for potting mixes or for hydroponics
- Discuss the problems associated with container growing
- Describe soilless growing of plants
- Identify the purposes for the components of potting mixes
- Describe the two main types of hydroponics
- Discuss the role of the medium in hydroponics
- List five types of media used for hydroponics
- Identify three types of hydroponic systems using medium culture
- Explain the importance of nutrient solutions used in hydroponics
- Name three advantages and three disadvantages of hydroponic production
- Describe aeroponic plant production

KEY TERMS

Aeroponics
Aquaponics
Chelating Agents
Coarse Aggregates
Coir
Conductivity Meter
Expanded Clay
Hydroponics
Media
Medium Culture
Nutrient Film Technique
Perched Water Table
Potting Mixes
Rockwool
Soilless

CONTAINER GROWING

Billions of container plants are produced annually, including fruit, shade and ornamental trees, shrubs, forest seedlings, vegetable seedlings, bedding plants, herbaceous perennials, and vines. Most container plants are produced in **soilless** media, representing soilless culture. However, most are not hydroponics because the soilless medium often provides some of the mineral nutrients through slow-release fertilizers, cation exchange, and decomposition of the organic medium itself. Most soilless media for container plants also contain organic materials such as peat or composted bark, which provide some nitrogen to the plant. Greenhouse growth of plants in peat bags is often termed **hydroponics**, but technically it is not, because the medium provides some of the mineral nutrients. Peat has a high cation exchange capacity and must be amended with limestone to raise the pH.

One of the most demanding ways to grow a plant is to grow it in a container. A containerized plant requires constant attention to watering, fertilizing, and other practices. Despite this, more and more plants are being grown in containers. Not only are greenhouse growers growing flowers in pots, more and more nurseries grow shrubs, evergreens, and trees in containers. The container grower has complete control over soil conditions, making it easier to grow a large, uniform crop of quality plants. More recently, the business of landscaping the interiors of buildings with potted plants (Figure 8-1) has grown rapidly. Apartment dwellers and even homeowners now garden in containers (Figure 8-2).

Growing plants in containers differs from growing plants in the ground in one key way: the plant's root system is confined to a small soil volume that must supply all the plant's water and nutrient needs. This means the container grower waters and fertilizes far more than those who grow plants in fields. Additionally, container growing introduces some specific concerns, namely: the naturally poor drainage of potted soil, types of potting soil,

Figure 8-1 Potted plants in a commercial greenhouse destined for market.

Figure 8-2 Container gardening in a lot.

soil sterilization, soluble salts and alkalinity, soil temperature, and water pollution.

Potting Mixes

Good **potting mixes**, or media, have a high holding capacity for both air and water. To accomplish this, the mix needs large particles that can absorb water. The mixes contain varying amounts of three materials:

1. Soil is the main part of soil-based mixes. It cannot be used alone, but must be mixed with the other two materials. Soil helps a mix to hold water and nutrients, and helps buffer the medium from rapid changes. However, the fine particles retard drainage and lower aeration, limiting its use in shallow containers. Soils used in potting mixes should be loamy with good structure and be free of pesticide residues.

2. **Coarse aggregates** are large, inorganic particles used to create large pores in the mix. Coarse aggregates include coarse sand, perlite, vermiculite (expanded mica), shredded plastics, and other materials.

3. Organic amendments hold water and may help porosity. Shredded sphagnum peat is most common. Many growers shred and compost tree bark or sawdust. Rice hulls, shredded coconut hulls, and wood chip/sludge compost are also being used.

Growers combine these materials into various mixes to suit their needs (Figure 8-3). Standard soil-based mixes follow the model set by the John Innes mixes developed in England in the 1930s. These mixes consist of loam, peat, a coarse aggregate, and fertilizers. Later, soilless mixes based on mixing peat with a coarse aggregate were developed. Table 8-1 compares the components of some mixes. Growers use mixes based on composted hardwood bark chips or pine bark. Bark chip mixes have very high porosity and seem to suppress many harmful soil organisms.

Figure 8-3 Numerous pots filled with potting mixes in a commercial greenhouse.

Table 8-1 Comparison of some common potting mixes. Numbers represent parts.

Component	Mix 1	Mix 2	Mix 3	Mix 4
Sand	1	—	—	1
Peat	3	1	1	—
Perlite	—	—	1	—
Vermiculite	—	1	—	—
Composted bark	—	—	—	2

Materials such as peat, perlite, vermiculite, and composted sawdust are superior to soil in aeration and drainage. One or more of these artificial materials with or without soil is a more desirable mixture for plant production. Some examples include perlite, vermiculite, and sphagnum peat moss.

+ Perlite is volcanic rock heated to about 1,800°F, causing it to expand and become porous. In the bag, it is sterile and neutral in reaction (pH 7.0). It does not decay, but handling will fracture it. Perlite is lightweight and will hold three to four times its weight in water. It is excellent for rooting cuttings as a seed-germinating medium and as an additive to soil mixes. It has little nutrient value.

+ Vermiculite is a mica compound heated to about 1,400°F to form a platelike structure enabling it to retain both water and fertilizer. Excessive handling can destroy this structure. Moist vermiculite, when compressed, does not expand, so its water-holding capacity is reduced. It is sterile in the bag and absorbs fertilizer much the same as clay. It also contains potassium and magnesium in an available form. Growers use horticultural vermiculite, sizes 2, 3, or 4.

+ Sphagnum peat moss is the preferred peat for artificial mixes. The horticultural grades mix well with soil or vermiculite and perlite. Because it is generally very acid, pulverized limestone should be added to reduce acidity. Muck, often referred to as peat, has a considerable adsorptive capacity for phosphorus. Phosphorus deficiencies in plants can occur if special precautions are not taken. Muck that has decomposed to a very fine texture is not as good as sphagnum peat moss in providing good aeration and drainage—two important factors in growing plants.

A typical growing medium would contain the following components in proper proportions:

+ Shredded sphagnum peat moss
+ Horticultural vermiculite
+ Dolomitic limestone
+ 20 percent superphosphate (pulverized)
+ Potassium nitrate or calcium nitrate
+ Trace elements

Standard soil mixes for plant growers are available; for example, Jiffy Mix, Redi-earth, and PRO-MIX.

Peat Pellets. One dramatic advance for sowing seeds and growing plants is the use of peat pellets and fiber blocks. The small, compressed, net-enclosed sphagnum peat pellet with fertilizer added, when moistened, expands into a pellet $1\,^3/_4$ inches in diameter and $2\,^1/_8$ inches high. Seeds are then pressed into the peat. After the plant has grown, everything is transplanted, preventing transplant shock.

Fiber Blocks. Two major types of fiber blocks are sold. In one type, the block is shaped and sized, then dried and used directly as taken from the package. The second type is slightly compressed and dried, and must be watered and allowed to expand for a few minutes to reach full size before use. Fiber blocks are light, clean, and easy to use. Because roots penetrate the block, the whole block is planted. Manufacturers control the content of blocks, giving a uniform, clean, disease-free container.

Container Drainage

A pot of soil is by nature poorly drained because of the shallow soil profile. Compare soil in a pot with soil in the ground. Capillary action "pulls" water into the drier soil below a wetting front. In a deep soil profile, then, lower layers of soil pull water downward.

In a pot, the soil column ends abruptly at the bottom of the pot. The bottom layer of soil has no capillary connection to deeper soil and the last bit of water cannot drain away after watering. Thus, in spite of drainage holes, a layer of soil on the bottom of the pot remains saturated after drainage ceases. This layer is called a perched water table (Figure 8-4). As a result, potted soil is wetter and has less air after drainage than the same soil in the ground.

The difficulty is the short water column—no "depth" to pull water down and no heavy mass of water pushing down by gravity. Therefore, the taller the pot, the less severe the problem. A 6-inch pot filled with a standard greenhouse mix has an air-filled porosity of about 20 percent. Some greenhouse containers are no more than an inch deep. With the same mix, such a pot has a porosity of perhaps 2 percent.

Figure 8-4 A perched water table in a pot causes poor drainage. Highly porous soils and fairly tall pots help solve the problem.

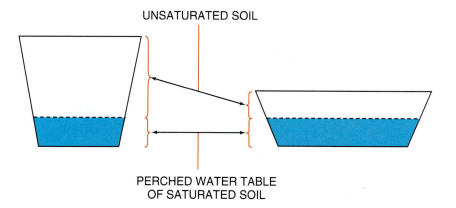

UNSATURATED SOIL

PERCHED WATER TABLE
OF SATURATED SOIL

Because of poor drainage, a potting mix must be highly porous, with very large pore spaces. Large pore spaces apply less capillary force to hold the water in the pot. There are two approaches to making a porous mix. One is to mix materials into a field soil to make a porous soil-based potting mix. The second method omits soil altogether, making a soilless potting mix of other materials.

The key to making potting mixes (Figure 8-5) is to use large particles to create large pore spaces. For example, one simple soilless mix is half sand and half peat. When fine sand is used to prepare the mix, the percentage of total soil volume filled with air after watering is about 5 percent. The same mix with coarse sand has an air capacity of 16 percent. Thus, the mix with coarse sand holds more than three times as much air as the fine sand mix.

Soil Sterilization

Soil-based mixes contain weed seeds, insects, nematodes, and parasitic fungi and bacteria. Of special concern are fungi that destroy young seedlings or cause root rots. To kill these organisms, some growers treat the soil with chemicals. The soil must "air out" for several weeks after such a treatment to avoid injuring crops planted in the mix. More commonly, soil

Figure 8-5 Potting mix ready to fill pots in a commercial greenhouse.

is sterilized by heat. Many growers use steam, normally at a temperature of 212°F. However, this high temperature may cause three problems:

1. Kills the bacteria that convert ammonia to nitrates (nitrification). This break in the nitrogen cycle can cause a buildup of ammonia to harmful levels.
2. Raises the solubility of manganese, which can result in toxic levels.
3. Creates a "biological vacuum," destroying organisms that could compete with pathogens if the soil were reinfected.

Several methods avoid these problems. First, lime can be used in a potting mix to maintain the pH between 6.0 and 7.0. At this level, manganese is insoluble, reducing toxicity problems. Second, potting mixes should be used soon after treatment, before high levels of ammonia can build up. Third, "live steam" (temperature, 212°F) can be replaced by "aerated steam" (temperature, 160–180°F). Special devices inject air into live steam to lower the temperature. This temperature reduces ammonia and manganese problems and allows some helpful organisms to survive while killing pathogens.

Soluble Salts and pH

To grow potted plants successfully, a great deal of fertilizer (fed through irrigation water or incorporated into the potting mix) must be poured into a small soil volume. Most irrigation water contains dissolved salts, and most fertilizers are salts. Therefore, a troublesome problem of growing in pots is the buildup of soluble salts.

Controlling soluble salts in potted plants means frequent testing of both irrigation water and the medium itself. Growers send samples of potting media to testing laboratories. However, they also monitor it themselves with a **conductivity meter.** The more ions dissolved in water, the more easily an electrical current will pass. Therefore, measuring the conductance of a standardized mixture of medium and water also measures its soluble salt concentration. Conductivity meters provide growers with critical and timely information needed to manage soluble salts and fertility. These other practices also help growers avoid problems:

+ Saline water can be treated with special devices, but the process can be expensive for large amounts of water. A common water softener merely replaces calcium ions with sodium ions and does not lower salinity. Some California nurseries, contending with salty irrigation water, have installed very large—and expensive—systems to remove salts from their water supply.
+ All pots should have good drainage. When watering, enough water is added so that some water leaks from the drainage holes. Drainage water leaches out excess salts.

A major difficulty encountered by container growers is dramatic changes in soil pH. In many areas of the country, high levels of dissolved carbonates—lime—in irrigation water raise pH of the potting mix. In other areas, the water has so little dissolved lime that leaching calcium from the potting mix during watering lowers pH. For most containerized crops, pH should be between 5.8 and 6.5.

The carbonate content of water, termed alkalinity, should be measured by a testing laboratory. If it is too high, the most common treatment injects acid into the water supply to neutralize the alkalinity. Water could also be treated to remove the ions, and some have even suggested mixing irrigation water with rainwater collected from the greenhouse roof. Crops irrigated with low-alkalinity water may need to be treated with lime to counteract falling pH.

Soil Temperature

Plants growing above ground in containers, especially when exposed to hot sunlight, suffer large swings in soil temperature that can damage roots.

More severe is the difficulty of overwintering containerized plants in cold climates. Plant roots are less hardy than plant tops, and a potted root system exposed to subzero temperatures will be damaged. Northern nurseries must protect container plants over winter by covering them with straw or some protective structure. Similarly, landscape plants in containers, like urns by the doorstep or trees in planters on city streets, experience the same freezing. Northern gardeners and landscapers should use only the root-hardiest plants for this purpose or plant only annuals that do not need to survive the winter. Insulating containers with rigid insulation also helps.

Water Pollution

The public is concerned about water pollution; this issue strongly affects nurseries and greenhouses. For example, container nurseries have pots sitting on the ground and fertilize through overhead sprinklers. Most of this water—as much as 70 percent—lands between, rather than in, the pots. Because a container field is watered and fertilized daily, the potential for water pollution is great.

One answer, where feasible, is to use drip irrigation (Figure 8-6). First little water or nutrients are wasted this way. Second, greater use of slow-release fertilizers added to the pots could reduce the amounts added to irrigation water. Third, pots should be set on a sealed, graded surface that prevents leaching into the soil. Water runs off into sealed holding ponds, where the water can be pumped for reuse (Figure 8-7).

HYDROPONICS

Hydroponics is the technique of growing plants in a medium other than soil, using a feeding mixture of essential plant nutrients dissolved in water. The method has proved valuable in areas where the soil is unsuitable or infertile, or where soil-borne diseases inhibit the growth of vegetables.

Figure 8-6 Drip irrigation (between pots with tags) being used for a large number of potted plants on a table.

Figure 8-7 Pots set on a sealed, graded surface that prevents leaching into the soil. Water runs off into holding ponds.

Techniques

The two main types of hydroponics are static solution culture and medium culture (Figure 8-8). Static solution culture does not use a solid medium for the roots, just the nutrient solution. The three main types of hydroponics culture include:

1. Static solution culture
2. Medium culture
3. Aeroponics

The medium culture method has a solid medium for the roots and is named for the type of medium, such as sand culture, gravel culture, or rockwool culture. The two main variations for each medium are subirrigation and top irrigation. For all techniques, most hydroponic reservoirs are now built of plastic but other materials have been used, including concrete, glass, metal, vegetable solids, and wood. The containers should exclude light to prevent algae growth in the nutrient solution.

Hydroponic Requirements

Vegetables such as tomatoes, lettuce, and spinach work well when grown in hydroponics. Plants can be spaced closer using this method. In hydroponic culture, there is no reserve. The nutrient solution that bathes the

Figure 8-8 A diagram comparing static solution with media hydroponic culture.

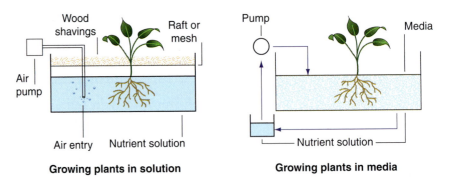

Growing plants in solution   **Growing plants in media**

roots in hydroponics is depleted of nutrients by the plants. The solution can be added to or changed to maintain the best concentrations for plant growth. Normally, it is changed weekly. Hydroponics does avoid several soil problems associated with soil texture, weeds, and disease. But, hydroponic growers face other problems. They must keep in balance all chemicals and physical conditions that surround the roots.

A general solution, such as the modified Hoagland solution, is good for tomatoes, cucumbers, and other fruiting crops. Many premixed, complete nutrient solutions are available. Hobby growers and beginning hydroponic growers often start with a commercial solution and mix their own only after other culture problems have been solved. Mixing a nutrient solution can be even more expensive than buying commercial formulations because many chemicals are expensive, especially when purchased in small quantities.

All essential plant nutrients must be accounted for in a nutrient solution. Any hydroponic solution must contain nitrogen, phosphorus, potassium, calcium, magnesium, iron, manganese, boron, zinc, copper, and molybdenum. Lack of an element, too little of an element, or too much of an element will mean inferior or dead plants. Commercial solutions include directions about application. Still, different environmental conditions, different varieties, and different seasons will create different demands. Some demands cannot be met by one particular blend or concentration of nutrients.

Static Solution Culture

In static solution culture, plants are grown in containers of nutrient solution, such as glass quart jars (typically in-home applications), plastic buckets, Styrofoam containers, tubs, or tanks. The solution is usually gently aerated but may be unaerated. If unaerated, the solution level is kept low enough that enough roots are above the solution so they get adequate oxygen (Figure 8-8). A hole is cut in the lid of the reservoir for each plant. Each reservoir may hold one to many plants. Size of the reservoir is increased as plant size increases. When containers are used, they need to be covered with aluminum foil, black plastic, or other material to exclude light to eliminate the formation of algae. The nutrient solution is either changed on a schedule, such as once per week, or when the concentration drops below a certain level as determined with an electrical conductivity meter. Whenever the solution is depleted below a certain level, either water or fresh nutrient solution is added. An automated system can be used to maintain the solution level. In raft solution culture, plants are placed in a sheet of buoyant plastic that is floated on the surface of the nutrient solution so the solution level never drops below the roots (Figure 8-9).

Medium Culture

In **medium culture** the nutrient solution constantly flows past the roots. It is much harder to automate than the static solution culture because sampling and adjustments to degree and nutrient concentrations can be

Figure 8-9 Demonstration of plants grown in static solution hydroponics.

made in a large storage tank that serves potentially thousands of plants. Computerized systems are used for this culture system. Continuous flow systems can be the nutrient film technique, passive subirrigation, ebb and flow/flood and drain subirrigation, or top irrigation.

Nutrient Film Technique. A popular variation of continuous flow is the nutrient film technique (NFT) whereby a very shallow stream of water containing all the dissolved nutrients required for plant growth is recirculated past the bare roots of plants in a watertight gully, also known as a channel. Ideally, the depth of the recirculating stream should be very shallow, little more than a film of water; hence, the name nutrient film. This ensures that the thick root mat, which develops in the bottom of the channel, has an upper surface that, although moist, is in the air. Subsequently, an abundant supply of oxygen is provided to the roots of the plants. A properly designed NFT system is based on using the right channel slope, the right flow rate, and the right channel length.

The main advantage of the NFT system over other forms of hydroponics is that the plant roots are exposed to adequate supplies of water, oxygen, and nutrients. In all other forms of production a conflict exists between the supply of these requirements as excessive or deficient amounts of one results in an imbalance of one or both of the others. A disadvantage of NFT is that it has very little buffering against interruptions in the flow; for example, when a power outage occurs.

Passive Subirrigation. Passive subirrigation, also known as passive hydroponics or semihydroponics, is a method where plants are grown in an inert porous medium that transports water and fertilizer to the roots by capillary action from a separate reservoir as necessary, reducing labor and providing a constant supply of water to the roots. In the simplest method, the pot sits in a shallow solution of fertilizer and water or on a capillary mat saturated with nutrient solution. The various hydroponic media available, such as expanded clay and coconut husk, contain more air space than more traditional potting mixes, delivering increased oxygen to the roots. Additional advantages of passive hydroponics are the reduction of root rot and the additional ambient humidity provided through evaporation.

Ebb and Flow/Flood and Drain Subirrigation. In the ebb and flow/flood and drain method, a tray above a reservoir of nutrient solution is either filled with growing medium (clay granules being the most common) and planted directly, or pots of medium stand in the tray. At regular intervals, a simple timer causes a pump to fill the upper tray with nutrient solution, after which the solution drains back down into the reservoir. This keeps the medium regularly flushed with nutrients and air.

Top Irrigation. In top irrigation, nutrient solution is periodically applied to the medium surface. This may be done manually once per day in large containers of some media, such as sand. Usually, it is automated with a pump, timer, and drip-irrigation tubing to deliver nutrient solution as frequently as 5 to 10 minutes every hour (Figure 8-10).

Media

Those who practice hydroponics must select the growing medium to use. Different media are appropriate for different growing techniques. Types of growing media include dihydro, expanded clay, rockwool, coir, perlite, vermiculite, sand, gravel, brick shards, or polystyrene packing peanuts.

Diahydro. Diahydro is a natural sedimentary rock medium that consists of the fossilized remains of diatoms. Diahydro is high in silica (87 to 94 percent).

Expanded Clay. Expanded clay aggregate consists of small, round, baked spheres of clay. These are inert and are suitable for hydroponic systems in which all nutrients are carefully controlled in water solution. The clay pellets are pH-neutral and provide no nutrient value. The clay is formed into round pellets and fired in rotary kilns at 1,200°C. This causes the clay to expand, like popcorn, and become porous. Manufacturers consider

Figure 8-10 Hydroponically grown tomatoes near the end of their production cycle. A drip system delivers the nutrient solution to the top of the growing media in the long white containers.

Organoponics

Organoponics might be considered a hydroponic system converted to organic cultivation by replacing the inorganic fertilizer with compost, but it is more or less than this. Organoponics draws its name from the field of hydroponics in which plants are grown without soil, but the system seems to have a couple of meanings depending on use and location. Its use is prevalent in urban agriculture in countries such as Cuba and where commercial fertilizers are costly or not available. One scheme focuses on the collection and use of human urine as a fertilizer for crops. Another scheme focuses on collection and composting of readily available organic matter such as food wastes, crop wastes, leaves, and grass.

If the scheme uses human urine, it is collected and stored in capped containers to minimize ammonium losses. In some cases the urine is fermented and the container must be kept at least partially uncapped in order not to inhibit the aerobic action and the development of beneficial actinomycetes (filamentous or rod-shaped microorganisms). Sediments in urine contain some of the phosphorus in compounds such as $MgPO_4$, $MgHPO_4$, NH_4HPO_4, and $NaHPO_4$. These can be potentially collected and applied to crops with a high phosphorus demand.

As with any fertilizer, application of the urine must be uniform and in adequate quantities. Application techniques have been kept simple, involving manual watering cans and/or buckets, or hoses connected to elevated tanks. The concentrated or diluted urine, as the case may be, is applied to an open furrow or shallow hole and then the fresh earth is quickly covered over with a hoe. Commercial fertilizer applicators, involving a hose and tube applicator attached to a small tank carried on shoulder straps, have also been used.

When the scheme relies instead on readily available organic matter, compost is developed as plants grow and liquid fertilizer comes from the compost. Plants are grown in small containers placed in household areas such as patios, balconies, and alleyways. The containers are filled mostly with some readily available organic matter such as food wastes, leaves, and grass. A small amount of soil (2 to 5 inches) is added to the top of the container. The contents of the container are then treated with small but continual doses of a liquid organic fertilizer that encourage the decomposition process. Liquid fertilizer is made by adding water compost and allowing the mixture to ferment. (Sometimes this is called compost tea.[1]) As the organic matter begins to break down in the container, it releases nutrients and acts as a soillike medium to support the plants. Within 8 to 10 months, the organic matter mixture will convert into soil. Using this form of organoponics, it is possible to create soil and grow plants at the same time.

Compost for the production of liquid fertilizer can be derived from regular composting or from the "vermicompost" derived from the action of earthworms such as California red worms. Worms break down the organic matter, leaving behind compost rich in nutrients—nitrogen, potassium, magnesium, and calcium.

---

[1]Compost tea, a liquid solution or suspension made by steeping compost in water.

expanded clay to be an ecologically sustainable and reusable growing medium because of its ability to be cleaned and sterilized.

Rockwool. Rockwool is one of the most widely used media in hydroponics. It is made from basalt rock heat-treated at high temperatures then spun back together like cotton candy. It can be purchased as cubes, blocks, slabs, and granulated or flock. Rockwool is an excellent inert substrate for both drainage and recirculating systems. Rockwool is also lightweight and self-contained, which allows plants to be grown at different densities in different stages. Its light weight permits setting up to be quick and inexpensive. Rockwool provides a favorable root environment, thus minimizing plant stress. Root temperature can also be controlled, thus giving substantial energy savings. Rockwool initially causes an increase in pH level. The disadvantages of rockwool include transport costs, minor skin irritations when working with it, and inhalation of dry particles. Wetting the rockwool before handling it lessens its hazards.

Coir. Coco peat, also known as coir or coco, is the leftover material after the fibers have been removed from the outermost shell of the coconut. Coco peat is not only a high-quality product, but also an environmentally friendly product. Coir comes in bags and in slabs. This medium combines the organic nature of soil with the qualities of rockwool. Buffering capability of the coir substrate and its spongelike structure allow nutrients needed to ensure high yields to be stored in the coco. Coconut fibers have sufficient capillary action to retain enough water and nutrients. Quality coir can be used a number of times and makes an excellent soil amendment after use.

Perlite. Perlite is a volcanic rock that has been superheated into very lightweight, expanded glass pebbles. It is used loose or in plastic sleeves immersed in the water. It is also used in potting soil mixes to decrease soil density. If not contained, it can float if ebb-and-flow feeding is used.

Vermiculite. Vermiculite is another mineral that has been superheated until it has expanded into light pebbles. Vermiculite holds more water than perlite, and can draw water and nutrients in a passive hydroponic system.

Sand. Sand is cheap and easily available. It is heavy, does not always drain well, and it must be sterilized between use (Figure 8-11).

Figure 8-11 Sorghum being grown in sand with a drip system to supply nutrients. A legume (clover) is planted between the rows.

Gravel. Gravel like that used in aquariums or any small, washed gravel can be used for a medium. Gravel is inexpensive, easy to keep clean, drains well, and will not become waterlogged. Gravel is heavy, and if the system fails to provide continuous water, the plant roots may dry out.

Brick Shards. Brick shards have properties similar to gravel. They possibly alter the pH and require extra cleaning before reuse.

Polystyrene Packing Peanuts. Polystyrene packing peanuts are inexpensive, readily available, and have excellent drainage. They can be too lightweight for some uses. Polystyrene packing peanuts are mainly used in closed tube systems. Polystyrene peanuts must be used; biodegradable packing peanuts will decompose into sludge.

Nutrient Solutions

Plant nutrients are dissolved in the water used in hydroponics and are mostly in inorganic and ionic form. Primary among the dissolved cations (positively charged ions) are Ca^{2+} (calcium), Mg^{2+} (magnesium), and $K+$ (potassium); the major nutrient anions in nutrient solutions are NO_3^- (nitrate), SO_4^{2-} (sulfate), and $H_2PO_4^-$ (phosphate).

Many recipes (formulations) for hydroponic solutions are available. Commonly used chemicals for the macronutrients include potassium nitrate, calcium nitrate, potassium phosphate, and magnesium sulfate. Various micronutrients are typically added to hydroponic solutions to supply essential elements; among them are Fe (iron), Mn (manganese), Cu (copper), Zn (zinc), B (boron), Cl (chlorine), and Ni (nickel). **Chelating agents** are sometimes used to keep Fe soluble. Many variations of the Hoagland solution are widely used.

Plants change the composition of the nutrient solutions upon contact by depleting specific nutrients more rapidly than others, removing water from the solution, and altering the pH by excretion of either acidity or alkalinity. Nutrient solutions must be monitored to prevent salt concentrations from becoming too high, nutrients becoming too depleted, or pH drifting far from the desired value.

Hydroponics fertilizers and other types of formulas for hydroponics have changed drastically during the last 10 years. Many of these changes have resulted in significant increases in plant growth rates, plant resistance to diseases and pests, and plant yields.

Commercial Use

Because of its arid climate, Israel has developed advanced hydroponic technology and marketed the system to other countries. Additionally, other countries have developed their own hydroponic systems (Figure 8-12).

The largest commercial hydroponics facility in the world is Eurofresh Farms in Willcox, Arizona (www.eurofresh.com). Eurofresh Farms has over 300 acres of greenhouses.

Some commercial hydroponic operations use no pesticides or herbicides, preferring integrated pest management techniques. Certifying

Figure 8-12 Large-scale hydroponic cucumber production in a greenhouse near Dnipro-petrovsk, Ukraine.

hydroponic crops as "organic" presents some challenges as nutrients used to make the hydroponic solutions do not come from "organic" sources as defined, and some definitions of organic require soil.

Hydroponics saves an incredible amount of water, using as little as 5 percent the amount as a regular farm production to produce the same amount of food. Nevertheless, hydroponic operations must control their water runoff and use.

The environment in a hydroponics greenhouse is tightly controlled for maximum efficiency. Computerized climate control systems take the guesswork out of the critical growing parameters, like temperature, humidity, light, irrigation, ventilation, and carbon dioxide levels within the greenhouse. With modern hydroponic systems, growers can produce premium foods anywhere in the world, regardless of temperature and growing seasons.

Advantages and Disadvantages

Advantages of hydroponics include:

+ Allows greater control over the root zone environment than soil culture
+ Over- and underwatering prevented
+ Accomplished in remote areas that lack suitable soil, such as Antarctica, space stations, space colonies, or atolls
+ Soil-borne diseases virtually eliminated
+ Weeds virtually eliminated
+ Fewer pesticides may be required
+ Edible crops not contaminated with soil
+ Water requirement less than with traditional irrigation of soil-grown crops

Figure 8-13 Squash plants being grown with aeroponics. Plants revolve from a track (above) and are sprayed with a nutrient solution as they move through a chamber (below).

- Provides plants more nutrition while at the same time using less energy and space
- Allows for easier fertilization as it is possible to use an automatic timer to fertilize the plants
- Provides plants with balanced nutrition because the essential nutrients are dissolved into the water-soluble nutrient solution

Disadvantages of hydroponics include:

- Quick loss of plants from failure of pumps, the clogging of the system, or springing a leak
- Costly to start up
- Disposal of a medium or sterilization and reuse of a medium
- As an intensive monoculture, increased risk of water-borne diseases such as *Pythium* becoming disastrous when nutrient solution is recirculated
- Requires intensive management and highly technical skills to operate

Aeroponics

Aeroponics is defined as a system where roots are continuously or discontinuously in an environment saturated with fine drops (a mist or aerosol) of nutrient solution (Figure 8-13). The method requires no substrate and entails growing plants with their roots suspended in a deep air or growth chamber with the roots periodically wetted with a fine mist of a nutrient solution. The main advantage of aeroponics is excellent aeration.

Aeroponic techniques are successful, but are not used on a commercial scale. Aeroponics is used in laboratory studies of plant physiology. Additionally, aeroponic techniques have been studied by the National Aeronautics and Space Administration for use in a zero-gravity environment.

AQUAPONICS

In **aquaponics,** waste from an aquaculture facility provides a food source for the growing plants and the plants provide a natural filter for the fish. This creates a miniecosystem in which both plants and fish can thrive. Aquaponics is the ideal answer to a fish farmer's problem of disposing of nutrient-rich water and a hydroponic grower's need for nutrient-rich water.

Commercially, aquaponics is in its infancy, but as the technology develops and is refined, it has the potential to be a more efficient and space-saving method of growing fish, vegetables, and herbs. By incorporating aquaponics, hydroponic growers can eliminate the cost and labor involved in mixing a fertilizer solution, and commercial aquaculturists may be able to drastically reduce the amount of filtration needed in recirculating fish culture.

Currently, a limited number of commercial aquaponic facilities operate. Interest in this intensive method of food production will continue.

The combination of aquaculture and hydoponics is quite new. Research in aquaponics began in the 1970s and continues.

On a hobby scale, aquaponics has the potential to catch on quickly. A home aquarium, with ornamental or food fish, can be combined with a minigarden growing herbs, vegetables, or flowers.

SUMMARY

One of the most demanding ways to grow a plant is to grow it in a container. A containerized plant requires constant attention to watering, fertilizing, and other management practices. Despite this, more and more plants are being grown in containers. Good potting mixes, or media, have a high holding capacity for both air and water. To accomplish this, the mix needs large particles that can absorb water. The mixes contain varying amounts of three materials: soil, course aggregates, and organic amendments.

Hydroponics is the technique of growing plants in a medium other than soil, using a feeding mixture of essential plant nutrients dissolved in water. The two main types of hydroponics are static solution culture and medium culture. Static solution culture does not use a solid medium for the roots, just the nutrient solution. The three main types of hydroponic culture include static solution culture, continuous flow solution culture, and aeroponics.

The medium culture method has a solid medium for the roots and is named for the type of medium, such as sand culture, gravel culture, or rockwool culture. The two main variations for each medium are subirrigation and top irrigation.

Aeroponics is a system in which roots are continuously or discontinuously in an environment saturated with fine drops (a mist or aerosol) of nutrient solution.

In aquaponics, waste from an aquaculture facility provides a food source for the growing plants and the plants provide a natural filter for the fish. This creates a miniecosystem in which both plants and fish can thrive.

Review

True or False

1. Those who use containers for growing plants fertilize and water more than those who grow plants in fields.

2. Good potting mixes eliminate air and allow water to pass through quickly.

3. Clay in potting mixes increases aeration and drainage.

4. Conductivity meters monitor the pH of nutrient solutions.

5. Environmentally friendly, biodegradable packing peanuts can be used for a medium in a closed tube hydroponic system.

Short Answer

6. Why is container growing often considered soilless?

7. Identify three types of hydroponic systems using medium culture.

8. List five types of media used for hydroponics.

9. Name three advantages and three disadvantages of hydroponic production.

10. How are nutrients supplied to the plant in aeroponic systems?

11. What are the two main types of hydroponic systems?

Discussion

12. Why is drainage a problem in container-grown plants and what can be done to prevent drainage problems?

13. Compare the media used for potting mixes to that used for hydroponics.

14. Discuss the problems associated with container growing.

15. Describe soilless growing of plants in its broadest sense.

16. Identify the purposes for the components of potting mixes.

17. Discuss the role of the medium in hydroponics.

18. Explain the importance of nutrient solutions used in hydroponics.

Knowledge Applied

1. Grow a garden in a container with a soilless medium you mix (http://aggie-horticulture.tamu.edu/extension/container/container.html). Track water use and plant production.

2. Create a simple static flow hydroponic system. Use Figure 8-8 as a guide and use the Internet to research some ideas and designs. Keep the system low cost and use easily available supplies.

continues

Knowledge Applied, *continued*

3. Using Internet resources, develop a report on container gardening.

4. Compare the contents of several commercial potting mixes.

5. Visit a commercial hydroponics operation or a greenhouse where containers are used. Report on your visit. Include some discussion of how the operations address some of the unique challenges of either production system. If a physical visit is not possible, make a virtual visit to such Internet sites as these: The Miracle Farm Web Tour (www.astralstar.com/hydrogreen.html), Crop King (www.cropking.com/commercial.shtml), or American Hydroponics (www.amhydro.com/comm/index.html).

Resources

Bridgewood, L. (2003). *Hydroponics: Soilless gardening explained.* Wiltshire, UK: Crowood Press.

Douthwaite, K. (1999). *Growing fruits in containers (Success with).* Sidney, Australia: Murdoch Books.

Jones, J. B., Jr. (2004). *Hydroponics: A practical guide for the soilless grower* (2nd ed.). Boca Raton, FL: CRC Press.

Marken, B., & Editors of the National Gardening Association. (1998). *Container gardening for dummies.* Indianapolis, IN: Wiley Publishing, Inc.

Mason, J. (2000). *Commercial hydroponics.* New York: Simon & Schuster, Inc.

Phillips, S., & Sutherland, N. (2002). *The container gardening encyclopedia.* Berkeley, CA: Thunder Bay Press.

Plaster, E. J. (2008). *Soil science and management* (5th ed.). Albany, NY: Delmar Cengage Learning.

Internet

Internet sites represent a vast resource of information. The URLs (uniform resource locators) for World Wide Web sites can change. Using one of the search engines on the Internet such as Google or Yahoo!, find more information by searching for these words or phrases: hydroponics, commercial hydroponics, soilless production, container planting, container gardening, commercial hydroponics, growing media, aeroponics, and aquaponics.

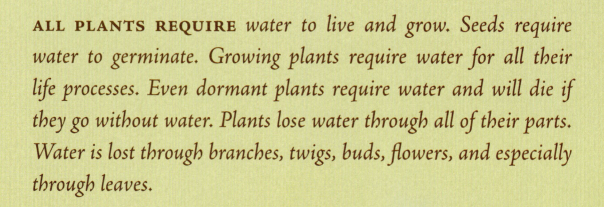

Chapter 9
Irrigation

ALL PLANTS REQUIRE water to live and grow. Seeds require water to germinate. Growing plants require water for all their life processes. Even dormant plants require water and will die if they go without water. Plants lose water through all of their parts. Water is lost through branches, twigs, buds, flowers, and especially through leaves.

After completing this chapter, you should be able to—

- Describe the unique characteristics of water

- List the role of water in plants

- Explain evapotranspiration and the factors that affect it

- Discuss how evapotranspiration is used to determine the water needs of plants

- Describe how water moves through plants

- Name four types of irrigation used to fill the water needs of plants

- Suggest three methods for determining the moisture content of soil

- Identify and explain wilt in plants

- Explain translocation and transpiration

- Explain the concept of hygroscopic water

- Discuss the effects of too little or too much water on a crop

- Compare the evapotranspiration values for two different crops

- List four factors that influence the type of irrigation system used

- Explain a water budget

KEY TERMS

Aquifer
Bipolar
Capillary Water
Drip Irrigation
Electrical Conductivity
Emitter
Evapotranspiration
Fallow
Flood Irrigation
Furrow Irrigation
Gravity Water
Humidity
Hydrophytes
Hygroscopic Water
Irrigation
Mesophytes
Percolation
Precipitation
Reactant
Runoff
Solvent
Sprinkler Irrigation
Stomata
Translocation
Transpiration
Turgor
Water Budgeting
Wilt
Xerophytes

CHARACTERISTICS OF WATER

Water—the stuff of life—covers three-fourths of the earth's surface and represents a major component of the bodies of plants and animals. A 1-ton load of fresh-cut alfalfa would weigh only about 520 pounds with all the water removed.

This miracle liquid forms from two gases, hydrogen (H) and oxygen (O). Two atoms of hydrogen and one atom of oxygen combine to form water, H_2O. The water molecule is bipolar, having charged poles like a magnet, giving water unique properties.

Depending on the temperature, water exists in three forms. It is a liquid between 32° and 212°F. It is a gas vapor at temperatures above 212°F, and becomes a solid (ice) at temperatures below 32°F.

Of all the naturally occurring substances, water has the highest specific heat. This makes it a good coolant in biological systems and makes it resist rapid temperature changes. Specific heat is the amount of heat required to raise the temperature of a substance 1°C.

Water is the universal solvent, dissolving almost everything. It is powerful enough to dissolve rocks yet gentle enough to hold an enzyme in a fragile plant cell. As a solvent it acts as a medium for biochemical reactions carrying products of metabolism and nutrients.

Within any temperature zone, the availability of water is the most important factor in determining which plants will grow and how productive they will be. Indeed, the future development of crop lands depends on the availability of water.

Earth's water is always in movement, and the water cycle, also known as the hydrologic cycle, describes the continuous movement of water on, above, and below the surface of the earth. Water can change states among liquid, vapor, and ice at various places in the water cycle. To learn more about the water cycle, visit the U.S. Geological Survey Web site at http://ga.water.usgs.gov/edu/watercycle.html.

ROLE OF WATER IN PLANTS

The vital functions that water performs in plants include—
+ Necessary constituent of all living plant cells and tissues
+ A biochemical medium and solvent as nutrients from the soil and some organic compounds move in solution from their site of uptake, production, or storage to sites of use
+ A chemical reactant or product in many metabolic processes, including photosynthesis
+ Responsible for cell turgor
+ Functioning of the stomata and normal plant turgidity
+ A coolant and temperature buffer

Plants can actually be classified on the basis of their water needs. Xerophytes grow in dry places, mesophytes grow in moderately wet areas, and **hydrophytes** thrive in very wet or even flooded conditions. Xerophytes conserve or store water. The best-known xerophytes are the cacti. Sorghum and the millets are the best examples of field crops that are xerophytic. Mesophytes include most crop plants such as pine, corn, tomato, soybean, and peach. The water lily is a true hydrophyte, but rice is the major crop plant that is most like a hydrophyte.

PRECIPITATION

Precipitation—rain, snow, sleet, hail—is important to crops. Precipitation is the weather element of greatest interest in the weather forecast. Fifteen years ago the National Weather Service started using probabilities in forecasts to indicate their thoughts about precipitation occurring in a given area. These probabilities can be useful in making plans. Few things are certain. The meteorologists use probabilities to indicate the degree of confidence they have about whether precipitation will occur at a location. People understand there is a greater chance for rain with a 60 percent chance than with a 30 percent chance.

A forecast is for measurable precipitation at a particular location in a specified period. To be measurable, the amount must be 0.01 inch or more. The forecast is based on two factors: the meteorologists' idea of whether a rain- or snow-producing system will occur in the forecast area, and their opinion as to how much of the area might be covered by the expected storm. Multiplying these two factors together produces the probability forecast. These forecasts are now made by computer and may actually be handled differently, but the idea is the same. A 30 percent chance of rain means that there is a 70 percent chance that the location will not receive rain. It also means that rain can be expected in an area three times for every ten times the forecaster predicts a 30 percent chance of rain.

A forecast does not say anything about the amount of precipitation. Because heavy thunderstorms cover a very small area, it would not be at all surprising to receive rain totaling 3.28 inches with a 20 percent chance forecast. It is also possible to receive 0.04 inch when the forecast was for 80 percent chance. The chance of precipitation indicates little about the amount of moisture expected.

The probability forecast gives a person a strong basis for making many management decisions—planting, harvesting, and **irrigation**.

Another term used in predicting precipitation is normal precipitation. It occurs in the outlooks issued by the National Weather Service. Outlooks are separated from forecasts by the meteorologist as they are for a longer period of time and thus more general and less accurate than the short-term forecasts. For example, the National Weather Service regularly issues 6- to 10-day and 30-day outlooks. These outlooks predict precipitation by comparison with the normal precipitation expected in an area at that time of year. A prediction that is much above normal may still not mean that a location is going to receive a lot of rain. It all depends on the normal. It is defined by the World Meteorological Organization simply as a

30-year average that is recalculated every 10 years. It is used by international agreement so that climate in one area can be compared with that in other parts of the world. Normal precipitation varies throughout the year, and throughout the state. Figures 9-1 through 9-3 show monthly distribution of normal precipitation amounts and temperature for three areas across

Figure 9-1 Normal precipitation and temperature of Pocatello, Idaho.

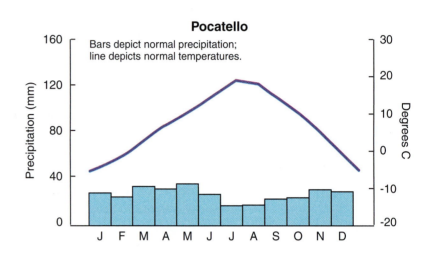

Figure 9-2 Normal precipitation and temperature of Albany, New York.

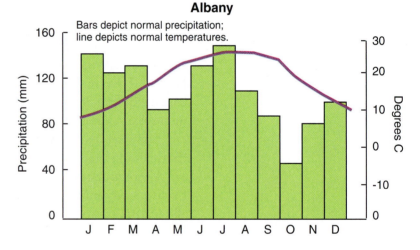

Figure 9-3 Normal precipitation and temperature of Dodge City, Kansas.

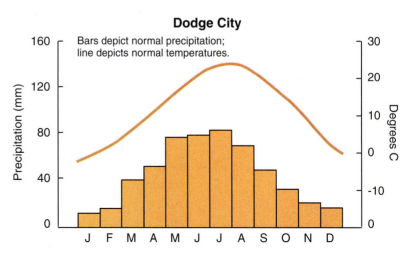

the United States. When the outlook calls for normal, above normal, or below normal, a comparison with the appropriate normal period will give an idea of what amount to expect.

WATER LOSS IN PLANTS

Water loss is directly proportional to the surface area exposed to air. Most of the surface area of a growing plant is leaf area. Leaves also have small holes in them called stomates. Stomates allow gases to pass in and out of the leaf. Water passes in and out of leaves as water vapor, along with other gases. These stomates can close to conserve water, but many of the life processes of the plant slow or even stop when the stomates close.

Desert plants frequently have very small leaves or even thorns in place of leaves. They may also have fleshy stems that store water. Fleshy stems and thorns or very small leaves are adaptations that help plants survive in hot, dry climates.

Plants take up most of their water through small hair roots. When they lose water more rapidly than they take it up, they **wilt**. Life processes slow and growth may even stop. If a plant remains wilted for too long, it will be damaged and may die. Some of the more common forms of damage include yellowing, leaf drop, and stunting. These usually result in yield loss and less profit for the farmer.

Water stress is more important to plants at some times than others. Germinating seed can be killed very quickly by lack of water. Lack of water will also cause many plants to drop flower buds, flowers, or even small fruit. Pollen may not develop fully if plants lack water. Corn, for example, may have many missing grains on the ears if adequate water is not available during pollination. Fruit of many plants such as peaches will be smaller if water is lacking while the fruit is expanding.

Most plants show indications that they need water before permanent damage occurs. Leaves develop a greenish blue tint and may wilt or roll (corn). Pumpkins and other plants with large leaves often wilt during the day and recover at night. When they remain wilted overnight, they need water immediately if yields are not to be reduced.

Farmers thus try to be sure that their plants will have the water they need. Farmers grow crops that are adapted to the area. They space individual plants far enough apart to allow the roots to spread out and gather moisture. They break up the soil so plant roots can grow down to subsurface moisture. Farmers avoid deep plowing during the summer, which will damage roots and allow moisture to escape. Sometimes fields are allowed to remain **fallow** (plowed but not seeded) for a year to allow moisture to collect in the lower soil layers. Controlling weeds reduces moisture lost to these plants. Both plastic and organic mulches reduce moisture lost directly to evaporation and also from competition by weeds. Early planting also assures that plants will grow and mature before summer droughts reduce yields.

Sometimes the only way to produce normal crops is to provide water artificially—to irrigate. There are many different kinds of irrigation systems. Some use overhead sprinklers, some use ditches or channels to carry water to the rows of plants, and some use drip or trickle hoses to supply water to individual plants. All of these systems are people's attempts to supply plants with a steady supply of water.

Too much moisture can be just as bad as too little moisture. In addition to moisture, seeds and plants also require oxygen and carbon dioxide to germinate and grow. Seeds and the roots of plants in waterlogged soil may not receive the gases they need to maintain life processes. Roots die, diseases set in, and both roots and seed may rot. Plants yellow and become stunted. Even large trees can be killed by having their roots covered by water for too long.

Farmers usually avoid soils that are subject to flooding while crops are growing. It may be possible to plant after spring floods. It may also be possible to avoid minor moisture problems by planting on raised beds or ridges.

Some crops, such as rice, tolerate excessive moisture better than other crops and may even flourish in soil covered with water for some or all of their growing season.

Excessive moisture can limit the usefulness of land for agricultural production, just as too little moisture can. Land with excess moisture usually is drained or relegated to nonagricultural uses.

WATER FOR PLANTS

Moisture for plant growth is stored in the soil. It takes about 500 pounds of water to produce 1 pound of dry plant material. About 5 pounds (or 1 percent) of this water becomes an integral part of the plant. The remainder (the other 495 pounds) is lost through the stomata of the leaves in the course of transpiration. This water must be stored by the soil and then provided to the plant.

Capillary water is used by the plants. This water moves freely in the soil and can move up and down, or horizontally. An example of capillary action can be demonstrated by dipping the end of a napkin into a glass of water and observing the water as it moves up through it. Forms of soil water not available to plants include:

+ Gravity water that is lost to drainage
+ Hygroscopic water that bonds to the soil particles

MOVEMENT OF WATER

Water moves through the plant by translocation and transpiration.

Translocation

The process of translocation is made possible by the "solvent of life," plain water. Water enters the plant through the root system and is transported

The Cost of Water

The cost of irrigation water varies widely throughout the United States and the world. Individual crops require a varying amount of water to reach marketable size, stage, or maturity. For example, perennial crops require 36 inches or more of usable (available) water. In many areas, rainfall is not sufficient to produce a crop.

A rainfall report is given in inches of water precipitated. This means that enough rain was received in an area to cover it with a depth of 1 inch of water. If this were measured over a given area of 1 acre then "1 acre inch of water" would have been received. Each time it rains during the season, the additional precipitation (in inches) is added to the total. For example, during a rainy season, 12 inches of rain is received. This would mean that enough water was received to cover each acre with 1 foot of water, one acre-foot.

Typically, water is measured in acre-feet. In irrigation districts, the landowners are charged a given amount for each acre-foot of water they use. For example, a grower uses 1.75 acre-feet of water; the cost is $50.00 per acre-foot; the total cost per acre paid by the landowner is $87.50.

In irrigated areas, growers supplement the natural precipitation (rain or snow) with irrigation water. A crop may require 32 inches of water to reach maturity, and it has rained 4 inches. Thirty-two inches minus 4 inches, or 28 inches, of water needs to be added. This would be the case for 100 percent efficiency of water use. This is seldom the case. The best is 85–90 percent, but 65–75 percent is common. So, water needed for the crop would be more than 28 inches of water and would be closer to 30–38 inches of water. Depending on the location, the grower will have to pay for this water to produce the crop.

throughout the plant. Translocated water moves up through the sieve tubes from the roots though the xylem to the rest of the plant. The water carries nutrients and metabolites. These metabolites are the products of the chemical reactions or metabolism of the plant.

Water also moves through the phloem sieve tubes to transport sugar produced by the leaves during photosynthesis to all parts of the plant. Roots have no other means to nourish themselves, and would die without translocated sugars manufactured in the leaves.

By being translocated, water adds turgor or stiffness to the cells. This pressure helps to provide support to leaves and new tissue. Further, translocated water also enables a plant to—

+ Carry on transpiration
+ Buffer temperature changes
+ Stabilize the pH in its metabolism
+ Carry on most chemical reactions in its processes (most of the reactions require water)
+ Maintain the volume of the cytoplasm (the cellular material and fluids surrounding the nucleus), which is mostly water

Transpiration

Plant transpiration is evaporation of water from leaf and plant surfaces. Transpiration is the last step in a continuous water pathway from soil, into plant roots, through plant stems and leaves, and out into the atmosphere. Water conditions drive the system by pulling the water uphill through the entire pathway. Because water in this pathway also carries nutrients, transpiration is an essential process in plant life.

Respiration and photosynthesis both give off water (H_2O) and heat. Transpiration cools the plant—plant sweat! About 90 percent of all water that enters the plants from the roots is given off during transpiration and the other 10 percent becomes involved in chemical processes or is tied up in the plant's structure.

The lower surface of the leaf is dotted with special porelike structures called "stomata." By the action of "guard cells," openings occur in the stomata during the daylight hours to permit the free exchange and release of water vapor. The stomata close at night or when the plant is water-stressed. Water vapor moves freely from the leaf through the open stomata.

WATER IN THE SOIL

Irrigators must learn to convert water to productive crops in the most efficient manner possible. Applying only enough water to meet full **evapotranspiration** (ET) of the crop is one key to efficient water use (ET is also called crop water use). Because ET is directly related to yield, the goal for irrigation management is to supplement normal precipitation with just enough water to meet full ET unless the water supply is inadequate. If irrigation along with rainfall is insufficient to meet ET demand, yield reduction is likely. Irrigating too much can cause percolation of excess water below the root zone, conveying nitrate nitrogen and other agrichemicals to the groundwater.

There are immediate short-range operating costs to the irrigator for either excess irrigation or less-than-full irrigation. For each acre-inch of excess irrigation, operating costs can increase for—

1. Nitrogen loss of 5 pounds or more per acre
2. Yield loss or extra fertilizer to compensate for nitrogen leaching
3. Extra energy for pumping

These factors can increase operating costs for each acre for each excess inch of irrigation. For each acre-inch of irrigation less than full ET demand by the crop, yields can be reduced, even though pumping costs are less. The net effect is an increase of operating costs.

The long-range costs of overpumping an aquifer, excessive percolation, or runoff of water and chemicals are more difficult to assess. One way to avoid these long-run or short-run costs is to match irrigation to crop needs. ET information can serve a key role in irrigation management decisions.

What Is ET?

Water from precipitation or irrigation can enter the soil where it comes into contact with the crop root system. Evapotranspiration (ET) is the water removed from soils by soil evaporation and plant transpiration. Soil evaporation is a direct pathway for water to move from soil to the atmosphere as water vapor. Over the course of an irrigation season, soil evaporation is 20–30 percent of total ET. Soil evaporation rates are highest after irrigation or rainfall. At those times, the soil surface is wet and the water readily evaporates. As the soil dries, the soil evaporation rates decline.

Both evaporation and transpiration are driven by a tremendous drying force the atmosphere exerts on soil and plant surfaces. Figure 9-4 shows the relative forces that exist in water as it is drawn through the plant directly from the soil. Water moves from higher to lower pressure. The bars (1 bar = 15 psi) noted in Figure 9-4 are negative pressure, or tension, terms. Water is drawn or pulled by more negative pressure as it moves from the soil, through the plant, and into the atmosphere.

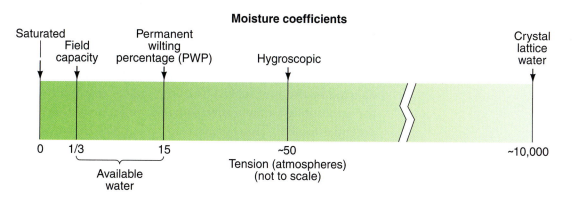

Figure 9-4 The relative forces that exist in water as it is drawn through the plant directly from the soil.

Figure 9-5 The yield response of a plant to watering is not totally linear.

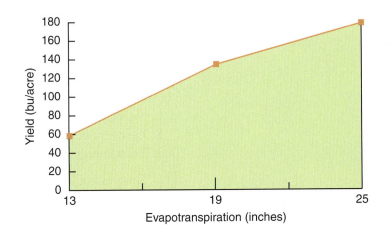

ET and Crop Yield

ET is important to irrigation management because crop yield relates directly to ET. This linear or straight-line relationship is shown in Figure 9-5. Because yield increases linearly with ET, maximum yield will not be reached unless the maximum ET level is reached. Irrigators who are working to achieve maximum yields need to apply water to meet the crops' ET demand. Applying extra water beyond ET demand will not translate into extra yield. A particular crop variety responding to a particular climate has only so much capacity to transpire water.

The goal for irrigation managers is to convert water to ET and ultimately to yield. However, the curved lines in Figure 9-5 represent inefficiencies in irrigation systems. All water cannot convert into ET and yield. More irrigation water is required to reach the same yield level from an inefficient system than a more efficient one. By improving system efficiency, the yield-irrigation curves can come closer to the yield-ET straight line.

Factors Affecting ET

Evapotranspiration can be affected by the weather, the crop type, the crop growth stage, the crop variety, the crop population, surface cover and tillage, and the availability of soil water.

Weather. The power of the atmosphere to evaporate water is the driving force for soil evaporation and crop transpiration. Weather factors that have major impact on this evaporative power include air temperature, humidity, solar radiation, and wind. High air temperatures, low humidity, clear skies, and high winds cause a large evaporative demand by the atmosphere. The crop may or may not be able to satisfy the atmosphere's evaporative demand, but the weather factors set the potential for ET. This potential for ET, governed by weather factors, is the starting point for estimating ET from weather data. Day-to-day variations in weather cause day-to-day variations in daily potential ET. The actual crop ET (or crop water use) also responds to these weather variations.

Crop Type. Different crops use different amounts of water over the course of the growing seasons. Table 9-1 shows total growing season ET requirements for corn, alfalfa, wheat, beans, and potatoes. Crop planting times and water use patterns are somewhat different among the crops listed. Alfalfa is harvested three to four times each season and is unique because it is always in the vegetative stage. Winter wheat requires 2–3 inches of water from emergence to dormancy. Differences in water use among corn, sorghum, and soybeans are mainly related to planting time and days to maturity. The range in ET values from year to year is due to differences in weather patterns.

Table 9-1 Examples of Seasonal Crop Water Use (ET) in the Midwest

Crop	Inches/year
Corn	23–28
Soybeans	20–25
Dry Beans	15–16
Sorghum	18–23
Winter Wheat	16–18
Alfalfa	31–36
Sugar Beets	24–26

Crop Growth Stage. During the course of the growing season, ET from crops depends not only on the potential ET demand from the atmosphere, but also on the crops' stage of growth. ET is related to leaf surface area, so small plants transpire less water than large ones. Because of growth patterns of different crops, maximum ET occurs at different times during the calendar year. Generally, crops reach maximum ET just prior to their reproductive growth stage. As crops continue through reproductive processes and approach maturity, ET decreases (Tables 9-2 through 9-6). The average maximum ET for corn, sorghum, soybeans, and wheat is approximately 0.30–0.35 inch per day. Individual daily ET could reach 0.45–0.50 inch per day.

Table 9-2 Average Water Use for Corn in Inches per Day

Temp. °F	Week After Emergence																	
	1	2	3	4	5	6	7	8	9	10	11	12	13	14	15	16	17	18
50-59	.01	.02	.03	.04	.05	.06	.08	.09	.09	.10	.10	.10	.09	.07	.06	.05	.04	.03
60-69	.02	.03	.04	.06	.08	.09	.11	.12	.13	.15	.14	14	.14	.11	.09	.07	.06	.04
70-79	.03	.04	.05	.07	.10	.12	.15	.16	.17	.19	.19	.18	.17	.14	.11	.09	.07	.05
80-89	.03	.05	.07	.09	.13	.15	.18	.20	.22	.24	.23	.22	.21	.17	.14	.11	.09	.06
90-99	.04	.06	.08	.11	.15	.18	.21	.24	.26	.28	.27	.26	.25	.20	.17	.13	.11	.07
Corn Growth Stages	Three Leaf		Eight Leaf				First Tassel		Silk		Blister		Early Dent				Kernel	

Table 9-3 Average Water Use for Wheat in Inches per Day

Temp.	Week After Emergence													
°F	1	2	3	4	5	6	7	8	9	10	11	12	13	14
50–59	.02	.03	.05	.06	.08	.09	.10	.10	.09	.09	.07	.05	.03	.02
60–69	.03	.05	.07	.09	.12	.13	.15	.14	.13	.13	.10	.07	.05	.03
70–79	.04	.07	.10	.12	.17	.17	.19	.19	.18	.17	.13	.10	.07	.04
80–89	.05	.08	.12	.16	.20	.22	.24	.24	.22	.21	.16	.12	.08	.04
90–99	.06	.10	.15	.18	.24	.26	.29	.28	.26	.25	.19	.15	.10	.05

Table 9-4 Average Water Use for Alfalfa in Inches per Day

Temp.	First Four Weeks After Growth Starts in the Spring				First Three Weeks After the First and Second Cuttings			First Three Weeks After Third Cutting			Weeks Not Covered by Earlier Charts for the Respective Months				
°F	1	2	3	4	1	2	3	1	2	3	May	June	July	Aug	Sept.
50–59	.04	.05	.07	.07	.06	.07	.09	.05	.06	.09	.07	.09	.09	.09	.07
60–69	.04	.07	.09	.11	.10	.12	.14	.09	.10	.13	.11	.14	.14	.13	.10
70–79	.05	.09	.12	.14	.12	.15	.18	.12	.14	.17	.14	.18	.18	.17	.13
80–89	.06	.11	.15	.19	.14	.18	.22	.15	.18	.21	.18	.22	.22	.21	.16
90–99	.08	.13	.18	.21	.18	.22	.27	.17	.21	.25	.21	.27	.27	.25	.19

Table 9-5 Average Water Use for Potatoes in Inches per Day

Temp.	Week After Emergence																
°F	1	2	3	4	5	6	7	8	9	10	11	12	13	14	15	16	17
50–59	.02	.03	.04	.06	.07	.08	.08	.09	.09	.09	.09	.07	.07	.06	.05	.04	.04
60–69	.03	.05	.06	.08	.10	.12	.12	.13	.13	.13	.13	.11	.10	.09	.07	.05	.05
70–79	.04	.06	.09	.11	.13	.16	.18	.18	.18	.17	.17	.14	.13	.11	.09	.07	.07
80–89	.05	.07	.11	.14	.16	.19	.21	.22	.22	.21	.21	.18	.16	.14	.11	.09	.08
90–99	.06	.09	.13	.16	.19	.23	.24	.26	.26	.25	.25	.21	.19	.17	.13	.11	.10
Potato Growth Stages			7 in.		Budding		Full Cover										

Table 9-6 Average Water Use for Field Beans in Inches per Day

Temp.	Week After Emergence																
°F	1	2	3	4	5	6	7	8	9	10	11	12	13	14	15	16	17
50–59	.02	.03	.05	.06	.07	.08	.09	.10	.09	.09	.08	.06	.04	.02	.05	.03	.02
60–69	.04	.04	.07	.08	.11	.12	.13	.15	.14	.13	.12	.09	.06	.03	.07	.04	.02
70–79	.05	.06	.09	.11	.14	.16	.17	.19	.19	.17	.15	.11	.07	.04	.09	.05	.03
80–89	.06	.07	.11	.13	.18	.20	.22	.24	.23	.22	.19	.14	.09	.04	.11	.06	.03
90–99	.07	.09	.13	.16	.21	.24	.25	.27	.26	.25	.23	.17	.11	.05	.13	.08	.05
Field Beans Growth Stages		Second Trifoliate		First Flower					Seed Filling				Leaves Yellowing				

Crop Variety. The relative maturity range of a particular variety has the most impact on seasonal crop ET. For example, at the same location, a corn variety with maturity of 120 days will use more water than an 85-day variety. However, if both varieties are able to mature fully, the grain produced for each inch of ET is approximately equal.

Longer-season corn varieties use more water, but they also produce more grain if the heat units and water supply are available. The difference in water use is because of total days of water use, not a difference in daily water use.

Crop Population. Plant population in a sparsely planted crop like corn can influence crop ET. Dryland farmers may grow 12,000–15,000 plants per acre, while a neighboring irrigator may use 25,000–30,000 plants per acre. Irrigators often wonder whether or not decreasing corn population will result in less ET and irrigation requirements. Irrigation research demonstrates that savings in transpiration from fewer plants per acre have been used up by increases in evaporation. Less shading from fewer plants resulted in more evaporation.

ET requirements decreased when irrigated corn populations were less than 18,000 plants per acre. Dryland farmers can take advantage of lower ET requirements with lower populations. Irrigators need to plant for higher populations to optimize yields.

Surface Cover and Tillage. The amount of soil surface cover influences soil evaporation. When the soil surface is wet, evaporation depends on the amount of radiant energy at the soil surface. Lowest evaporation rates occur from shaded and mulched soil surfaces. Crops shade more and more of the soil as they grow, but soil evaporation continues. However, crop residues can reduce soil evaporation by 1–3 inches during the irrigation season.

Availability of Soil Water. Research shows soil water content cannot be considered alone as the single factor controlling whether crop ET is reduced below its potential rate. The ability of soil to transmit water to plant roots and the actual evaporative demand for a given day also are important.

For progressively drier soil, actual crop ET is below the demand rate. The drier soil is unable to transmit water to the roots fast enough to satisfy higher potential ET demand. When the soil is very dry, plants may not experience reductions in ET, if the potential ET demand is low. Controlling factors are the potential ET demand and the soil's ability to transmit water to the roots. This transmitting ability is different for every soil, and for a given soil it depends on the water content.

Crops including corn, winter wheat, and determinate soybeans need water during reproductive and grain-filling growth stages. Indeterminate soybeans, which flower over a longer time, need water especially during grain filling. When crops do not receive water to meet their ET demand, grain yields can be reduced. Limited water applications targeted to critical growth stages can be very effective for production (Table 9-7).

Table 9-7 Grain Yields, Water Use, and Irrigation Water Use Efficiencies (IWUE) for Continuous Corn Grown in the Midwest

Item	Dryland	Limited Irrigation	Full Irrigation
Irrigation	0.06	0.013	0.8
ET (in)	13.5	19.1	25.3
Grain yield (bu/ac)	59.0	135.0	178.0
Grain/ET (bu/ac-in)	4.3	7.1	7.1
IWUE (bu/ac-in)	—	12.7	5.5

IWUE = Irrigation Water Use Efficiency. IWUE (Limited) = Limited Irrigation Yield – Dryland Yield Limited Irrigation – Dryland Irrigation (0). IWUE (Full) = Full Irrigation Yield – Limited Yield Full Irrigation – Limited Irrigation.

Estimating ET

Automated weather stations measure and record the air temperature, relative humidity, incoming radiation, and wind speed. Weather information from each station is collected daily by a computer. These data are used to calculate potential ET, which estimates crop water use for the region around the weather station. The calculation is based on years of research that related these weather factors to evaporative demand.

The crop ET is calculated from potential ET. Growth stages of crops near the station are based on growing degree days accumulated since crop emergence. The growth stages combined with the potential ET from the weather station give crop ET estimates. Field research has furnished the relationships between potential ET from the weather stations and crop ET throughout the growing season.

Crop estimates assume soil water does not limit crop ET. The increased soil evaporation rates that occur immediately after rain and irrigation are not included in the estimates. These specific adjustments to crop ET vary from field to field and cannot be included in regional estimates.

Regional crop ET estimates are an excellent starting point for tabulating water use from a particular irrigated field. Periodic checks of soil moisture in each irrigated field are necessary to confirm the water use from that field.

Sources for ET Information

Computerized sources, newspapers, radio and television stations, telephone recordings, and extension services give regional ET information. Usually, ET is reported in units of inches per day. To find the total crop ET over several days, multiply the reported ET by the number of days. Some examples of Web sites providing ET information include:

+ K-State Research and Extension (www.oznet.ksu.edu/irrigate/ ET/ETinfo.htm)
+ Oklahoma Evapotranspiration Model (http://agweather.mesonet. org/models/evapotranspiration/default.html)
+ Texas Evapotranspiration Network (http://texaset.tamu.edu)

Using ET Information

A bank account provides an analogy to matching irrigation and rainfall amounts to crop ET. The soil is the bank for water. Rainfall and irrigation are deposits to the account, and ET is the withdrawal from the account. This approach has been called checkbook irrigation scheduling. ET estimates are a key component for tracking how much water the crops are using, when to irrigate, and how much to apply.

IRRIGATION

Rainfall usually reduces the requirement for irrigation. Not all rainfall can be considered effective and many crops are grown in areas where the rainfall is not sufficient for a productive crop. Also, a small amount of precipitation has little effect on irrigation water management. Only rainfall sufficient enough to fill much of the root zone will place an irrigation schedule in a start-over mode.

When rainfall is sufficient, it can be lost from the soil in two ways—runoff and deep percolation. Runoff occurs when the rainfall rate exceeds the soil's infiltration rate. Low rates of rainfall over an extended period of time will often replenish the soil without runoff. Water is lost by deep percolation when rainfall exceeds the soil's total water-holding capacity. Because rainfall is highly variable, measurements should be taken near the fields that are scheduled for irrigation.

In arid areas and in even some quite wet areas, irrigation is used to ensure a productive crop. The type of irrigation system varies.

Irrigation Systems

Flood, furrow, sprinkler, and drip (above or below the surface) irrigation are the four basic methods of applying irrigation water. Selecting the appropriate method for a site depends on land slope, soil water intake rates and water-holding capacity, water tolerance of the crop, and the effect of wind.

Physical features play an important role in identifying irrigation options. For example, if an adequate water supply is not available at reasonable cost, irrigation may be impractical, limited in its use, or confined to a system such as drip. Slopes greater than 10 percent may prevent the use of some sprinkler systems. Soils that absorb water slowly, tend to crust and seal, or are erosive should not be irrigated with sprinkler systems. Selection and cost-effectiveness of irrigation systems is also affected by crop height, row spacing, plant sensitivity to spray, the need for frost or other environmental protection, and the permanency of the crop.

Flood Irrigation. Flood or border irrigation is used where the land is level. This method is often used for orchards and vineyards and for hay, pasture, and cereal grains. The land must be leveled and graded. A uniform slope of 0.1–0.4 feet per 100 feet (0.1–0.4 percent slope) is used for most crops. Water enters through a head ditch or biplane and is released into the individual checks (areas bounded by levees running downslope) by siphons, gates, or valves.

Figure 9-6 Furrowed field ready for irrigation.

Furrow Irrigation. This is one of the most widely used methods for crops. Water runs down the furrows between plant rows (Figure 9-6). Water moves to all parts of the soil by capillary action or gravity. This system is efficient in water use, but is expensive because of high labor costs. Costs include forming the furrows, forming and maintaining the irrigation ditches, hiring help to maintain the ditches, and irrigation cycles using gates and siphons.

An advantage of **furrow irrigation** is uniform crop maturity, which is possible because water is more uniformly applied than in **flood irrigation**.

Sprinkler and Drip Systems. Sprinkler and drip systems are widely used. **Sprinkler and drip irrigation** systems have several common components, such as the pumping unit, control head, mainline and perhaps submain pipe(s), and laterals. The basic water distribution hardware for both systems is similar. The primary difference is in the physical characteristics of these components and the equipment that applies the water—either sprinklers or **emitters.**

In drip or trickle irrigation systems (Figure 9-7), emitters are grouped into two categories: (1) line-source and (2) point-source. Line-source emitters are an integral part of a lateral, as was previously mentioned. Point-source emitters can have single or multiple outlets and are generally classified as drippers, bubblers, or misters (foggers), depending on the method of final water application.

Compared with sprinkler nozzles, emitters have very small openings, usually pinhole size. Different emitters have different internal flow characteristics that determine how sensitive they are to pressure changes and particles in the water. Because of the small openings in emitters, water filtration using a 30-200 mesh screen is normally required. However, some emitters are self-cleaning or can be taken apart and cleaned.

Figure 9-7 Soil cut away to expose a drip irrigation line in a tomato field. (Courtesy USDA ARS)

Emitters normally operate at pressures of 5–40 psi, with flow rates of 0.5–15 gallons per hour. Emitter spacing depends on the discharge rate and soil type because most of the water distribution is through the soil. The low pressure requirements of emitters result in more sensitivity to pressure losses along a lateral line or an elevation gradient. Pressure-compensating emitters may be necessary to achieve uniform water application. Because the pressure and discharge requirements of emitters are much less than those of sprinklers, drip laterals and mainlines are usually smaller, and annual operating costs of these systems tend to be lower.

Sprinklers are found on hand lines, wheel lines, and center pivots (Figures 9-8 and 9-9). Three basic types of sprinklers are used: rotating sprinklers, stationary spray-type nozzles, and perforated pipe. Rotating sprinklers, such as the slowly rotating, impact-driven sprinkler, are most commonly used. Perforated pipe is the simplest of sprinklers, consisting of only a pipe with numerous holes through which water sprays.

Figure 9-8 Center pivot irrigation is used to irrigate many crops. A crop will be sprinkled for 10 to 12 hours, then the water is turned off and the line rolls forward to the next setting.

Figure 9-9 Aerial view of over 100 center-pivot sprinklers controlled by a central computer irrigating wheat, alfalfa, potatoes, and melons along the Columbia River near Hermiston, Oregon. (Courtesy USDA ARS)

Rotating impact sprinklers come in many sizes and variations to meet different design conditions. Some sprinklers operate at pressures as low as 5–30 psi, while large, big gun types require pressures exceeding 80 psi. Sprinkler discharge ranges from a few gallons of water per minute to 1,000 gallons per minute for a big gun. Wetting diameters can range from only a few feet to 500 feet.

Pressure, discharge, and wetting diameter are the most significant characteristics in selecting a particular sprinkler, but nozzle size, jet angle, wind, sprinkler overlap, and sprinkler rotation speed are also important. These characteristics determine water application rates, sprinkler spacings, and water droplet sizes. For uniform water application, sprinklers are generally spaced so that 50–60 percent of their wetted areas overlap. Because annual operating costs of irrigation increase with increasing pressure and discharge requirements, the large gun-type sprinkler will require from one and a half to two times as much fuel or electric energy as smaller sprinklers to apply equal amounts of water.

Irrigation System Efficiency

Another important factor to consider is irrigation system efficiency. With sprinklers, for example, not all of the water sprayed into the air reaches the ground. Normally, 15 percent is lost directly to evaporation, but the actual amount will vary with droplet size and wind speed. Thus, if 1 inch of water is needed to replenish soil moisture for the crop, and the sprinklers are rated and spaced to apply 1 inch per hour, more than an hour would actually be needed to apply 1 inch of water to the soil. Drip systems are more efficient because they are not affected by evaporation. Under most on-farm circumstances, drip systems are rated at 90 percent efficient and sprinkler systems about 75 percent efficient. Nonuniform application and water losses in the distribution line may also affect efficiency.

Timing Irrigation

A soil's water-holding capacity indicates both the amount of water available for plant use and the maximum allowable depletion. To determine when irrigation is needed, the irrigator must know how much water is in the soil reservoir at any given point in time. There are a number of ways to make this estimate. The most common ones are discussed here and can be used regardless of soil type.

Feel Method. The feel method is one of the oldest ways to determine amount of soil moisture. As water content changes, the feel of soil changes. With experience, some people can calibrate their sense of touch to estimate available soil water. Tables 9-8 and 9-9 describe the appearance and feel of sandy, loamy, and clayey soils at various moisture content levels.

The only instrument needed for employing this technique is an inexpensive hand probe long enough to reach down through the full root profile. The probe-feel method allows for only approximate soil moisture values.

Table 9-8 Available Water Capacity (AWC) Characteristics of Soil

Profile (Inches)	Depth/Texture Class	AWC (Inches)		
		Per Inch	Per Zone	Cumulative
0–12	Loam	.21	2.52	2.52
12–18	Sandy loam	.16	.96	3.48
18–24	Sand and Gravel	.04	.24	3.72
24–60	Sand and Gravel	.03	1.08	4.80

Table 9-9 Feel Chart for Determining Moisture in Medium to Fine Soil

Degree of Moisture	Feel	Percent of Field Capacity
Dry	Powder dry.	0
Low (critical)	Crumbly, will not form a ball.	Less than 25
Fair (usual time to irrigate)	Forms a ball, but will crumble upon being tossed several times.	25–50
Good	Forms a ball that will remain intact after being tossed five times; will stick slightly with pressure.	50–75
Excellent	Forms a durable ball and is pliable; sticks readily; a sizable chunk will stick to the thumb after soil is squeezed firmly.	75–100
Too wet	With firm pressure, some water can be squeezed from the ball.	In excess of field capacity

Electrical Conductivity. Electrical resistance blocks measure soil water more precisely than the feel method. Small blocks made from plaster of paris (gypsum), nylon, fiberglass, or other material containing electrodes are buried in the soil with wires extending to the surface. An electrical meter touching the wires measures the change in electrical resistance during wetting and drying cycles. The blocks, which can be used only one season, are placed at four locations or stations in a field. At each station, three blocks are buried at 0.5-, 1.5-, and 2.0-foot depths. The electrical conductivity method is recommended on finer-textured soils.

Tensiometers. Tensiometers are mechanical instruments that measure soil water suction, as shown in Figure 9-10. They can be used successfully for scheduling irrigations if the relationship between soil water suction and available water for a particular soil is known. Tensiometers are recommended for use on loamy sands and fine sands. Their placement and the number of stations in a field are the same as for electrical resistance blocks. Tension readings in centibars for various soils are listed in Table 9-10.

Neutron Moisture Probe. A neutron probe measures soil water content using a radioactive source. The instrument itself and the access tubes, which require installation, are quite expensive. The neutron probe method is used

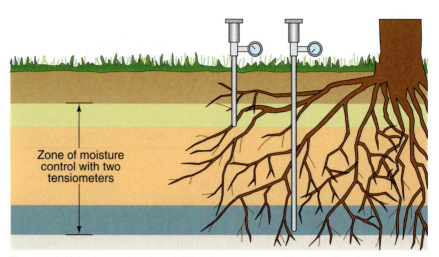

Figure 9-10 Using two tensiometers at different depths can provide information about the water in a plant's root zone.

Table 9-10 Soil Water Deficit Estimates in Inches per Foot for Various Soil Tensions

| Soil Texture | Soil Tension (Centibars) | | | | | | |
| | 10 | 30 | 50 | 70 | 100 | 200 | 1500* |
	Deficit in/ft						
Coarse sands	0	0.1	0.2	0.3	0.4	0.6	0.7
Fine sands	0	0.3	0.4	0.6	0.7	0.9	1.1
Loamy sands	0	0.4	0.5	0.8	0.9	1.1	1.4
Sandy loams	0	0.5	0.7	0.9	1.0	1.3	1.7
Clay loams	0	0.2	0.5	0.8	1.0	1.6	2.4

*1500 centibars is the permanent wilting point, and the soil water deficit value is equal to the soil's total available water-holding capacity.

mostly for research and by some irrigation consulting firms that measure large numbers of fields. It is not recommended for small-scale irrigation scheduling.

Evaporation Pans. Evaporation pans may be used to calculate the amount of water used by the crop. The pans can serve wide areas, but calibration is required based on a crop coefficient to relate evaporation rate to actual water use. The coefficient is the ratio of potential evapotranspiration to pan evaporation.

Computer Programs. Computer programs have been developed for scheduling irrigation, and they use meteorological data to calculate water use. The computer forecasts the timing and amount of irrigation water necessary for optimum crop production. Computer irrigation scheduling is available from computer networks that provide scheduling programs for a service fee. Similar programs can be purchased for personal computers and handheld calculators.

Water Budgeting. This is a method used to balance the available soil moisture. Rainfall and irrigation amounts represent credit entries.

Evapotranspiration is a debit entry. Rainfall and irrigation should be measured. Evapotranspiration can be estimated from evaporation pan data, from weather data obtained at a nearby weather station, or from tables available through local agencies that calculate crop water use.

The critical value in **water budgeting** is soil water-holding capacity. Fifty percent of this value is the maximum allowable depletion level, and 50 percent is the maximum application amount. A good starting point for a water budget system is when the soil water level is at field capacity. This allows the center-pivot irrigator to follow a replacement schedule whereby any water used by the plant is replaced through irrigation or rainfall.

Water budget scheduling also provides water management for the gravity irrigator. Plant water use can be estimated from any of the sources previously noted. Because the soil profile is normally filled with each irrigation, soil water-holding capacity can serve as a guide to determine amount of water available. When water used equals the maximum allowable depletion, irrigation is needed.

Humid Versus Subhumid Climates

The high probability of precipitation makes scheduling irrigation in humid regions difficult. Chances of overirrigation are high because of the greater likelihood of summer rains. Rainfall usually supplies the major part of a crop's water requirement in these regions and should be used to the full extent.

Because the chances are good of rainfall occurring shortly after irrigation, complete recharge of the root zone is not advisable. This not only reduces irrigation costs, but also minimizes nutrient leaching.

If the available soil water is permitted to reach the maximum allowable depletion level following rain, a portion of the field will likely experience considerable water stress before the irrigation cycle is completed. This problem can be minimized by (1) starting the system when half of allowable depletion occurs and (2) applying half the normal water rate (which should allow the abbreviated cycle to be completed in half the normal time). Such a procedure re-establishes the irrigation cycle without subjecting part of the crop to excessive water stress.

Arid and Semiarid Regions. Early precipitation or irrigation at planting normally meets the crop's water demand during the spring in arid and semiarid regions. Then as reserve moisture is used, irrigation becomes the primary means for plant growth. In most cases, scheduling in the arid areas provides the greatest return during the spring and fall periods.

Arid and semiarid regions have low probabilities of precipitation. Irrigators should plan to fill the soil profile during irrigation rather than leave room for possible rain. Water use can average above 0.25 inch a day during the summer, which means over 1 inch of soil moisture depletion in only 4 days. Irrigation should keep ahead of crop demand rather than risk falling behind trying to catch up to crop demand. The exception to this occurs in the fall, when crop water demand is on its downward trend.

SUMMARY

Water performs vital functions in plants. Based on their origin, plants have different water requirements. Precipitation and irrigation supply water to plants. Many crops are grown under some type of irrigation system because normal precipitation is too little or too unreliable. Soil type can affect a plant's ability to take up the water in the soil. Capillary water is used by plants. Through the process of transpiration, plants loose water. Water moves through plants by translocation. This process also moves the nutrients around in the plant.

To be efficient, irrigation must supply only enough water to meet evapotranspiration of the crop. Several factors affect the rate of evapotranspiration. Irrigation systems are some variation of flood, furrow, sprinkler, or drip. Methods to determine the soil moisture content are used to schedule of irrigation.

Review

True or False

1. Hygroscopic water is the only water available to plants.

2. Plants lose water through transpiration.

3. Water is a bipolar solvent.

4. Soil moisture content can be determined by feeling the soil.

5. Hydrophytes thrive in the desert.

Short Answer

6. Name four general types of irrigation.

7. List four factors affecting evapotranspiration.

8. Give the two processes that move water through a plant.

9. Name four vital functions water performs in plants.

10. List four ways to estimate the water content of a soil.

11. Water lost to drainage is called _____ water. Water used by plants is called _____ water, and _____ water bonds to the soil particles.

Discussion

12. Discuss the effects of too little or too much water on a crop.

13. Explain the concept of a water budget or scheduling irrigation.

14. Compare the daily water use of two crops during their stages of growth, germination, seedling, vegetative growth, blooming, and harvest.

15. Define evapotranspiration.

16. Describe what happens when plants wilt.

Knowledge Applied

1. Using an electronic spreadsheet, create a computerized graph of the crop water use data in Tables 9-2 through 9-6. Label all axes. (Try a three-dimensional graph using week, temperature, and water use.)

2. Search the Internet and find normal precipitation and temperature data for your area. Print out the information you find and compare it with the type of weather you experience for a month. Keep a weather log.

continues

Knowledge Applied, *continued*

3. Use colored food dye to show the translocation of water. Put white carnations in warm water and add a colored dye. The flowers will change color. Measure the length of the stems and record the length of time it takes the dye to reach the blooms. Observe the progress of the dye through the flower. Calculate how fast water is translocated through the stems. Express your answer in centimeters per unit of time.

4. Visit areas using each of these methods of irrigation: flood, furrow, sprinkler, and drip.

5. Make a collection of sprinkler heads and drip emitters; share it with the class and demonstrate how each works.

6. Use a tensiometer; make readings for two weeks and graph the readings.

Resources

Decoteau, D. R. (2004). *Principles of plant science: Environmental factors and technology in growing plants.* Englewood Cliffs, NJ: Prentice Hall.

Hoffman, G.J., Evans, R. G., Jensen, M. E., Martin, D. L, & Elliot, R. L. (2007). *Design and operation of farm irrigation systems* (2nd ed.). St. Joseph, MI: American Society of Agricultural Engineers.

Kirkham, M. B. (2004). *Principles of soil and plant water relations.* New York: Academic Press.

United States Department of Agriculture. (1994). *Irrigating efficiently.* Beltsville, MD: Author.

Vorst, J. J. (1998). *Crop production* (3rd ed.). Champaign, IL: Stipes Publishing Company.

Internet

Internet sites represent a vast resource of information. The URLs (uniform resource locators) for World Wide Web sites can change. Using one of the search engines on the Internet such as Google or Yahoo!, find more information by searching for these words or phrases: crop water use, evapotranspiration, translocation, transpiration, irrigation, irrigation systems, precipitation, crop water use, and crop weather.

Chapter 10
Soil and Organic Farming

ORGANIC FARMING *is becoming increasingly popular. Organic growers use different methods and materials to maintain soil structure and fertility and to control diseases and pests. Methods and materials used by organic farmers can include: using recycled and composted crop wastes and animal manures, soil cultivation at the right time, crop rotation, green manures, and mulching. Additionally, organic growers control diseases, pests, and weeds through careful planning and crop choice, resistant crops, cultivation practices, crop rotation, pest predators, genetic diversity, and natural pesticides.*

After completing this chapter, you should be able to—

- Describe organic and sustainable agriculture

- Explain why growers choose organic farming

- Discuss how organic farming benefits the soil

- Discuss the pros and cons of using manure to supply plant nutrients

- Explain the importance of the carbon:nitrogen (C:N) ratio

- List the Best Management Practices for using manure

- Describe the benefits of composting

- Explain the use of green manures

- Discuss why crop rotation is especially important to the organic grower

- Describe the weed, pest, and disease control options used by organic growers

Allelopathy
Best Management Practices
Composting
Cover Crop
Crop Rotation
Green Manure
Mesophilic
Mulching
Organic Standards
Thermophilic

ORGANIC FARMING

Organic farming is a type of sustainable agriculture that also prohibits the use of synthetic substances, including inorganic fertilizers and pesticides. A major theme shared by organic farms is promoting healthy soil by controlling erosion and keeping organic matter levels high. Buyers believe organic products are safer, more nutritious or flavorful, or support the process of organic farming.

Organic growers add nitrogen by the use of manures, composts, legumes, and organic nitrogen fertilizers. Phosphorus and potassium come from manures and mineral fertilizers such as rock phosphate. Organic farms tend to rely more on natural nutrient cycles than do conventional farms. Crop rotations figure centrally in many organic operations. Weed control tends to rely on crop rotation, cultivation, and sometimes flaming or mulches.

According to a 1980 United States Department of Agriculture (USDA) study, successful organic farms come in all sizes and crops. This and other studies point to soil benefits including reduced erosion, increased soil organic matter content, higher populations of earthworms, richer soil flora, and others. In a 2000 review of studies comparing conventional, sustainable, and organic systems in horticultural crops, the comparative production and profit of organic systems were highly variable; results depended greatly on the specific sites and practices. Profitability for organic production tended to rely on the higher prices offered for organic produce (Figure 10-1).

State-sponsored programs to certify organic production have grown in recent years, and in 1990 the Organic Foods Production Act directed the USDA to set up a federal program. The final rules for that program were published in 2000. The rules set certain **organic standards**, prohibit the use of many substances on organic land including sewage sludge, and provide a list of allowed and disallowed synthetic materials. It also sets labeling requirements. These rules and state standards define what can be sold as organic and help the consumer purchase organically grown foods. The National Organic Program (NOP), through the USDA Agricultural Marketing Service, maintains a website: www.ams.usda.gov/nop/NOP/NOPhome.html.

Organic food is becoming popular in Europe and the United States. However, for food to be sold as organic, it must bear a symbol that proves that it is truly organic. This is obtained through a certification organization. The certifying procedure is quite complex and is potentially expensive.

Methods and Materials

Methods and materials used by organic farmers include keeping and building good soil structure and fertility, using recycled and composted crop wastes and animal manures, correct soil cultivation at the right time, **crop rotation**, **green manures**, and **mulching**.

Figure 10-1 Organic farmers aim to produce healthy crops, such as these cauliflowers, and sell at a higher price. (Photo courtesy of USDA ARS)

International Organic Standards

The International Federation of Organic Agriculture Movements (IFOAM) produced a set of international organic standards that was developed by individuals from many countries. These standards give guidelines about what organic farming is and how it should be practiced on the farm.

International standards are also used to help countries set their own standards, which take into account different farming systems. Many countries have an organic standards authority that enforces national standards and awards a symbol to farms that have followed the standards. This symbol then allows farmers to market certified organic produce. This ensures that people know that the food they buy is organic.

The main principles of organic farming were developed by IFOAM in 1992. These principles include:

+ To produce food of high nutritional quality in sufficient quantity
+ To interact in a constructive and life-enhancing way with all natural systems and cycles
+ To encourage and enhance biological cycles within the farming system involving microorganisms, soil flora and fauna, plants, and animals
+ To maintain and increase long-term fertility of soils
+ To use, as far as possible, renewable resources in locally organized agricultural systems
+ To work, as far as possible, within a closed system with regard to organic matter and nutrient elements, thus reducing external inputs
+ To work, as far as possible, with materials and substances that can be reused or recycled, either on the farm or elsewhere
+ To give all livestock living conditions that will allow them to perform the basic aspects of their natural behavior
+ To minimize all forms of pollution that may result from agricultural practices
+ To maintain the genetic diversity of the agricultural system and its surroundings, including the protection of plant and wildlife habitats
+ To allow agricultural producers a living according to fundamental human rights; to cover their basic needs; and to obtain an adequate return and satisfaction from their work in a safe working environment.
+ To consider the wider social and ecological impact of the farming system

Further information may be obtained by visiting the IFOAM Web site: www.ifoam.org.

Control of Pests and Disease

Organic growers control diseases, pests, and weeds through careful planning and crop choice, resistant crops, cultivation practice, crop rotation, useful predators that eat pests, increasing genetic diversity, and natural pesticides.

Organic farmers use a combination of techniques to allow them to work together for the maximum benefit. For example, the use of green manures and careful cultivation together provide better control of weeds than if the techniques were used alone. Organic farming also involves careful use of water resources and good animal husbandry.

Why Farm Organically?

Those who practice organic farming claim it provides long-term benefits to people and the environment. Organic farming aims to:

+ Increase long-term soil fertility
+ Control weeds, pests, and diseases without harming the environment (Figure 10-2)
+ Ensure that water stays clean and safe
+ Use resources that the farmer already has, so the farmer needs less money to buy farm inputs
+ Produce nutritious food, feed for animals, and high-quality crops to sell at a good price

Proponents of organic agriculture claim that modern, intensive agriculture causes many problems, including:

+ Artificial fertilizers and herbicides easily wash from the soil and pollute rivers, lakes, and water courses.
+ Prolonged use of artificial fertilizers results in soils with a low organic matter content, which is easily eroded by wind and rain.

Figure 10-2 Corn gluten meal (the yellow substance) applied for organic weed control tests in Lane, Oklahoma. (Photo courtesy of USDA ARS)

+ Dependency on artificial fertilizers leads to greater amounts needed every year to produce the same yields of crops.
+ Artificial pesticides stay in the soil for a long time and enter the food chain, where they build up in the bodies of animals and humans, causing health problems.
+ Artificial chemicals destroy soil microorganisms, resulting in poor soil structure and aeration and decreasing nutrient availability.
+ Pests and diseases become more difficult to control as they become resistant to artificial pesticides. The numbers of natural enemies decrease because of pesticide use and habitat loss.

CROP NUTRITION

An organic farmer manages the soil to produce a healthy crop. This involves considering soil life, soil nutrients, and soil structure. Feeding the soil with manure or compost feeds the whole variety of life in the soil, which then turns this material into food for plant growth. This also adds nutrients and organic matter to the soil. Green manures also provide nutrients and organic matter

Manure

Ironically, for many farms today, manure has become a waste disposal problem. This is an abrupt change, for throughout history people have long relied on animals as a source of soil nutrients. However, many farms do still use this resource. Properly used, manure offers many benefits; improperly used, it poses many problems.

Benefits of Manure. When handled and applied correctly, manure benefits growers several ways:
+ Manure is a fertilizer with good amounts of nitrogen and potash. Phosphorus and calcium are present, as are lesser amounts of sulfur and magnesium. Most manures also have traces of several micronutrients. Table 10-1 provides examples of the nutrient content of several manures.

Table 10-1 Sample nutrient composition of several manures, on an as-is basis (not dried or composted), in pounds nutrient per ton. Actual composition of manures varies widely and should be measured

| Animal | Pounds/Ton | | | | | |
	N	P_2O_5	K_2O	S	Ca	Mg
Dairy cattle	10	4	8	1	6	2
Beef cattle	11	8	10	1	3	2
Poultry	23	11	10	3	36	6
Swine	10	3	8	3	11	2
Sheep	28	4	20	2	11	4
Horse	13	5	13	—	—	—

+ Manure adds organic matter to the soil. Organic solids make up 20–40 percent of manure. This matter decays readily because of its high nitrogen content. Nitrogen tie-up occurs only if the manure includes a lot of straw or wood shavings used as animal bedding.
+ Manure has longer lasting effects than an equivalent amount of chemical fertilizer. Improved yields may continue years after manure stops being added to the soil.

Problems of Manure. Manure can also pose problems for the environment. The growth in recent years of large feeding operations, generating large amounts of manure in a small area, raises the potential for environmental side effects. These include:

+ Excessive application of phosphorus to land. Although manure is not high in phosphorus, it builds up over repeated applications. With high levels of phosphate in soil, runoff elevates phosphate levels in surface waters, with results detailed later in the chapter.
+ Excessive rates of manure application to land. This occurs most commonly in situations in which more animals are being raised than there is land available to safely absorb their manure.
+ Leaching of nitrates under animal confinement areas.
+ Runoff of nitrates and organic materials into lakes and streams.
+ Large spills from manure lagoons.
+ Generation of gaseous air pollutants such as hydrogen sulfide (H_2S), which has human health effects; methane (CH_4), a greenhouse gas; and ammonia (NH_3), which can dissolve in local surface waters.

Under the Federal Clean Water Act of 1972, large feedlots are considered to be point sources of water pollution, so their operations can be regulated. For growers today, the goal is to make the most efficient use of manures for profit while minimizing environmental problems.

Content of Manure. Manure includes both solids and liquids, which, for the most part, are the feces and urine of the animal. The solid part may also include bedding. As Figure 10-3 shows, the solid part of the manure contains most of the phosphate. Most of the potash is in the liquid part. Urine holds about half the nitrogen in manure, primarily

Figure 10-3 Most of the potash in manure is contained in the urine, and the phosphate is contained primarily in the feces. Nitrogen is distributed equally between the two parts.

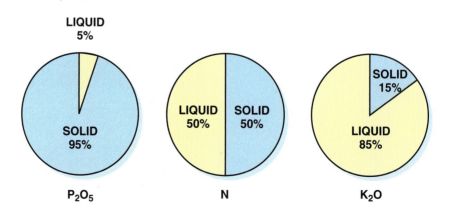

in the form of urea and similar compounds. The rest of the nitrogen is contained in the animal feces.

Several factors determine the amount of nutrient in manure, including the type of animal. In general, sheep and poultry manure have high nitrogen content; the manure of cattle, pigs, and horses has lower nitrogen content. The amount and type of bedding also influences nutrient content because it thins out the manure. If manure has a high carbon:nitrogen (C:N) ratio, nitrogen will be trapped in the soil. The amount and type of rations and the age and health of the animal are also important factors.

Table 10-1 lists average values of nutrient content for several animal manures. To change these values to the standard percentages used in commercial fertilizers, divide by 20. This operation changes pounds per ton to percent. For example, poultry manure contains 25, 11, and 10 pounds per ton of nitrogen, phosphate, and potash, respectively. Dividing these values by 20, the nitrogen-phosphate-potash levels become 1.2-0.5-0.5. This is much weaker than commercial fertilizer, mainly because manure is largely water and organic carbon. Manure must be applied in quantities of tons per acre rather than pounds per acre. Part of the secret to using manure is to keep its nutrient value intact and prevent large losses.

Nutrient Losses from Manure. Urine contains about 50 percent of the nutrient value of manure. If this part of the manure is lost, most of the potassium and much of the nitrogen also will be lost. Urine is lost when it seeps into the ground through barn floors or in feedlots. A great deal of urine simply drains away from manure heaps.

Sharp nitrogen losses occur if the manure begins to decay before it is spread. As much as 90 percent of the nitrogen can be lost within 3 weeks if manure is poorly handled. The losses occur when urea changes to ammonia gas during decay. The loss is most rapid when it is warm and the concentration of urea is highest. Water in the manure dilutes the urea and slows the change. The following storage conditions promote nitrogen loss:

+ High air temperatures speed decay, with resulting nitrogen loss.
+ Heat in a manure pile during decay also speeds up losses.
+ As manure dries out, urea becomes more concentrated. Therefore, as manure dries out during storage, ammonia enters the air rapidly.
+ Freezing also speeds up losses. As the water in the pile begins to freeze, urea is concentrated in the remaining unfrozen water. The higher concentration speeds up losses.

Nutrient losses continue after manure is spread in the field. Ammonia continues to escape unless the manure is mixed quickly into the soil. Runoff and leaching increase the loss. Spreading manure on frozen, sloping land increases the chances of manure being lost to runoff.

Decay organisms respiring in a manure pile change organic carbon to carbon dioxide gas. Organic matter decay explains why manure piles shrink over time. Because the organic matter would be a desirable addition to the soil, it can be considered a loss as well.

Figure 10-4 Much of the nitrogen and potash can be easily lost during manure storage.

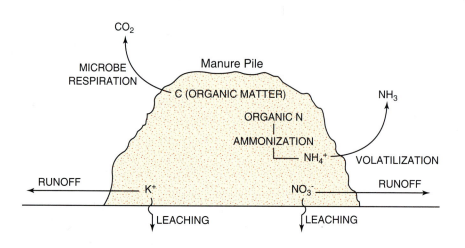

Figure 10-4 summarizes the ways in which nutrients are lost from manure.

Handling Manure. The best way to handle manure is to spread it immediately on unfrozen ground and then plow it into the soil. In this way, the grower prevents the loss of ammonia gas that occurs during storage and in the field. However, in some regions, this technique is not practical in every season.

If manure cannot be mixed into the soil immediately, it should be stored properly and then applied when it can be plowed into the soil. The actual loss of nitrogen varies with handling and storage systems. Piles in an open lot, exposed to sun, rain, and air movement, will lose about half their nitrogen.

Long-term storage in lagoons is even worse. The better method is short-term storage of solid or liquid manures in proper storage structures. Good storage facilities have concrete floors and walls and a roof to stop drainage losses and slow down the drying of the manure.

Liquid manure handling systems, except for large lagoons, are the best way of saving nutrients. In these systems, growers store liquid animal wastes in concrete pits or tanks. The manure is about 90 percent or more liquid and can be handled by pumps. The liquid manure can be spread on fields by machinery or even by gun irrigation.

Freshly spread liquid manure should be plowed into the soil immediately. In warm weather, 20 percent of the nitrogen volatilizes within six hours. The best system uses a tank that knifes the liquid into the ground. Knifing stops the loss of gaseous ammonia and reduces odors.

Table 10-2 summarizes nitrogen losses from various methods of application.

Best Management Practices. A grower's task is to make the most efficient use of animal manures without inflicting damage on the environment. To this end, a number of **Best Management Practices** have been proposed:

+ Test manure and soil for nutrient levels. Charts like Table 10-1 merely suggest possible levels; actual tests are needed to measure exact nutrient content. With tests, the grower can reduce fertilization by an amount equivalent to the nutrients in the manure.

Table 10-2 Nitrogen losses from animal manures as affected by methods of handling and storage.

Method	Nitrogen Loss (percent)
Solid Systems	
Daily scrape and haul	15–35
Manure pack	20–40
Open lot	40–60
Deep pit (poultry)	15–35
Liquid Systems	
Anaerobic deep pit	15–30
Aboveground storage	15–30
Earthen storage pit	20–40
Lagoon	70–80

Source: *Utilization of Animal Manures as Fertilizer; Sutton, Nelson, and Jones, Purdue University*

+ Base manure application rates on phosphorus needs rather than nitrogen. The latter is more commonly done, but this leads to heavy phosphorus soil loading in many soils. Basing application rates on phosphorus will often result in manure being spread more thinly over more acres of land, which could be difficult for operations lacking the needed acres. In such situations, reducing phosphorus in animal rations reduces its level in manure, so it can be spread more thickly.
+ Incorporate all manures into the soil as soon as possible.
+ Where there is not enough land for safe spreading of all the manure, consider composting and selling the excess.

Compost

Composting is defined as a method of causing the decay of organic matter contained in a pile above the ground. Composting can be done on a household or garden scale or for larger-scale commercial uses as well. Examples of important composting operations include composting of yard wastes by cities and landscape companies, other municipal solid wastes, food-processing wastes, or sewage sludge. Nurseries and greenhouses compost organic materials like wood chips for use in potting soils, and farmers may compost manure and bedding as well as animal carcasses. The compost that results from these operations may be used as a soil amendment by the operator. In some cases it may be given away to homeowners for their gardens or sold as a commercial product. In the process, a variety of wastes are put to good use.

Composting offers three benefits over simply spreading uncomposted organic matter on the soil:

1. Reduces the weight and volume of organic material, making it easier to handle and ship
2. Reduces the C:N ratio of materials like wood chips or leaves, eliminating the problem of nitrogen tie-up when compost is added to soil. Table 10-3 shows the C:N ratio of some common materials.

Table 10-3 Carbon-nitrogen (C:N) ratios of several materials. The lower the ratio, the richer its nitrogen content.

Material	C:N Ratio
Soil microflora, average	8
Sewage sludge	7
Garden soil	12–15
Young alfalfa	13
Mature alfalfa	25
Compost	15–20
Rotted manure	20
Lawn clippings, fresh	31
Corn stover	60
Straw of leaves	60
Newspaper	120
Pine needles	225
Sawdust	400–600

3. Heat generated in a compost pile kills most plant or human pathogens and weed seeds. Of course, the pile must be mixed a few times to ensure that the outer edges of the pile are pulled into the hot interior (Figure 10-5).

A properly prepared compost pile mixes carbonaceous and nitrogenous materials to achieve a C:N ratio of about 30:1. At this ratio, nitrogen is preserved because it remains immobilized; at a lower ratio, much will be free to leach or volatilize. At much higher ratios, the process is very slow.

The pile is also kept moist but not wet, at a moisture level of about 50 percent. The pile must be moist to permit rapid decay, but excess moisture creates anaerobic conditions.

Anaerobic piles decay very slowly, generate unpleasant odors, and produce "sour compost" containing organic acids and other chemicals that damage plants.

Figure 10-5 Turning compost windrows to replenish oxygen and mix the organic material for efficient composting. (Photo courtesy of USDA ARS)

In commercial composting, wastes may be shredded before they are piled in long windrows and moistened. The composting process now follows three stages. In the first, **mesophilic**, stage, organisms that prefer moderate temperatures begin the decay process, and the temperature begins to rise. In the second, **thermophilic**, stage, heat-loving microbes replace mesophilic ones and the temperature rises to around 150°F. During this stage, temperatures are monitored, and whenever they begin to drop, the pile is turned to bring in fresh oxygen and organic matter, and temperatures rise again. The bulk of decay occurs during the thermophilic stage. When temperatures drop for good, a second mesophilic stage takes one or two months, during which time stabilization of the compost occurs.

Well-prepared compost is well decayed and stabilized. It is low in heavy metals and soluble salts, with a pH between 5.0 and 8.0. Particles should be around a millimeter in size, with little or no foreign matter, such as pop can tabs. The C:N ratio should be between 15 and 25:1, and the compost should be free of the toxic residues of anaerobic decay. It should also be mostly organic matter, with little soil mixed in. Table 10-4 shows average composition of composts.

Green Manures

Some crops are planted to be turned into the soil rather than for harvest. Green manure may be incorporated into four broad systems of production. First is traditional green manure, planted as a main crop for later plow-down. This method is particularly useful to reduce erosion and weed growth on land that has been idled (Figure 10-6). Otherwise, the loss of a paying crop for one season seldom justifies the practice for most growers. However, it is a standard and important practice for nursery growers to renew organic matter between crops of trees and shrubs.

Figure 10-6 A cover crop like this mustard can be disked into soil as green manure to act as a natural fumigant for weeds and diseases. (Photo courtesy of USDA ARS)

Table 10-4 Average Analysis of One Wood Shaving/ Turkey Manure Compost Product.

Component	Pounds/Ton
Total nitrogen	44.00
Phosphate (P_2O_5)	68.00
Potash (K_2O)	38.00
Calcium carbonate	160.00
Magnesium	8.00
Sulfur	12.50
Sodium	5.56
Iron	6.80
Aluminium	4.24
Manganese	0.73
Copper	0.59
Zinc	0.50
Organic matter	1,000.00

Courtesy of Agri-Brand Compost, Holden Frams, Inc.

Two types of plants can be grown as green manures. Legumes such as clover or vetch are useful because, in addition to the organic matter they leave behind, the nitrogen they fix supplies later crops. To get the most bulk of organic matter at the least cost, or if nitrogen additions are undesirable, grasses like oats or rye may be used. Sudan grass, a tropical grass that grows to 6 feet, develops the greatest amount of green matter and is particularly popular with nursery growers.

Where winter erosion is a problem, a cover crop may be planted in the fall after the main crop is harvested. The green cover protects soil during the fall, winter, and early spring when it is most prone to erosion. The cover crop is then plowed down the following spring. Winter rye and other winter crops work quite well for this purpose.

A cover crop may also be planted immediately after harvest of a rapidly maturing crop, like oats following peas in the same season. The crop could be fall plowed. If left until the following spring, this method also provides winter protection of the soil.

Lastly, a cover crop may be planted between rows of the main crop. This cover is sometimes called a companion crop or living mulch. The cover crop may be planted later in the development of the main crop to reduce yield losses from competition.

CHOICE OF CROPS

Choice of crops is important to organic farmers. Each crop or crop variety has its own specific needs. In some places a crop will grow well and in others it will not. Crops are affected by:

+ Soil type
+ Rainfall
+ Altitude
+ Temperature
+ Type and amount of nutrients required
+ Amount of water needed

These factors affect how a crop grows and yields. If a crop is grown in a climate to which it is not suited, it is likely to produce low yields and be more susceptible to pests and diseases. This then creates the need to use nonorganic sources of fertilizer for the crop and to control pests and diseases.

Successful organic farmers learn to grow the crops and varieties that are suited to the local conditions, geography, and climate. Organic farmers tend to choose local crop varieties suited to the local conditions.

CROP ROTATIONS

Growing the same crops in the same site year after year reduces soil fertility and can encourage a buildup of pests, diseases, and weeds in the soil. Crops should be moved to a different area of land each year and not returned to the original site for several years. For vegetables, a three- to four-year rotation is usually recommended as a minimum.

Crop rotation means having times in which the fertility of the soil is being built up and times in which crops are grown that remove nutrients. Crop rotation also helps a variety of natural predators to survive on the farm by providing diverse habitats and sources of food for them.

For the organic grower, a typical four-year rotation could include a cycle with corn and beans, a root crop, and cereals with either of the following:

1. Grass or bush fallow (a fallow period during which no crops are grown)
2. A legume crop in which a green manure—a plant grown mainly for the benefit of the soil—is grown

Some conventional farmers do not rotate crops because it means planting some less-profitable crops. Often a farmer has no use for certain crops in common rotations. For instance, a farmer who feeds no animals has little use for hay unless a buyer can be found for it. However, crop rotation has important benefits for those who practice it:

+ Aids the control of diseases and insects that rely on one plant host, reducing a grower's pesticide bill.
+ Helps control weeds. Many weed species grow best in certain crop types, so alternating crops suppresses the weeds. Some rotations suppress weeds by allelopathy, in which one plant emits chemicals from the roots that suppress growth of other plants. For instance, soybeans planted into wheat residues suffer fewer weed problems because of the allelopathic effects of the wheat.
+ Supplies nitrogen if certain legumes like alfalfa are in the rotation. This can lower a farmer's fertilizer bill.
+ Improves soil organic matter and tilth. Deep-rooted crops like alfalfa also improve subsoil conditions.
+ Reduces erosion compared with continuous row crops, as long as the rotation includes small grains or hay.

Generally, crop rotations involve some combination of three kinds of crops: row crops, small grains, and forages. The specific crops and crop sequence vary from place to place (Figure 10-7).

WEED CONTROL

Weed control presents a special challenge to organic farmers. Organic farmers avoid the use of herbicides, which they claim leave harmful residues in the environment. Additionally, beneficial plant life such as host plants for useful insects may also be destroyed by herbicides. In organic farming systems, the aim is not necessarily the elimination of weeds but their control. Weed control means reducing the effects of weeds on crop growth and yield using one or more of the following methods:

+ Crop rotation
+ Hoeing

Figure 10-7 Sunflowers and proso millet plots in the alternative crop rotation plots being compared with the old wheat-fallow system. (Photo courtesy of USDA ARS)

- Mulches, which cover the soil and stop weed seeds from germinating
- Hand-weeding or the use of mechanical weeders
- Planting crops close together within each bed to prevent space for weeds to emerge
- Green manures or cover crops to outcompete weeds
- Soil cultivation carried out at repeated intervals and at the appropriate time, when the soil is moist. Care should be taken that cultivation does not cause soil erosion.
- Animals as weeders to graze on weeds

Mulches. Mulching means covering the ground with a layer of loose material such as compost, manure, straw, dry grass, leaves, or crop residues (Figure 10-8). Green vegetation is not normally used as it can take a long time to decompose and can attract pests and fungal diseases. Alternative mulching materials include black plastic sheeting or cardboard. These materials do not add nutrients to the soil or improve its structure.

Mulches should always be applied to a warm, wet soil. Mulch applied to a dry soil keeps the soil dry. Too thick of a mulch will prevent air flow and encourage pests. To allow the germination of planted seeds through the mulch, a layer of less than 4 inches should be used.

Organic growers consider weeds to have some useful purposes, such as providing protection from erosion, food for animals and beneficial insects, and food for human use.

Figure 10-8 Rows of no-till cotton planted directly into this unplowed cornfield. Abundant crop residue between rows creates a mulch that helps prevent moisture loss and weed growth. (Photo courtesy of USDA ARS)

PEST AND DISEASE CONTROL

Organic growers view pests and diseases as part of nature, suggesting that in the ideal system a natural balance exists between predators and pests.

If the system is imbalanced, then one population can become dominant because it is not being preyed upon by another. The aim of natural control in organic methods is to restore a natural balance between pest and predator and to keep pests and diseases down to an acceptable level. The aim is not to eradicate them altogether.

Organic growers avoid chemical control of pests, believing that pesticides do not solve the pest problem. Their preference for natural control is based on three reasons:

1. Safety for people—artificial pesticides can quickly find their way into food chains and water sources, creating health hazards for humans.
2. Cost—natural pest and disease control is often cheaper than applying chemical pesticides because natural methods do not involve buying commercial materials.
3. Safety for the environment—harmful effects that chemical pesticides can have on the environment include killing useful insects that eat pests, staying in the environment and in the bodies of animals and causing problems for many years, and developing pests resistant to the chemical pesticides, thus requiring more and stronger chemicals.

Natural Control

Organic farmers control pests and diseases using a variety of methods, for example:

+ Growing healthy crops that suffer less damage from pests and diseases
+ Choosing crops with a natural resistance to specific pests and diseases
+ Timely planting of crops to avoid the period when a pest does the most damage
+ Companion planting with other crops that pests will avoid, such as onion or garlic
+ Trapping or picking pests from the crop
+ Identifying pests and diseases correctly, preventing the grower from wasting time or accidentally eliminating beneficial insects
+ Using crop rotations to help break pest cycles and prevent a carryover of pests to the next season
+ Providing natural habitats to encourage natural predators that control pests
+ Recognizing insects and other animals that eat and control pests (Figure 10-9)

Through careful planning and using all of the other techniques available, organic growers believe it should be possible to avoid the need for any crop spraying. If pests are still a problem, natural products can be used to manage pests, including sprays made from chillies, onions, garlic, or neem.

Figure 10-9 A tray of ladybird beetles (family Coccinellidae) from the Agricultural Research Service museum's extensive collection of biological control agents. (Photo courtesy of USDA ARS)

Even with these natural pesticides, organic growers limit their use as much as possible and use only the safest ones. Organic growers should check with national and international organic standards to see which natural products are allowed or recommended.

GENETIC DIVERSITY

Within a single crop many genetic differences may exist between plants; for example, they may vary in height or ability to resist diseases. Traditional crops grown by organic farmers contain greater genetic diversity than modern-bred crops. Traditional varieties have been selected over many centuries to meet the requirements of farmers. Although many are being replaced by modern varieties, seeds are often still saved locally.

Crops that have been bred by modern breeding methods tend to be very similar, and if one plant is prone to disease, all the other plants are as well. Although some modern varieties may be very resistant to specific pests and diseases, they are often less suited to local conditions than traditional varieties. Therefore, it can be dangerous to rely too much on any one of them.

In organic systems, some variation or "genetic diversity" between the plants within a crop is beneficial. Growing a number of different crops rather than relying on one is also very important. This helps to protect against pests and diseases and acts as insurance against crop failure in unusual weather such as drought or flood. It is important to remember this when choosing which crops to grow.

An organic farmer should:

+ Grow a mixture of crops in the same field (mixed cropping, inter-cropping, strip cropping)
+ Grow different varieties of the same crop
+ Use as many local crop varieties as possible
+ Save the seed of local and improved crop varieties rather than relying on buying seed from outside the farm every year. Exchange of seed with other farmers can also help to increase diversity, and ensure the survival of the many traditional crop varieties that are being lost as they are replaced by a few modern varieties.

Organic growers are opposed to genetically engineered plants.

USE OF WATER

In arid lands the careful use of water is as much a part of organic growing as is any other technique. As with other resources, organic farmers should try to use water that is available locally, avoiding using water faster than it is replaced naturally.

Methods organic growers use to use water carefully include:

+ Use of terracing, rainwater basins or catchments, and careful irrigation
+ Addition of organic matter to the soil to improve its ability to hold water
+ Use of mulches to hold water in the soil by stopping the soil surface from drying out or becoming too hot

ANIMALS AND ORGANIC GROWING

In an organic system, the welfare of the animals is considered very important. Animals should not be kept in confined spaces where they cannot carry out their natural behavior, such as standing and moving around, which cannot be done in an inadequate amount of space. Animals must be kept from damaging crops. Of course, food for animals should be grown organically.

Types and breeds of livestock should be chosen to suit local needs and local conditions and resources. This ensures that livestock are healthier, better able to resist diseases, and to provide good yields for the farmer.

Finally, livestock provide manure used to maintain soil fertility.

SUSTAINABLE AGRICULTURE

Increasing concern for long-term farm productivity and the effect of agricultural practices on the environment led to the concept of sustainable agriculture. The American Society of Agronomy in 1989 declared that "a sustainable agriculture is one that, over the long term, enhances environmental quality and the resource base on which agriculture depends; provides for basic human food and fiber needs; is economically viable; and enhances the quality of life for farmers and society as a whole."

Those who research or practice sustainable agriculture have several concerns:

+ Depleting agriculture's resource base
+ Declining soil productivity due to erosion and loss of organic matter and nutrients
+ Depleting fertilizer sources like phosphate rock
+ Increasing cost and declining availability of energy

A feared consequence of conventional agriculture is a degraded environment—pollution of water by agricultural chemicals, nutrients, and siltation. Stability of the farm economy and community further motivates sustainable agriculture.

Sustainable agriculture, then, is a philosophy and collection of practices that seeks to protect resources while ensuring adequate productivity. It strives to minimize off-farm inputs like fertilizers and pesticides and to maximize on-farm resources like nitrogen fixation by legumes. Top yields are less a goal than optimum and profitable yields based on reduced input costs.

Soil and water management are central components of sustainable agriculture. Techniques include crop rotation, conservation tillage, cover cropping, and nutrient management.

SUMMARY

Methods and materials used by organic farmers include keeping and building good soil structure and fertility, using recycled and composted crop wastes and animal manures, correct soil cultivation at the right time, crop rotation, green manures, and mulching. Additionally, organic growers control diseases, pests, and weeds through careful planning and crop choice, resistant crops, cultivation practice, crop rotation, useful predators that eat pests, increasing genetic diversity, and natural pesticides.

Manure and compost provide a double benefit to growers—they contain nutrients to promote crop growth and organic matter to improve the soil. Applying these materials to land is a good alternative to other disposal methods.

Manure is highest in carbon, nitrogen, and potash. It also contains phosphates and secondary and trace elements. Proper handling of manure reduces nutrient losses and lowers the chance of polluting surface or groundwater. Manure is best spread on unfrozen land as soon as possible after collection and then tilled into the soil. If this is not practical, the grower should store the manure in sealed, covered pits. Some liquid systems also work well to preserve nutrients.

Composting is a way to reduce the C:N ratio of organic materials and to kill harmful organisms.

Any nutrient source—fertilizer, manure, or compost—can harm the environment if used improperly. Nutrients and pathogens can wash into surface waters or leach into groundwater, causing pollution and human and animal health problems. Using organic fertilizers in the suggested ways and rates and avoiding erosion are important ways to reduce these problems.

Review

True or False

1. Composting reduces the C:N ratio of organic materials.

2. A national program sets organic standards.

3. Green manure means that it is fresh from the barn or corral.

4. Nitrogen loss from manure is negligible.

5. Composting increases the weight and volume of organic material used to make the compost.

6. Heat generated during composting kills weed seeds.

Short Answer

7. Identify NOP and IFOAM.

8. Name four crops likely to be used for a green manure.

9. To reduce nitrogen loss from manure, what is the best way to spread it on a field?

10. List three ways nutrients are lost from manure.

11. Why should manure applications to fields be based on phosphorus needs rather than nitrogen?

12. Identify four factors organic growers should consider when choosing crops to grow.

13. Name four benefits of crop rotation for the organic grower.

14. List four acceptable methods used to control weeds in an organic field.

15. Identify five natural control methods for controlling pests and diseases on an organic operation.

Critical Thinking/Discussion

16. Discuss factors that will determine how much of the nutrient value of manure as it is excreted by an animal will end up being used by crops in a field on which it has been spread.

17. What happens if a compost pile becomes too dry? Too wet?

18. Why can heavy phosphorus loading occur on lands receiving manures?

19. Discuss the pros and cons of using manure or compost for fertilizing fields.

20. Why is genetic diversity important to the organic grower?

21. What weed, pest, and disease control options are acceptable to organic growers?

22. Why is crop rotation especially important to the organic grower?

Knowledge Applied

1. Visit a composting or manure storage facility.

2. Build and manage a compost pile.

3. Proper manure management involves land application at the correct rate, which means calibrating the spreader. Describe this process after reading the instructions on this Web site: www.ext.colostate.edu/pubs/crops/00561.html.

4. Visit an organic farming or gardening operation and report back to the class.

5. Compare organically grown vegetables with conventionally grown vegetables in a presentation.

6. Observe the effects of nitrogen tie-up by growing corn in pots using two different soil mixes. Grow one group of plants in a normal but unsterilized soil mix. Grow another group in a mix that is half fresh sawdust. Note differences in crop appearance or growth.

7. Develop a presentation on organic production certification rules for the National Organic Program. Use this Web site as the source of your information: www.ams.usda.gov/nop/NOP/NOPhome.html.

Resources

Coleman, E. (1995). *The new organic grower* (Revised ed.). Chelsea, VT: Chelsea Green Publishing Co.

Fossel, P. V. (2007). *Organic farming: Everything you need to know.* St. Paul, MN: MBI Publishing Company.

Krishna, K. R. (2003). *Agrosphere: Nutrient dynamics, ecology, and productivity.* Enfield, NH: Science Publishers.

Lampkin, N. (2003). *Organic farming* (2nd ed.). Suffolk, UK: Old Pond Publishing.

Newton, J. (2004). *Profitable organic farming* (2nd ed.). Indianapolis, IN: Wiley-Blackwell.

Plaster, E. J. (2008). *Soil science and management* (5th ed.). Albany, NY: Delmar Cengage Learning.

Poincelot, R. P. (2004). *Sustainable horticulture: Today and tomorrow.* Englewood Cliffs, NJ: Prentice Hall.

Zimmer, G. F. (2000). *The biological farmer.* Austin, TX: Acres U.S.A., Publishers.

Internet

Internet sites represent a vast resource of information. The URLs (uniform resource locators) for World Wide Web sites can change. Using one of the search engines on the Internet such as Google or Yahoo!, find more information by searching for these words or phrases: organic farming, organic standards, organic food, composting, nutrient value of manure, crop rotation, green manure, natural pest control, natural weed control, sustainable agriculture, and biological farming.

PART 3

Fundamentals of Plant Growth and Propagation

Chapter 11
Temperature

NOT ALL PLANTS are alike. They vary greatly as to the temperature they tolerate or even require to germinate, grow, and survive the winter.

Temperature affects the productivity and growth of a plant, depending upon whether the plant is a warm- or cool-season crop. If temperatures are high and day length is long, cool-season crops such as spinach will flower. Temperatures that are too low for a warm-season crop such as tomato will prevent fruit set. Adverse temperatures also cause stunted growth and poor-quality vegetable production. For example, bitterness in lettuce is caused by high temperatures.

Sometimes temperatures are used in connection with day length to manipulate the flowering of plants.

After completing this chapter, you should be able to—

- List five cool-season crops and five warm-season crops
- Discuss the effect temperature has on plant dormancy
- Give the germination temperatures for five plants
- Identify and explain the hardiness zones in the United States
- Discuss temperature differences across the United States
- Describe the use of growing degree days
- Calculate the growing degree days
- Discuss the growing degree day differences across the United States
- Define vernalization
- Discuss the damage caused by frost
- Identify the differences in first and last frost across the United States
- List the effects temperature may have on plants
- Discuss the concept of thermoperiod

KEY TERMS

Cool-Season
Growing Degree Day (GDD)
Hardening
Hardiness Zone
Optimal Temperature
Thermoperiod
Vernalization
Warm-Season

BIOLOGICAL TEMPERATURE RANGE

Plants vary in their ability to tolerate temperature differences. Lettuce and turnips germinate well and grow at 50°F, while okra does not come up at all. Lettuce and turnips can be planted while it is cool, but not okra. Every plant has a temperature below which it cannot grow and an optimum temperature for growth. If it gets too warm, growth again ceases. Broccoli and lettuce, for example, do not grow well during the heat of the summer, but cotton does very well. Table 11-1 lists the days needed for germination at various temperatures for some common seeds.

Table 11-1 Days Needed for Germination at Various Temperatures

Plant	Temperature (°F)					
	41°	50°	59°	68°	77°	95°
Cabbage	—	15	9	6	5	—
Corn	—	22	12	7	4	3
Lettuce	15	7	4	3	2	—
Okra	—	—	27	17	13	6
Tomato	—	43	14	8	6	9
Turnip	—	5	3	2	1	1

Common agricultural plants are divided into cool-season and warm-season plants. **Cool-season** plants survive mild spring frosts and may be planted early in the spring or in the fall. **Warm-season** plants are usually killed by frosts and require much warmer temperatures to grow properly. They are planted later in the spring. Some cool-season crops include:

Wheat
Oats
Barley
Rye
Fescue
Cabbage
Collards
Lettuce
Turnips
Onions

Warm-season crops include:

Cotton
Corn
Soybeans
Grain sorghum
Bermuda grass
Beans
Southern peas
Tomatoes
Peppers

Cantaloupe
Squash
Watermelon

Some plants are able to adapt to low temperatures. Tropical plants and annuals (plants that live only one year) adapt very little and are normally killed by winters. Native evergreen trees and shrubs are able to stop growing until warm temperatures return. Spinach and winter wheat remain alive all winter and even grow a little during warm periods.

Hardening

Cool-season vegetables can be made to adapt to somewhat cooler temperatures by gradually exposing the young transplants to stress. The young plants are allowed to wilt slightly before watering or they are grown under temperatures 10 degrees below normal. The plants become tougher and less likely to be killed or injured by the environment after they are set into the field or garden. This process is called **hardening**.

Temperature and Dormancy

Some plants drop their leaves and appear to be dead but are only resting or dormant. Bulbs and some smaller plants (johnsongrass) are killed to the ground by cold temperatures but emerge from the underground parts in the spring. Those plants that are killed by winter temperatures often overwinter as seeds and thrive without individual plants surviving more than a few months. Many of the most troublesome weeds survive in this manner.

Those plants that survive by going dormant in the winter may still be killed in some years. They are able to survive only to a certain temperature. If it gets lower than that temperature, they will be injured and may be killed. The lowest temperatures that a dormant plant can survive depend on many things. These include how long the cold lasts, how windy it is, what the recent temperature has been, how old and healthy the plant is, and even how much moisture is in the soil. A period of warm winter temperatures followed by a sudden extreme cold spell can be deadly to many plants.

Not all parts of a dormant plant can survive the same temperature. Flower buds of peaches, for example, are often killed by temperatures that do not kill the entire tree. Roots of many trees can be killed by temperatures that do not kill tree trunks or branches. Sometimes small fruit trees shipped through the mail arrive with their roots frozen. These trees leaf out and begin to grow and then die because their roots are dead.

Seed Germination

Temperature determines seed germination. The temperature can be too cold or too hot and, as Table 11-2 shows, seeds have an **optimal temperature** for germination.

Thermoperiod

Thermoperiod refers to a daily temperature change. Plants respond to and produce maximum growth when exposed to a day temperature that is about 10–15 degrees higher than a night temperature. This allows the

Table 11-2 Temperature Requirements of Vegetable Crops

| Crop | Temperature Requirements (°F) | | | | |
| | Germination | | | Growth | |
	Min.	Opt.	Max.	Day	Night
Asparagus	50	75	95	65–80	55–70
Broccoli	45	85	95	65–75	55–60
Brussel sprouts	45	85	95	65–75	55–60
Cabbage	40	85	100	55–75	45–55
Cauliflower	40	80	100	55–75	45–55
Celery	40	70	85	60–75	60–65
Collard	40	80	100	55–75	55–60
Cucumber	60	95	105	70–85	65–70
Eggplant	60	85	95	70–85	65–70
Endive	35	75	85	70–85	65–70
Lettuce	35	75	85	50–70	45–55
Muskmelon	60	90	100	70–85	65–70
Okra	60	95	105	75–85	60–75
Onion	35	75	95	60–75	60–75
Pepper	60	85	95	70–80	65–70
Pumpkin	60	95	100	70–85	60–70
Summer squash	60	85	95	70–85	60–70
Sweet potato	—	—	—	75–85	65–70
Tomato	50	85	95	65–80	60–65
Watermelon	60	95	105	65–80	60–65

plant to photosynthesize (build up) and respire (break down) during an optimum daytime temperature and to curtail the rate of respiration during a cooler night. Temperatures higher than needed cause increased respiration, sometimes above the rate of photosynthesis. This means that the products of photosynthesis are being used more rapidly than they are being produced. For growth to occur, photosynthesis must be greater than respiration. Too low temperatures can produce poor growth. Photosynthesis is slowed down at low temperatures. Because photosynthesis is slowed, growth is slowed, and this results in lower yields.

Not all plants grow best under the same temperature range. For example, snapdragons grow best at nighttime temperatures of 55°F and the poinsettia at 62°F. Florist cyclamen does very well under very cool conditions, while many bedding plants prefer a higher temperature.

In some cases, a certain number of days of low temperature are needed by plants in order to grow properly. This is true of crops growing in cold regions of the country. Peaches are a prime example. Most varieties require 700 to 1,000 hours below 45°F but above 32°F before they break their rest period and begin growth. Lilies need six weeks of temperatures at 33°F before blooming.

Plants can be classified as either hardy or nonhardy depending upon their ability to withstand cold temperatures. Winter injury can occur to nonhardy plants if temperatures are too low or if unseasonably low temperatures occur early in the fall or late in the spring. Winter injury may also occur because of desiccation or drying out. Plants need water during the winter. When the soil is frozen, the movement of water into the plant is severely restricted. On a windy winter day, broadleaf evergreens can become water-deficient in a few minutes and the leaves or needles then turn brown. Wide variations in winter temperatures can cause premature bud break in some plants and consequent fruit bud freezing damage. Late spring frosts can ruin entire peach crops. If temperatures drop too low during the winter, entire trees of some species are killed by freezing and splitting plant cells and tissue.

CLIMATIC CLASSIFICATION

The United States is divided into **hardiness zones** for growing plants. These zones are shown in Figure 11-1. Hardiness zones determine the types of plants that will grow based on the average annual minimum temperatures; Table 11-3 provides the actual minimal temperatures for the zones. The U.S. Department of Agriculture (USDA) plant hardiness map divides North America into 11 hardiness zones. Zone 1 is the coldest. Zone 11 is the warmest, a tropical area found only in Hawaii and southernmost Florida. In between, the zones follow a fairly predictable pattern across the continent, although a closer look will reveal scattered patterns of variations. Generally, the colder zones are found at higher latitudes and higher elevations.

Today, the USDA map, which was last updated and released in 1990 (based on weather records from 1974–1986), is generally considered the standard measure of plant hardiness throughout much of the United States. Hence we have the USDA plant hardiness zones.

The average minimum temperature is not the only factor in figuring out whether a plant will survive. Soil types, rainfall, daytime temperatures, day length, wind, humidity, and heat also play their roles. For example, although both Phoenix and Portland are in the same zone (8), the local climates are dramatically different—even within a city, a street, or a spot protected by a warm wall. Microclimates within an area affect how plants grow. The zones are a good starting point, but growers still need to determine what will and will not work in a specific area.

Climate and temperature are such an important part of raising crops that a variety of agricultural weather stations track the weather and report on it across the United States. Figure 11-2 shows the climate features of the United States and the location of weather (climate) stations across the country.

Helping Plants Tolerate Cold Nights

Researchers at the Photosynthesis Research Unit in Urbana, Illinois, are working to help warm-weather plants deal with chilly temperatures. Chilly evenings can cause biochemical chaos in warm-weather plants like tomatoes and corn, resulting in reduced yields. But Agricultural Research Service (ARS) scientists know that manipulating certain plant enzymes could keep crop biochemistry on an orderly schedule. If the researchers succeed, the same tactic may also expand the geographic area in which some warm-weather plants are grown. Crucial processes that normally take place at night in the plant may shut down when temperatures dip below 50°F. When the weather warms up in the morning, the nighttime processes resume, but they may clash with different, but equally important processes set to occur during the day. ARS researchers have determined that two enzymes play a key role in turning on and off central biochemical activities. Their aim is to manipulate the enzymes sucrose phosphate synthase and nitrate reductase to override nature's obsession with temperature.

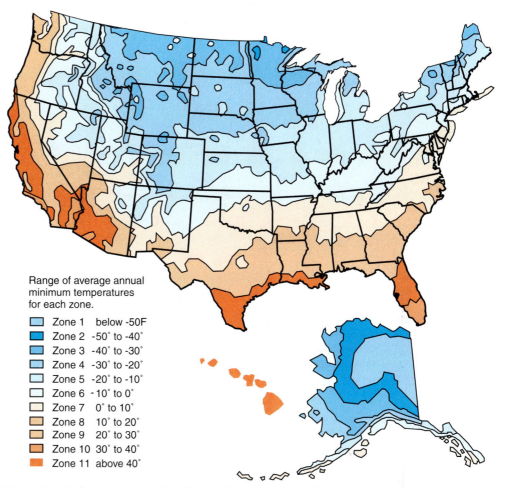

Range of average annual minimum temperatures for each zone.

- Zone 1   below -50F
- Zone 2   -50° to -40°
- Zone 3   -40° to -30°
- Zone 4   -30° to -20°
- Zone 5   -20° to -10°
- Zone 6   -10° to 0°
- Zone 7   0° to 10°
- Zone 8   10° to 20°
- Zone 9   20° to 30°
- Zone 10   30° to 40°
- Zone 11   above 40°

Figure 11-1 Hardiness zones in the United States. *(Image courtesy of USDA)*

Table 11-3 Approximate Range of Average Annual Minimum Temperatures

Zone	Centigrade	Farenheit
Zone 1	below −45°	below −50°
Zone 2	−45° to −40°	−50° to −40°
Zone 3	−40° to −34°	−40° to −30°
Zone 4	−34° to −29°	−30° to −20°
Zone 5	−29° to −23°	−20° to −10°
Zone 6	−23° to −18°	−10° to 0°
Zone 7	−18° to −12°	0° to +10°
Zone 8	−12° to −7°	+10° to +20°
Zone 9	−7° to −1°	+20° to +30°
Zone 10	−1° to +4°	+30° to +40°
Zone 11	above +4°	above +40°

GROWING DEGREE DAYS

Growth and development of many plants and cold-blooded animals depend upon the amount of heat present in or around the organism. Growers may use this fact to help monitor development of plants, pests,

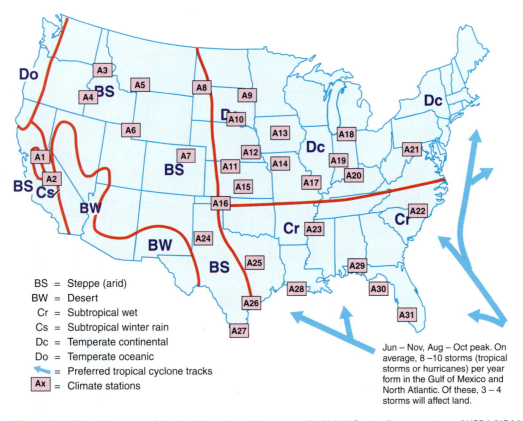

BS = Steppe (arid)
BW = Desert
Cr = Subtropical wet
Cs = Subtropical winter rain
Dc = Temperate continental
Do = Temperate oceanic
= Preferred tropical cyclone tracks
Ax = Climate stations

Jun – Nov, Aug – Oct peak. On average, 8 –10 storms (tropical storms or hurricanes) per year form in the Gulf of Mexico and North Atlantic. Of these, 3 – 4 storms will affect land.

Figure 11-2 Climate features and location of weather stations across the United States. *(Image courtesy of USDA/NOAA Joint Agricultural Weather Facility)*

and diseases during the course of a growing season through use of a simple, temperature-derived index, **growing degree days** (**GDD**). The basic concept is that development will occur only if a certain temperature threshold is reached or exceeded. This developmental threshold is referred to as the base temperature (T_{base}; Table 11-3). The base temperatures are determined experimentally and are different for each organism. Table 11-4 gives the reported base temperatures for GDD computations for different crops and insects.

Table 11-4 Base Temperatures for Growing Degree Days for Crops and Insects

Base Temperature (°F)	Crop or Insect
40	Wheat, barley, rye, oats, flaxseed, lettuce, asparagus
44	Corn rootworm
45	Sunflower, potato
48	Alfalfa weevil
50	Sweet corn, corn, sorghum, rice, soybeans, tomato; Black cutworm, European corn borer
52	Green cloverworm

Calculation of Growing Degree Days

Calculation of GDD may take one of several forms, depending on the application. All forms of GDD have the common principle that the biological process will not begin until Tbase is obtained.

To calculate GDD, the mean temperature for the day first needs to be calculated. This is usually done by taking the high and low temperatures for the day, adding them together, and dividing by 2. If the mean temperature is at or below Tbase, then the GDD value is zero. If the mean temperature is above Tbase, then the GDD amount equals the mean temperature minus Tbase. For example, if the mean temperature was 75°F, then the growing degree day amount equals 10, using a Tbase = 65°F. In equation form:

GDD = Ta – Tbase   if Ta is greater than Tbase
GDD = 0   if Ta is less than or equal to Tbase

Where Tbase = temperature base and Ta = average temperature. This can be expressed another way:

GDD = (high temperature for the day minus the low temperature for the day)/2 – Base

After the base temperature is reached, most biological processes continue until the air temperature falls below the threshold once again. Many of these thresholds lie between 40° and 50°F, such as 50°F for corn growth. The GDD on a given day are then calculated with observed daily temperature data relative to the base temperature. The state of development of the biological process in question is usually correlated with the accumulation of daily GDD through the growing season.

The optimum method of calculating GDD is through use of a thermograph or electronic digital temperature sensor, as these instruments can provide near-continuous temperature data during the course of a day. Many agricultural weather services calculate and publish the GDD for use during the growing season. These are available in newspapers, newsletters, and at various weather sites on the Internet.

Use of Growing Degree Days

Figure 11-3 shows the GDD in all of the United States during a recent growing season. For some dramatic comparisons, this figure shows areas in the northern United States with total GDD of less than 2,000, while parts the southern United States show GDD in excess of 6,000.

Reliability of a GDD rating system increases with the number of years that local GDD data have been collected. Depending on the region, accumulated GDD can differ significantly within even a rather small area. In some areas GDD may vary by 200 units over a distance of less than 20 miles. GDD values may have to be adjusted depending on the location of one's farm relative to the weather station at which the data were recorded.

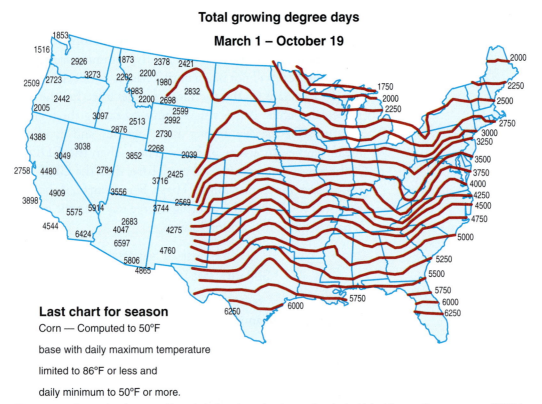

Figure 11-3 A map showing a specific period of total growing degree days in the United States. *(Image courtesy USDA/ NOAA Joint Agricultural Weather Facility)*

Modified Growing Degree Days

Modified GDD are similar to GDD with several temperature adjustments. If the high temperature is above 86°F, it is reset to 86°F. If the low is below 50°F, it is reset to 50°F. Once the high and/or low temperature has been modified as needed, the average for the day is computed and compared with the Tbase (usually 50°F). Modified GDD are typically used to monitor the development of corn, the assumption being that development is limited once the temperature exceeds 86°F.

An Example in Crop Production

All corn hybrid maturity rating systems are related to temperature affects rather loosely. Presently, the most widely used maturity rating systems in the United States is based on these two facts: (1) a corn plant must accumulate a certain amount of heat in order to complete its life cycle, and (2) the total amount of heat needed will be relatively constant for a given hybrid. Under the system, the quantity of heat being added is determined from daily temperatures and usually expressed as GDD.

The daily high temperature cannot be higher than 86°F. The daily low temperature cannot be lower than 50°F, and the base is 50°F. A base is used in order to keep accumulated GDD numbers relatively

Figure 11-4 Graph of the GDD (growing degree days) for corn.

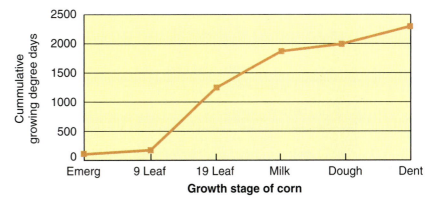

small. It is set at 50°F because corn makes little or no growth below that temperature. The lower cutoff for the daily temperature prevents calculation of negative values. The upper cutoff for the daily temperature is 86°F because at that temperature, corn growth rate begins to decline rapidly because of excessive respiration and moisture stress.

Figure 11-4 gives examples of how daily GDD are determined using the previously presented equation and cutoffs. The accumulated GDD that identify a corn hybrid's maturity rating is merely the sum of daily GDD values over a given length of time. The period that seed corn producers normally use to rate their hybrids is from planting to physiological maturity.

Because GDD characterize several important relationships between corn hybrids and the local growing season, the GDD maturity rating system can be a valuable decision-making aid that is easy to use. For instance, having access to daily GDD information for his area, a farmer can choose hybrids rated to take advantage of the full growing season based on the intended planting date. If planting must be delayed substantially, the grower can substitute hybrids having a maturity rating within the range of accumulated GDD left after the new intended planting date. Also, by calculating daily GDD values during the growing season, the farmer who knows the hybrid's maturity rating can often schedule harvesting much more accurately. Figure 11-5 shows the GDD information provided for some corn hybrids.

Although the GDD maturity rating system is the most common one in the United States, not all seed producers use the same set of standards to determine GDD. For instance, some may use temperature cutoffs other than 50°F and 86°F in calculating daily GDD.

VERNALIZATION

Vernalization is the promotion of flowering by cold treatment given to plants or imbibed seeds. Plants must be exposed to a given temperature for a certain length of time to become and remain vernalized.

AGRONOMIC RATINGS

	MANAGEMENT					CHARACTERISTICS											DISEASE						INSECTS		GRAIN COMPOSITON		
						PLANTING			GROWTH						HARVEST												
	Hybrid	Relative Maturity*	GDUs to Mid-Pollination	GDUs to Black Layer	Yield for Maturity	Planting Rate	Emergence	Seedling Growth Rate	Root Strength	Stalk Strength	Drought Tolerance	Plant Height	Ear Height	Staygreen	Drydown	Test Weight	Eyespot	Anthracnose Leaf Blight	Northern Leaf Blight 1	Northern Leaf Blight 2	Gray Leaf Spot	Southern Leaf Blight	First-Brood European Corn Borer	Second-Brood European Corn Borer	Protein	Oil	Starch
	DK 512	101	1275	2520	1	H	3	4	4	5	3	M-T	M	4	4	6	4	3	4	3	7	4	7	7	6	4	5
	DK 527	102	1265	2555	1	H	2	3	6	5	3	M	M	4	5	2	3	3	4	3	6	2	6	7	5	6	4
EARLY	DK 554	105	1290	2630	2	M-H	5	6	3	4	4	M	M	4	4	6	4	2	4	5	7	6	7	6	6	4	5
	DK 559	105	1330	2680	1	M-H	2	2	3	5	4	M-T	M	5	2	6	4	2	4	4	7	5	6	7	5	5	5
	DK 560	106	1325	2645	1	H	2	3	4	5	4	M-T	M	4	2	3	4	2	2	2	7	5	5	6	5	4	5
	DK 561SR	106	1300	2655	1	H	4	2	5	4	4	M	M	7	2	6	3	3	3	2	7	3	5	7	NA	NA	NA
	DK 566	106	1300	2660	1	H	2	2	5	4	4	M	M	6	2	5	3	3	3	2	6	2	5	7	6	5	5
	DK 569	106	1290	2670	1	M-H	4	6	5	5	3	M-S	M	6	3	6	3	2	2	2	7	2	6	6	5	4	6
MEDIUM	DK 574GR	107	1330	2710	1	M-H	3	3	3	4	4	M	M	4	2	5	4	4	4	5	7	3	7	6	NA	NA	NA
	DK 580	108	1330	2710	1	M-H	3	3	3	4	4	M	M	4	2	5	3	2	4	4	7	3	7	6	6	6	5
	DK 586	108	1315	2710	1	H	2	2	4	4	4	M-T	M	4	3	4	4	2	1	3	8	2	6	6	5	5	5
	DK 591	108	1370	2750	1	H	4	4	3	5	3	M-T	M-H	4	2	5	4	2	3	3	7	3	7	6	4	4	6
	DK 592SR	109	1370	2750	1	M-H	4	4	3	5	3	M-T	M	4	2	6	5	4	4	6	7	2	7	8	3	4	7
	DK 604	110	1380	2740	1	M-H	2	2	4	5	3	M-T	H	3	1	5	3	3	3	3	7	5	7	7	5	4	6
FULL	DK 616	111	1360	2770	1	M-H	2	2	3	4	3	M-T	M	3	3	6	4	2	4	4	7	5	6	7	4	4	6
	DK 618	111	1355	2760	1	H	2	2	2	2	3	M-T	M	2	5	4	6	3	3	3	5	2	6	7	5	6	5
	DK 626	112	1420	2800	1	M-H	2	1	5	6	4	T	M-H	2	3	6	4	2	2	4	6	3	6	7	5	4	5
	DK 642	114	1445	2845	1	M-H	2	3	4	5	5	T	M	3	4	6	4	2	3	4	6	4	6	7	6	5	5
	SPECIALTY																										
	DK 512WX	102	1265	2500	1	H	3	4	3	5	3	M-T	M	4	4	6	3	2	4	3	5	4	7	9	5	4	5
	DK 580WX	108	1320	2700	1	M-H	3	3	3	4	4	M	M	4	2	5	4	3	3	5	8	3	7	7	4	5	5
	DK 631W	113	1380	2835	1	M-H	2	3	3	4	4	S	M-L	5	4	2	3	4	2	5	9	5	6	7	5	7	4

Corn Legend

Rating Scale: 1–2 = Excellent; 3–4 = Very Good; 5–6 = Good; 7–8 = Fair; 9 = Poor

Disease and Insect Ratings: 1 = Excellent Resistance; 9 = Poor Resistance

Code/Symbol: NA = Not Available

* *Note:* Although maturity of one hybrid relative to another remains reasonably constant, the number of actual calendar days from planting or emergence to any given point of plant development varies with temperature, day length, rate and date of planting, soil fertility, and other environmental factors.

Grain composition ratings are based on tests conducted by DEKALB, using standard methods and indicate relative strengths of commercial hybrids. Grain quality measurements are highly sensitive to environmental conditions before and during grain development. Actual values observed in any season may differ from those on which these ratings were based. Each unit difference in the oil, protein, and starch ratings reflects 0.25% oil, 0.5% protein, and 1.0% starch, respectively.

Figure 11-5 Growers selecting hybrid corn use information provided by seed dealers to select varieties suited to their needs and the number of growing degree days (units). (Information courtesy Seed Resource Guide, provided by Dekalb Genetics Corporation, Dekalb, IL.)

Vernalization may be reversed by exposure to warm temperatures before the required time has passed. This process is called devernalization. The plant site for the perception of vernalization is the areas of active cell division—meristems.

Likely, a vernalization stimulus exists in plants. Vernalized plants grafted to nonvernalized plants induce flowering in some species. Plants that do not require vernalization grafted onto plants that do require vernalization induces flowering in some species. Extracts from vernalized plants applied to nonvernalized plants induces flowering in some species

The growth regulator, gibberellin, shows evidence of playing a part in vernalization.

TEMPERATURE STRESS

Temperature affects many essential plant growth processes including most biochemical reactions in the plant. Generally, with higher temperature the rate of reaction increases. Photosynthesis is slower at lower temperatures, but the rate increases up to a certain point as the temperature goes up. This varies by plant. Plant growth functions and rates, such as absorption of minerals and water, are determined by temperature. All plants have an optimal temperature at which they function best. Plants also have a minimum temperature tolerance below which the plant may be injured and in some cases killed. Sunscald of plant's leaves is much more likely to occur at high temperatures. High temperatures cause desiccation or may directly kill protoplasm of cells. Plants should be selected according to the growth conditions and climate of the area. Of course, the exception is in greenhouses, where temperature conditions can be controlled.

A = Freeze occurs after specified dates

B = Freeze occurs before specified dates

⌐ = Spring freezes occur south of this line in less than half the years

Figure 11-6 Average times for the last spring frost in the areas of the United States. (*Source: National Climatic Data Center*)

A = Freeze occurs after specified dates

B = Freeze occurs before specified dates

⌐⌐⌐ = Autumn freezes occur south of this line in less than half the years

Figure 11-7 Average times for the first fall frost in the areas of the United States. (*Source: National Climatic Data Center*)

High- and Low-Temperature Injury

Both high and low temperatures can injure plants. Low-temperature damage usually results from frost injury. High-temperature injury is often linked to light and water effects.

Freezing can cause damage to crops. Figures 11-6 and 11-7 show the average times for the last spring frost and first fall frost in the areas of the United States.

Warm spring temperatures bring renewed tender growth. Sometimes this new growth is too early. For example, in some areas, apricots bloom in February and their flowers are usually killed by frosts. Plums bloom a little later, then peaches, and finally apples. Apples are much more likely to bear fruit than apricots. Table 11-5 lists the average dates for the first and last frosts in the 11 hardiness zones defined by the USDA.

Tomatoes that are planted too early, wheat that begins to grow too soon, and even early leaves on some trees may be injured or killed by late frosts. Sometimes, for example, new, unfolding leaves on oak trees will be killed. This will not necessarily kill the tree. The tree will regrow its leaves but this requires more use of its supply of stored food. If the tree is weak or the emerging leaves freeze too many times, the tree may die. Table 11-6 describes light, moderate, and severe freezes and the damage they can cause.

Table 11-5 Average Dates for First and Last Frost in the United States

Zone 1	Average dates Last Frost = 1 Jun / 30 Jun Average dates First Frost = 1 Jul / 31 Jul ***Note:*** Vulnerable to frost 365 days per year
Zone 2	Average dates Last Frost = 1 May / 31 May Average dates First Frost = 1 Aug / 31 Aug
Zone 3	Average dates Last Frost = 1 May / 31 May Average dates First Frost = 1 Sep / 30 Sep
Zone 4	Average dates Last Frost = 1 May / 30 May Average dates First Frost = 1 Sep / 30 Sep
Zone 5	Average dates Last Frost = 30 Mar / 30 Apr Average dates First Frost = 30 Sep / 30 Oct
Zone 6	Average dates Last Frost = 30 Mar / 30 Apr Average dates First Frost = 30 Sep / 30 Oct
Zone 7	Average dates Last Frost = 30 Mar / 30 Apr Average dates First Frost = 30 Sep / 30 Oct
Zone 8	Average dates Last Frost = 28 Feb / 30 Mar Average dates First Frost = 30 Oct / 30 Nov
Zone 9	Average dates Last Frost = 30 Jan / 28 Feb Average dates First Frost = 30 Nov / 30 Dec
Zone 10	Average dates Last Frost = 30 Jan or before Average dates First Frost = 30 Nov / 30 Dec
Zone 11	Free of frost throughout the year

Table 11-6 Frost Classification and Damage

Frost	Damage
Light freeze[1]	29°F/32°F Tender plants killed with little destructive effect on other vegetation
Moderate freeze	25°F/28°F Wide destruction on most vegetation with heavy damage to fruit blossoms and tender semi-hardy plants
Severe freeze	24°F and colder Heavy damage to most plants

[1] For light freeze, damage depends on length of frost duration.

SUMMARY

The rate of photosynthesis increases with temperature to a point. Respiration rapidly increases with temperature. Transpiration increases with temperature. All plants have optimal temperatures. Germination depends upon the temperature. Flowering may be partially triggered by temperature. As for sugar storage, low temperatures reduce energy use and increase sugar storage. After a period of low temperature, dormancy will be broken and the plant will resume active growth. Some plants depend on a thermoperiod to be productive.

Plants are adapted to various temperature ranges across the United States. To help determine when to plant and what to plant, the United States is divided into hardiness zones and the weather across the United States is monitored closely. Calculating growing degree days provides a method of determining how much heat is required or is available for the plant to grow.

Plants can be injured or killed by temperature extremes. High-temperature injury is often associated with the light and water effects. Freezing temperatures can damage crops to varying degrees depending upon the severity of the frost.

Review

True or False

1. Temperature determines seed germination.

2. Growing degree days is the same as the average daily temperature.

3. Vernalization is the same as scarification.

4. Growing temperatures across the United States are similar.

5. Zone 1 of the USDA hardiness zones never experiences frost.

Short Answer

6. List five cool-season and five warm-season crops.

7. If the low temperature for the day was 45°F and the high was 80°F and the Tbase is 44°F, calculate the growing degree days.

8. List the hardiness zones in the United States.

9. What is the optimal germination temperature for eggplant, lettuce, cabbage, tomato, and onions?

10. _____ affects many essential plant growth processes.

Critical Thinking/Discussion

11. Compare the growing degree days in Florida with those in Montana, New York, and Oregon.

12. Explain the concept of hardiness zones.

13. What is a thermoperiod? Give an example.

14. Discuss the damages that can be caused by varying severities of frost.

15. Describe vernalization and dormancy.

Knowledge Applied

1. Place three containerized bean plants on a thermostatic heating pad set to 70°F; place three more plants next to these, just off of the heating pad. Make sure all of the plants get the water and light they need. After two weeks, compare the size of the plants in the two groups.

2. During the growing season, track the daily high and low temperatures. Plot this information and use it to calculate the growing degree days.

continues

Knowledge Applied, *continued*

3. Develop a presentation on the growing conditions of your area. Find out which of the hardiness zones your area is in, and find out the average dates for the first and last frosts. Using this information, make recommendations for potential new crops that could be grown in your area.

4. Design and conduct an experiment on the effects of freezing. Using potted plants, subject them to brief periods of freezing in the freezer compartment of a refrigerator. Report on your findings.

5. Find a site on the Internet that tracks and predicts weather. Based on this general information, predict the weather for your own area. Compare the accuracy of your predictions with the actual weather. Track your predictions and comparisons using a journal that is kept on a word processor.

Resources

Acquaah, G. (2004). *Horticulture: Principles and practices* (3rd ed.). Englewood Cliffs, NJ: Prentice Hall.

Connors, J. J., & Cordell, S. (Eds.). (2003). *Soil fertility manual.* Tucson, AZ: Potash & Phosphate Institute.

Hay, R., & Porter, J. (2006). *The physiology of crop yield* (2nd ed.). Ames, IA: Blackwell Publishing.

Janick, J. (Ed.). (1989). *Classic papers in horticultural science.* Englewood Cliffs, NY: Prentice Hall.

McMahon, M., Kofranek, A. M., & Rubatzky, V. E. (2006). *Hartmann's plant science* (4th ed.). *Growth, development, and utilization of cultivated plants.* Englewood Cliffs, NJ: Prentice Hall Career & Technology.

Simpson, B. B., & Ogorzaly, M. C. (2000). *Economic botany: Plants in our world* (3rd ed.). New York: McGraw-Hill.

Stern, K. R. (2005). *Introductory plant biology* (10th ed.). New York: McGraw-Hill.

Taiz, L., & Zeiger, E. (2006). *Plant physiology* (4th ed.). Sunderland, MA: Sinauer Associates, Inc.

Electronic Mailing List

USDA-reports@usda.mannlib.cornell.edu

Internet

Internet sites represent a vast resource of information. The URLs (uniform resource locators) for World Wide Web sites can change. Using one of the search engines on the Internet such as Google or Yahoo!, find more information by searching for these words or phrases: plant growth and temperature, growing degree days, frost damage, plant dormancy, germination, thermoperiod, and weather and crop production.

Some examples of an Internet search:

- National Weather Service Climate Prediction Center: www.cpc.noaa.gov/products /analysis_monitoring/cdus/degree_days/
- Oregon State University Extension, Environmental Factors Affecting Growth: http://extension.oregonstate.edu/mg/botany/heat2.html
- West Virginia University Extension Service, Plant Growth and Development as the Basis of Forage Management: www.caf.wvu.edu/~forage/growth.htm
- University of Arizona, Biology, Seed Germination: http://biology.arizona.edu/sciconn /lessons2/Roxane/teach_sec.htm

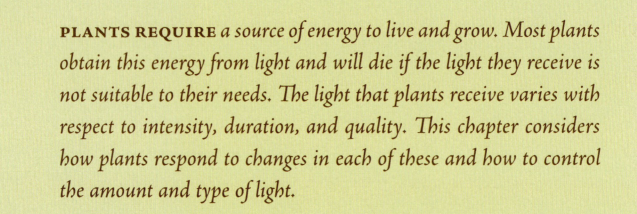

Chapter 12
Light

PLANTS REQUIRE *a source of energy to live and grow. Most plants obtain this energy from light and will die if the light they receive is not suitable to their needs. The light that plants receive varies with respect to intensity, duration, and quality. This chapter considers how plants respond to changes in each of these and how to control the amount and type of light.*

After completing this chapter, you should be able to—

- Describe the nature of light

- Explain absorption and reflection

- Describe light quality and light intensity

- Define the words used to describe the amount of sunlight a plant receives

- Define long-day, short-day, and day-neutral

- Give three examples of plants that are long-day, short-day, and day-neutral

- Define photoperiodism

- List factors to consider when using artificial illumination

- Describe light absorption and photomorphogenesis

- Name all the processes that light can affect

NATURE OF LIGHT

To understand how plants use light and how important light is to plants, some understanding of the nature of light is necessary. Light is a form of radiant energy. All radiant energy travels in waves, and the distance between consecutive crests of the waves is called wavelength (Figure 12-1). The human eye is sensitive to wavelengths of about 400 to 700 nanometers (nm). This portion of radiant energy is called white light. White light is made up of the colors red, orange, yellow, green, blue, and violet. Each of these colors has a different wavelength. Red is the longest wavelength and violet is the shortest. Ultraviolet and infrared are on the ends of the visible spectrum.

Figure 12-1 The distance between consecutive crests of waves is called wavelength.

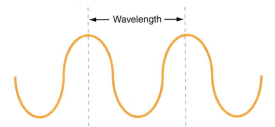

When light energy is trapped it is changed to another form of energy; for example, a dark color absorbs light, and the energy becomes heat energy. Not all light is absorbed when it strikes an object. Some is reflected. The reflected portion of the spectrum is the color of an object. A red object reflects red light; a green object reflects green and so on. Reflection and absorption of light are important concepts in understanding the trapping of light in photosynthesis.

Photosynthetic organisms contain the pigment chlorophyll that traps light energy and changes it to chemical energy. Chlorophyll a is the most common type; it has the chemical formula of $C_{55}H_{72}O_5N_4Mg$. The photosynthetic parts of plants appear green because chlorophyll reflects most of the green portion of the spectrum. Because certain wavelengths are absorbed by the chlorophyll, the spectrum produced is called an absorption spectrum. Violet and red wavelengths are greatly absorbed, as are some blue wavelengths. These absorbed wavelengths are changed from light energy to chemical energy.

SUNLIGHT

Plants require a minimum duration of light to survive. The life process of plants requires considerable time and some of them can be carried out only when it is light. The length of the days usually does not limit plant growth in some areas, but it may in some latitudes. The duration of light, or how long it lasts, varies with the time of the year and how far one is from the equator. Days are longer during the summer than the winter everywhere except right on the equator. Summer days become relatively longer and winter days relatively shorter as one moves away from the equator (Figure 12-2).

Figure 12-2 As the earth revolves around the sun, the length of day and intensity of light changes. As the earth tips on its axis, the days become longer or shorter and the angle at which the light strikes the earth changes.

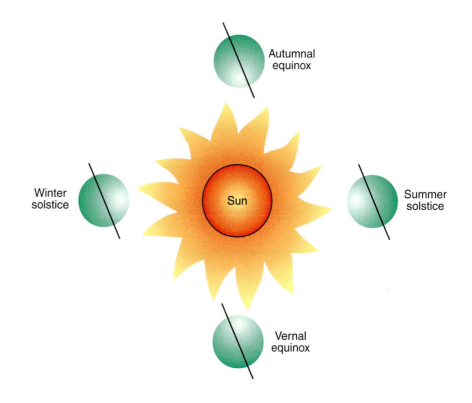

Light Quality and Intensity

The intensity of light can be similarly important. Most agricultural plants require six or more hours of full sunlight on sunny days if they are to develop respectable yields. The more shade they are exposed to, the lower their yield will be. Not only will yields be reduced but the characteristics of the plants will also change.

As tomato plants are grown in more and more shade, they produce fewer flowers, the stems elongate, and the leaves become smaller. Less branching occurs. Examination of the internal leaf cell structure shows that leaves grown in heavy shade have fewer cells and a thinner protective coating.

Other plants respond in similar fashion. If a plant grown in heavy shade is suddenly exposed to bright sunlight, the leaves may be unable to adjust to the bright light and they may die. This frequently occurs if houseplants are taken outside in the summer and placed directly in the sun.

Some plants grow well in the shade. Dogwoods and redbuds, for example, frequently grow beneath larger trees. Hostas, azaleas, and many other plants prefer shade to bright sunlight. Some of these plants will grow in bright sunlight but are easier to grow and more attractive when grown in shade or semishade.

Sometimes it is necessary to garden where areas of the garden are shaded for all or part of the day. All of the common vegetables prefer full sunlight, but some will tolerate shade better than others. Leafy vegetables tolerate shade better than rooting vegetables, and rooting vegetables do better in shade than fruiting vegetables. Table 12-1 describes the amount of solar radiation that plants receive.

Many plants can be observed to determine whether or not they are receiving enough light. An African violet that is grown in proper lighting

Table 12-1 Solar Radiation Description

Classification	Description
Shade	Usually an area under a closed tree canopy that receives no direct sun at all
Partial shade	Usually an area under a lone or limited number of trees receiving only one to five hours of dappled sun during the day
Partial sun	Usually refers to an area that receives approximately three to five hours of full sun per day
Full sun	Refers to an area receiving six or more hours of full sun per day

will have leaves held parallel to the surface of its container. If the light intensity is too high, the leaves bend down, and if the light intensity is too low, the leaves bend up toward the light.

Many plants are able to orient their leaves parallel to light that comes in greater intensity from one direction. This movement in response to light is called phototropism. Houseplants commonly show phototropic responses to light because of their exposure to uneven light. This can be prevented by turning the plant one-quarter turn every one or two days.

With small, rapidly growing plants the entire stem may bend toward the light. If the light intensity is low enough, the stems will elongate and the leaves will not develop their full size. This is a problem when seedlings are grown in shady locations. Seedlings can also shade each other, causing elongated stems and small leaves. The resulting plants or transplants are not as sturdy and are more easily damaged if exposed to severe conditions such as being transplanted into a garden or field.

Light and Germination

Seeds usually germinate underground. The young seedling lives on energy stored in the seed until it emerges above ground and receives light. The shoot grows up away from gravity and the root grows down toward gravity. This response to gravity is called geotropism. When the shoot reaches the light, it toughens, develops a green pigmentation (chlorophyll), and begins to manufacture food from minerals, water, and air, using sunlight for energy (photosynthesis). A seedling unable to reach the light soon dies. It may first send a long, thin, white shoot with almost no leaves toward any visible light or up into the air until its stored food supply is exhausted. A white stem with little leaf growth is said to be etiolated.

Light Absorption and Photomorphogenesis

Light absorption occurs in an array of pigment molecules that act like a radio or a television antenna in collecting radiant energy that falls on its surface. The energy is transferred at enormous speed to a processing center.

For television or radio, an antenna collects energy, and energy processing results in sound or a picture. In the plant, the collected energy is used to do chemical work that ultimately results in more plant material. The energy processing occurs in what is called a reaction center—a complex of proteins and pigments. Light energy collected in the antenna and delivered to the "reaction center" is used to make a chemical change—to push an electron away from one molecule into a neighboring molecule. This happens with dazzling speed—less than a picosecond (10^{-12}), which is one millionth of one millionth of a second. A quick reaction occurs when light strikes many molecules, but the usual result is for the electron to bounce back to its point of origin and no chemical change is accomplished.

Photosynthesis is unique because the energized electron is captured and stabilized with high efficiency. Part of the efficiency may come from the series of slightly slower steps that move the electron on to other neighboring molecules. Only recently, with the development of laser technology, could scientists measure changes that occur so rapidly. Now the picosecond measurements are allowing new understanding of the result of light absorption by a molecule. Part of the efficiency may depend on the way in which the complex molecules of the reaction center are fitted together.

Figure 12-3 outlines a pathway to stable products of photosynthetic energy conversion that are used in yet another complex process of conversion of carbon dioxide (CO_2) from the atmosphere into sugar molecules. **Photomorphogenesis** occurs as these molecules are converted into all of the other kinds of molecules—proteins, nucleic acids, fats, and so forth— that are the substance of the new plant.

The first step in this CO_2 fixation process is made possible by the enzyme **ribulose phosphate carboxylase**. It is the limiting reagent of photosynthetic CO_2 fixation. If there is more enzyme or if the enzyme works more efficiently, there is more CO_2 fixation and more plant production. It catalyzes the series of reactions indicated in the upper pathway, which also use compounds produced by light and which ultimately fix the CO_2 into the stable products of a growing plant. The alternative is for the enzyme to react with oxygen (O_2) instead of CO_2, as shown in the lower pathway of Figure 12-3, and the result is a lowering of CO_2 fixation and a decrease in productivity. This undesirable alternative often occurs in crop plants.

Scientists are learning the precise structure and the chemical details of functioning of this enzyme. There is a good possibility that the powerful tools of biotechnology—molecular biology and genetic engineering—will allow us to regulate the choice between these two alternative reactions and improve crop productivity at its most fundamental level.

Figure 12-3 Pathway to stable products of photosynthetic energy conversion.

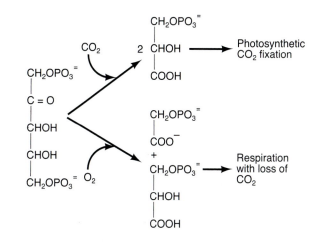

PHOTOPERIODISM

Many plants have adapted to changes in day length and respond to the new length. A response to the length of daylight is called photoperiodism. One of the main responses of plants to duration of daylight is flowering.

In general, plants have been divided into three different types according to their length of day requirements for flower initiation. These are long-day, short-day, and day-neutral types. In reality, these three types are misnamed, because it is the length of the uninterrupted night, the dark period, that initiates the floral buds. A long-day plant is really a short-night plant. The day period can be interrupted by a short dark period without affecting the plant, but if the dark period is broken by a short period of light, flowering will not take place. Table 12-2 lists the common and scientific names for some long-day, short-day, and day-neutral plants.

One interesting example of photoperiodism may be observed with the cocklebur. A cocklebur plant that begins growth in the long days and short nights of June will grow to a height of five to six feet before flowering. A cocklebur that germinates in September develops two leaves and then produces several seed pods at the six-inch height. This is the effect of the long night in initiating flower buds immediately after germination.

The chrysanthemum is also a short-day/long-night plant. By covering it with a black cloth in late afternoon, night length can be increased, and the plant can be brought into blossom at any time during the year. The plant must remain in a short-day or long-night environment until it reaches the desirable height for floral bud initiation, or the flowering plant will be very small.

Some plants are valuable only when they flower at the right time. Poinsettias are sold almost entirely for Christmas. Turning a light on in a greenhouse and interrupting the long night for only a short period of time can delay flowering enough to make the entire crop worthless. A streetlight near the greenhouse can have the same delaying effect.

Sometimes flowers must be avoided if a crop is to have value. Spinach is planted in the fall or early in the spring while nights are long; it is harvested before the change in the photoperiod causes flower bud formation or bolting.

There is variation in how much or little change in the photoperiod is needed to initiate flowers in different cultivars of a plant species. For example, we have Thanksgiving, Christmas, or Easter cactus depending on the time of the year the plant normally blooms.

Both long- and short-day varieties of some plants have been developed. Soybeans and onions are two examples. Different varieties are grown in the north and the south because of the differences in length of day. Soybean varieties grown in the wrong day length bloom too early or too late, depending on whether they are northern varieties being grown in the south or southern varieties grown in the north. Onions produce bulbs rather than leaves when the day length is correct. If the wrong variety of

Table 12-2 Partial List of Long-Day, Short-Day, and Day-Neutral Plants

Common Name	Scientific Name
Long-Day Plants	
Barley, winter	*Hordeum vulgare*
Bent grass	*Agrostis palustris*
Canary grass	*Phalaris arundinacea*
Chrysanthemum	*Chrysanthemum maximum*
Clover, red	*Trifolium pratense*
Coneflower	*Rudbeckia bicolor*
Dill	*Anethum graveolens*
Henbane, annual	*Hyoscyamus niger*
Oat	*Avena sativa*
Orchard grass	*Dactylis glomerata*
Ryegrass, early perennial	*Lolium perenne*
Spinach	*Spinacia oleracea*
Timothy, hay	*Phleum pratensis*
Wheatgrass	*Agropyron smithii*
Wheat, winter	*Triticum aestivum*
Day-Neutral Plants	
Bluegrass, annual	*Poa annua*
Buckwheat	*Fagopyrum tataricum*
Corn (maize)	*Zea mays*
Cucumber	*Cucumis sativus*
Fruit and nut tree species	
Globe-amaranth	*Gomphrina globosa*
Grapes	*Vitis spp.*
Kidney bean	*Phaseolus vulgaris*
Pea	*Pisum sativum*
Strawberry, everbearing	*Fragari x Anansia*
Tomato	*Lycopersicon lycopersicum*
Short-Day Plants	
Chrysanthemum	*Chrysanthemum x morifolium*
Cocklebur	*Xanthium strumarium*
Cotton, Upland	*Gossypium hirsutum*
Orchid	*Cattleya trianae*
Poinsettia	*Euphorbia pulcherrima*
Rice, winter	*Oryza sativa*
Soybean	*Glycine max*
Strawberry	*Fragaria x Ananasia*
Tobacco, Maryland Mammoth	*Nicotiana tabacum*
Violet	*Viola papilionacea*

Managing Solar Radiation

Most often, climate is controlled by adjusting the thermostat in the home. But the landscaping around the home has a large effect on temperature, air movement, and humidity. Vegetation is beneficial even in the winter to block the winds that remove heat from the home.

The most obvious use of trees for landscaping is to shade windows, walls, and the roof, reducing solar heat gain that penetrates to the interior of the home. Vegetation can be used in other ways to cool the home. For example, the area surrounding the home has a direct impact on air temperature. Paved surfaces, especially blacktop, absorb huge amounts of heat from the sun and radiate it back into the surrounding air. By shading these outdoor areas, the air temperature can be kept down. This results in lower indoor temperatures. A shaded grass lawn would be the best option to surround the home.

The cooling effect of vegetation is also enhanced by the process of evapotranspiration. A single tree absorbs as much as 600,000 Btus (British thermal units) of solar radiation every day and evaporates water into the air. This action maintains lower temperatures, which pushes the cooler air toward the ground and moves the warmer air up (by convection) and away from the home. Trees literally are nature's air conditioners.

During the heating season (winter), the heat gain from the sun is very beneficial. The ideal situation is to place trees so they provide shading during the cooling season, but not during the heating season. One way of doing this is by using deciduous trees—trees that loose their leaves during the winter. Even bare-branched trees can block 30–60 percent of the desired sunshine in the winter.

The best approach to managing solar radiation is to have a good understanding of the sun's path during the summer and the winter. During the summer, the sun rises far to the northeast, is very high in the sky during the middle of the day, and sets in the northwest; this provides about 13 hours of daylight at the summer solstice. At the winter solstice, the sun rises in the southeast, is much lower in the sky during the middle of the day, and sets in the southwest, which provides about nine hours of sunlight. With this understanding, trees can be located to get the best use of their cooling ability.

onion is grown, it may produce a bulb that is very small because the plant has not grown enough, or it may produce a huge plant and then not have time to produce a bulb before winter.

ARTIFICIAL ILLUMINATION

All artificial light looks about the same but really varies in quality. Incandescent bulbs emit very heavily in the red end of the spectrum. Fluorescent lights produce more blue light. Many plants grow poorly, if at all, under incandescent light and only a little better under fluorescent light. Mixing lights in a ratio of 3 watts of fluorescent light to 1 watt of incandescent light provides a quality of lights under which most plants grow much better. There are also special lights designed for plant growth. These produce a quality of light that is more like natural sunlight.

Light Color and Plant Growth

Light quality refers to the color or wavelength reaching the plant surface. Sunlight can be broken up by a prism into respective colors of red, orange, yellow, green, blue, indigo, and violet. On a rainy day, raindrops act as tiny prisms and break the sunlight into these colors, producing a rainbow. Red and blue lights have the greatest effect on plant growth. Green light is least effective to plants as they reflect green light and absorb none of the light. This reflected light makes them appear green to the human eye. Blue light is primarily responsible for vegetative growth or leaf growth.

Red light, when combined with blue light, encourages flowering in plants. Fluorescent light or cool white is high in the blue range of light quality and is used to encourage leafy growth. Such light would be excellent for starting seedlings. Incandescent light is high in the red or orange range, but generally produces too much heat to be a valuable light source. Fluorescent grow lights have a mixture of red and blue colors that attempt to imitate sunlight as closely as possible. However, these special lights are costly and generally not of any greater value than regular fluorescent lights.

Growing Plants under Artificial Lights

Light is probably the most essential factor for indoor or greenhouse plant growth (Figure 12-4). The growth of plants and the length of time they remain active depend on the amount of light they receive. Light is necessary for all plants because they use this energy source to photosynthesize. When examining light levels, three aspects of light are considered:

+ Intensity
+ Duration
+ Quality

Light intensity influences the manufacture of plant food, stem length, leaf color, and flowering. A geranium grown in low light tends to be spindly

Figure 12-4 Natural lighting in a greenhouse.

and the leaves will be light green in color. A similar plant grown in very bright light would tend to be shorter, better branched, and have larger, dark green leaves. Indoor plants can be classified according to their light needs, such as high, medium, and low light requirements. The intensity of light a plant receives indoors depends on the nearness of the light source to the plant. Light intensity decreases rapidly as plants are moved away from the source of light. The location of the windows will affect the intensity of natural sunlight that plants receive. Southern exposures have the most intense light, eastern and western exposures receive about 60 percent of the intensity of southern exposures, and northern exposures receive 20 percent of a southern exposure. A southern exposure is the warmest, eastern and western are less warm, and a northern exposure is the coolest.

Other factors that can influence the intensity of light penetrating a window are the presence of curtains, trees outside the window, weather, seasons of the year, shade from other buildings, and the cleanliness of the window. Reflective (light-colored) surfaces inside the building will increase the intensity of light available to plants. Dark surfaces will decrease light intensity.

Day length or duration of light received by plants is also of some importance but generally only to those house plants that are photosensitive. Poinsettia, kalanchoe, and Christmas cactus bud and flower only when day length is short (11 hours of daylight or less). Most flowering houseplants are indifferent to day length.

Low light intensity can be compensated by increasing the time (duration) the plant is exposed to light, as long as the plant is not sensitive to day length in its flowering response. Increased hours of lighting allow the plant to make sufficient food to survive and/or grow. However, plants require some period of darkness to develop properly and thus should be illuminated for no more than 16 hours. Excessive light is as harmful as too little light. When a plant gets too much direct light, the leaves become pale and sometimes sunburned, turn brown, and die. During the summer months, plants need protection from too much direct sunlight.

Additional lighting may be supplied by either incandescent or fluorescent lights. Incandescent lights produce a great deal of heat and are not very efficient users of electricity. If artificial lights are to be used as the only source of light for growing plants, the quality of light (wavelength) must be considered. For photosynthesis, plants require mostly blues and reds, but for flowering, infrared light is also needed. Incandescent lights produce mostly red light, and some infrared light, but are very low in blue light. Fluorescent lights vary according to the phosphorus used by the manufacturer. Cool white lights produce mostly blue light and are low in red light. Foliage plants grow well under cool white fluorescent lights, and these lights are cool enough to position quite close to plants. Blooming plants require extra infrared, which can be supplied by incandescent lights or special horticultural-type fluorescent lights.

SUMMARY

Light affects plants based on its intensity, duration, and quality. Growers exercise little control over the quality when the source of light is the sun, but greenhouse growers can change the quality by using artificial light. Normal plant growth requires white light or sunlight. Chlorophyll absorbs the red and blue portions of the light spectrum and appears to be green because the leaf reflects green light. Light quality is important; it must contain the essential wavelengths that are represented by the colors of the visible spectrum.

Light intensity provides energy for photosynthesis. The rate of photosynthesis is affected by the availability of water, CO_2, and sunlight. In absence of light, plants will grow until their food reserves are exhausted. The growth will be abnormal and result in an elongated plant. Phototropism in plants is the tendency for the plant to lean in the direction of the greatest light intensity. Plants vary in the intensity of light they need. Some plants such as impatiens grow better in the shade with indirect light, and others such as zinnias grow best in full sunlight.

Photoperiodism is the growth response to the length of the dark period. This affects whether a plant is growing vegetatively or is in the flowering stage. In a greenhouse, plants can be forced to bloom by controlling the duration of light. Types of plants include short-day, long-day, and day-neutral plants. In a greenhouse, light can be controlled to influence blooming or vegetative growth by using a black cover to shorten the days or by adding artificial light in the evening hours to simulate long days.

Review

True or False

1. Plants appear green because they absorb all of the green wavelengths from the visible spectrum of light.

2. The chemical formula for chlorophyll is $C_{72}H_{55}O_4N_4Mg$.

3. Some plants prefer shade.

4. Photoperiodism is the response of a plant to a photograph.

5. Light duration can be altered in a greenhouse.

Short Answer

6. Name the colors in white light.

7. Which colors of white light do plants absorb?

8. Name three long-day, three short-day, and three day-neutral plants.

9. List three factors to consider when placing plants under artificial light.

10. What colors of light encourage flowering?

Critical Thinking/Discussion

11. Discuss the effect shade can have on plants.

12. Explain photomorphogenesis.

13. Describe photoperiodism and give examples of it.

14. Explain absorption and reflection as it relates to plants.

15. Define light wavelength.

Knowledge Applied

1. Design and conduct an experiment on the type of light used to grow plants. Compare growth under incandescent and fluorescent lights. If possible, create some red and green filters to put over the lights. Report on the design and results of your experiment.

2. Grow a potted plant in normal light. Then place it in total darkness and check on it each day. Report on your findings each day. After the plant changes color from the darkness, put it back in full light and observe the changes. Report your findings.

3. Visit a greenhouse and ask what types of lighting schemes they use for the plants they are growing.

4. Use a prism to show the colors in sunlight, fluorescent, and incandescent lighting.

5. Develop a presentation using computer software to explain a nanometer and a picosecond. Tell how and where these measures are used and give some comparisons as to their size.

Resources

Asimov, I. (1966). *Understanding physics.* Cutchogue, NY: Buccaneer Books, Inc.

Cromer, A. H. (1994). *Physics for the life sciences* (2nd ed.). New York: McGraw-Hill.

Janick, J. (Ed.). (1989). *Classic papers in horticultural science.* Englewood Cliffs, NJ: Prentice Hall.

McMahon, M., Kofranek, A. M., & Rubatzky, V. E. (2006). *Hartmann's plant science: Growth, development, and utilization of cultivated plants* (4th ed.). Englewood Cliffs, NJ: Prentice Hall Career & Technology.

Internet

Internet sites represent a vast resource of information. The URLs (uniform resource locators) for World Wide Web sites can change. Using one of the search engines on the Internet such as Google ore Yahoo!, find more information by searching for these words or phrases: nature of light, light wavelength, light color and plant growth, light and plant growth, light requirements for plant growth, absorption and reflection of light, photoperiodism in plants, photomorphogenesis, greenhouse lighting, and chlorophyll.

Some examples of an Internet search:

✦ Iowa State University—Photomorphogenesis: www.public.iastate.edu/~bot.512 /lectures/Photo.htm

✦ Indiana University—Photomorphogenis: www.bio.indiana.edu/~hangarterlab/courses /b373/lecturenotes/photomorph/photomorph1.html

✦ Wikipedia—Photoperiodism: http://en.wikipedia.org/wiki/Main_Page

✦ The Physics Hypertextbook—Nature of light: http://hypertextbook.com/physics /waves/light/

Chapter 13
Photosynthesis

PHOTOSYNTHESIS IS the "reaction of life." The conversion of the sun's energy into a form that man and other living creatures can use is done almost entirely by plants through photosynthesis. An easy way to remember what photosynthesis accomplishes is to remember the root meanings of the words "photo" (light) and "synthesis" (to put together). In the process of photosynthesis, a plant uses the energy of light to put together chemicals that are useful to itself and, consequently, to other living things. Photosynthesis is the original source of all important fuels including oil, coal, wood, and natural gas. It is also the source of all foods because all animals are dependent on plants at some point in the food chain.

After completing this chapter, you should be able to—

- Discuss the importance of photosynthesis and carbohydrate production to the global ecology and to agricultural crop plant production

- Describe the general process of carbon fixation by plants

- List five important steps or control points in photosynthesis

- Name five environmental and agricultural management factors and practices that affect photosynthesis

- Explain the differences between the light and dark reactions

- Describe the difference between the C-3 and the C-4 pathways in photosynthesis

- Write the general chemical reaction for photosynthesis and explain it in words

- List four requirements of photosynthesis

- Explain why plants are considered a carbon sink

- Name the products of photosynthesis

- Detail a photosystem

KEY TERMS

ATP
Calvin Cycle
Carbohydrates
Carbon Sinks
Chlorophyll
Chloroplast
Dark Reaction
Electron Transport
Energy
Enzyme
Grana
Light Reaction
NADPH
Oxaloacetic Acid
Photoelectric Effect
Photophosphorylation
Photosynthates
Photosynthesis
Photosystems
Phototropism
Ribulose Biphosphate
Starches
Sugars
Thylakoid
Wavelength

NATURE OF LIGHT

White light is separated into the different colors (**wavelengths**) of light by passing through a prism. The order of colors is determined by the wavelength of light (Figure 13-1). Visible light is one small part of the electromagnetic spectrum. The longer the wavelength of visible light, the more red the color appears. Likewise, the shorter wavelengths are toward the violet side of the spectrum. Wavelengths longer than red are referred to as infrared, and those shorter than violet are referred to as ultraviolet.

Figure 13-1 A field of sunflowers capturing sunlight.

Light behaves both as a wave and as a particle. Wave properties of light include the bending of the wave path when passing from one material (medium) into another. The particle properties are demonstrated by the **photoelectric effect**. Zinc exposed to ultraviolet light becomes positively charged because light **energy** forces electrons from the zinc. These electrons can create an electrical current. Sodium, potassium, and selenium have critical wavelengths in the visible light range. The critical wavelength is the maximum wavelength of light (visible or invisible) that creates a photoelectric effect. Albert Einstein developed a theory in 1905 that light was composed of particles—photons—whose energy is inversely proportional to the wavelength of the light. Light thus has properties explainable by the wave model and also by the particle model.

STRUCTURE OF THE PLANT PHOTOSYSTEM

The **thylakoid** is the structural unit of **photosynthesis**. Photosynthetic eukaryotes have these flattened sacs/vesicles containing photosynthetic chemicals. Thylakoids are stacked like pancakes in stacks known collectively as **grana** (Figure 13-2). The areas between grana are referred to as stroma. Although the mitochondrion has two membrane systems, the chloroplast has three, forming three compartments. The **chloroplast** contains the **chlorophyll**.

Chlorophyll is a complex molecule. Several modifications of chlorophyll occur among plants and other photosynthetic organisms. All photosynthetic organisms have chlorophyll a. Accessory pigments absorb energy that chlorophyll a does not absorb. Accessory pigments include chlorophyll b (also c, d, and e in algae and protistans), xanthophylls, and carotenoids (such as beta-carotene). Chlorophyll a absorbs the majority of its energy from the violet-blue and reddish orange-red wavelengths, and little from the intermediate (green-yellow-orange) wavelengths. Carotenoids and chlorophyll b absorb in the green wavelength.

The action spectrum of photosynthesis is the relative effectiveness of different wavelengths of light at generating electrons. If a pigment absorbs light energy, one of three things will occur; energy is dissipated as heat; the energy may be emitted immediately as a longer wavelength, a phenomenon known as fluorescence; or energy may trigger a chemical reaction, as in photosynthesis. Chlorophyll only triggers a chemical reaction when it is associated with proteins embedded in chloroplast.

Figure 13-2 Thylakoid is the structural unit of photosynthesis.

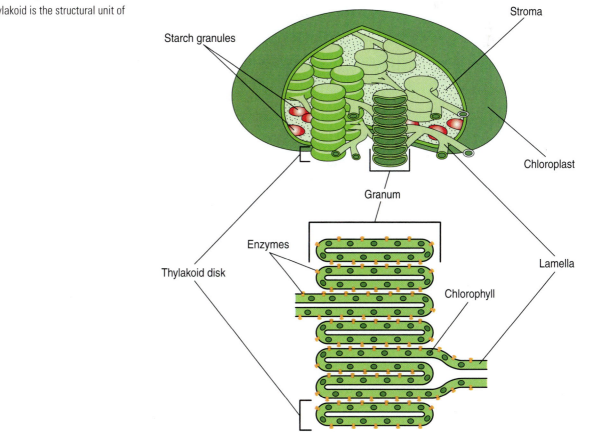

CHANGING LIGHT ENERGY INTO CHEMICAL ENERGY

Photosynthesis accomplishes three important processes in plants. First, it transforms light energy from the sun into chemical energy, which can be transported and stored in plants. Second, it "fixes" carbon (from carbon dioxide in the atmosphere) into a solid form. This is the most important component in all organic (living) matter. Finally, photosynthesis produces oxygen as an important by-product.

Energy Transformation

The earth receives vast quantities of energy from the sun each day in the form of solar radiation (light). Virtually all of this would be reradiated into the darkness of space, without serving any purpose but to heat the surface of the earth, except plants capture a portion of this energy. This is done through photosynthesis, in which radiation energy is changed into a useful form. This useful form of energy is called chemical energy. One energy-yielding chemical that photosynthesis produces happens to be sugar ($C_6H_{12}O_6$). Humans use many other chemicals that contain energy. These chemicals may be in solid, liquid, or sometimes gaseous form and are easy to transport and handle. These are useful to us much more so than radiation energy, which is diffuse and difficult to contain. Figure 13-3 summarizes how energy is transformed during photosynthesis.

Requirements and Products of Photosynthesis

The requirements for photosynthesis are—
- A living healthy plant
- An ample supply of carbon dioxide (CO_2) from the atmosphere
- Water (H_2O) from the soil or atmosphere
- Light, usually from the sun

The products of photosynthesis are **carbohydrates**, such as **sugars** and **starches** (CHOs), other complex compounds referred to collectively as **photosynthates**, water (H_2O), and oxygen (O_2).

Photosynthesis is a collection of many complex reactions. These reactions can be summarized into one basic reaction expressed as follows:

$$6CO_2 + 12H_2O \xrightarrow[\text{Light Energy}]{\text{Chlorophyll}} C_6H_{12}O_6 + 6H_2O + 6O_2$$

$$(\text{Carbon Dioxide} + \text{Water} \longrightarrow \text{Sugar} + \text{Water} + \text{Oxygen})$$

This expression is read: Six molecules of carbon dioxide plus 12 molecules of water combine, in the presence of a green plant and light energy, to form one molecule of sugar plus six molecules of water and six molecules of oxygen.

Figure 13-3 How energy is transformed during photosynthesis.

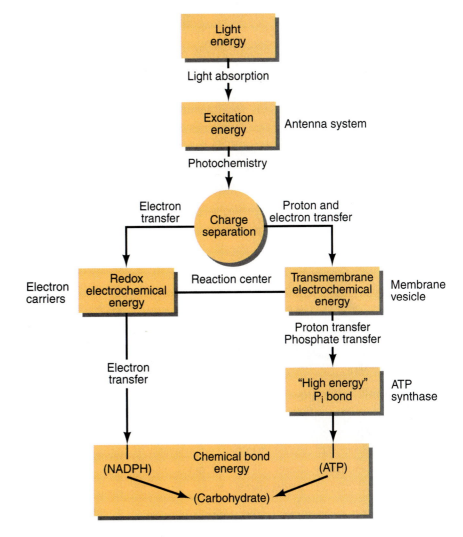

Water appears on both sides of the equation, so often the reaction for photosynthesis is expressed as:

$$6CO_2 + 6H_2O \xrightarrow[\text{Light Energy}]{\text{Chlorophyll}} C_6H_{12}O_6 + 6O_2$$

In this equation, six waters have been subtracted from each side of the equation, leaving an expression of the net photosynthesis reaction.

Carbon Fixation

By this process of photosynthesis, vast quantities of carbon are "fixed" from atmospheric carbon dioxide and converted into carbohydrates by green plants. An estimated 620,000,000,000 tons of carbon are converted by plants from the earth's atmosphere annually. This carbon would be equivalent to enough coal to fill 97 railroad cars every second of every hour of every day all year long. Most of this energy is used

Miracle of Hybrid Corn–Improving on Photosynthesis

The widespread use of hybrids, coupled with improved cultural practices by farmers, has more than tripled corn grain yields in the United States over the past 70 years from an average 35 bushels per acre in the 1930s to the current 115 bushels per acre. No other major crop anywhere in the world even comes close to equaling that sort of success story.

A major contributing factor to these increasing yields has been continued genetic improvement in the performance potential of corn hybrids.

Hybrid corn is the first generation grown from seed produced by crossing unrelated parents, which are generally inbreds or single crosses. In 1908, G. H. Shull of Carnegie Institute proposed the development of inbreds by self-pollination—pollen from the tassel placed on the silks of the same plant. This procedure is called "selfing-or-inbreeding."

The genetic variation within progenies (descendents) is reduced by half after each generation of selfing so that within six to seven generations, inbreds (pure lines) result that are uniform and transmit their characteristics consistently to their next generation. The inbreds are then tested in crosses with one another to find the best cross or hybrid that could be grown by the farmer.

As is true of most scientific proposals, acceptance of hybrid corn and the necessary developmental work were slow in coming. The primary reason for this delay was the apparent economic impracticality of producing hybrid seed because of the low plant vigor and low seed yield of the inbred parents.

In 1922, D. F. Jones, University of Kentucky, proposed using double-cross hybrids to facilitate economical seed production. The concept involved crossing inbreds to produce single crosses, which were subsequently crossed to make double crosses. After yield testing, the best-performing double-cross hybrids could then be selected for production of seed that would be sold to farmers.

Another decade passed before corn-breeding research developed superior double-cross hybrids. From that time on, however, use of hybrids expanded rapidly so that by the early 1940s the majority of corn grown in the United States was hybrid.

Between the early 1930s and early 1950s, hybrid seed was usually produced in fields isolated from other corn by at least 40 rods. The female and male single-cross parents were planted in alternating blocks of six and two rows, respectively. Tassels of corn in the female rows were removed by hand just prior to silk emergence so that pollen from corn in the male rows pollinated the female parent ears to produce the hybrid seed.

The seed was harvested "on the ear" and dried on forced-air driers at a temperature of about 110°F. Off-type and poor-quality ears were discarded before shelling. The seed was then cleaned and graded by shape ("flats" and "rounds") and by size to facilitate planting specific populations with various planter plates. It was also treated with a fungicide to control seedling diseases.

During this period, most of the inbreds and hybrids were developed by scientists in the U.S. Department of Agriculture and at land-grant universities. Also, each state usually published a list of recommended hybrids and "certified" their production as to proper removal of tassels, isolation, purity, and so on.

During the 1940s, however, many seed companies started their own corn-breeding programs. This resulted in a trend by companies to produce only "closed-pedigree" hybrids; the inbred parents were not made public. This trend continued so that by the late 1950s only a few open-pedigreed hybrids were still being marketed.

The early 1950s marked the introduction of two significant cultural practices relative to corn: increased fertilizer rates (especially nitrogen) and higher plant densities. Corn breeders responded quickly to these changes by developing hybrids that possessed both (1) greater yield potential at high fertility and high plant population levels and (2) increased stalk rot resistance and strength for better stability.

Around 1960, an even more important cultural practice change took place; a shift to using single-cross hybrids, which are generally superior to double-crosses for yield and other performance characteristics. Seed of a single-cross hybrid is usually more expensive because its inbred parents are lower-yielding and more susceptible to environmental stresses than are the single-cross

continues

Miracle of Hybrid Corn–Improving on Photosynthesis, *continued*

seed parents used in producing a double-cross hybrid. However, most farmers found that they were justified in paying extra for single-cross seed.

Today, about 80 percent of the hybrid-seed corn sold in the United States is single-cross, with the remaining seed being double-, three-way, and modified (related-line parents) crosses. These hybrids that are other than single-cross hybrids are used more frequently in marginal corn-growing areas because of lower seed cost and the lower yield potential in these areas.

The majority of corn-breeding research today is conducted by private seed companies. Over the past 10 years, these companies have continued to commit more and more staff and resources to corn hybrid improvement. Although scientists at public institutions are also involved in breeding research, their efforts are now directed to more basic areas, such as development of germplasm, resistance to pests and stress mechanisms, and evaluation of selection methods. In spite of this shift to basic research, publicly-developed hybrids are still widely used by seed companies in commercial hybrids.

The ultimate goal of a corn-breeding program is to develop a new, commercial hybrid possessing such desirable characteristics as

improved yield, standability, disease and insect resistance, and stress tolerance. Accomplishment of this goal requires the coordinated efforts of breeders, pathologists, entomologists, physiologists, and others, and involves the following phases:

1. Selection and development of source germplasm.
2. Development of improved inbreds.
3. Testing of inbreds in experimental hybrids.

Hybrid corn is a success story that along with other improved cultural practices, has enabled American farmers to triple their corn yields over the past 70 years. Current research and data indicate that by using conventional plant-breeding methods, hybrid performance improvement will continue and probably at an accelerated rate. Development of additional genetic tools and breeding methods should contribute to even higher yield, standability, and stress resistance over time.

The greatest factor ensuring future availability of improved corn hybrids is the intense competition that exists among seed companies. These companies must continue to offer high-quality seed of competitive or superior hybrids in order to maintain or increase their market share.

by plants themselves in conducting their life processes and by other creatures using plants as a food source.

Oxygen Production

Although oxygen is considered a by-product of the photosynthetic process, plants produce enough of it to be considered the primary source of oxygen worldwide. Without the rain forests of the tropics and the vast populations of phytoplankton in the oceans, humankind would not have the oxygen that is so necessary to sustain respiration processes and animal life.

Stages of Photosynthesis

Photosynthesis is a two-stage process. The first stage is the light-dependent process. This stage requires the direct energy of light to make energy-carrier molecules that are used in the second stage. The light-independent process (or dark reaction) occurs when the products of the light reaction are used to form carbon to carbon (C-C) covalent bonds of carbohydrates. The dark reactions can usually occur in the dark, if the energy carriers from the light process are present. The light reaction occurs in the grana and the dark reaction in the stroma of the chloroplasts.

Light Reaction

In the light-dependent processes, light strikes chlorophyll a in such a way as to excite electrons to a higher energy state. In a series of reactions the energy is converted (along an electron transport-like process) into ATP (adenosine triphosphate) and NADPH (nicotinamide adenine dinucleotide phosphate). Water is split in the process, releasing oxygen as a by-product of the reaction. The ATP and NADPH are used to make C-C bonds in the dark reaction.

In the light-independent process, carbon dioxide from the atmosphere is captured and modified by the addition of hydrogen (H) to form carbohydrates. The general formula of carbohydrates is $(CH_2O)_n$. The incorporation of carbon dioxide into organic compounds is known as carbon fixation. The energy for this comes from the first phase of the photosynthetic process. Living systems cannot directly use light energy, but can through a complicated series of reactions convert it into C-C bond energy that can be released by glycolysis (breakdown of glucose) and other metabolic processes.

Photosystems are arrangements of chlorophyll and other pigments packed into thylakoids. Eukaryotes have photosystem II plus photosystem I. Photosystem I uses chlorophyll a, in the form referred to as P700. Photosystem II uses a form of chlorophyll a known as P680. Both "active" forms of chlorophyll a function in photosynthesis because of their association with proteins in the thylakoid membrane.

Photophosphorylation is the process of converting energy from a light-excited electron into the pyrophosphate bond of an ADP molecule. This occurs when the electrons from water are excited by the light in the presence of P680. The energy transfer is similar to the electron transport occurring in the mitochondria. Light energy causes the removal of an electron from a molecule of P680 that is part of photosystem II. The P680 requires an electron, which is taken from a water molecule, breaking the water into hydrogen ions (H^+) and oxygen (O) ions. These O ions combine to form the diatomic O_2 that is released. The electron is "boosted" to a higher energy state and attached to a primary electron acceptor, which begins a series of redox reactions, passing the electron through a series of electron carriers, eventually attaching it to a molecule in photosystem I.

Light acts on a molecule of P700 in photosystem I, causing an electron to be boosted to a still higher potential. The electron is attached to a different primary electron acceptor (i.e., a different molecule from the one associated with photosystem II). The electron is passed again through a series of redox reactions, eventually being attached to $NADP^+$ and H^+ to form NADPH, an energy carrier needed in the light-independent reaction. The electron from photosystem II replaces the excited electron in the P700 molecule, so a continuous flow of electrons from water to NADPH occurs. This energy is used in carbon fixation. In photosystem II, the pumping of H ions into the thylakoid and the conversion of ADP + P into ATP is driven by electron gradients established in the thylakoid membrane. Figure 13-4 illustrates the reactions in both photosystems.

Dark Reaction

Carbon-fixing reactions are also known as the dark- or light-independent reactions. Land plants must guard against drying out (desiccation) and so have evolved specialized cells known as stomata to allow gas to enter and leave the leaf. The **Calvin Cycle** occurs in the stroma of chloroplasts.

Figure 13-4 Reactions to photosystem I and II.

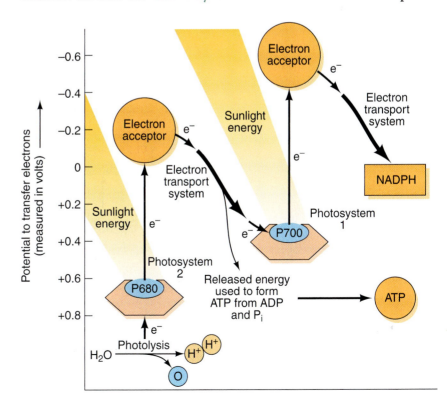

Carbon dioxide is captured by the chemical **ribulose biphosphate** (RuBP). RuBP is a five-carbon chemical. Six molecules of carbon dioxide enter the Calvin Cycle, eventually producing one molecule of glucose. The first stable product of the Calvin Cycle is phosphoglycerate (PGA), a three-carbon chemical.

The energy from ATP and NADPH energy carriers generated by the photosystems is used to attach phosphates to the PGA. Eventually 12 molecules of glyceraldehyde phosphate (GAP, a three-carbon chemical) are produced—two of which are removed from the cycle to make a glucose. The remaining GAP molecules are converted by ATP energy to reform six RuBP molecules, and thus start the cycle again. As in metabolic pathways like the Krebs cycle, each step is catalyzed by a different reaction-specific **enzyme**. These reactions are summarized in Figure 13-5.

C-4 Pathway

Some plants have developed a preliminary step to the Calvin Cycle (which is also referred to as a C-3 pathway); this preamble step is known as C-4. While most carbon-fixation begins with RuBP, C-4 begins with a new molecule, phosphoenolpyruvate (PEP), a three-carbon chemical that is converted into **oxaloacetic acid** (OAA, a four-carbon chemical) when carbon dioxide is present with PEP. The OAA is converted to malic acid and then transported from the mesophyll cell into the bundle-sheath cell, where OAA is broken down into PEP plus carbon dioxide. The carbon dioxide then enters the Calvin Cycle, with PEP returning to the mesophyll cell. The resulting sugars are now adjacent to the leaf veins and can be transported readily throughout the plant.

Figure 13-5 Sugar and starch synthesis.

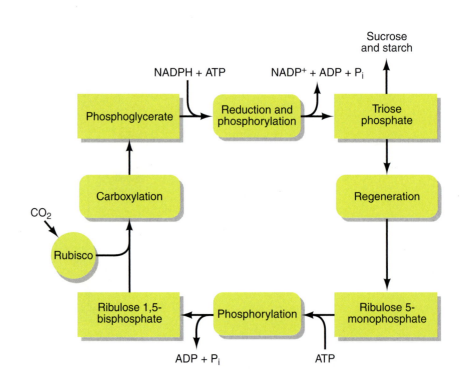

The capture of carbon dioxide by PEP is mediated by the enzyme PEP carboxylase, which has a stronger affinity for carbon dioxide than does RuBP carboxylase. When carbon dioxide levels decline below the threshold for RuBP carboxylase, RuBP is catalyzed with oxygen instead of carbon dioxide. The product of that reaction forms glycolic acid, a chemical that can be broken down by photorespiration, producing neither NADH nor ATP, in effect dismantling the Calvin Cycle.

The C-4 plants, which often grow close together, have had to adjust to decreased levels of carbon dioxide by artificially raising the carbon dioxide concentration in certain cells to prevent photorespiration. The C-4 plants evolved in the tropics and are adapted to higher temperatures than are the C-3 plants found at higher latitudes. Common C-4 plants include crabgrass, corn, and sugar cane.

Chloroplasts

Photosynthetic chemical activity takes place within plant organelles called chloroplasts. Chloroplasts are present mostly in leaves, but are also present in green stems and unripe fruit. In order to take place, photosynthesis depends on the presence of the chemical chlorophyll (which is green) within the chloroplasts (Figure 13-6).

Figure 13-6 Chloroplasts.

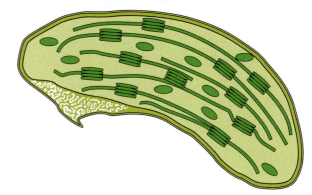

Efficiency of Photosynthesis

Light energy is not converted very efficiently to chemical energy during photosynthesis (usually between 0.1 and 3.0 percent), but it is going on constantly when the needed precursors for photosynthesis are present. As previously explained, this adds up to a tremendous annual yield of energy and plant material.

FACTORS AFFECTING PHOTOSYNTHESIS

Six factors affect photosynthesis: light quality, light intensity, light duration, carbon dioxide concentration, temperature, and water availability.

Light Quality (Wavelength)

Chlorophyll reacts only to certain wavelengths of light. Specifically, light in the blue and red wavelength range (340–700 nm) are the types of light most necessary for photosynthesis.

Light Intensity

Plants react differently to different levels of light; some plants are sun-loving and others prefer shade. As a general rule, however, the brighter the light, the more efficient the photosynthesis. In absence of light, plants will grow until their food reserves are exhausted. The growth will be abnormal and result in an elongated plant. Phototropism in plants is the tendency to "lean" in the direction of the greatest light intensity (Figure 13-7.) Because plants vary in the intensity of light they need, some grow better in shade with indirect light and others grow best in full sunlight.

Figure 13-7 An explanation of phototropism.

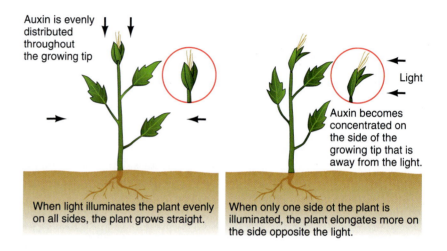

Auxin is evenly distributed throughout the growing tip

Light

Auxin becomes concentrated on the side of the growing tip that is away from the light.

When light illuminates the plant evenly on all sides, the plant grows straight.

When only one side of the plant is illuminated, the plant elongates more on the side opposite the light.

Light Duration (Day Length)

Photosynthetic production is directly proportional to the length of the day. The longer the day, the more photosynthesis takes place, and the more the plant grows. Photoperiodism is the growth response to the length of the dark period and affects whether a plant is growing vegetatively or is in the flowering stage. Plants can be forced to bloom in a greenhouse by controlling the duration of light. A plant's reaction to the duration of light can be classified as short-day, long-day, and day-neutral (see Table 12-2).

Carbon Dioxide Concentration

This has a profound effect on photosynthesis. Many plant scientists consider carbon dioxide availability to be the limiting factor in the photosynthetic process in most cases. Raising the CO_2 level from the usual atmospheric concentration of 0.03 percent to about 0.10 percent can double the rate of photosynthesis. Of course, it is impossible to raise the CO_2 level in most crop-growing situations. Greenhouse growers of roses, carnations, and other crops significantly increase their production by enriching the CO_2 concentration in the greenhouse.

Temperature

Except under very low light conditions, photosynthesis increases with temperature. In fact, for every 18°F increase in temperature, the level of photosynthesis will double. However, in some species of plants, excessively high temperatures will slow photosynthesis by causing stomata to close and gas exchange to cease.

Water Availability

Plants suffering water stress (lack of water) will close their stomata and photosynthetic activity will slow or cease.

THE CARBON CYCLE

Figure 13-8 shows the carbon cycle. Plants may be viewed as **carbon sinks**, removing carbon dioxide from the atmosphere and oceans by fixing it into organic chemicals. Plants also produce some carbon dioxide by their respiration but this is quickly used by photosynthesis. Plants also convert energy from light into chemical energy of C-C covalent bonds. Animals are carbon dioxide producers that derive their energy from carbohydrates and other chemicals produced by plants through the process of photosynthesis.

The balance between the plant and animal removal and generation (respectively) of carbon dioxide is equalized also by the formation of carbonates in the oceans. This removes excess carbon dioxide from the air and water (both of which are in equilibrium with regard to carbon dioxide). Fossil fuels, such as petroleum and coal, as well as more recent fuels such as peat and wood generate carbon dioxide when burned. Fossil fuels are formed ultimately by organic processes and also represent a tremendous carbon sink. Human activity has greatly increased the concentration of carbon dioxide in air. This led some scientists to suggest the theory of global warming—an increase in temperatures around the world, or the greenhouse effect. The increase in CO_2 and other pollutants in the air has also led to acid rain in some areas, where water falls through polluted air and chemically combines with carbon dioxide, nitrous oxides, and sulfur

Figure 13-8 The carbon cycle.

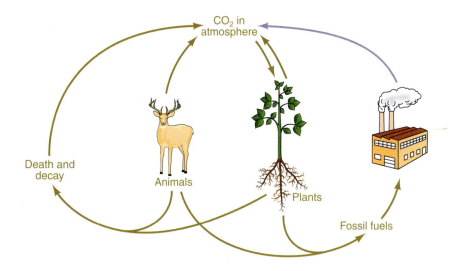

Figure 13-9 Some of the products of photosynthesis—fruits, seeds, and vegetative tissues. (Courtesy USDA ARS)

oxides, producing rainfall with pH as low as 4. This results in fish kills and changes in soil pH, which can alter the natural vegetation and uses of the land.

STORAGE OF THE PRODUCTS OF PHOTOSYNTHESIS

The glucose manufactured during photosynthesis does not accumulate to any marked degree in most green plant cells as glucose. Instead, it is used as a source of chemical energy by the plant or it is transformed into other biochemicals. Some of the glucose is converted to sucrose; some is converted to starch, which is stored in plant cells. Starch is found in seeds and roots and tubers. Some of the glucose is converted to cellulose and used in the construction of plant cell walls. Another portion of the glucose goes through a series of biochemical transformations and is converted into fats and oils. These are often stored in the seeds. Some of the breakdown products of glucose react with nitrogen and sulfur to form amino acids, which combine and form proteins. These concentrate in the active growing portions of the plant and in the seed (Figure 13-9).

SUMMARY

Photosynthesis makes plants unique. Because of their ability to produce their own energy directly, plants are self-sufficient; they are autotrophic. All other forms of life, including animals, fungi, bacteria, and even viruses, derive their sustenance from other living creatures. Plants are self-sufficient in terms of energy, and can survive independently of other living things.

Photosynthesis captures the light energy. The thylakoid is the structural unit of photosynthesis; they are stacked in grana. Light strikes chlorophyll and excites electrons to a higher energy state. Electrons move through an electron-transport process. Water is split and oxygen is released, and the energy stored in ATP and NADPH is used to make carbon-to-carbon bonds and eventually glucose. Photosynthesis has a series of light and dark reactions. Six factors can affect photosynthesis. Because plants convert tons of carbon, they are considered a carbon sink.

Review

True or False

1. Photosynthesis is a collection of a very few complex reactions.

2. The incorporation of carbon dioxide into organic compounds is known as carbon fixation.

3. Photosynthetic chemical activity takes place within plant organelles called chloroplasts.

4. Plants are self-sufficient, or autotrophic.

Short Answer

5. List five important steps or control points in photosynthesis.

6. Name four requirements of photosynthesis.

7. Give the three important processes photosynthesis accomplishes in plants.

8. List the six factors that affect photosynthesis.

9. Glucose manufactured during photosynthesis is converted to _____ or sugar, or _____ and is found in seeds, roots, and tubers, or it is converted to _____, which is used in the construction of plant cell walls.

Critical Thinking/Discussion

10. Explain the difference between light and dark reactions.

11. Define carbon fixation.

12. Why are plants considered a carbon sink?

13. Write the general chemical reaction for photosynthesis and explain how it is read.

14. What is the greenhouse effect?

Knowledge Applied

1. Grow plants under green light, "grow lights," and regular fluorescent tubing. Compare the differences in growth. Do a simple laboratory write-up including procedure, results, and possible explanations for your observations.

2. Place a bean plant under a box with one side cut out. Check the bean plant the next day to see which direction it is leaning. Turn the box 180 degrees and observe what happens to the direction the plant leans after two more days.

3. Select three coleus or geranium plant; make sure the foliage is dry and the soil is well watered. Place a clear dry plastic bag over each, and tie them off around the base of the stalk. Place one plant in the sun, another in an enclosed or dark area, and the other in open shade. Record your observations after 0.5 hour, after another 0.5 hour, then again in 24 hours. Based on your observations, discuss the effect of temperature and light on transpiration.

continues

Knowledge Applied, *continued*

4. Make a list of the different forms of chemical energy used by living things, including humans. Try to include examples of solids, liquids, and gases. Write the chemical formulas for these forms of energy.

5. Obtain a small elodea plant (aquarium or water plant). Put it in water. Put a glass funnel upside down over the water plant. Be sure that all the air is out of the funnel. Then place a jar or test tube full of water over the funnel. Put the plant in bright sunlight and observe it for several days. What happens? Start the experiment again, but this time put the plant in a dim light. What happens?

6. Create a computer presentation that completely explains all of the steps in photosynthesis. Include the light and dark reactions. Use graphics and chemical formulas to explain the steps.

Resources

Janick, J. (Ed.). (1989). *Classic papers in horticultural science.* Englewood Cliffs, NJ: Prentice Hall.

Preece, J. E., & Read, P. E. (2005). *The biology of horticulture: An introductory textbook* (2nd ed.). Hoboken, NJ: Wiley Publishing.

Raven, P. H., Evert, R. F., & Eichhorn, S. E. (2004). *Biology of plants* (7th ed.). New York: W. H. Freeman.

Schmidt, D., Allison, M. M., Clark, K. A., Jacobs, P. F., & Porta, M. A. (2005). *Guide to reference and information sources in plant biology* (3rd ed.). Portsmouth, NH: Libraries Unlimited.

Simpson, B. B., & Ogorzaly, M. C. (2000). *Economic botany: Plants in our world* (3rd ed.). New York: McGraw-Hill.

Stern, K. R. (2005). *Introductory plant biology* (10th ed.). New York: McGraw-Hill.

United States Department of Agriculture. (1986). *Research for tomorrow. The yearbook of agriculture.* Washington, DC: Author.

Internet

Internet sites represent a vast resource of information. The URLs (uniform resource locators) for World Wide Web sites can change. Using one of the search engines on the Internet such as Google or Yahoo!, find more information by searching for these words or phrases: photosynthesis, carbon fixation, light reactions, dark reactions, respiration, photosystem, photoelectric effect, chloroplast, and carbon cycle.

Chapter 14
Respiration

RESPIRATION IS *necessary in all living cells. It is the controlled expenditure of an organism's energy reserves to sustain its life processes. Plants are well known for photosynthesis but they must also respire in order to survive. Photosynthesis occurs only in plant cells containing chlorophyll and only during the daylight hours. Respiration occurs in all of a plant's living cells 24 hours a day.*

After completing this chapter, you should be able to—

- Describe the steps of respiration
- Compare respiration with photosynthesis
- List three differences between respiration and photosynthesis
- Give the generalized reaction for respiration
- List four factors that affect the rate of respiration
- Name the three biochemical pathways involved in respiration
- Discuss the importance of respiration
- Explain how respiration influences crop production
- Give the generalized reaction for fermentation
- Define aerobic and anaerobic

KEY TERMS

Aerobic
Anaerobic
Degradation
Electron Transport System
Fermentation
Glycolysis
Oxidation
Pyruvic Acid
Reduction
Respiration
Tricarboxylic Acid (TCA)

SUGAR INTO ENERGY

Aerobic respiration is the opposite of photosynthesis. Aerobic means that it occurs in the presence of atmospheric oxygen. The generalized reaction for respiration is:

$$C_6H_{12}O_6 + 6H_2O + 6O_2 \longrightarrow 6CO_2 + 12H_2O + energy$$
(Glucose)   (ATP)

This reaction shows glucose as the main fuel for respiration. This is accurate, but many other chemicals such as starches, fats, proteins, and organic acids can also be consumed in respiration.

Respiration occurs in all living cells in the mitochondria (Figure 14-1). Respiration uses up stored energy and gives off heat. Respiration is the opposite of photosynthesis. Photosynthesis captures energy from the sun, uses up carbon dioxide, and produces oxygen. Respiration, on the other hand, releases energy, evolves carbon dioxide, and uses up oxygen. Although the actual biochemical pathways are not closely related, the net effect of respiration is opposite of the net effect of photosynthesis. Table 14-1 compares photosynthesis and respiration.

In many ways, respiration and combustion are similar processes. When a carbohydrate such as sugar is burned, the resulting by-products

Figure 14-1 A mitochondrion.

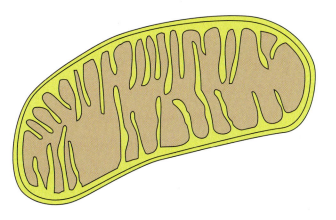

Table 14-1 Comparison of Photosynthesis and Respiration

Photosynthesis	Respiration
Produces food	Uses food for plant energy
ATP produced through photophosphorylation	ATP produced through oxidative phosphorylation
Occurs in cells that contain chloroplasts	Occurs in all cells
Oxygen released	Oxygen used
Water used	Water produced
Carbon dioxide used	Carbon dioxide produced
Occurs in sunlight	Occurs in dark as well as light

are carbon dioxide, water, and heat; these same by-products result from respiration. The difference is the rate of the reactions. Respiration is a controlled, stepwise release of energy that is put to useful purposes by the plant, as it occurs. In contrast, combustion releases energy all at one time. The unit of energy released by respiration is ATP (adenosine triphosphate).

Factors Affecting Respiration

Respiration is affected by temperature, oxygen, soil conditions, light, growth rate, and metabolic activity of the tissues.

Temperature. As temperature increases, so does the rate of respiration. At normal temperatures, respiration increases two to four times for each 18°F rise in temperature.

Oxygen Concentration. Because oxygen is required for respiration, abnormally low concentrations of oxygen in an environment will result in a lower rate of respiration.

Soil Conditions. Because compacted or water-logged soils exclude air, and therefore oxygen, respiration in a plant's root system is inhibited under such conditions.

Light. Because plants grown in low light photosynthesize at a reduced rate, the level of available carbohydrates (sugars, among others) and rate of respiration are also reduced.

Growth Rate and Metabolic Activity of Tissues. Maintenance respiration provides energy maintenance of existing cells and tissues. Growth respiration provides energy for production of new cells and tissues. When growing plants reach full development, the rate of photosynthesis and the rate of respiration have reached their maximum. After this time the rates decline, so age and the availability of glucose affect the rate of respiration.

Respiration Pathways

Metabolic pathways for respiration are shown in Figures 14-2, 14-3, and 14-4. The first pathway is that of glycolysis (Figure 14-2). The net outcome of glycolysis is pyruvic acid, ATP, and water. Glycolysis is reversible. Pyruvic acid from glycolysis enters the tricarboxylic acid (TCA) pathway (Figure 14-3). The net outcome of the TCA pathway is the release of carbon dioxide and electrons. These electrons enter the electron transport system where they are handed off through a series of oxidation (combining with oxygen) and reduction (removing oxygen) reactions (Figure 14-4). Finally, oxygen is reduced to water, and 36 units of ATP are formed.

The potential energy in the glucose molecule is released in glycolysis (not an oxygen-requiring step), the TCA cycle, and the electron transport chain. The occurrence of cellular respiration in an organism can be demonstrated in several ways. Some energy is given off as heat. The amount of glucose used can be determined, and the amount of oxygen consumed can be measured.

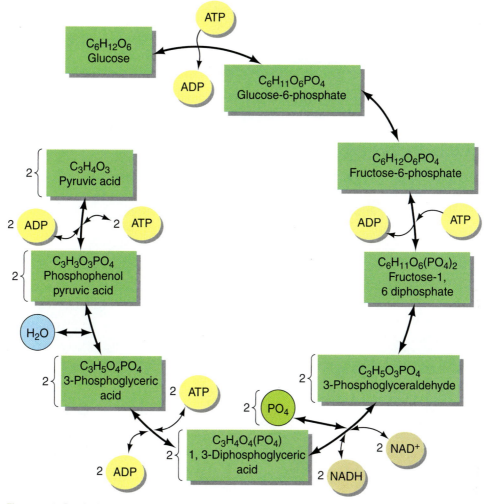

Figure 14-2 Respiration pathway of glycolysis.

NECESSITY OF RESPIRATION

Cellular respiration is the mechanism by which chemical energy is converted from one form to a more usable form, with a release of a certain amount of heat. Energy is defined as the capacity to do work; students can be directed to envision energy as being necessary for biological work. Biological work includes the simple maintenance of life, and energy comes from food/nutrient molecules that are broken down. Stored (potential) energy is transformed to be used to fuel biological processes. Nearly all of the energy used to maintain life is ultimately derived from the sun.

Plants need energy to perform many of the essential functions of life such as growth, repair, movement, reproduction, and transport. Plants and other primary producers transform energy from solar into storage products that are used as energy sources or food by consumer organisms. Energy needed to do biological work comes from the potential energy stored in chemical bonds in these food or nutrient sources. Respiration is the conversion of chemical bond energy in nutrient molecules into forms directly usable by the cell, such as ATP. Respiration can be aerobic

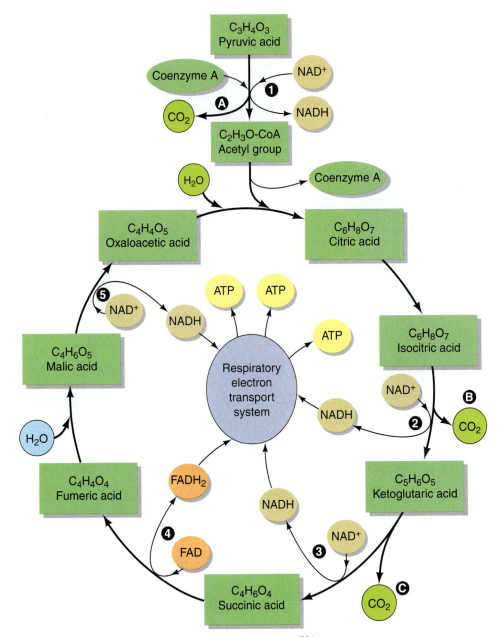

Figure 14-3 Respiration pathway of the tricarboxylic acid cycle or TCA.

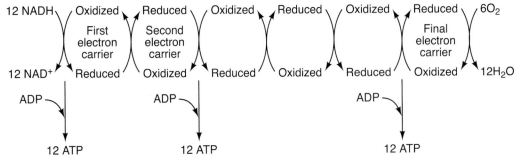

Figure 14-4 Respiration pathway of the electron transport system.

(oxygen) or **anaerobic** (without atmospheric oxygen). In aerobic respiration, a molecule of glucose is broken down completely to carbon dioxide (CO_2) and water, with the assistance of enzymes.

HOW PLANT RESPIRATION AFFECTS AGRICULTURE

Growers need an understanding of respiration for better crop management and postharvest handling of crops.

Crop Management

Because respiration is taking place 24 hours a day, plants needs to be able to photosynthesize enough during the day to provide for respiration during both the day and the night. If this is not accomplished, no energy will be available for growth and the plant will stagnate or die. For this reason, most plants grow best when nighttime temperatures are about 9°F lower than daytime temperatures. Nighttime temperatures in a greenhouse kept too warm will cause respiration rates to be high and stall the growth of a crop. Temperatures that are too cool will also cause stagnation of the crop because respiration will be too low and growth processes will not take place.

Postharvest Handling of Crops

Respiration is closely tied to the **degradation** (the breakdown of the tissues, or rotting) of harvested crops. By slowing respiration rates after harvest, we can improve the quality of the foods or the goods reaching the consumer (Figure 14-5). Two strategies to slow respiration are:

+ Refrigeration—Cooling a fruit, vegetable, or floral crop helps to slow the rate of respiration and the associated deterioration processes.
+ Increasing atmospheric concentration of nitrogen—This pushes oxygen out of the air surrounding the crop and brings the rate of respiration to a very low level. This is especially useful in storing many fruit crops, such as apples, over a long period of time.

Figure 14-5 Postharvest handling of a perishable crop requires an understanding of respiration.

Many Uses of Starch through the Ages

Starch, a product of photosynthesis, has found many uses over the years—and not always as a food.

Leonardo Da Vinci, a pioneer in aviation, proposed to stiffen the wing fabric of flying machines with starch.

The use of starch as a hair powder is said to have originated in France in the sixteenth century. Near the end of the eighteenth century, large quantities of starch were used for this purpose.

Starch is a common ingredient of face powder, as are kaolin, magnesium carbonate, light precipitated chalk, zinc and titanium oxides. Rice starch, because of the small size of its granulates, is often used for this purpose. Recently, some manufacturers have added magnesium and zinc stearates to face powders. These ingredients help to keep the powder from being wet from perspiration. Wet face powder loses part of is power to reflect light and becomes less visible. In order to give the desired "bloom" to the skin, the powder must be visible.

Holland was noted for the high quality of wheat starch produced there in the Middle Ages. Then and later, starch was used mostly in the laundry for stiffening fabrics and was an expensive luxury afforded only by church dignitaries, aristocrats, and other wealthy persons. Apparently, starch was first used in England during the reign of Elizabeth I, who is reported to have appointed a laundry starcher as a special court official. Later, the manufacture of starch from wheat was prohibited more than once in England because wheat was considered to be more useful as food.

In the seventeenth century laundry starches were colored. Yellow starch was ultrafashionable for men. The Cavaliers often used green starch; the Roundheads used blue starch.

Fermentation

Anaerobic respiration occurs without atmospheric oxygen. This is fermentation and the general reaction is:

Glucose → Pyruvic Acid → Acetaldehyde + CO_2 → Alcohol (Ethanol)

Yeast and lower plants carry on **fermentation** following glycolysis. Many higher plants have a limited ability to produce alcohol, but the alcohol is so toxic that it can destroy the tissues producing it. One exception is rice seedlings growing submerged in oxygen-depleted water.

Fermentation is important in making silage.

SUMMARY

Carbohydrates made during photosynthesis are of value to the plant when they are converted to energy. This energy is used in the process of building new tissues, or growth of the plant. The chemical process by which sugars and starches produced by photosynthesis are converted to energy is called oxidation. It is similar to the burning of wood or coal to produce heat. Controlled oxidation in a living cell is known as respiration. As with photosynthesis, respiration is actually a total of many reactions, some of which are complex. These reactions can be summarized as glucose plus water and oxygen yields carbon dioxide, water, and energy in the form of ATP.

Glycolysis, tricarboxylic acid cycle, and the electron transport system are the three main parts of respiration. Glycolysis is the breakdown of glucose to pyruvic acid, which then enters the tricarboxylic acid cycle. The tricarboxylic acid cycle produces electrons that enter the electron transport system and produce ATP. Energy derived from respiration is used for maintenance, growth, and reproduction.

An understanding of respiration is helpful for managing crop production and postharvest crops. Fermentation or anaerobic respiration occurs without atmospheric oxygen.

Review

True or False

1. Aerobic respiration produces alcohol.

2. Pyruvic acid is the end product of glycolysis.

3. Both photosynthesis and respiration produce ATP.

4. Aerobic respiration occurs only at night.

5. After harvest, slowing respiration improves the quality of vegetables for the consumer.

Short Answer

6. List four factors that affect the rate of respiration.

7. Give the generalized reaction for aerobic respiration.

8. Name the three biochemical pathways involved in aerobic respiration.

9. Respiration occurs in the _____ of the plant cell.

10. What is another name for anaerobic respiration?

Critical Thinking/Discussion

11. Discuss the importance of aerobic respiration.

12. Compare respiration with photosynthesis.

13. Describe what happens in each of the three biochemical pathways used in aerobic respiration.

14. Explain how knowledge of respiration is used in crop production.

15. Define aerobic and anaerobic.

Knowledge Applied

1. Using some type of presentation software, develop a presentation that explains how photosynthesis and respiration result in the recycling of materials.

2. Develop an experiment demonstrating how reducing the oxygen level can improve the storage of some crops.

3. Devise an experiment for making a small amount of silage, or use yeast to produce alcohol from sugar. Then develop a report to explain aerobic and anaerobic respiration.

4. Using presentation software, develop a presentation tracing the flow of energy from sunlight to the energy used by animals for growth and reproduction.

5. Glucose is a high-energy compound. Develop a demonstration or a report to show how much energy is in glucose.

Resources

Fageria, N. K., Baligar, V. C., & Clark, R. B. (2006). *Physiology of crop production.* Binghamton, NY: Haworth Food and Agricultural Products Press.

Hamrick, D. (2003). *Ball redbook: Vol. 2. Crop production* (17th ed.). Batavia, IL: Ball Publishing.

Hay, R., & Porter, J. (2006). *The physiology of crop yield* (2nd ed.). Ames, IA: Blackwell Publishing.

Janick, J. (Ed.). (1989). *Classic papers in horticultural science.* Englewood Cliffs, NJ: Prentice Hall.

Simpson, B. B., & Ogorzaly, M. D. (2000). *Economic botany: Plants in our world* (3rd ed.). New York: McGraw-Hill.

Internet

Internet sites represent a vast resource of information. The URLs (uniform resource locators) for World Wide Web sites can change. Using one of the search engines on the Internet such as Google or Yahoo!, find more information by searching for these words or phrases: plant respiration, fermentation, respiration pathways, electron transport system, tricarboxylic acid pathway, glycolysis, and cellular respiration.

Chapter 15
Basics of Plant Growth

PLANT GROWTH OCCURS *because the number of plant cells increase through division and the size of the cells increase as they become differentiated into specialized tissues and organs. Understanding plant growth requires a basic knowledge of genetics and how DNA controls the processes of cells. Heredity and environment interact to determine growth and growth patterns.*

After completing this chapter, you should be able to—

- Name three regions of growth on plants
- Discuss five assumptions about cells
- Name three functions of cells
- Describe the structure of DNA
- Explain how DNA directs the growth of plants
- Discuss the relationship between DNA, genes, and chromosomes
- Explain how plant growth regulators and hormones act
- List five ways plant growth regulators are used on agronomic crops
- Describe the sequencing of bases in DNA
- Discuss the relationship between DNA and RNA
- List four general examples of plant growth regulators
- Diagram a growth curve for a plant
- Discuss the interaction of heredity and environment on plant growth

KEY TERMS

Abscisic Acid
Adenine
Alleles
Bases
Chromosome
Codons
Cytosine
Differentiation
Ethylene
Gametes
Genes
Genome
Guanine
Hormone
Metabolites
Mitosis
Morphogenesis
Nucleotides
RNA
Senescence
Thymine

REGIONS OF GROWTH

A most important part of plants that is not readily visible is the region or point of growth. Growth usually takes place in two ways. One is the enlargement of a single seed, leaf, stem, root, or fruit. This type of growth is simply due to the intake of water and other substances by previously formed cells. The second type of growth actually involves the formation of new cells in a specialized part of the plant. If these specialized parts of the plant are destroyed, then growth in that portion of the plant is stopped.

The three parts of the plant where new cells are formed—

1. Tips of stems and roots
2. Axils of leaves
3. Cambium layer in stems and roots

The tips of stems (terminal buds) and roots are responsible for increase in length of these structures. Buds in leaf axils are responsible for the formation of new stems, leaves, and flowers. The cambium layer of dicots gives rise to cells that increase the diameter of stems, roots, and tree trunks. This layer lies just beneath the bark of tree trunks and is responsible for the formation of growth rings in trees. The growing point of lawn grasses is located near the ground and is not removed in mowing. The grasses continue to grow after each mowing.

CONCEPTS AND COMPONENTS OF GROWTH

A meristem is a region of plant growth (see Chapter 4). Cells of meristems actively divide at certain times. As the root or stem grows, new cells are produced. These cells mature, increase in size, and develop into specialized tissues. Changes in cells that result in the formation of specialized parts of the plant are called **differentiation**. For example, in a young root, early differentiation results in the formation of different types of tissue for conduction of water and food, storage of food, protection, and the uptake of water and minerals.

Some seed plants live for years. During their lives, they continue to produce new tissues and organs year after year—for example, cones, leaves, and flowers.

CELLS

All living material is made of cells or the chemical products of cells (Figure 15-1). This concept of the cell as the fundamental unit of life is the basis for an understanding of living organisms such as the plants.

Modern cellular biology makes six assumptions—

1. All living material is made up of cells or the products of cells.
2. All cells are derived from previously existing cells; most cells arise by cell division, but in sexual organisms they may be formed by the fusion of gametes.

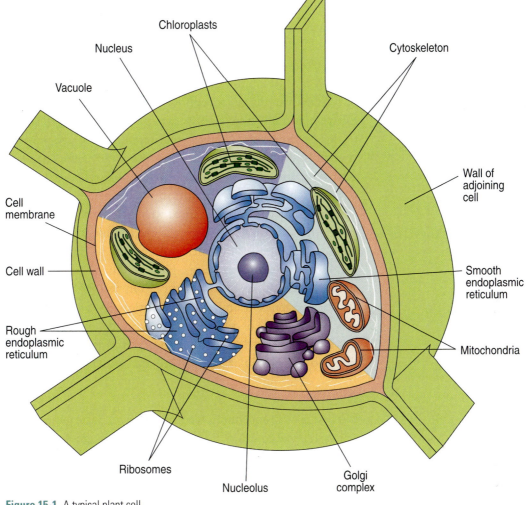

Figure 15-1 A typical plant cell.

3. A cell is the most elementary unit of life.
4. Every cell is bounded by a plasma membrane, an extremely thin skin separating it from the environment and from other cells.
5. All cells have strong biochemical similarities.
6. Most cells are small, about 0.001 cm (0.00004 inches) in length.

The three general functions of most cells include:

+ Maintenance
+ Synthesis of cell products
+ Cell division

These functions require the cell to take in nutrients and excrete waste products. The nutrients are used either as building blocks in synthesizing large molecules, or they are oxidized—burned—producing energy for powering the cell's activities. Because synthesis, maintenance, and mechanical and electrical activity all require energy, a major chemical activity in nearly all types of cells is the energy-linked conversion of **metabolites**. Adenosine triphosphate (ATP) is the universal energy-transfer molecule. ATP is constantly used and regenerated by the energy-yielding chemical reactions described in Chapter 13.

Morphogenesis

All organisms, regardless of their complexity, begin as a single cell. By repeated cell growth and mitosis, or division, the organism eventually develops into an adult, containing thousands of billions of cells. This process of development is called **morphogenesis**. Because many types of cells exist in plants, morphogenesis involves not only cell growth but differentiation into specialized types of cells. This differentiation is controlled by the genes. The information needed to program and guide the growth is contained within the **chromosomes**. Size, shape, and chemical activity of the cells are governed to some extent by the function of the tissue in which they are found.

Each cell contains the same total genetic information that was present in the fertilized egg. The cells are not identical because in different types of cells, groups of genes are controlled—switched on and off—by various biochemical processes. Each cell manufactures the proteins and structures needed for it to function. On the average, only about 10 percent of the genes of any cell are functional—which genes, in particular, vary with the type of cell. Although morphogenesis has been scientifically described in great detail for a number of organisms, all of the processes involved at the cellular level are still not understood.

Mitosis

Mitosis is the reproduction of cells in which the genetic material of the cell is duplicated exactly. The cells simply divide and produce new cells like themselves. Daughter cells have the same genetic makeup as parent cells. The six steps of mitosis are—

1. Interphase
2. Prophase
3. Metaphase
4. Anaphase
5. Telophase
6. Interphase

These steps of cell division were covered in Chapter 4.

Meiotic Cycle

The meiotic cycle is covered in detail in Chapter 17. This cycle is known as a reduction division because the cells produced from this division have one-half the number of chromosomes (1n) typical for the species. The meiotic cycle produces the pollen and the egg in the ovary.

BASIC GENETICS

Genes are the basic unit of inheritance. Genes are carried on the chromosomes in the **gametes**—eggs or pollen (sperm). Different forms of the same gene—same location on the chromosome—are called **alleles**. Genes contain the blueprint or code that determines how the plant will look and interact with its environment. The number of chromosomes varies from species to species but is consistent for a species.

Genes are made of DNA (deoxyribonucleic acid). Resemblances and differences among related individuals are primarily because of genes. Genes cause the production of enzymes, which control chemical reactions in the plant, thus affecting plant development and function. For normal plant development and function, genes must occur in pairs. Genes are a part of the chromosomes that reside in the nucleus of cells. Chromosomes in the nucleus of a particular cell contain the same genetic information as the chromosomes in every cell of the plant. So the chromosomes in the cells of the root are the same as the chromosomes in the cells of the stem. The genes on the chromosomes, however, know their function in specific tissues.

In the normal cell of a plant, chromosomes occur in distinct pairs. Table 15-1 shows the chromosome numbers for some common plants.

Table 15-1 Chromosome Numbers in Some Common Plants

Common Name	Chromosome Number
Barley	14
Rye	14
Corn (Maize)	20
Tomato	24
Durum wheat	28
Wheat	42
Oats	42
Spelt	42

Description of Genes

The complete set of instructions for making a plant is called its **genome**. It contains the master blueprint for all cellular structures and activities for the lifetime of the cell or organism. Found in every nucleus of cells, the genome consists of tightly coiled threads of DNA and associated protein molecules, organized into structures called chromosomes.

All genes are arranged linearly along the chromosomes. The nucleus cells contain two sets of chromosomes, one set given by each parent. Chromosomes contain roughly equal parts of protein and DNA; chromosomal DNA contains an average of 150 million bases. DNA molecules are among the largest molecules now known. Chromosomes can be seen under a light microscope.

If unwound and tied together, the strands of DNA would stretch more than five feet but would be only 50 trillionths of an inch wide. For each organism, the components of these slender threads encode all the information necessary for building and maintaining life, from simple bacteria to remarkably complex plants. Understanding how DNA performs this function requires some knowledge of its structure and organization.

Gregor Mendel—Founder of Modern Genetics

Modern genetics was founded by Gregor Johann Mendel, a cigar-smoking Austrian monk who gardened. He conducted plant-breeding experiments with garden peas from 1857 to 1865. In his monastery at Brunn (now Brno, Czechoslovakia), Mendel applied a clear mind and a powerful curiosity to discover some basic principles of heredity.

In 1866 Mendel published a report covering eight years of research. He published this in the proceedings of a local scientific society. For 34 years his findings went unrecognized and ignored. Finally in 1900, 16 years after Mendel's death, three European biologists independently duplicated his findings. This led to the rediscovery of the original article published by Mendel 34 years earlier.

The essence of Mendelism is that inheritance is by particles or units, now called genes. These units (genes) are present in pairs, one member of each pair having come from each parent. Each gene maintains its identity generation after generation. So Mendel's work with peas laid the basis for two of the general laws of inheritance:

- The law of segregation
- The law of independent assortment of genes

Other genetic principles have been added, but to honor the work of Mendel, all the phenomena of inheritance are generally known as Mendelism or Mendelian genetics.

Modern genetics was founded by an amateur not trained in science and who did his work merely as a hobby.

Mendel's original article was published in German. An English translation of it can be found on the Internet at www.mendelweb.org/MWNotes.html. The site includes photographs and hypertext links to other sources of information on genetics.

Structure of DNA

In plants and animals, a DNA molecule consists of two strands that wrap around each other to resemble a twisted ladder whose sides, made of sugar and phosphate molecules, are connected by rungs of nitrogen-containing chemicals called **bases** (Figure 15-2). Each strand is a linear arrangement of repeating similar units called **nucleotides**, which are each composed of one sugar, one phosphate, and a nitrogenous base.

Figure 15-2 DNA strands divide, and bases attach themselves to the new strands to form identical genes for new cells.

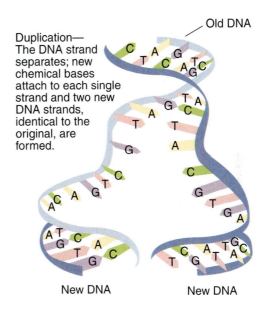

Duplication— The DNA strand separates; new chemical bases attach to each single strand and two new DNA strands, identical to the original, are formed.

Old DNA

New DNA   New DNA

Four different bases are present in DNA—**adenine** (A), **thymine** (T), **cytosine** (C), and **guanine** (G). The particular order of the bases arranged along the sugar-phosphate backbone is called the DNA sequence. The sequence specifies the exact genetic instructions required to create a particular organism with its own unique traits.

The two DNA strands are held together by weak bonds between the bases on each strand, forming base pairs. Genome size is usually stated as the total number of base pairs. Some genomes contains over 3 billion of these base pairs.

Each time a cell divides into two daughter cells, its full genome is duplicated; for horses and other complex organisms, this duplication occurs in the nucleus. During cell division the DNA molecule unwinds and the weak bonds between the base pairs break, allowing the strands to separate (Figure 15-3). Each strand directs the synthesis of a complementary new strand, with free nucleotides matching up with their complementary bases on each of the separated strands. Strict base-pairing rules are adhered to. Adenine will pair only with thymine (an A-T pair) and cytosine with guanine (a C-G pair). Each daughter cell receives one old and one new DNA strand. The cells adherence to these base-pairing rules ensures that the new strand is an exact copy of the old one. This minimizes the incidence of errors (mutations), which may greatly affect the resulting organism or its offspring.

Figure 15-3 DNA structure coiled inside a replication chromosome.

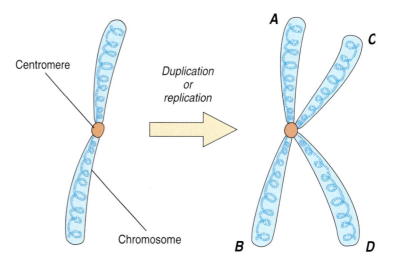

How the Code Works

Each DNA molecule contains many genes, the basic physical and functional units of heredity. A gene is a specific sequence of nucleotide bases whose sequences carry the information required for constructing proteins, which provide the structural components of cells and tissues as well as enzymes for essential biochemical reactions.

All living organisms are composed largely of proteins. Proteins are large, complex molecules made up of long chains of subunits called amino acids. Twenty different kinds of amino acids are usually found in proteins. Within the gene, each specific sequence of three DNA bases (**codons**) directs the protein-synthesizing machinery of the cell to add specific amino acids. For example, the base sequence ATG codes for the amino acid methionine. Because three bases code for one amino acid, the protein coded by an average-sized gene (3,000 base pairs) will contain 1,000 amino acids. The genetic code is thus a series of codons that specify which amino acids are required to make up specific proteins.

The protein-coding instructions from the genes are transmitted indirectly through messenger ribonucleic acid (**mRNA**), a transient intermediary molecule similar to a single strand of DNA. For the information within a gene to be expressed, a complementary RNA strand is produced (a process called transcription) from the DNA template in the nucleus. This mRNA is moved from the nucleus to the cellular cytoplasm, where it serves as the template for protein synthesis. The protein-synthesizing machinery of the cell then translates the codons into a string of amino acids that will constitute the protein molecule for which it codes (Figure 15-4).

Control of Growth

The important factors affecting plant growth and development include heredity, hormones, nutrition, and environment.

Heredity. Hereditary, or genetic, factors control the general species characteristics of the individual and set limits on size and rate of growth.

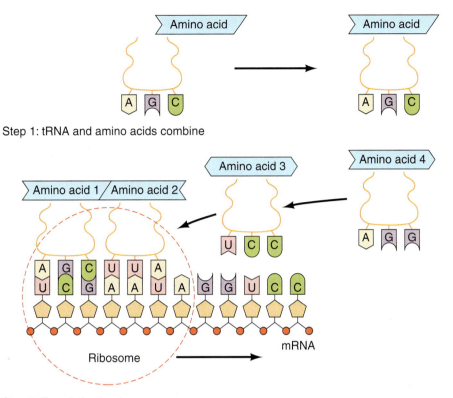

Step 1: tRNA and amino acids combine

Step 2: Translation begins

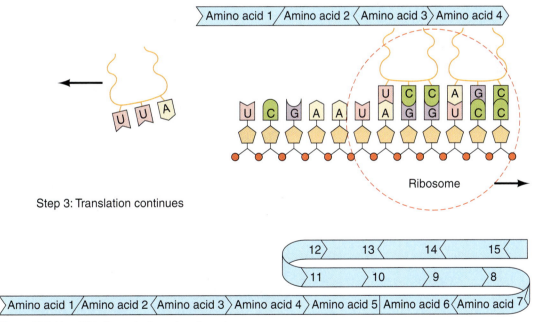

Step 3: Translation continues

Figure 15-4 How DNA directs the formation of protein.

The genetic structure, through DNA and RNA patterns, acts by regulating protein synthesis, especially the manufacture of enzymes, as well as cell division, cell enlargement, the incorporation of substances into the cell walls, and the production and activity of the hormones. The gene action, in turn, is controlled by various growth regulators, particularly **hormones** and nutrients.

So-called plant hormones are organic chemicals produced in small amounts at one place in the plant that cause some physiological action in another. The several classes of plant hormones include the auxins, cytokinins, gibberellins, abscisic acid, and ethylene. Cytokinins are especially important in cell division. Elongation is promoted by auxins and gibberellins. The bending of stems toward light is caused by auxins in higher concentration on the dark side, inducing more cell elongation.

Cell and organ differentiation are usually regulated by the interaction of several hormones. The initiation of roots by auxins and of buds by cytokinins depends on the presence of opposing hormones in the proper amounts. Other growth-related activities regulated by hormones include seed germination, flower and fruit development, and leaf enlargement.

Nutrition and Environment. Plants require all the essential ingredients of photosynthesis to construct the necessary compounds and structures (Figure 15-5). Water is especially important, because cell enlargement is a result of internal water pressure (turgor) extending the walls. In periods of drought, plants tend to have smaller leaves. Calcium interacts with auxins and cytokinins in regulating cell division and elongation. Nitrogen is involved in the structure of chlorophyll, proteins, auxins, and cytokinins.

The intake and use of nutrients and the activities of hormones and other regulators are affected greatly by the external environment, particularly temperature and light. Certain wavelengths of light affect the activity of a pigment called phytochrome, which in turn interacts with hormones

Figure 15-5 Fruits and vegetables with consistent flavor, size, and color are available because of plant breeding and genetic selection. *(Photo courtesy of USDA ARS)*

in regulating flowering, leaf expansion, stem elongation, sleep movement of leaves, and seed germination.

All physiological activities are directly related to temperature, with warmer temperatures favorable to more growth. Cold temperatures are required for some seeds to germinate, some buds to begin growing, and some plants to flower.

Temperature, light, water, nutrients, and soil are all environmental factors influencing plant growth and development as they interact with the genetic makeup of plants. Chapters 7, 9, 11, and 12 discuss how the environment—internal or external—influences the growth and development of plants.

Plant Hormones and Regulators

Plant growth regulators (PGRs) are organic compounds, other than nutrients, that modify plant physiological processes. PGRs, called biostimulants or bioinhibitors, act inside plant cells to stimulate or inhibit specific enzymes or enzyme systems and help regulate plant metabolism. They normally are active at very low concentrations in plants. The importance of PGRs was first recognized in the 1930s. Since that time, natural and synthetic compounds that alter function, shape, and size of crop plants have been discovered.

Specific PGRs are used to modify crop growth rate and growth pattern during the various stages of development, from germination through harvest and postharvest preservation. Growth-regulating chemicals that have positive influences on major agronomic crops can be of value. Some of these uses include:

+ Preventing lodging in cereals
+ Preventing preharvest fruit drop
+ Synchronizing maturity to facilitate mechanical harvest
+ Hastening maturity to decrease turnover time
+ Reducing labor requirements

Classes of PGRs. PGRs may be naturally occurring, plant-produced chemicals called hormones, or they may be synthetically produced compounds. Most PGRs, natural and synthetic, fall into one of the following classes:

+ Auxins primarily control growth through cell enlargement. Some instances of auxin-induced cell division do occur. They may act as both stimulators and inhibitors of growth, and cause different plant parts—shoots, buds, and roots—to respond differently. Auxins also stimulate differentiation of cells, the formation of roots on plant cuttings, and the formation of xylem and phloem tissues.
+ Gibberellins control cell elongation and division in plant shoots. They have been shown to stimulate RNA and protein synthesis in plant cells.
+ Cytokinins act in cell division, cell enlargement, senescence, and transport of amino acids in plants.

For the specific regulation of many plant processes and the differentiation of cells into specific plant parts, a variety of ratios and concentrations of these three plant hormone classes are required rather than a single hormone acting alone.

Other naturally occurring regulators of plant growth and plant metabolic activity can be classed as inhibitors and ethylene.

+ Inhibitors represent a wide assortment of internally produced chemical compounds, each of which inhibits the catalytic action of a specific enzyme. Because a plant cell may contain as many as 10,000 different enzymes, there are a wide variety of inhibitors acting inside the cell.

+ Ethylene is internally produced by plants and has a multitude of effects on cell processes. It interacts with auxins to regulate many metabolic processes. Several chemical compounds that release ethylene after being sprayed on plants are currently commercial PGR products.

A wide assortment of plant growth–promoting products is being marketed with claims made for beneficial effects on crop growth and yields. Typically, these products are supposed to—

+ Promote germination and/or emergence
+ Stimulate root growth
+ Promote mobilization and translocation of nutrients within plants
+ Increase stress tolerance and improve water relations in plants
+ Promote early maturity
+ Increase disease resistance
+ Retard senescence
+ Improve crop yields and/or quality

Usually, the claims are made for plant hormone products or products that affect the concentrations and ratios of plant hormones internally.

MEASURING AND MODELING PLANT GROWTH

Growth can be measured as increases in fresh weight, dry weight, volume, length, height, or surface area. As its overall size increases, a plant's form and shape change as directed by genetic factors. Figures 15-6, 15-7, and 15-8 illustrate some typical growth curves for some common plants.

Figure 15-6 Typical growth curves for annuals.

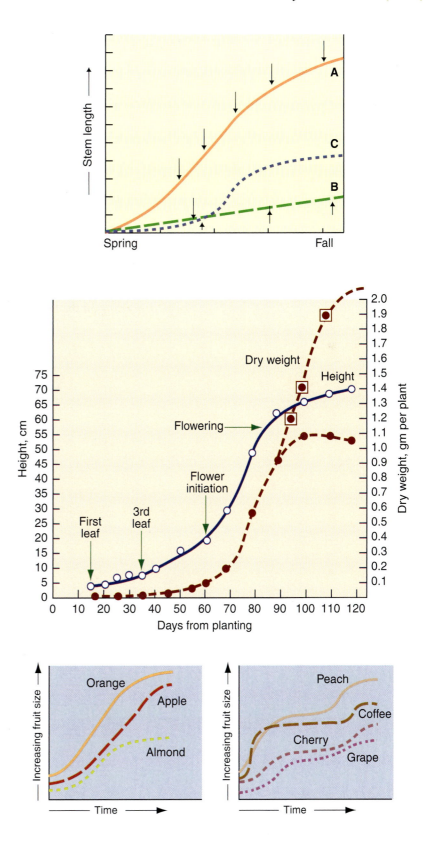

Figure 15-7 Typical growth curve for field barley.

Figure 15-8 Typical growth curves for several fruits.

SUMMARY

Plant growth is a product of living cells, with all of their wide-ranging metabolic processes. Meristematic tissue produces cells that differentiate into the tissues and organs of plants. Cell function and ultimately plant structure and function are under control of DNA. Genes are the basic unit of inheritance and they are made of DNA. The sequence of the bases adenine, thymine, cytosine, and guanine in DNA are the codes for plant function and structure. This base sequence determines the structure of messenger RNA (mRNA) and transfer RNA(tRNA) for protein synthesis. Even though inherited traits are set by the genes, hormones, environment, and nutrition affect plant growth and development. Plant growth regulators and hormones can be used to control a variety of plant growth patterns and rate during various stages of development. Finally, plant growth can be measured a variety of different ways, depending on the need.

Review

True or False

1. The complete set of instructions for making an organism is called its genome.

2. The possible number of distinct assortments of genes in forming gametes is not over 100.

3. Differences in genetic makeup are often referred to as genetic variation.

4. The environment has little effect on the expression of the gene pair(s) controlling a trait.

5. Alanine is one of the four bases in DNA.

Short Answer

6. Name three functions of all cells.

7. List five basic assumptions about all cells.

8. Identify four general examples or types of plant growth regulators.

9. Name five ways plant growth regulators are used on agronomic plants.

10. What are the complementary bases for the following sequence: adenine, adenine, cytosine, thymine, adenine?

11. List the three parts of plants where new cells continue to be formed.

Critical Thinking/Discussion

12. Diagram and label a growth curve for an annual plant.

13. Describe the relationship between DNA, genes, chromosomes, and RNA.

14. Give an example of how heredity and the environment interact to alter plant growth and development.

15. Discuss three ways that plant growth regulators or hormones act on plants or plant cells.

Knowledge Applied

1. Obtain prepared microscope slides and view the chromosomes in the nucleus of some plant cells.

2. Draw or make a three-dimensional representation of DNA.

3. Using the Internet and/or other research methods, develop a report about any work that is being done to map the genome of plants.

4. To understand the randomness of the genetic process, have five people roll a pair of dice 12 times. Make a table that tracks how many times each person rolls two ones, two twos, two threes, two fours, two fives, and two sixes.

5. Using potted plants, design an experiment that shows the effects of one of the plant growth regulators. Keep a record of your experimental design, materials, methods, and results. Present your findings to a group using photographs, charts, and tables to explain your experiment.

Resources

Asimov, I. (1962). *The genetic code.* New York: New American Library.

Beck, C. B. (2005). *An introduction to plant structure and development.* Cambridge, UK: Cambridge University Press.

Janick, J., ed. (1989). *Classic papers in horticultural science.* Englewood Cliffs, NJ: Prentice Hall.

McMahon, M., Kofranek, A. M., & Rubatzky, V. E. (2006). *Hartmann's plant science: Growth, development, and utilization of cultivated plants* (4th ed.). Englewood Cliffs, NJ: Prentice Hall Career & Technology.

Preece, J. E., & Read, P. E. (2005). *The biology of horticulture: An introductory textbook* (2nd ed.). Hoboken, NJ: Wiley Publishing.

Simpson, B. B., & Ogorzaly, M. C. (2000). *Economic botany: Plants in our world* (3rd ed.). New York: McGraw-Hill.

Thomas, L. (1974). *The lives of a cell: Notes of a biology watcher.* New York: The Viking Press.

Internet

Internet sites represent a vast resource of information. The URLs (uniform resource locators) for World Wide Web sites can change. Using one of the search engines on the Internet such as Google or Yahoo!, find more information by searching for these words or phrases: DNA, chromosomes, genes, plant growth regulators, genetic mapping, plant growth, plant environment, mRNA, and mitosis.

Chapter 16
Vegetative Growth

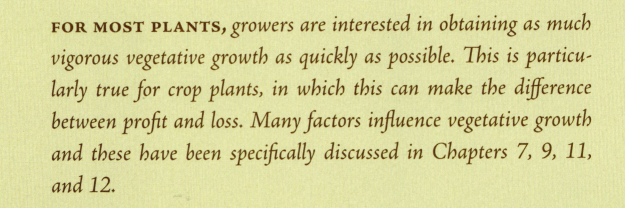

FOR MOST PLANTS, *growers are interested in obtaining as much vigorous vegetative growth as quickly as possible. This is particularly true for crop plants, in which this can make the difference between profit and loss. Many factors influence vegetative growth and these have been specifically discussed in Chapters 7, 9, 11, and 12.*

After completing this chapter, you should be able to—

- Describe the growth of vegetative organs, or stems, leaves, and roots

- Discuss the coordination of vegetative growth and how environmental factors or agricultural management practices affect growth rates

- List differences between annuals, biennials, and perennials

- Name the three stages, or phases of plant development

- List factors that lead to "natural" plant death

- Identify the six steps for germination to occur

- Describe the scarification process

- Explain seed dormancy

- List five factors affecting plant growth

- Describe senescence in plants and fruit

- Discuss how light possibly affects seeds during germination

- Identify the role of water in germination

GROWTH

Growth is an irreversible increase in volume and/or weight. Plant growth occurs by an increase in cell numbers and cell size. Cell division and enlargement involves the synthesis of new cellular materials.

Plant growth has two aspects—primary and secondary. Primary growth takes place in young, herbaceous organs, resulting in an increase in length of shoots and roots. Secondary growth follows primary growth in some plants and results in an increased girth as layers of woody tissue are laid down. Monocots and herbaceous dicots typically exhibit only primary growth.

The formation of new cells takes place in regions known as meristems. At the tip, or apex, of each stem and root is an apical meristem, where cells are actively dividing (see Chapter 4). Each apical meristem produces three other meristems (protoderm, ground meristem, and procambium) called primary meristems. Tissues derived from the primary meristems are called primary tissues and include epidermis, cortex, pith, and primary xylem and phloem (vascular tissues). The elongation of cells produced by the primary meristems accounts for most of the increased length of stems and roots.

SEED GERMINATION

Growth begins with germination. A seed certainly looks dead. It does not seem to move, to grow, nor do anything. In fact, even with biochemical tests for the metabolic processes we associate with life (such as respiration), the rate of these processes is so slow that it would be difficult to determine whether there really was anything alive in a seed.

Indeed, if a seed is not allowed to germinate (sprout) within a certain length of time, the embryo inside will die. Each species of seed has a certain length of viability. Some maple species have seeds that need to sprout within 2 weeks of being dispersed, or they die. Some seeds of lotus plants are known to be up to 2,000 years old and still can be germinated.

Assuming the seed is still viable, the embryo inside the seed coat needs something to get its metabolism activated to start the embryo growing. The process of getting a seed to germinate can be simple or complicated.

Germination involves six steps—
1. Water imbibition
2. Enzyme activation
3. Hydrolysis and catabolism of storage compounds
4. Initiation of embryo growth
5. Anabolism and formation of new cell structures
6. Emergence of seedling

Four environmental factors affect germination: water, oxygen, light, and heat (Figure 16-1).

Figure 16-1 Requirements for optimum plant growth.

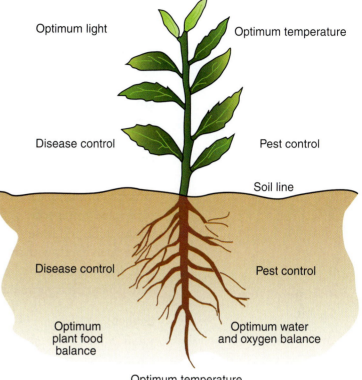

Water

The first step in the germination process is the imbibition or absorption of water. Even though seeds have great absorbing power because of the nature of the seed coat, the amount of available water in the germination medium affects the uptake of water. An adequate, continuous supply of water is important to ensure germination. Once the germination process has begun, a dry period will cause the death of the embryo.

During seed germination, water functions as a media of transport. It also functions in hydrolysis of storage compounds, for turgor pressure, for cell growth, and to re-establish spatial relationships. The general enzymes activated during germination are important in the catabolism of starch, lipids, and proteins. These include—

* Amylase for starch
* Lipase for lipids
* Protease for protein

The basic unit of starch is glucose. Two main types of starch are present in seeds—amylose and amylopectin. Amylose is a straight chain of glucose that can be 100 percent hydrolyzable to glucose. Amylopectin is a branched chain and is only 60 percent hydrolyzable to glucose. During germination, glucose yields energy in the form of adenosine triphosphate. Lipids are broken down to fatty acids, and proteins are broken down to amino acids. This all occurs because of the activation of enzymes.

Light

Light is known to stimulate or to inhibit germination of some seeds. The light reaction involved here is a complex process. In some plant species, a pigment called phytochrome, or Pfr, mediates seed germination. Light in the red wavelength induces Pfr formation. Pfr induces formation of growth regulators, which stimulate formation of hydrolytic enzymes active in germination. Some crops that have a requirement for light to assist seed germination are ageratum, begonia, browallia, impatiens, lettuce, and petunia.

Conversely, calendula, centaurea, annual phlox, verbena, and vinca are examples of plants with seeds that will germinate best in the dark. Other plants are not specific at all. When sowing light-requiring seeds, do as nature does, and leave them on the soil surface. If they are covered at all, cover them lightly with fine peat moss or fine vermiculite. These two materials, if not applied too heavily, will permit some light to reach the seeds and will not limit germination.

Oxygen

In all viable seeds, respiration takes place. The respiration in nongerminating seeds is low, but some oxygen is required. The respiration rate increases during germination, so the medium in which the seeds are placed should be loose and well aerated. If the oxygen supply during germination is limited or reduced, germination can be severely retarded or inhibited.

Heat

A favorable temperature is another important requirement of germination. It not only affects the germination percentage but also the rate of germination. Some seeds will germinate over a wide range of temperatures; others require a narrow range. Many seeds have minimum, maximum, and optimum temperatures at which they germinate. For example, tomato seed has a minimum germination temperature of 50°F and a maximum temperature of 95°F, but an optimum germination temperature of about 80°F. When germination temperatures are listed, they are usually the optimum temperatures unless otherwise specified. Generally, 65–75°F is best for most plants. This often means the germination flats may have to be placed in special chambers or on radiators, heating cables, or heating mats to maintain optimum temperature.

Germination will begin when certain internal requirements have been met. A seed must have a mature embryo, contain a large enough endosperm to sustain the embryo during the germination process, and contain sufficient growth regulators—auxins, gibberellins, cytokinins—to initiate the process.

Some seeds do not have a dormant period. Other seeds exhibit some type of dormancy that must be overcome before germination can occur.

Seeds Lacking True Dormancy

Common vegetable garden seeds generally lack any kind of dormancy. The seeds are ready to sprout. All they need is some moisture to get their biochemistry activated, and temperatures warm enough to allow the chemistry of life to proceed. Seeds taken from the wild, however, are frequently endowed with deeper forms of dormancy.

Seeds with Truly Dormant Embryos

Several mechanisms permit seeds to be truly dormant:

+ Thick seed coat
+ Thin seed coat
+ Insufficient development
+ Inhibitors

Thick Seed Coat. Many kinds of seeds have very thick seed coats. These obviously keep water out of the seed, so the embryo cannot get the water needed to activate its metabolism and start growing. The lotus seeds are an example of this. An outstanding example from the northern temperate zone is the Kentucky coffeetree (*Gymnocladus dioica*). The seed coat is perhaps 2 mm thick.

Scarification allows thick-coated seeds to germinate. This occurs naturally through a variety of processes such as the freezing action of water, the pounding along a river or seacoast, or action by an animal. A very common example of a way to scarify a seed coat is observed in strawberry and raspberry. The thick seed coat is designed to be swallowed by a bird. The animal digests the fruit pulp, but the seed coat passes through the digestive system, still protecting the viable embryo inside, but weakened enough to allow sprouting. The seed is deposited with a little organic fertilizer in the environment and can now sprout!

Thin Seed Coat. A thin seed coat is so thin that it is no barrier to water. Some other kind of dormancy mechanism is needed. A pigment molecule in the seed can absorb light and cause a change in the behavior of the embryo. The pigment is phytochrome. Like chlorophyll, it is made of a chromophore and is associated with proteins. This pigment is different from chlorophyll, however, in one critical way: it exists in two interconvertible forms.

One form of phytochrome, named Pfr, is the form of the phytochrome found in plant cells that are exposed to red (660 nm) or common white light. This form of phytochrome is biologically very active and plays a role in all systems when a plant needs to know if the lights are "on" or "off." In lettuce (*Lactuca sativa*) seeds, Pfr causes the seeds to begin to germinate. Lettuce seeds germinate only when placed in white or red light. They will not germinate if they are buried in deep soil.

The other form of phytochrome, named Pr, is formed when phytochrome is exposed to far-red (730 nm) light. This form is biologically inactive or inhibits responses. Thus, if lettuce seeds are placed in far-red light, they do not germinate.

Insufficient Development

If a seed's embryo is not completely developed, some additional maturation may be needed before the seed can sprout. This happens in seeds with little-to-no storage material invested in the seed. Examples include orchid seeds. They are the size of dust and have almost nothing but a very immature embryo inside. Such a seed needs an association with fungi in the soil or other environments to feed the developing embryo until the embryo is mature enough to actually penetrate the seed coat. These seeds are also likely to have a very brief viability. The fungal association must be established rapidly or the embryo dies.

Inhibitors Present

Many plant species invest chemicals in the developing seeds, and these chemicals inhibit the development of the embryos. They keep the embryos dormant. Obviously, the seed must have some way to eliminate these chemicals before they can sprout.

Abscisic Acid. Many temperate-zone species that use inhibitors use abscisic acid. This chemical induces dormancy in the embryo. The chemical is produced in abundance in the late summer and early fall. The seeds in the fruits become dormant, so even if they are dispersed in autumn, they cannot sprout. During the winter, enzymes in the seeds degrade the abscisic acid. By spring the abscisic acid is gone and the seed can sprout.

Seeds of these species can be forced to sprout early. The seeds are put in moist soil and refrigerated for about 4 weeks. This is sufficient time to degrade the abscisic acid. Then the planted seeds are placed in a warm greenhouse. The seeds assume winter is over, spring has come, and they begin to sprout. This process is called **vernalization**.

Phenolic Compounds. Plants that live in deserts have a different problem. There is no cold, moist, winter to allow vernalization of abscisic acid. These plants instead use more potent toxins, **phenolic compounds**, to keep their seeds dormant until the proper season for germination. Phenolic compounds are freely water-soluble; the plant is living in a desert. Deserts typically have very long dry seasons and a short wet season accompanied by flash floods and so on. When the rains come, the phenolic compounds are leached from the seeds.

The Process

The following describes the germination of corn—seed to seedling. The first step in germination is the absorption of water by the corn kernel. The dormant kernel starts to swell and the chemical changes needed for the growth process get under way. The absorbed water activates enzymes in the scutellum and aleurone layer that break down the food reserves needed to initiate growth in the embryo axis region. Endosperm starches are converted to sugars, which are readily available to the embryonic plant.

During germination, the radicle elongates and is the first structure to break through the seed coat. Next comes the coleoptile, which surrounds

Figure 16-2 Lima bean germination.

(a) Lima bean seed   (b) Germination   (c) After germination

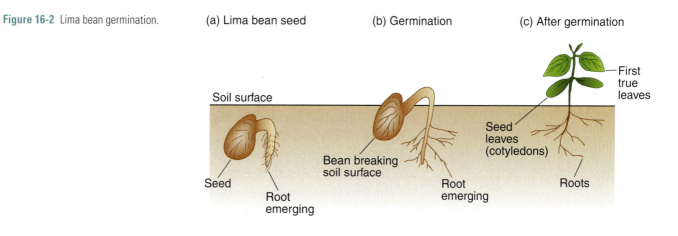

the plumule. Then the two to five seminal roots emerge. Under optimum temperatures, this stage may be reached in 45 days. Under cool soil conditions, the process will take much longer (Figure 16-2).

Initially, the radicle grows in whatever direction the kernel tip is pointed (except up). The other roots grow at varying angles from the horizontal, depending on temperature and moisture conditions. This initial root system serves to help anchor, provide moisture, and eventually absorb nutrients for the young plant.

Emergence is normally accomplished by a combination of coleoptile growth and mesocotyl (first internode) elongation. The mesocotyl is the structure between the scutellar and coleoptilar nodes or crown. The deeper the kernel is planted, the greater the length of the mesocotyl.

The crown area, from which the nodal or **permanent roots** eventually develop, will grow from 1 to 1.5 inches below the soil surface with little variation because of planting depth. Thus, if a kernel is planted 1 inch deep, emergence will be accomplished entirely by growth of the coleoptile. If planting depth is greater than that, the mesocotyl becomes responsible for elevating the coleoptile close enough to the soil surface to make emergence possible.

The coleoptile pushes up through the soil until it reaches light, at which point the tip opens and the first true leaves of the plumule emerge. Seeds or kernels fail to emerge when they are planted deeper than the mesocotyl can elongate or when the coleoptile tip ruptures below the ground.

The time from planting to emergence is influenced most by temperature. Under ideal conditions, emergence can take less than a week, but typically one to two weeks are required. Under very cool conditions, emergence may not occur for two to three weeks. Soil moisture, soil compaction, crusting, planting depth, and so forth also influence time to emergence.

The Seedling. The first leaf blade to emerge has a rounded tip. All others are pointed. (This has importance for leaf number identification later on.) All of the first leaf blade, most of the second, and parts of the third and fourth leaf blades are visible on a newly emerged seedling. Remnants of the coleoptile are seen at the soil surface.

The radicle and seminal roots of the seed root system have been taking up water and nutrients from the soil. Leaves are now carrying on photosynthesis, and the young plant is nearly independent of the seed as a food source.

At or shortly after emergence, the nodal (crown) root system begins to develop and soon becomes dominant as the role of the seed roots system diminishes. Some authors refer to seminal and nodal roots as adventitious roots. Establishment of the nodal roots is very important in order to provide the water and nutrients the corn plant needs for normal growth. Inability of a plant to grow away from many seedling problems often is associated with poor establishment of this root system.

The structure that initiates new leaves is the apical meristem or growing point. At seedling stage, it is below the soil surface because the internodes have not yet begun to elongate. Thus, if the above-ground leaves are destroyed, additional leaves will still emerge unless the growing point has become diseased, frozen, or destroyed in some other manner.

Plants exhibit either hypogeal or epigeal emergence. In **hypogeal emergence**, the growing point remains below the soil surface for a time after emergence. In **epigeal emergence,** the growing point is above the soil surface immediately upon emergence.

ROOTS

Because the roots are unseen they tend to be forgotten for most plants, but they are essential for growth. The structure of roots was described in Chapters 3 and 4. As a reminder, roots serve four important functions:

1. Anchoring the plant in the soil
2. Absorbing water and nutrients
3. Conducting water and dissolved minerals and organic materials
4. Storing food materials

During growth, the root and shoot system tend to maintain a balance. As the top of the plant grows, the leaf area increases, as does water loss through transpiration. This increased water loss is made up by water absorption from an increasing root system.

SHOOT GROWTH

Plant shoot growth is either determinate or indeterminate. In **determinate** growth, after a certain period of vegetative growth, flower clusters form at the shoot terminals so most growth to the shoot elongation stops. Many vegetables grow this way.

Plants with **indeterminate** growth bear the flower clusters laterally along the stem and in the axils of the leaves, so the shoot terminals remain vegetative. The shoot continues to grow until it is stopped by **senescence** or some environmental influence. Pole beans and grapevines are examples of indeterminate growth.

Plantibodies

What if human antibodies could be produced by field crops?

In some new cornfields, the ears are ears of gold. The kernels growing on a few acres could be worth millions—not to grocers or ranchers but to drug companies. This corn is not bred for sweetness, but it is a strain genetically engineered by a biotechnology company to secrete human antibodies. The first of these antibodies from mutant corn seeds will be injected into cancer patients. If the treatment works as intended, the antibodies will stick to tumor cells and deliver radioisotopes to kill them.

Using antibodies as drugs is not new, but manufacturing them in plants is, and the technique could be a real boon to the many biotechnology firms that have spent years and hundreds of millions of dollars trying to bring these promising medicines to market. So far most have failed, for two reasons.

First, many early antibody drugs either did not work or provoked severe allergic reactions. They were not human but mouse antibodies produced in vats of cloned mouse cells. In recent years, geneticists have bred cell lines that churn out antibodies that are mostly or completely human. These chimeras seem to work better than the mouse cells only.

Second, the new drug may be effective but it will not be cheap. Cost is the second barrier faced by these medicines. Cloned animal cells make inefficient factories: 10,000 liters of them eke out only a kilogram or two of usable antibodies. So some antibody therapies, which typically require a gram or more of drug for each patient, may cost more than insurance companies will cover. Low yields also increase the expense and risk of developing antibody drugs.

This is where "plantibodies" come in. By transplanting a human gene into corn reproductive cells and adding other DNA that intensifies the production of the foreign protein in the cells, a biotechnology company has created a strain that it claims yields about 1.5 kilograms of pharmaceutical-quality antibodies per acre of corn. Enough antibodies to supply the entire U.S. market could be grown on just 30 acres.

Currently, the developmental process takes about a year longer in plants than in mammal cells, but start-up costs are far lower, and in full-scale production, proteins can be made for orders of magnitude less cost.

Plantibodies might reduce another risk as well. The billions of cells in fermentation tanks can catch human diseases; plants do not. So even though a biotechnology company must ensure that its plantibodies are free from pesticides and other kinds of contaminants, it can forgo expensive screening for viruses and bacterial toxins.

Corn is not the only crop that can mimic human cells. Biotechnology companies are also cultivating soybeans that contain human antibodies against herpes simplex virus type 2, a culprit in venereal disease, in the hope of producing a drug inexpensive enough to add to contraceptives. Another biotechnology company is testing an antitooth decay mouthwash made with antibodies extracted from transgenic tobacco plants. Still another company has modified tobacco to manufacture an enzyme called glucocerebrosidase in its leaves. People with Gaucher disease pay up to $160,000 a year for a supply of this crucial protein, which their bodies cannot make.

Amazingly, transgenic plants can accurately translate the subtle signals that control human protein processing. The future can only hold more promise as more transgenic plants produce unique biochemicals for human use and health.

Growth Patterns

Shoot growth patterns of plants fall into three classifications: annual, biennial, and perennial.

Annuals. Annuals complete their life cycle in less than 1 year and must be planted again. Examples include corn, wheat, and annual flowers. Shoot growth starts after germination and continues in a fairly uniform pattern until stopped by frost or some senescence-inducing factor. Flowering followed by fruit and seed production occurs during the summer. Figure 16-3 shows a typical life cycle of an annual. Annuals are herbaceous (leaves and stems that die down at the end of the growing season , not woody).

Biennials. Biennials complete their cycle in two growing seasons, not necessarily two years but more than one year. Energy stored by the plant during its first year is used in the reproductive phase during the second year. Examples include celery, parsnip, sugar beets, and asparagus. Figure 16-4 graphically illustrates the growth patterns of biennials. Stem growth is limited during the first growing season. Then the plant remains alive but dormant through the winter. Exposure to cold temperatures triggers

Figure 16-3 The typical life cycle of an annual.

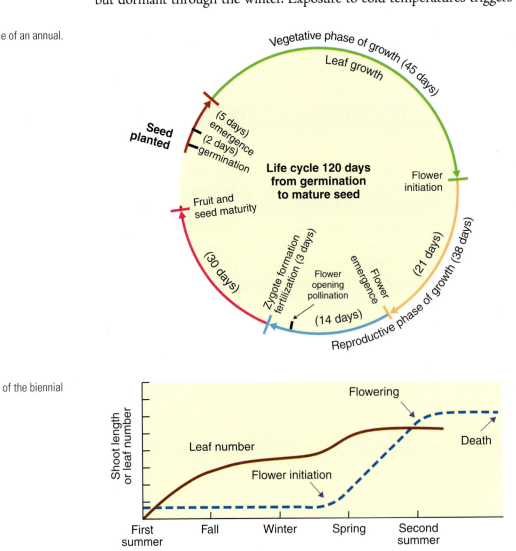

Figure 16-4 The growth curves of the biennial plant.

Figure16-5 Typical growth pattern of the perennial plant.

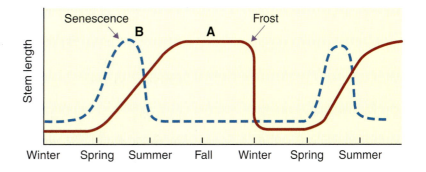

hormonal changes, causing stem elongation. Flowering, fruit formation, and seed set occur during the second growing season. Senescence and death of the plant follow shortly. Biennials are herbaceous.

Perennials. Perennials continue to grow for more than two years. The plant continues to develop vegetatively during and after the reproductive phase. Examples include Bermuda grass and all shrubs and trees. Shoot growth resumes each spring from latent or adventitious buds at the crown of the plant. Figure 16-5 shows a typical growth pattern for a perennial. Perennials can be herbaceous or woody.

In temperate zones, all trees exhibit intermittent annual vegetative growth. These patterns fall into four categories:

1. A single flush of terminal shoot growth during the growing season followed by the formation of a resting terminal bud—oaks, hickories, many conifers, and most fruit trees are examples.
2. A flush of growth followed by shoot tip abortion at the end of the growing season—the shoot the following season starts from a lateral bud. This causes the zigzag pattern of shoot development seen in elm, birch, willow, beech, and honey locust trees.
3. Recurrent flushes of terminal growth with terminal bud formation at the end of each flush—examples of this type of growth include pines in the southeastern United States, some subtropical evergreens, citrus, and Persian walnut.
4. A sustained growth flush for varying periods, producing new growing points that develop as late-season leaves—these plants end the season with the formation of a distinct terminal bud. Examples include the tulip tree and the sweet gum tree.

FACTORS AFFECTING PLANT GROWTH AND DEVELOPMENT

Many factors affect plant growth and development. These are covered in other chapters, for example:

+ Soil fertility in Chapter 7
+ Water use and requirements in Chapter 9
+ Temperature in Chapter 11
+ Light in Chapters 12 and 13
+ Pests such as weeds, insects, and disease in Chapter 18
+ Plant growth regulators and genetic engineering in Chapters 18 and 19

PHASE CHANGES

Seedling plants go through phases of their life similar to animals—embryonic growth, juvenility, maturity, senescence, and death.

Juvenility

For some plants, the juvenile stage may have a very different physical appearance. A primary factor for judging that a plant is in the juvenile stage is its inability to form flowers even though all environmental conditions are conducive for flowering. The onset of flowering and fruiting indicates the end of the juvenile stage. Another important indication of phase change from juvenility to maturity is a loss of or reduced ability to form adventitious roots.

Length of the juvenile stage varies from days to years. For most crops, the juvenile stages last a varying number of days during the growing season. For some forest trees, the juvenile stage can last as long as 30–40 years.

Breeders of fruit and nut crops are eager to bring seedlings through the juvenile stage as rapidly as possible so they will flower and fruit, and the breeders can see the results of their work. Plant propagators like to maintain the juvenile stage as long as possible so that high percentages of cuttings taken from them will root rapidly.

Once the mature phase is reached, a plant does not easily revert to the juvenile stage. Treatments with plant hormones such as gibberellins or grafting tissues from mature plant parts back onto juvenile plants can cause the mature tissue to revert.

Sexually propagated plants do not go through "true" juvenile stage.

Maturity

When a plant reaches maturity it is able to produce flowers, fruit, and seeds. In other words, the plant is able to reproduce. The mature plant may have a different physical appearance. Length of the mature phase may range from days to years.

Flowering. Some annuals mature and can flower in only a few days or weeks after sprouting. Some forest and fruit trees require years before flowering. Once mature, the plant can be induced to flower by becoming sensitive to conditions in the environment, such as photoperiodism (see Chapter 12). After flowering, pollination and fertilization follow.

Fruit Growth and Development

Sometimes tissues associated with the fruit begin to grow once the fruit has set. Food materials from other parts of the plant move into these developing tissues. Hormonal substances such as the auxins, gibberellins, ethylene, and cytokinins may be involved in fruit growth, as they are in fruit set (see Chapter 15, Figure 15-8 for fruit growth curves).

Senescence and Death

Senescence is considered to be a terminal, irreversible deteriorative change in living organisms that leads to cell breakdown and death. It is nonpathogenic. Senescence is an obvious period of physical decline; it is particularly noticeable toward the end of the life cycle of annual plants. Senescence can also occur in leaves, seeds, flowers, or fruits (organ senescence).

Plants show senescence in different ways. In annuals, the entire plant dies at the end of the growing season. This occurs after and probably because of the fruit and seed production. In herbaceous perennials, the tops of the plants die at the end of the growing season, but the shoot grows again in the spring and the roots can live for many years. In deciduous woody perennials, the leaves senesce, die, and fall off each year but the shoot and root remain alive for many years.

Generally, senescence is considered to be caused by inherent physiological changes in the plant, but it can also be caused by a disease attack or environmental stress. Also, as perennials age they become more vulnerable to attacks by fungi, bacteria, or viruses.

Plant senescence is hastened by the transfer of stored nutrients to the flowers, fruits, and seeds. Senescence can be postponed in many plants by picking off the flowers before seeds start to form.

Abscission. During leaf senescence, DNA, RNA, proteins, chlorophyll, photosynthesis, starch, auxins, and giberellins decrease. Abscisic acid is involved in leaf and fruit **abscission**—shedding leaves or fruit.

Fruits. As fruits near their maximum size, significant changes lead to the end product—a fruit with its distinctive color and flavor (Figure 16-6). These changes are all associated with the maturing and eventual senescence of the fruit. All fruits on a plant do not mature at the same time. Most fruits reach maturation and ripen on the tree, vine, or bush. They should be picked at this time. If not picked, they enter a stage of senescence and deterioration.

Figure 16-6 Ripe fruit—Red Delicious apples—the end result of plant development. (Photo courtesy of USDA)

Some fruits are climacteric and others are nonclimacteric. In both types, the respiration rate of the cells (see Chapter 14) slowly decreases as maturation proceeds. In climacteric fruits, the respiration rate rises abruptly as the fruit ripens and reaches a peak and then declines. Senescence follows, leading to the eventual death of all the fruit cells. In nonclimacteric fruits, the respiration rate gradually declines during ripening with no obvious peak.

SUMMARY

The shoots are the desired product of many kinds of plants; for example, hay crops, pastures, lawns, foliage plants, and shade trees. Growth begins with germination. The goal of plant science is to produce the shoot system in a fast and economical manner. Plants may be annual, biennial, or perennial in their growth patterns. The annuals and the biennials are herbaceous. The perennials can be either herbaceous or woody. The shoot system develops through the activities of the meristem that is found in the terminal and lateral buds and in the cambium layer. Growth patterns are controlled by genetic factors as modified by environmental stresses. Important environmental factors influencing plant growth include light, temperature, water, and the gases in the atmosphere. Following germination, plants go through juvenile, mature, and senescence stages. Juvenile plants cannot reproduce. Mature plants reproduce and eventually senesce. Across the plant world, the times for these stages vary widely, from days to years.

Review

True or False

1. The first leaf blade to emerge has a rounded tip.

2. A juvenile plant can reproduce.

3. Abscisic acid is involved in leaf and fruit abscission.

4. The first step in the germination of a seed is the absorption of water.

5. The first structure to break through the seed coat is the radicle.

Short Answer

6. Identify the six steps necessary for germination to occur.

7. Scarification allows thick-coated seeds to _____.

8. What is a general optimum temperature for seeds to germinate?

9. Plant growth occurs by an increase in cell _____ and cell _____.

10. In what region of the plant does the formation of new cells occur?

Discussion

11. List the differences between annuals, biennials, and perennials.

12. Discuss the ways that plants show senescence and death.

13. Besides having a thick seed, what are some other ways seeds are truly dormant?

14. Why do plant breeders want to bring plants through the juvenile stage quickly?

15. How is emergence accomplished in seed germination?

Knowledge Applied

1. Germinate seeds from plants such as some of the vegetable crops. Devise a method for charting the growth of your plants. Include such data as the high and low temperatures and the amount of light.

2. Generate a list of plants based on those around the school or your homes. Group them into list of annuals, biennials, and perennials. Make a table of these plants and give their common name, scientific name, classification (annual, biennial or perennial), and use.

3. Conduct a germination test and report on the results.

4. Make a seed collection. Identify the seeds as monocot or dicot and annual, biennial, or perennial. Also, give the common and scientific names of all the seeds.

5. Make a chart of the fresh fruits found in a grocery store. Classify the fruits using the information in Chapter 3. Also on the chart, indicate how each fresh fruit is stored.

Resources

Hay, R., & Porter, J. (2006). *The physiology of crop yield* (2nd ed.). Ames, IA: Blackwell Publishing.

Kirkham, M. B. (2004). *Principles of soil and plant water relations.* New York: Academic Press.

McMahon, M., Kofranek, A. M., & Rubatzky, V. E. (2006). *Hartmann's plant science: Growth, development, and utilization of cultivated plants* (4th ed.). Englewood Cliffs, NJ: Prentice Hall Career & Technology.

Raven, P. H., Evert, R. F., & Eichhorn, S. E. (2004). *Biology of plants* (7th ed.). New York: W.H. Freeman.

Simpson, B. B., & Ogorzaly, M. C. (2000). *Economic botany: Plants in our world* (3rd ed.). New York: McGraw-Hill.

United States Department of Agriculture. (1992). *New crops, new uses, new markets. 1992 yearbook of agriculture.* Washington, DC: Author.

Internet

Internet sites represent a vast resource of information. The URLs (uniform resource locators) for World Wide Web sites can change. Using one of the search engines on the Internet such as Google or Yahoo!, find more information by searching for these words or phrases: annuals, perennials, biennials, scarification, seed dormancy, germination, plant growth rates, plant emergence, and vernalization.

Chapter 17
Plant Propagation

PLANTS CAN BE produced from seed (sexual reproduction) or by vegetative propagation (asexual reproduction). The world's major food crops produce seeds that people may consume directly, as corn and rice, or process into food, such as bread from wheat and vegetable oil from soybeans. People use many plant parts other than seeds for food, such as leaves (lettuce and cabbage), stems (asparagus), flowers (broccoli), and roots (carrots). Many plants are not grown from seeds. Irish potatoes, for example, are raised from tubers and sweet potatoes from slips. There are also other means of asexual propagation, such as grafting, layering, and rooting of cuttings. The plant scientist needs to understand sexual and asexual propagation.

Feeding the growing population of the world requires continuous improvement in plant varieties. Older varieties are frequently replaced by new varieties that are higher yielding, more pest-resistant, and otherwise superior. The science of improving plant varieties is called plant breeding. It draws heavily on genetics and many of the basic sciences.

After completing this chapter, you should be able to—

- Discuss the differences between sexual and asexual plant propagation

- Describe the production of gametes and the processes of pollination and fertilization

- Differentiate between self-fertilization and cross-fertilization

- Explain how plants produce seeds

- Name four requirements for seeds to germinate and grow

- Explain dormancy

- List five ways plants may be propagated without seeds and give examples of each

- Describe micropropagation

- Explain the basic principles of genetics

- Identify how a mutation is helpful and how it may be created

- Discuss how and why people improve plants

KEY TERMS

Asexual
Budding
Bulbs
Corm
Cross-Pollination
Cuttings
Diploid
Division
Dominant
Dormancy
Double Fertilization
Embryo
Endosperm Nucleus
Explants
Fertilization
Grafting
Haploid
Heterozygous
Homozygous
Hybrid
Hypocotyl
Layering
Megaspore
Meiosis
Micropyle
Microspore
Mutagens
Mutation
Ovary
Ovule
Plantlets
Plumule
Pollen
Pollen grain
Pollination
Radicle
Recessive
Rhizomes
Scion
Self-Pollination
Separation
Sexual
Tetraploids
Triploid
Viability

SEXUAL REPRODUCTION

Sexual reproduction in plants requires that flowers form, pollination and fertilization occur, seeds develop, and that the seeds grow into new plants. Several variations are possible in the process. A complete flower is one having all four main flower parts: sepals, petals, stamens, and one or more pistils. Figure 17-1 illustrates a complete flower.

Figure 17-1 A complete flower.

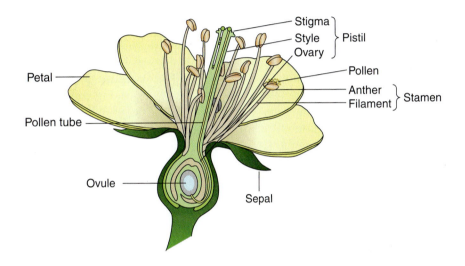

Flower Induction

Flowers having both pistil and stamen are called perfect flowers. Only the pistil and stamen are necessary for seed formation. The stamen is the male part of the flower. A stamen is composed of a filament and an anther. The anther produces **pollen grains**.

The pistil is the female part of a flower. It is made up of the style, the stigma, and the **ovary**, which contains the **ovules**. Most flowers have only one pistil.

The flower is the sexual reproduction unit that functions to produce and house gametes (sex cells) and to attract pollinators. The stalk portion (pedicel) of the flower ends in an enlargement, called the receptacle, from which arise the sepals, petals, stamens (male units), and carpels (female units). Sepals are often green, being the least evolutionarily modified from leaves. The petals are usually larger and colorful, serving to attract the pollinator. Stamens produce and house the **pollen**, which contains the male gametes. Carpels produce and enclose the ovules, which contain the female gametes. A carpel may develop into a simple pistil, or several carpels may fuse to form a compound pistil. Whether simple or compound, a pistil typically consists of an enlarged ovary, from which arises an elongated style topped by a pollen-receiving stigma. Plants use many agents for transporting pollen from one flower to another, including wind, insects, birds, and bats.

Meiosis

Sexual activities begin within the ovules and anthers of a flower. Diploid (2n) cells in the pollen sacs on the anther go through meiosis to produce haploid cells known as microspores. Each microspore nucleus divides by mitosis to form two haploid nuclei—the tube nucleus and the generative nucleus. The outer wall of the microspore hardens and the structure becomes the pollen grain (Figure 17-2).

Figure 17-2 The formation of a pollen grain.

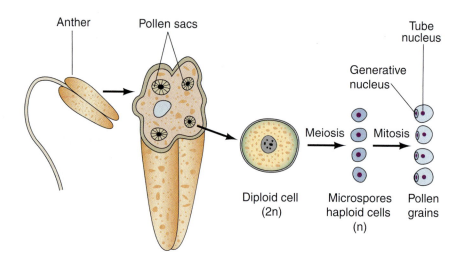

A similar process takes place within each ovule. A single diploid cell of the ovule undergoes meiosis to produce four haploid cells. Only one of these survives (Figure 17-3). This surviving cell is called a megaspore. In some flowers, the megaspore nucleus divides by mitosis to form two haploid nuclei. Mitosis continues until a total of eight haploid nuclei are produced. These nuclei and the cytoplasm around them represent the entire gametophyte generation. Of the eight nuclei produced, only three are important in reproduction—two polar nuclei in the center of the ovule and the true egg at one end. The remaining five haploid nuclei die (Figure 17-3).

Figure 17-3 The formation of an egg.

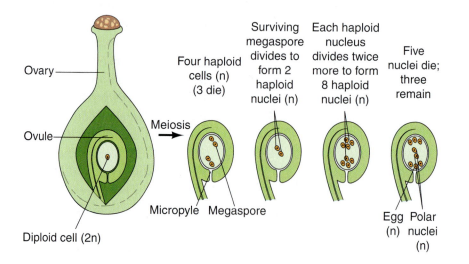

Pollination and Fertilization

Pollination is the transfer of pollen from an anther to a stigma. If pollen is transferred from the anther of a flower to the stigma of the same flower or another flower on the same plant, it is self-pollination. Pollination can be natural or it can be accomplished by people.

The transfer of pollen from the anther of one flower to the stigma of a flower on another plant of the same species is cross-pollination.

Many plants have anthers and stigmas very close together. Self-pollination occurs when pollen falls onto the stigma or is brushed from the anthers to the stigma by contact between the anther and stigma. Soybean and tomato are examples of this type of plant.

Corn, pine trees, and some other plants are cross-pollinated by wind, while many of the horticultural plants rely on insects for cross-pollination. Honeybees and other insects pick up sticky pollen from the anthers of flowers they visit. Pollen grains stick to the bees and are rubbed off on stigmas of flowers they visit later. Self-pollination and cross-pollination are illustrated in Figure 17-4.

After pollination, a small tube grows from the pollen grain down through the style into the ovary. A cell (gamete) from the pollen grain moves down through the pollen tube and is released into the ovary through a small opening called the micropyle. The pollen gamete unites with a cell (gamete) from the ovary. The union of gametes from pollen and ovary is called fertilization. Meanwhile, the other haploid sperm nucleus that was carried down the pollen tube to the ovule joins with the two polar nuclei to

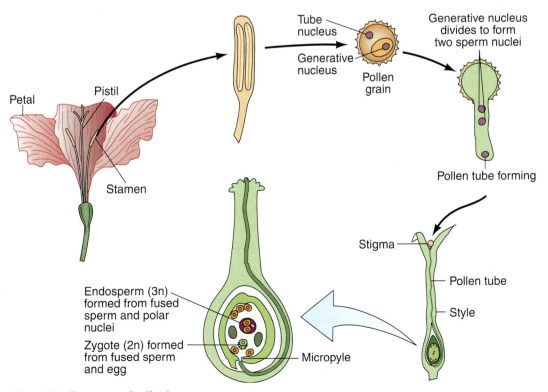

Figure 17-4 The process of pollination.

form a **triploid** (3n) structure called the **endosperm nucleus**. These two fusions are known as **double fertilization**.

After fertilization, a seed develops. The zygote (2n) develops into an **embryo** and the endosperm nucleus (3n) divides many times to form a mass of tissue called the endosperm. This is the food source for the embryo.

Some plants have separate male and female flowers on the same plant (imperfect flowers). Examples are corn, nut trees, cucumbers, watermelon, cantaloupe, and pumpkins.

Other plants have only male flowers on one plant and female flowers on another. American holly, spinach, and asparagus are examples of this type. The male plants flower and produce pollen that must be transferred to flowers on female plants. The male plants cannot produce seed or fruit.

Fruit Development and Maturation

Upon fertilization, the ovary begins developing into a fruit and the ovules into seeds. The function of the fruit is to aid in the dispersal of the seeds. Some fruits develop a fleshy, edible wall that attracts fruit eaters, which will disperse the seeds in their droppings; others develop a dry, hard wall that splits open, allowing the seeds to be shaken out by some disturbance. Some, such as the dandelion, develop light, feathery structures that are carried by the wind. Still others have hooks or barbs that stick to passing animals.

Seeds

As the fertilized egg within the ovule develops into an embryo, the ovule walls develop into a seed coat, forming the ovule into a seed. The seed serves as the unit of dispersal for the new plant. It also provides some protection from injury and drying and some nourishment for the young plant until it can make its own food. Chapter 3 provides a more complete discussion on fruits and seeds.

Seeds are living organisms in a dormant or inactive state. There are many kinds of seeds, but most of them have three main parts. These parts are the seed coat, a supply of stored food (cotyledons or endosperm), and an embryo. These parts are illustrated in Figure 17-5. The seed coat protects the seed and may be very thick and hard. The cotyledons, or endosperm, feed the embryo until the young plant can make its own food. The embryo is made up of three parts: the young shoot (**plumule**), the stem (**hypocotyl**), and the root (**radicle**).

Figure 17-5 Parts of a seed.

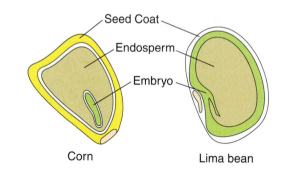

Seed Coat

Endosperm

Embryo

Corn   Lima bean

A seed will begin to grow (germinate) when conditions inside and outside the seed are suitable. The most critical factors for germination are moisture, oxygen, and temperature.

Water. Water helps soften the seed coat. The seed absorbs moisture and swells, and the embryo begins to grow by cell division. Water is necessary for movement of the stored food in a seed to the growing points of the embryo. The growing embryo breaks though the seed coat that has been softened by water.

Temperature. The temperature at which germination occurs may range widely. Some seeds, like cabbage, germinate best at low temperatures, and others, like okra or cotton, require warmer temperatures (see Table 11-2).

Oxygen. Germinating seeds need oxygen. Dormant seeds also require oxygen, but not as much as the germinating ones. Seeds planted in soil saturated or covered by water may swell because they have absorbed water, but fail to germinate because they do not get enough oxygen.

Light. Seeds of some plants require light for germination. The seeds of lettuce, for example, require exposure to light.

Dormancy

Seeds will not germinate while they are dormant. There are several kinds of dormancy. In some plants, seed dormancy prevents germination in the fall when the plants would be killed by low temperature. Seeds of some plants have a very hard seed coat that prevents absorption of moisture by the seed. If no moisture passes through the seed coat to reach the embryo, the seed cannot germinate. Examples of plants with hard seed coats are alfalfa, clover, honey locust, and black locust. These seeds do not germinate until the seed coat loses its resistance to moisture.

Some seed coats temporarily keep oxygen from reaching the embryo. This prevents growth of the embryo. Oat seeds are sometimes dormant because of oxygen exclusion from the seed.

Sometimes the seed coat may be too strong for the embryo to break through. Germination takes place only after the seed coat is weakened or broken. Examples of this type of seed are the common weeds pigweed, shepherd's purse, and peppergrass.

Another kind of dormancy is due to the embryo itself. It may not be fully developed, or chemical changes must take place in the embryo or in the stored food before the seed will germinate. Apple, hawthorn, basswood, dogwood, and pine have this type of dormancy.

There are several ways to overcome dormancy and force seeds to germinate. Seeds may be soaked in hot water for several hours. Some seeds are soaked in acid to soften the seed coat. Another way to break dormancy caused by hard seed coat is scarification. In scarification the seed coat is carefully scratched by some mechanical process to allow moisture or oxygen to enter the seed. A common way to overcome dormancy caused by the embryo is to place seeds in modified cold storage. Seeds can be placed in a plastic bag containing moist sand or peat moss and held at a temperature

Table 17-1 Seed Viability

Years	Seed
1	Onion, parsnip
2	Pepper, soybean, okra
3	Bean, carrot, broccoli
4	Beet, squash, eggplant, pumpkin, cabbage, tomato, kale, turnip, watermelon
5+	Flowers: zinnia, marigold, petunia; Vegetables: cucumber, melon, lettuce; Grains: wheat, corn, alfalfa; Trees: black locust

of 40°F for one to four months. This is a form of artificial aging of seed; the temperature is about that inside a refrigerator.

The viability (ability to germinate) of seeds depends on the kind of seed and conditions under which it is stored. Most seed remains viable longest under cool, dry conditions. Some seeds stored for 80 years have been found to be still viable. Table 17-1 shows the number of years seeds remain viable under favorable storage conditions.

ASEXUAL REPRODUCTION

Although the most common form of plant reproduction is from seed, some plants, like cultivated banana, do not produce seed. Other plants, such as most fruits, have seeds but the seed may not always produce plants or fruit identical to the parent plants. Asexual reproduction of these plants is used to ensure that new plants are identical to the parents. Asexual reproduction may also be used to dwarf plants or to maintain variegated or otherwise desirable specimens. Eight common types of asexual reproduction include:

1. Cuttings—A portion of a plant is removed and made to form roots. This is commonly used to propagate shrubs and house plants.
2. Grafting—A shoot or scion is removed from the desired plant and placed on another plant (the stock). This method is used with some fruit and nut trees.
3. Budding—A bud is removed from the desired plant and placed on the stock. This method is used with some fruit trees and ornamentals such as roses.
4. Layering—A portion of an attached shoot is partially buried underground where roots develop. The new plant may then be separated from the parent plant. Figs, raspberries, blackberries, and many ornamentals can be propagated this way.
5. Dividing—Plants like daylilies, mums, and daffodils form clumps of plants. These may be dug and cut apart to form new plants.
6. Rhizomes—Iris, Bermuda grass, johnsongrass, and other plants produce underground stems called rhizomes. The rhizomes can be dug, cut into sections, and the sections planted. New plants and roots grow from joints on the rhizomes.

7. Stolons—Strawberries produce stems that grow horizontally along the soil surface. New plants are produced at joints along the stolons.

8. Tillers or suckers—These are formed at the base of some trees like mimosa, sassafras, and some plums when the trees are cut. Blackberries, rabbit eye blueberries, raspberries, and many grasses also produce tillers. These tillers may be cut free from the main plant to form new plants.

Cuttings

Propagation by cutting is the most common method of asexual or vegetative reproduction. A cutting is any vegetative plant part such as stem, leaves, or roots that, when detached from the parent plant, is capable of reproducing a plant exactly like the parent. Stem cuttings may be herbaceous cuttings (nonwoody plants), softwood cuttings (normally woody plants; cuttings made from new growth before it hardens), and hardwood cuttings (from a woody plant after the stem has hardened).

Cuttings from roses and spring-flowering shrubs (softwood cutting) should be made in midsummer, when the shoot is still growing. Cuttings of some evergreens, holly, boxwood, yew, and juniper (hardwood cuttings) root best if taken from the plants in late fall or early winter. Most houseplants or herbaceous plant cuttings will root at almost any season.

Stem cuttings should be 6 inches long with three to four leaves retained at the terminal end. Cut the stem at a 45-degree angle immediately below a node. Only healthy, insect-free cuttings should be selected. Early morning is the best time to take cuttings. Keep the cut end moist until it is rooted.

Cuttings are taken with a sharp knife or razor blade to reduce injury to the parent plant. Dipping the cutting tool in rubbing alcohol or a mixture of one part bleach and nine parts water prevents transmitting diseases from infected plant parts to healthy ones. Flowers and flower buds are removed from cuttings to allow the cutting to use its energy and stored carbohydrates for root and shoot formation rather than fruit and seed production. To hasten rooting, increase the number of roots, or to obtain uniform rooting except on soft fleshy stems, a rooting hormone, preferably one containing a fungicide, should be used.

Cuttings are inserted into a rooting medium such as coarse sand, vermiculite, soil, water, or a mixture of peat and perlite. Selection of the correct rooting medium is important to get optimum rooting in the shortest time. In general, the rooting medium should be sterile, low in fertility, drain well enough to provide oxygen, and retain enough moisture to prevent water stress. The medium needs to be moist before inserting cuttings and kept evenly moist while cuttings are rooting and forming new shoots.

Stem and leaf cuttings can be placed in bright but indirect light. Root cuttings can be kept dark until new shoots appear.

Stem Cuttings. Numerous plant species are propagated by stem cuttings. Some can be taken at any time of the year, but stem cuttings of many woody plants must be taken in the fall or in the dormant season (Figure 17-6).

Figure 17-6 Stem tip cuttings.

Stem cutting with unwanted portions
(leaves, seed heads, and flowers) removed

Tip Cuttings. Tip cuttings are made by detaching a 2- to 6-inch piece of stem including the terminal bud. The cut is made just below a node. Lower leaves that would touch or be below the medium are removed before dipping in the rooting medium and placing in the media. At least one node must be below the surface (Figure 17-6).

Medial Cuttings. For medial cuttings the first cut is made just above a node, and the second cut just above a node 2–6 inches down the stem. Medial cutting must be positioned right side up. Axial buds are always above leaves.

Cane Cuttings. Canelike stems are cut into sections containing one or two eyes, or nodes. The ends are dusted with fungicide or activated charcoal. After drying several hours, the cuttings are laid horizontally with about half of the cutting below the media surface, eye facing upward. Cane cuttings are usually potted when roots and new shoots appear, but new shoots from dracaena and croton are often cut off and rerooted in sand (Figure 17-7).

Single Eye. The eye refers to the node. This is used for plants with alternate leaves when space or stock materials are limited. The stem is cut

Figure 17-7 Cane cutting.

Node or eye   Medium line

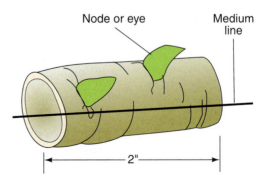

2"

about 1/2 inch above and 1/2 inch below a node. The cuttings are placed horizontally or vertically in the medium (Figure 17-8).

Double Eye. This is used for plants with opposite leaves when space or stock material is limited. Stems are cut about 1/2 inch above and 1/2 inch below the same node and inserted into the medium vertically with the node just touching the surface (Figure 17-9).

Heel Cutting. This method uses stock material with woody stems efficiently. Make a shield-shaped cut about halfway through the wood around a leaf and axial bud. The shield is inserted horizontally into the medium (Figure 17-10).

Leaf Cuttings. Leaf cuttings are used almost exclusively for a few indoor plants. Leaves of most plants will either produce a few roots but no plant, or they will just decay (Figure 17-11).

Figure 17-8 Single-eye cutting.

Figure 17-9 Double-eye cutting.

Figure 17-10 Heel cutting.

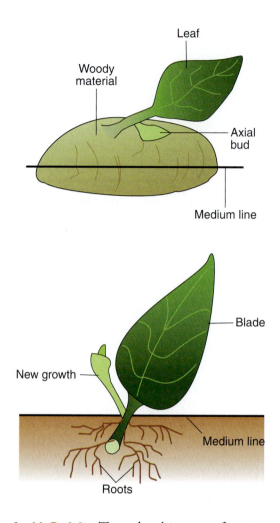

Figure 17-11 Leaf cutting.

Whole Leaf with Petiole. To make this type of cutting, the leaf and 0.5–1.5 inch of petiole is detached. The lower end of the petiole goes into the medium. One or more new plants will form at the base of the petiole. The leaf may be severed from the new plants when they have their own roots, and the petiole can be reused (Figure 17-12).

Figure 17-12 Leaf petiole cutting.

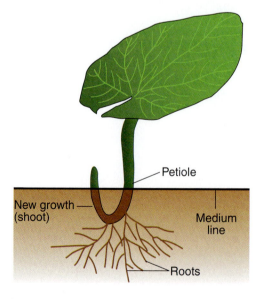

Whole Leaf without Petiole. This is used for plants with sessile or petioleless leaves. The cutting is inserted vertically into the medium. A new plant will form from the axillary bud. The leaf may be removed when the new plant has its own roots.

Split Vein. For a split vein cutting, a leaf from the stock plant is detached and its veins are slit on the lower leaf surface. The cutting is laid, lower side down, on the medium. New plants will form at each cut (Figure 17-13).

Leaf Sections. This method is frequently used with snake plant and fibrous rooted begonias. Begonia leaves are cut into wedges with at least one vein. A new plant will arise at the vein. Snake plant leaves are cut into 2-inch sections. Roots will form fairly soon, and eventually a new plant will appear at the base of the cutting. These and other succulent cuttings will rot if kept too moist (Figure 17-14).

Root Cuttings. Root cuttings are usually taken from two- to three-year-old plants during their dormant season when they have a large carbohydrate supply. Root cuttings of some species produce new shoots, which then form their own root systems; root cuttings of other plants develop root systems before producing new shoots (Figure 17-15, *A* and *B*).

Figure 17-13 A split vein cutting.

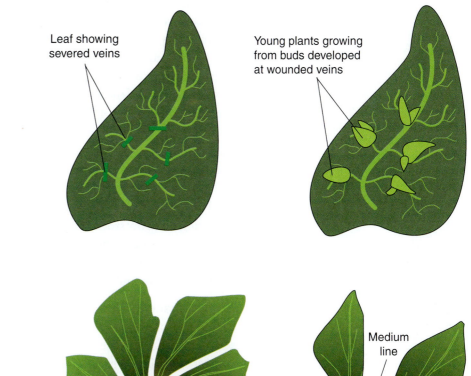

Leaf showing severed veins

Young plants growing from buds developed at wounded veins

Figure 17-14 Leaf section cutting.

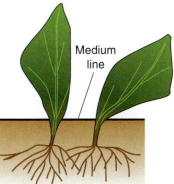

Medium line

Leaf sections cut into wedge-shaped pieces

Wedge-shaped sections showing root growth

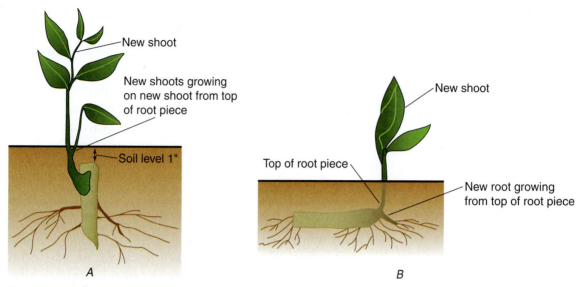

Figure 17-15 *A*, Fleshy root cutting. After the cutting is placed in soil, it is given no special care. *B*, Small, nonfleshy root cutting. This cutting is placed horizontally in the soil.

Grafting

Grafting and budding are methods of asexual plant propagation that join plant parts so they will grow as one plant. These techniques are used to propagate cultivars that will not root well as cuttings or whose own root systems are inadequate. One or more new cultivars can be added to existing fruit and nut trees by grafting or budding.

The portion of the cultivar that is to be propagated is called the scion. It consists of a piece of shoot with dormant buds that will produce the stem and branches. The rootstock, or stock, provides the new plant's root system and sometimes the lower part of the stem. The cambium is a layer of cells located between the wood and bark of a stem from which new bark and wood cells originate.

Four conditions must be met for grafting to be successful:
+ Scion and rootstock must be compatible
+ Each must be at the proper physiological stage
+ Cambial layers of the scion and stock must meet
+ Graft union must be kept moist until the wound has healed

Cleft Grafting. Cleft grafting is often used to change the cultivar or top growth of a shoot or a young tree (usually a seedling). It is especially successful if done in the early spring.

Bark Grafting. Unlike most grafting methods, bark grafting can be used on large limbs, although these are often infected before the wound can completely heal.

Whip or Tongue Grafting. This method is often used for material 1/4–1/2 inch in diameter. The scion and rootstock are usually of the same diameter, but the scion may be narrower than the stock. This strong graft heals quickly and provides excellent cambial contact (Figure 17-16).

Very little success in grafting will be obtained unless proper care is maintained for the following year or two. If a binding material such as

(A)
The scion before any cuts are made.

(B)
The first cut is made in the scion.

(C)
The second cut is made in the scion.

(D)
The root, before any cuts are made.

(E)
The first cut is made in the root.

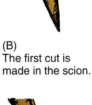

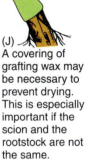

(F)
The second cut is made in the root.

(G)
The scion and root are positioned for joining.

(H)
The scion and root are pushed together. (Cambrium must match on at least one side.)

(I)
The two pieces are tied together.

(J)
A covering of grafting wax may be necessary to prevent drying. This is especially important if the scion and the rootstock are not the same.

Figure 17-16 Typical grafts.

strong cord or nursery tape is used on the graft, this must be cut shortly after growth starts to prevent girdling and later dying of the graft. Rubber budding strips have some advantages over other materials. They expand with growth and usually do not need to be cut as they deteriorate and break after a short time. It is also an excellent idea to inspect the grafts after a two- to three-week period to see if the wax has cracked, and if necessary reapply wax to the exposed areas. After this the union will probably be strong enough and no more waxing will be necessary.

Limbs of the old variety that are not selected for grafting should be cut back at the time of grafting. The total leaf surface of the old variety should be gradually reduced as the new one increases, until at the end of one or two years the new variety has completely taken over. Completely removing all of the limbs of the old variety at the time of grafting increases the shock to the tree and causes excessive suckering. Also, the scions may grow too fast, making them susceptible to wind damage.

Budding

Budding, or bud grafting, is the union of one bud and a small piece of bark from the scion with a rootstock. It is especially useful when scion material is limited. It is also faster and forms a stronger union than grafting.

Patch Budding. Plants with thick bark should be patch budded. This requires removal of a rectangular piece of bark before the insertion of a bud. This is done while the plants are actively growing so their bark slips easily.

Chip Budding. This budding method can be used when the bark is not slipping. Slice downward into the rootstock at a 45-degree angle through one-fourth of the wood (Figure 17-17).

T-Budding. This is the most commonly used budding technique. A T-shaped cut for inserting the bud is made when the bark is slipping (Figure 17-18).

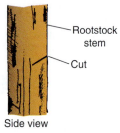

Side view

(a) A 45° cut is made in the rootstock about one-quarter of the way through the stem.

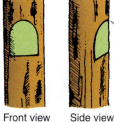

Front view   Side view

(b) A second cut is made starting about 1½ inches above the first cut and extended down to meet the first cut. *Cut deeply enough at the top so that an upside-down "U" form is made. A cut that is made too shallow results in an "A" shape.*

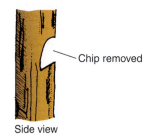

Side view

(c) The chip produced by the two cuts is removed.

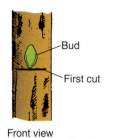

Front view   Side view

(d) The bud to be inserted is cut from the bud stick exactly as the chip was removed from the rootstock.

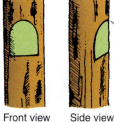

Side view   Front view

(e) Second cut on the bud.

Side view

(f) The bud is removed from the bud stick.

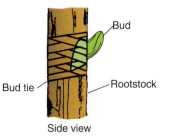

Side view

Note: Be sure to align the cambium of the bud and the rootstock on at least one side

(g) The bud is inserted in the rootstock and tied with rubber bud tie.

Figure 17-17 Chip budding procedure.

Figure 17-18 Making a T-bud.

Scion
The bud shield is cut and removed.

Stock
A T-cut is made through the bark.

The shield is inserted until tops of the shield and T-cut are even.

The bud is left exposed while remaining surfaces are wrapped tightly with rubber strips.

Layering

Stems still attached to their parent plants may form roots where they touch a rooting medium. Severed from the parent plant, the rooted stem becomes a new plant. This method of vegetative propagation, called **layering**, promotes a high success rate because it prevents the water stress and carbohydrate shortage that plague cuttings.

Some plants layer themselves naturally, but sometimes plant propagators assist the process. Layering is enhanced by girdling the stem where it is bent, by wounding one side of the stem, or by bending it very sharply. The rooting medium should always provide aeration and a constant supply of moisture.

Tip Layering. The shoot tip is inserted in a hole three to four inches deep and covered with soil. The tip grows downward first, then bends sharply and grows upward. Roots form at the bend and the recurved tip becomes a new plant (Figure 17-19). The tip layer plant can be removed and planted in the early spring or late fall. Tip layering can be used on purple and black raspberries and trailing blackberries.

Figure 17-19 Tip layering.

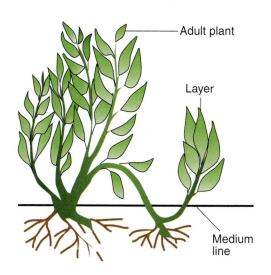

Adult plant

Layer

Medium line

Figure 17-20 Simple layering.

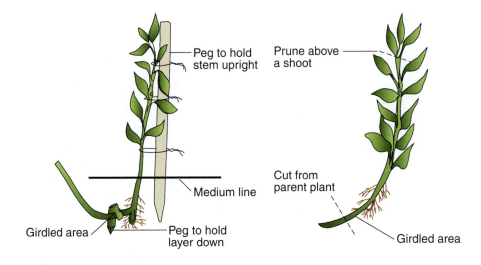

Peg to hold stem upright

Prune above a shoot

Medium line

Cut from parent plant

Girdled area

Peg to hold layer down

Girdled area

Simple Layering. A stem is bent to the ground and partially covered with soil, leaving the last six to twelve inches exposed. The tip is bent into a vertical position and staked in place (Figure 17-20). The sharp bend will often induce rooting, but wounding the lower side of the branch or loosening the bark by twisting the stem may help. Simple layering is used with rhododendron and honeysuckle.

Compound Layering. This method works for plants with flexible stems. The stem is bent to the rooting medium as for simple layering, but stem sections are alternately covered and exposed (Figure 17-21). The lower side of the stem sections to be covered is wounded. Compound layering works for heartleaf philodendron and pothos.

Mound (Stool) Layering. The plant is cut back to 1 inch above the ground in the dormant season. In the spring, soil is mounded over the emerging shoots to enhance their rooting (Figure 17-22). This type of layering works for gooseberries and apple rootstocks.

Air Layering. Air layering is used to propagate some indoor plants with thick stems, or to rejuvenate them when they become leggy. The stem is slit just below a node. Then the slit is pried open with a toothpick and the wound is surrounded with wet, unmilled sphagnum moss. Plastic or foil is wrapped around the sphagnum moss and tied in place (Figure 17-23).

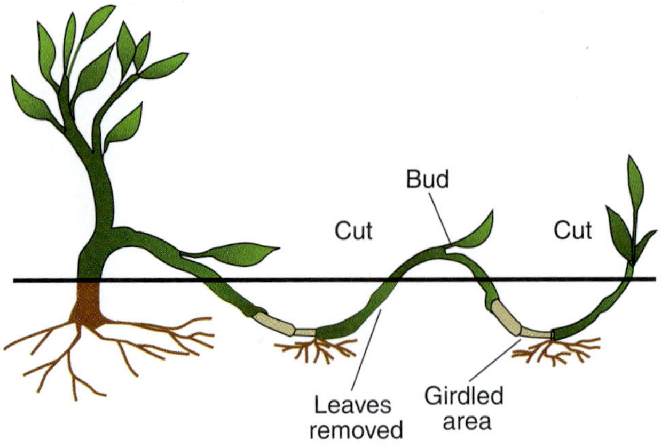

Figure 17-21 Compound layering.

Bud

Cut

Cut

Leaves removed

Girdled area

Figure 17-22 Mound (or stool) layering.

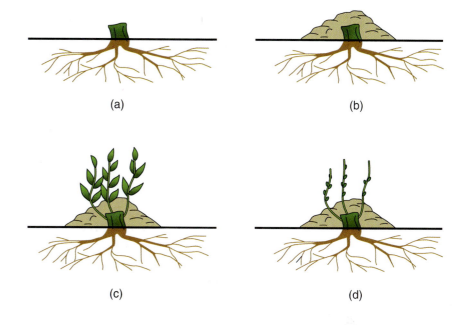

(a)

(b)

(c)

(d)

When roots pervade the moss, the plant is cut off below the root ball. Air layering works for dumb cane and rubber tree.

The following propagation methods can all be considered types of layering, as the new plants form before they are detached from their parent plants.

Stolons and Runners. A stolon is a horizontal, often fleshy, stem that can root, then produce new shoots where it touches the medium. A runner is a slender stem that originates in a leaf axil and grows along the ground or downward from a hanging basket, producing a new plant at its tip (Figure 17-24). Plants that produce stolons or runners are propagated by severing the new plants from their parent stems. Plantlets at the tips of runners may be rooted while still attached to the parent, or detached and placed in a rooting medium. Examples include the strawberry and spider plant.

Offsets. Plants with a rosetted stem often reproduce by forming new shoots at their base or in leaf axils. New shoots can be severed from the parent plant after they have developed their own root system

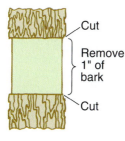

Cut
Remove 1" of bark
Cut

1. Prepare stem.

2. Soak Sphagnum moss and squeeze out excess water.

3. Pack damp moss over girdled area and tie.

4. Wrap with plastic and tape ends tightly

Figure 17-23 Air layering.

Figure 17-24 Propagation of stolons and runners.

(Figure 17-25). Unrooted offsets of some species may be removed and placed in a rooting medium. Some of these must be cut off, while others may be simply lifted off the parent stem. Offset propagation works for date palm, haworthia, bromeliads, and many cacti.

Separation. Separation is a term applied to a form of propagation by which plants that produce **bulbs** or **corms** multiply.

+ **Bulbs.** New bulbs form beside the originally planted bulb. Separate these bulb clumps every three to five years for largest blooms and to increase bulb population. The clump can be dug after the leaves have withered. Bulbs are then gently pulled apart and replanted immediately so their roots can begin to develop. Small new bulbs may not flower for two or three years, but large ones should bloom the first year. This works for the tulip and narcissus.

+ **Corms.** A large new corm forms on top of the old corm, and tiny cormels form around the large corm (Figure 17-26). After the leaves wither, the corms are dug and allowed to dry in indirect light for two or three weeks. Then the cormels are removed and the new corms are separated from the old corms. All new corms should be dusted with a fungicide and stored in a cool place until planting time. Crocus and gladiolus are propagated with corms.

Division. Plants with more than one rooted crown may be divided and the crowns planted separately. If the stems are not joined, they can be gently pulled apart. If the crowns are united by horizontal stems, the stems and roots need to be cut with a sharp knife to minimize injury. Divisions

Figure 17-25 Propagation by offsets.

Figure 17-26 Propagation by separation.

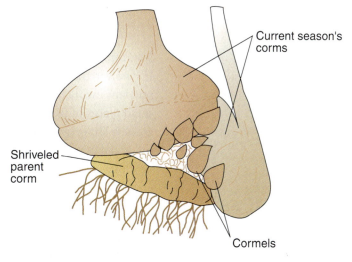

Gladiolus corm system

of some outdoor plants should be dusted with a fungicide before they are replanted. **Division** propagation can be used for the snake plant, iris, prayer plant, and day lilies. Commercial and garden potatoes are propagated by division (Figures 17-27 and 17-28).

Hosta root clump before division

Hosta root divisions

Figure 17-27 Propagation by rhizomes.

Figure 17-28 Propagation by division of the potato.

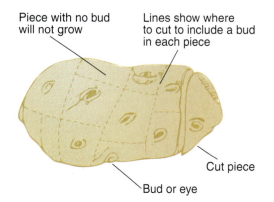

MICROPROPAGATION

Micropropagation, or tissue culture, developed in the 1970s and 1980s. It involves the use of very small pieces of plant tissue grown on sterile nutrient media under aseptic conditions in a small glass container. These small pieces of plant tissue are called **explants**. They are used to regenerate new shoot systems that can be separated for rooting and growing into full-sized plants. Some nurseries use this method to produce millions of new plantlets a year. The process requires trained technicians working under sterile conditions.

Micropropagation was first used on plants such as ferns, orchid, gerberas, carnations, tobacco, chrysanthemum, asparagus, gladiolus, strawberry, and many other herbaceous plants. After modification of the media, woody plants such as azaleas, roses, plums (Figure 17-29), apples, and many others were successfully propagated.

Figure 17-29 Genetically engineered plums in tissue culture; new lines providing resistance to plum pox virus. (Photo courtesy of USDA ARS)

Different parts of the plant can be taken as an explant, including entire seeds, shoot tips, and even single grains of pollen.

Besides being a method of propagation, micropropagation under aseptic conditions is used to eliminate pathogens and maintain germplasm. The media and the plant tissue are sterile; micropropagation must be done with laboratory skill and procedures to prevent contamination.

The nutrient media for growing excised plant parts includes mineral salts, sugar, vitamins, auxins, cytokinins, and sometimes organic complexes such as coconut milk, yeast extract, or banana puree. Many media have been developed for many different plants.

PLANT IMPROVEMENT

Plant breeding is the deliberate attempt to change the genetic architecture of a plant species and is now considered to be an applied branch of the field of genetics, although significant improvements were made in all the important crop plants long before anyone had an understanding of genetic principles. About 200 species of plants have been domesticated, and plant breeding has been practiced directly or indirectly on all of them.

Brief History

Einkorn and emmer wheat kernels that date back to 6750 BC have been found in Iraq. The primitive hulled wheats were largely replaced by the naked durum and poulard wheats in the Mediterranean region during the period 500 BC to 500 AD, and the common and club bread wheats were beginning to replace durum toward the end of this span. Excavations have shown that corn was used as food in Mexico as early as 5000 BC. Although primitive, this corn was unquestionably domesticated. No wild forms of corn have been found.

Rice sowing was an important religious ceremony in China by 3000 BC, and differences among rice varieties were recognized as early as 1000 BC in India. Evidence of the cultivation of these grass crops as food and of flower horticulture in such regions as China suggests that plant breeding was practiced by ancient civilizations. The primary objectives of early cultivators appear to have been to increase grain size and ease of harvesting and threshing, but indirect selection for yield and resistance to disease also must have been practiced.

These objectives changed very little during the Middle Ages, although rye replaced wheat as the most important grain crop in Europe during this period. Although few changes were made in plant-breeding methods during the eighteenth and nineteenth centuries, a number of significant discoveries were reported that revolutionized procedures and firmly established plant breeding as a scientific discipline in the early part of the twentieth century. Chief among these were Joseph G. Koelreuter's observations (1763) that plant hybrids often possessed unusual vigor. Charles Darwin's publications on evolution, inbreeding, and natural selection, which influenced plant breeder Luther Burbank, and Gregor Mendel's discovery of the basic laws of heredity in 1865 were other major happenings.

The realization that yield and quality of plant products are quantitatively inherited—a large number of genes contribute to these characteristics—and that controlled hybrids could be produced, led to the basic breeding and evaluation procedures that constitute modern plant breeding. One of the first researchers to apply these breeding theories was the American botanist George Shull. From two pure strains of corn, he developed a hybrid corn in 1909 whose yield was enormously greater than those of pure strains. Hybrid corn is now cultivated worldwide, and the development of other food crop hybrids is being researched.

Modern Practices

Yield is still the primary selection criterion in plant breeding. It is affected by almost all factors that influence the growth and health of a plant,

Improving Sunflowers

Oilseed researchers from Fargo, North Dakota, are working to improve commercial sunflower production. To do this, they are using wild sunflowers. Wild species of sunflowers might provide their cultivated cousins with genes for such valuable traits as drought tolerance or disease resistance. Until recently, embryos of most crosses between cultivated sunflowers and wild species would not develop into fertile plants.

Now scientists have broken the interbreeding barrier for at least nine difficult-to-cross species by first growing a meager portion of the hybrid embryos on a new tissue-culturing medium. Then the researchers treat the "rescued" embryos with the chemical colchicine to double the number of chromosomes in reproductive cells. Plants called amphiploids are produced from these male and female reproductive cells with doubled chromosomes. These amphiploids of interspecies hybrids are genetically compatible with other sunflowers used in breeding experiments.

including such inherited characteristics as nutrient uptake, photosynthetic efficiency, rate and period of growth, and the number, size, and density of seeds produced. Diseases, insects, and weather conditions can seriously affect yield, and a breeding program must consider resistance or avoidance of these external factors. Cultural practices and user requirements must also be incorporated into the breeding plan. Mechanized agriculture, for instance, requires that all plants in a field mature at virtually the same time.

Gene banks contain collections of seeds of wild relatives and unimproved and improved varieties of crops. They have been organized to preserve the genetic resources of major food crops for future use by plant breeders. Efforts of agricultural research in the 1960s to find ways for farmers in developing countries to produce far more food on the same amount of land led to early successes popularly called the Green Revolution, initiated by a plant scientist, Dr. Norman Borlaug.

Basic Genetics

Living organisms produce offspring (progeny) similar to themselves. However, the degree of resemblance between parents and their offspring is related to the method of reproduction and the genetic makeup of the parents. A soybean seed will grow into a plant almost identical to its parent, but an apple seed may produce a tree very different from the tree on which it grew. An apple tree grown from seed, for example, may have a different size, shape, hardiness, amount of fruit, or fruit quality than the tree from which it originated. This is because some plants contain a greater degree of genetic variation than others.

New variations that are not present in the parent plants may also suddenly appear. An apple tree or a part of an apple tree may produce red apples although the ancestors, or even other branches on the tree, produce green apples. This sudden genetic change is called a **mutation**.

Genetics is the science that studies natural variation and mutations. Plant breeders are trained in genetics and use natural variation and mutations to produce improved varieties of plants.

Some plants have only a few chromosomes in the cell nucleus, while others have 100 or more. Chromosomes occur in pairs in all higher plants. A plant with 26 chromosomes has 13 pairs of chromosomes in every cell nucleus. The two chromosomes that form a pair are identical in appearance, but each pair of chromosomes may have unique identifying characteristics. In corn, for example, chromosome 7 is very short and chromosome 4 is very long. Cytogenetics is the science that studies chromosome structure and function. Under the microscope, chromosomes appear as small, thick wormlike structures that are bent in different shapes (Figure 17-30).

Seeds that develop from pollination followed by fertilization produce a new plant with the same number of chromosomes as the parent plant or plants. Seeds develop only when a flower is fertilized with pollen from the same species.

Figure 17-30 Typical plant chromosomes.

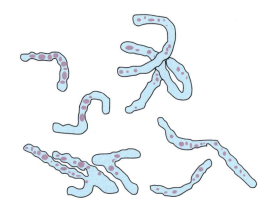

Fertilization occurs when the pollen nucleus unites with the egg nucleus to form a zygote. The zygote then grows by cell division called mitosis to form an embryo. During mitosis, an individual cell reproduces itself completely and then forms a new cell wall that divides the components in half. The result is two cells that are identical to the one parent cell. Once a seed germinates and begins to produce a new plant, cells duplicate themselves by the same process (mitosis), and this increase in cell numbers results in plant growth.

The pollen grain and egg cell are called gametes, and are formed by a special type of cell division called **meiosis**. During meiosis, the nucleus does not duplicate itself. Instead, chromosome pairs separate, with one chromosome of each pair moving to opposite ends of the cell. When a new cell wall forms, the result is two cells that contain one chromosome of each pair in their nucleus. These cells are the pollen grains and egg cells, and each contains one-half the number of chromosomes found in all other cells throughout the plant.

Fertilization occurs when the pollen gamete unites with the egg gamete. The chromosomes form pairs again; that is, chromosome 1 from the pollen pairs with chromosome 1 from the egg, and the process results in a new cell with a complete set of chromosomes. Thus, a cell having 13 pairs of chromosomes would contribute one-half of these (one chromosome from each pair) to the resulting gamete. When the gametes fuse, they would have 13 pairs, or a total of 26 chromosomes.

The chromosomes contain the genetic code of life. Chemically, they are composed of deoxyribonucleic acid (DNA). Groups of DNA molecules arranged in a specific sequence on a chromosome make up a gene. Each chromosome may contain several thousand genes, and the genes determine plant characteristics. For example, genes determine whether a plant is tall or short, has green or brown seed, or white or purple flowers.

Because chromosomes exist in pairs, plants have two genes for every characteristic, one derived from each parent. The two genes for a characteristic may be identical, and the plant is said to be **homozygous** for that characteristic. If the two genes for a characteristic are not alike, the plant is **heterozygous** for the trait. For example, if one parent has red flowers and the other parent has white flowers, then the progeny that result from crossing these two plants may have one gene for red flowers and one gene for white flowers.

The progeny from this cross may have red, pink, or white flowers, depending on how the genes for flower color function. When one gene completely dominates other genes, the relationship is **dominant-recessive**. If red is dominant to white, and a plant has one gene for red and one for white, the plant will have red flowers.

In some cases, genes are neither dominant nor recessive, but their effects are added together. If genes for red and white flower color are additive, two genes for red would produce red flowers, while a plant with one gene for red and one gene for white would produce pink flowers. A plant with two genes for white could have only white flowers. This situation is illustrated in Figure 17-31.

Sometimes the situation is more complicated because several pairs of genes may affect the same characteristic. For example, the height of sunflowers is genetically controlled, and several pairs of genes affect the height. Because several genes are involved, sunflowers can show a wide range in plant height, depending on the combination of genes present in the plant.

Self-Pollination

Any plant will have many thousands of genes, one-half of them derived from each parent. Although it seems unlikely that any two plants would be exactly alike, some plants produce progeny that are essentially identical

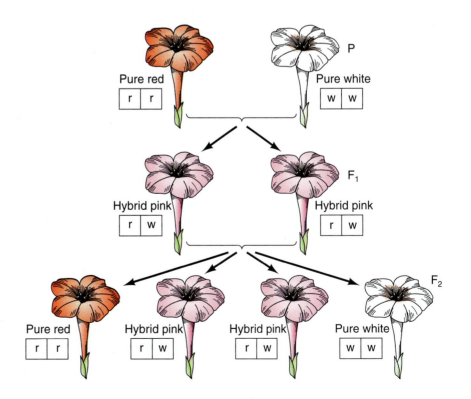

Figure 17-31 A plant with a gene for white and red flowers will produce pink flowers.

to each other and the parent. This occurs primarily in self-pollinated plants, in which both male and female gametes come from the same parent plant. For example, a self-pollinated plant may have two genes for red flowers or two genes for white flowers, but it is unlikely to have one gene for red flowers and one gene for white flowers. If a plant has two genes for red flowers, the gametes from that plant will all have one gene for red flowers, and when self fertilization occurs, the progeny can have only two genes for red flowers, the same as the parent. Plants that are homozygous will produce progeny identical to the parent when self-pollinated.

Cross-Pollination

In cross-pollination, gametes come from different parents and they may contain genes that differ, such as red and white flowers. Cross-pollinated plants may produce progeny that differ from each other and from the parents.

Plants that normally self-pollinate may occasionally be cross-pollinated for some reason (Figure 17-32). When this occurs, the progeny may have gene pairs that are unlike. As the progeny self-pollinate in future generations, more and more of the gene pairs in a particular line of progeny become alike. After seven or eight generations of self-pollination, any single plant will again have pairs of genes that are alike and it will produce progeny very much like itself. This process of self-pollination for several generations to produce genetically uniform plants is very important to plant breeders.

Plant breeders try to develop plant varieties with the best possible sets of characteristics. Some desirable characteristics include increased yield, early maturity, disease resistance, uniformity, and improved quality. A plant breeder uses self-pollination and selection for desirable traits to produce new varieties in which all plants exhibit the same improved characteristics.

The actual procedure a plant breeder uses depends on several factors; whether a plant is naturally self- or cross-pollinated is one of the most important factors.

Figure 17-32 Self- and cross-pollination.

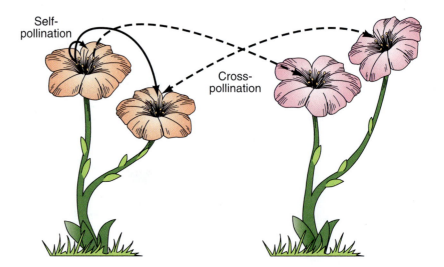

Cross-pollinated plants, such as apple trees and corn, can produce progeny that differ widely from each other. A plant breeder crosses varieties with different characteristics and selects the best progeny. This can be accomplished quickly for crops like corn because the plants mature in one growing season. With crops like apples, it is necessary to wait several years for the apples to fruit before selecting those with the desired characteristics.

Once a desirable apple tree is obtained, it can be quickly and easily propagated by grafting; it then becomes a new variety. Developing a new variety is more complicated with corn and other cross-pollinated crops that are not asexually propagated. These crops are grown from seeds, and remember that seeds are the result of sexual reproduction. Plant breeders must develop varieties in which all of the plants are genetically alike (true breeding).

Developing and maintaining a variety of a cross-pollinated crop that is genetically uniform is more complicated than for a self-pollinated crop. One way to accomplish this is by developing hybrid varieties.

Hybrids

Hybrid varieties are obtained by crossing two or more true breeding parents. True breeding corn lines can be developed by self-pollinating corn each year for seven to eight years. The product of this self-pollination is usually called an inbred line. At the end of this time, every corn plant of a line will have gene pairs that are alike, and all of the plants will be almost genetically identical. Also, the plants of the line will produce gametes and progeny that are nearly identical. Figure 17-33 shows how hybrid corn is grown.

Inbreeding, however, has some disadvantages. Inbred lines of corn are not very vigorous, the plants are often small, and grain yields are usually low. The lines are useful only because the gametes they produce are nearly identical. If two inbred corn lines, A and B for example, are crossed, the result is hybrid seed. Because all of the gametes from A are identical, and all of the gametes from B are identical, all of the resulting hybrid seeds will produce plants that are alike. The hybrid seeds will produce vigorous plants with desirable characteristics if the inbred lines were selected wisely.

Self-pollinated plants such as soybean, wheat, and tomato are usually grown as pure genetic varieties rather than hybrids. Here the problem is obtaining new genetic combinations. To accomplish this, genetic variation must first be created by artificially crossing different lines with desirable characteristics.

The first step in artificially cross-pollinating a plant that is normally self-pollinated is to prevent self-pollination. A plant is chosen as a female parent. A mature flower is selected that has not yet released its pollen. The anthers, and often the petals also, are carefully removed before any pollen has been released. This is called emasculation. Care must be taken not to damage the pistil during emasculation. The emasculated flower must be protected to prevent pollination by wind or insects; this is usually done by covering the flower with a small paper bag.

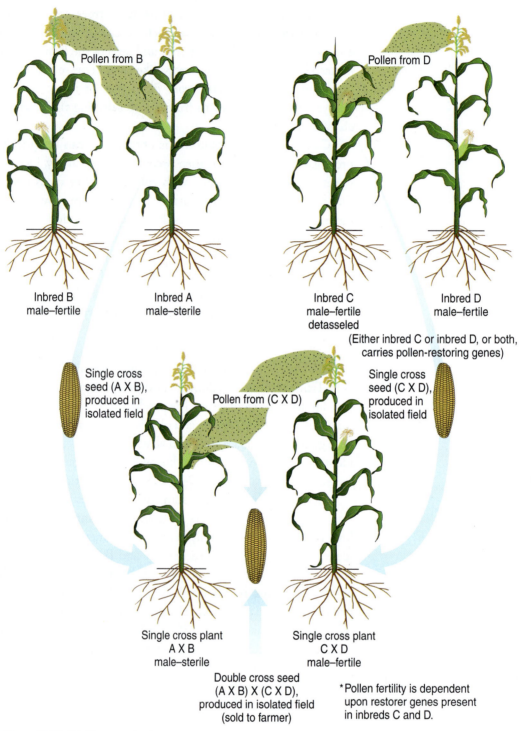

Figure 17-33 Hybrid corn production.

Second, the pollen is collected from the male parent. This may be done by taking the entire anther, shaking the pollen into a small container, or collecting the pollen on a small brush or pipe cleaner.

Third, pollinate the emasculated flower by touching the pollen to the stigma. Place the protective paper bag over the flower after pollination.

Fourth, return when the fruit or seed matures and collect the hybrid seed. If you can produce enough hybrid seeds by cross-pollinating, they can be used as a new variety. This method is used with plants that produce numerous seeds per pollination, such as corn, tomatoes, cucumbers, melons, and tobacco. This procedure is usually too expensive for crops in which only a few seeds are produced by each flower, such as soybean.

With soybean and similar crops, the initial cross is made as previously described. The plants grown from the hybrid seeds are allowed to self-pollinate for seven to eight generations. In each generation, plants showing the desired characteristics are selected as parent plants for the next year and all other plants are discarded. After several generations of self-pollination, any single plant will again have pairs of genes that are alike and will produce progeny like itself. Once a desirable line is identified, the seeds of this line are increased and the line is evaluated further. The new line may be released as a new variety.

Mutations

Sometimes a true breeding line may show a sudden change in a characteristic. For example, a gene that causes the skin of an apple to be green may change and begin producing red apples. These changes are called mutations.

Most mutations are undesirable, but rare favorable mutations have been found. Plant breeders attempt to use favorable mutations to develop new varieties. This is easiest in plants that are asexually propagated. Many of our apple varieties were first observed as mutations in a single bud of an apple tree. If the mutation produces a favorable effect, an entire variety can be propagated from one bud.

Incorporating a mutation into a new variety is more complex for plants that are sexually propagated. The mutation must be reproducible in the progeny of the plant, and a genetic evaluation is required to determine whether the mutation is dominant, recessive, or additive. Recessive mutations are the most common, and they are the most difficult to work with for the progeny to express the characteristic. Many recessive mutations that occur in a single gene of one chromosome go undetected.

Geneticists and plant breeders can increase the number of mutations by using radiation or chemicals called **mutagens**. Plant material can be treated with mutagens, and plants grown from the material are carefully observed for new characteristics. High doses of mutagenic agents may kill the plant, and low doses may not cause mutations. The proper dosage of the mutagenic agent produces mutations without killing the plant. A very small proportion of the mutations that result may be desirable and can be propagated. Many ornamental plants with variegated foliage have been developed this way.

The chemical colchicine is used to develop plants with four sets of chromosomes. These plants are called **tetraploids** (4n) and often have larger leaves, flowers, or fruits than the plants from which they were developed.

With the advances in biotechnology and genetic engineering, plant breeders and plant geneticists have new tools for creating new and better crops. Chapter 19 is completely devoted to a discussion of these new tools and how they are being used.

SUMMARY

Plants can be propagated either by seeds—sexually—or by vegetative means—asexually. Vegetative propagation includes such methods as cutting, grafting, budding, and layering. Asexual reproduction produces plants with identical genetic makeup. Micropropagation, or tissue culture, is a fairly new method of asexual reproduction. Micropropagation techniques also produce disease-free plants. Seeds are the product of pollination and fertilization in the flower, followed by fruit set and seed development. A seed contains a miniature plant—the embryo—and food storage in the form of endosperm or cotyledons is enclosed in the seed coat. For seeds to germinate and start the next generation, they require adequate moisture, proper temperature, oxygen, and in some cases light.

Plant breeding is modification of a plant's genetic makeup in a purposeful way. The general goal of plant breeding is to assemble into single varieties the best possible combination of genes that control desirable traits. The traits of importance include yield, local environmental adaptation, uniformity, quality, disease or insect resistance, and early maturity. Numerous plant-breeding methods can be used worldwide, but actual testing of experimental varieties must be done in the regions in which they are planned for use in order to ensure suitability.

Review

True or False

1. Pollination and fertilization are the same process.

2. Imperfect flowers are incapable of being pollinated.

3. Seeds are living organisms.

4. Stolons and runners represent a type of layering.

5. The small pieces of tissue used in micropropagation are called plantlets.

Short Answer

6. List four requirements for seed germination.

7. Name five examples of asexual plant propagation and give examples of each.

8. Identify two elements of sexual propagation that are produced by meiosis.

9. What are the three parts of a plant embryo?

10. What two methods of asexual propagation join plant parts so they will grow as one plant?

Critical Thinking/Discussion

11. Explain the difference between pollination and fertilization.

12. Discuss the basic principles of plant genetics.

13. Describe how a plant mutation could be useful in crop improvement.

14. Define the following genetic terms: homozygous, heterozygous, dominant, recessive, self-fertilization, cross-fertilization, haploid, diploid, and triploid.

15. Discuss how hybrids form and what they have done for production agriculture.

Knowledge Applied

1. Attempt an experiment using micropropagation or tissue culture. Although technical procedures for aseptic culture of plant cells, tissues, and organs are as diverse as the plant material on which they are practiced, a simplified general procedure can be followed in the home or classroom. All that is needed are a few basic supplies that can easily be obtained at the local grocery store. The procedures outlined in this section can be used to propagate various species of plants that are either easy (African violets, coleus, chrysanthemums) or difficult (orchids, ferns, weeping figs) to propagate. Figure 17-34 illustrates the process.

continues

Knowledge Applied, *continued*

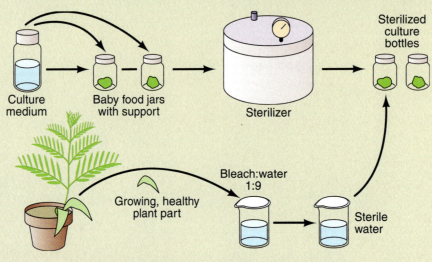

Figure 17-34 Plant disinfection and culture.

Medium preparation for 2 pints of medium: Mix the following ingredients in a 1-quart home canning jar:

1/8 cup sugar

1 teaspoon all-purpose soluble fertilizer mixture. (Check the label to make sure it has all of the major and minor elements, especially ammonium nitrate. If the latter is lacking, add 1/3 teaspoon of a 35-0-0 soluble fertilizer.)

1 tablet (100 mg) of inositol (myo-inositol), which can be obtained at most health food stores

1/4 of a pulverized vitamin tablet that has 1 to 2 mg of thiamine

4 tablespoons coconut milk (cytokinin source) drained from a fresh coconut. The remainder can be frozen and used later.

3 to 4 grains (1/400 teaspoon) of a commercial rooting compound that has 0.1% active ingredient indolebutyric acid (IBA)

Fill the jar with distilled or deionized water. If purified water is not available, water that has been boiled for several minutes can be substituted.

Shake the mixture and make sure all materials have dissolved.

Baby food jars with lids or other heat-resistant glass receptacles with lids can be used as individual culture jars. They should be half-filled with cotton or paper to support the plant material. The medium should be poured into each culture bottle to the point where the support material is just above the solution.

When all bottles contain the medium and have the lids loosely screwed on, they are ready to be sterilized. This can be done by placing them in a pressure cooker and sterilizing them under pressure for 30 minutes or in an oven at 320°F for four hours. After removing the bottles from the sterilizer, place them in a clean area and allow the medium to cool. If the bottles will not be used for several days, wrap groups of culture bottles in foil before sterilizing and

continues

Knowledge Applied, *continued*

then sterilize the whole package. Then the bottles can be removed and cooled without removing the foil cover. Sterilized water, tweezers, and razor blades, which will be needed later, can be prepared in the same manner.

Once the growing medium is sterilized and cooled, plant material can be prepared for culture. Because plants usually harbor bacterial and fungal spores, they must be cleaned (disinfested) before placement on the sterile medium. Otherwise, bacteria and fungi may grow faster than the plants and dominate the culture.

Various plant parts can be cultured but small, actively growing portions usually result in the most vigorous plantlets. For example, ferns are most readily propagated by using only 1/2 inch of the tip of a rhizome. For other species, 1/2–1 inch of the shoot tip is sufficient. Remove leaves attached to the tip and discard. Place the plant part into a solution of one part commercial bleach to nine parts water for 8 to 10 minutes. Submerge all plant tissue in the bleach solution. After this period, rinse off excess bleach by dropping the plant part into sterile water. Remember, once the plant material has been in the bleach, it has been disinfested and should be touched only with sterile tweezers.

After the plant material has been rinsed, remove any bleach-damaged tissue with a sterile razor blade. Then remove the cap of a culture bottle containing sterile medium, place the plant part onto the support material in the bottle, making sure that it is not completely submerged in the medium, and recap quickly.

Transferring should be done as quickly as possible in a clean environment. Therefore, scrub hands and countertops with soap and water just before beginning to disinfest plant material. Rubbing alcohol or a dilute bleach solution can be used to wipe down the working surface.

After all plants have been cultured, place them in a warm, well-lit (no direct sunlight) environment to encourage growth. If contamination of the medium has occurred, it should be obvious within 3 to 4 days. Remove and wash contaminated culture bottles as quickly as possible to prevent the spread to uncontaminated cultures.

When plantlets have grown to sufficient size, transplant them into soil. Handle as gently as possible because the plants are leaving a warm, humid environment for a cool, dry one. After transplanting, water the plants thoroughly and place in a clear plastic bag for several days. Gradually remove the bag to acclimate the plants to their new environment; start with 1 hour per day and gradually increase time out of the bag over a 2-week period until the plants are strong enough to dispense with the bag altogether.

Keep a log of your activities and report on your successes and failures.

Note: As an alternative to this do-it-yourself approach to micropropagation, the biological suppliers listed at the end of this chapter provide kits for tissue culture.

2. Try propagating plants with any of the other asexual methods described in this chapter. Keep a log of your activities and report on your successes and failures.

3. Develop a presentation on one of the early geneticist, such as George Shull, Charles Darwin, Luther Burbank, Joseph Koelreuter, or Norman Borlaug.

Resources

Asimov, I. (1962). *The genetic code*. New York: New American Library.

Brown, J., & Caligari, P. (2007). *An introduction to plant breeding*. Ames, IA: Blackwell Publishing.

Litz, R. E. (2005). *Biotechnology of fruit and nut crops*. Cambridge, MA: CABI Publishing.

Raven, P. H., Evert, R. F., & Eichhorn, S. E. (2004). *Biology of plants* (7th ed.). New York: W. H. Freeman.

Sleper, D. A., & Poehlman, J. M. (2006). *Breeding field crops* (5th ed.). Ames, IA: Blackwell Publishing.

Thomas, L. (1974). *The lives of a cell: Notes of a biology watcher*. New York: The Viking Press.

Toogood, A. (1999). *American Horticultural Society plant propagation: The fully illustrated plant-by-plant manual of practical techniques*. New York: DK Publishing.

United States Department of Agriculture. (1992). *New crops, new uses, new markets: The yearbook of agriculture*. Washington, DC: Author.

Wilson, E.O., Wright, A. B., & Ward, G. (2001). *The diversity of life*. London: Penguin Books.

Zimmerman, R. H., Griesbach, R. J., Hammerschlag, F. A., & Lawson, R. H. (2002). *Tissue culture as a plant production system for horticultural crops*. New York: Springer Publishing Company.

Equipment and Supplies

Carolina Biological Supply Company, Carolina Science and Math Catalog 66, 2700 York Road, Burlington, NC 27215-3398 (www.carolina.com).

Fisher-EMD, 4901 W. LeMoyne Street, Chicago, IL 60651 (www.fishersci.com).

Nebraska Scientific, 3823 Leavenworth Street, Omaha, NE 68105-1180 (www.nebraskascientific.com).

Internet

Internet sites represent a vast resource of information. The URLs (uniform resource locators) for World Wide Web sites can change. Using one of the search engines on the Internet such as Google or Yahoo!, find more information by searching for these words or phrases: sexual propagation, asexual propagation, pollination, fertilization, double fertilization, self-fertilization, cross-fertilization, micropropagation, genetics mutation, plant cuttings, heterozygous plants, homozygous plants, and hybrid.

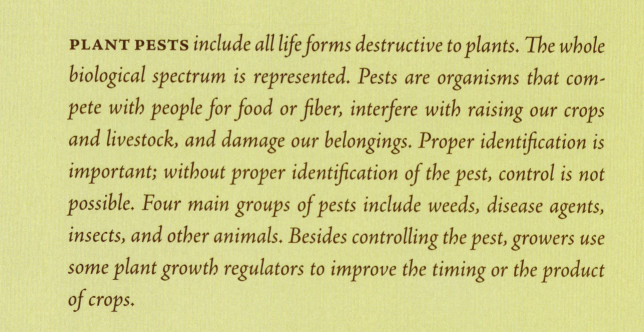

Chapter 18
Plant Pests

PLANT PESTS *include all life forms destructive to plants. The whole biological spectrum is represented. Pests are organisms that compete with people for food or fiber, interfere with raising our crops and livestock, and damage our belongings. Proper identification is important; without proper identification of the pest, control is not possible. Four main groups of pests include weeds, disease agents, insects, and other animals. Besides controlling the pest, growers use some plant growth regulators to improve the timing or the product of crops.*

After completing this chapter, you should be able to—

- Define a weed
- Describe how weeds reduce the yield and quality of desirable plants
- Explain the spread of weeds
- Name three types of life cycles exhibited by weeds
- Discuss the history of weed-control efforts
- List four general categories of weed control
- Describe a prohibited noxious weed
- Name the four different categories of plant growth regulators
- Explain how growth regulators are used in modifying plant growth
- Name three biotic and three abiotic causes of disease
- List four symptoms of plant disease
- Describe four ways that insects harm plants
- Select from a list the control practices used for insects and identify them as mechanical, cultural, biological, or chemical
- Name the five phases of disease development

WEEDS

Weeds are plants that grow where they are not wanted (Figure 18-1). A certain plant may be a weed when growing in one location, but may not be a weed when growing in another. Bluegrass growing in a flower bed is a weed. In the lawn, just a few feet from a flower bed, bluegrass would not be a weed.

Plants such as chickweed, henbit, dandelion, johnsongrass, and pigweed are considered weeds in most situations. This is because no useful function or value has been found for them.

Weeds Increase the Cost of Producing Food and Fiber

Weeds lower the yield of many crops by competing for water, nutrients, light, and space. Weeds are adapted to more diverse growing conditions than most crop plants and use a large share of the nutrients that are available. One wild mustard plant will remove as much nitrogen and phosphorus from the soil as two oat plants. Therefore, the amount of nutrients remaining for the crop plants is greatly reduced. This reduction of available nutrients results in stunting or a reduction in yield of the crop.

Weeds compete with desirable plants for light and water. Lack of sunlight or shading may cause crop plants to be yellow instead of their normal healthy green color. A soybean plant growing under a cocklebur canopy will grow abnormally and produce only a few healthy pods. Weeds also use large amounts of water. One pigweed plant may use as much water as a corn plant. In many areas, weeds reduce the soil moisture level to the point that moisture is the limiting factor in crop production.

The quality of many farm products is reduced because of weeds. Hay and grain for livestock is of inferior quality when contaminated with weeds or weed seeds. If cows eat weeds such as wild garlic or bitter sneezeweed, the milk will have an unpleasant odor and taste. Weeds can stain the lint of cotton, causing the lint to be graded lower.

Figure 18-1 Weeds competing with a field of pepper plants.

Weeds also interfere with harvest, causing delays and greater harvest loss. They often necessitate drying of harvested grain. This can add substantially to the cost of grain production (Figure 18-2).

Because crop yields or quality may be reduced by weeds, it is important that farmers receive a higher price for these products to help offset these losses. For example, if corn yields are reduced from 100 to 80 bushels an acre by weed competition, then the farmer would have to receive an increase of 25 percent in price to offset the loss in yield:

+ 100 bushels at $2.00 = $200
+ 80 bushels at $2.50 = $200

Weed competition affects everyone by increasing the price paid for food products.

Figure 18-2 Some common weeds and their seeds that interfere with crop production and harvest.

Weeds Are Harmful to Health

Some weeds have properties that are harmful to our health. An example is poison ivy, which is irritating to the skin of many people. Ragweed produces pollen that causes many people to suffer from hay fever. Other weeds such as jimsonweed, buttercup, wild cherry, and white snakeroot contain poisonous chemicals. These weeds often poison the livestock that eat them. Livestock producers must use control measures or fence livestock away from infestations of poisonous plants.

Spread of Weeds

Structural features of the seed and fruit aid in distribution of many weeds. Some seeds, in addition to being light and feathery, are often winged, chaffy, or may even resemble a parachute. Examples are thistles, dandelion, and milkweed. These seeds may be blown several miles by wind.

Weed seeds are readily carried from farm to farm by machinery, such as combines and hay-baling equipment. Plows and harrows may carry other plant parts such as **rhizomes**, **stolons**, and roots from one field to another. Weed seeds that float, such as curly dock, are carried along by runoff water from sloping fields. Others are carried by flooding ditches and streams, and are widely spread in the environment.

Some seeds pass through the digestive tracts of livestock and birds and remain viable. Others may cling to the hair or wool of domestic and wild animals. Examples are the cocklebur and beggartick. Weed seeds may also stick to the feet of animals, especially under muddy conditions.

Life Cycle of Weeds

Weeds are grouped into four types on the basis of their life cycle. These are **summer annuals**, **winter annuals**, **biennials**, and **perennials**.

Summer annuals are those plants that germinate in the spring and die in the fall after the first frost. Seeds mature and are left scattered on the ground. Some seeds may germinate in the early fall, but most lie dormant until spring. Examples are morning glory, pigweed, cocklebur, and crabgrass.

Winter annuals germinate in the fall or winter, produce their seeds in the spring or early summer, and then die. The seeds lie dormant in the soil during most of the summer months. Examples are wild mustard, chickweed, henbit, and annual bluegrass.

Biennial weeds produce leafy or vegetative growth from seeds during the first growing season, become dormant over the winter, and produce flowers and seed the second season. These weeds live for more than 12 but less than 36 months. Because two growing seasons are required for reproduction, they are called biennials. Examples are musk thistle, wild carrot, and common mullen.

Perennials live for three years or more and may actually persist almost indefinitely unless destroyed. Most of these weeds produce seeds each year after establishment. Examples are honeysuckle, dandelion, pokeweed, and Bermuda grass. Perennials are further classified on the basis of their method of reproduction. Simple perennials reproduce primarily by seed. Dandelion is an example a of simple perennial.

Bulbous perennials propagate by bulbs and bulblets as well as seeds. Wild onion and wild garlic are examples of bulbous perennials.

Creeping perennials are those that propagate by means of rhizomes (underground stems), stolons (aboveground stems), or spreading roots or tubers, as well as by seeds. Johnsongrass propagates by rhizomes and seeds, while Bermuda grass propagates by rhizomes, stolons, and seeds.

Methods of Weed Control

The ideal method of weed control would be to prevent them from becoming established. Once established, the grower must rely on mechanical (Figures 18-3 and 18-4), cultural, biological, and chemical methods to control them.

Mechanical methods of controlling weeds include—
+ Hand pulling
+ Hoeing
+ Cultivation (includes plowing and clean tillage)
+ Burning
+ Mowing
+ Smothering with plastic mulches (Figure 18-5)

Cultural methods of controlling weeds include—
+ Crop rotation
+ Crop competition
+ Weed-free crop seeds
+ Smother crops

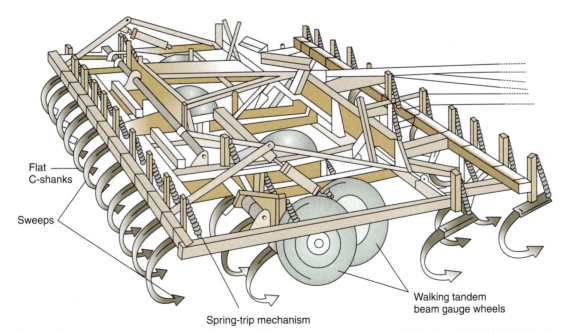

Flat
C-shanks

Sweeps

Spring-trip mechanism

Walking tandem
beam gauge wheels

Figure 18-3 Cultivators range from a garden hoe to small tillers to large pieces of equipment that are pulled behind tractors, as shown in the diagram—all are designed to mechanically destroy weeds.

Figure 18-4 Cultivating a field of okra. (Photo courtesy of USDA)

Figure 18-5 Plastic mulch being used to control weeds in a tomato field.

Biological weed control involves the use of natural enemies for the control of certain weeds. The objective of biological control is reduction, not eradication, of the weed population. Examples of success in biological weed control include use of the Klamath weed beetle to control Klamath weed, use of the puncture vine beetles and weevils to control puncture vine, or domestic geese to control a variety of weed seedlings.

Chemical control of weeds is the fourth method that can be used. Chemical weed killers (herbicides) are being used because they are economical and effective when used properly. Herbicides must be used carefully, always following label directions exactly. Intelligent use of herbicides requires positive identification of weeds and recognizing their stage of growth. Several methods are used for controlling weeds with herbicides—

+ Broadcast treatment
+ Band treatment
+ Spot treatment
+ Direct spray (in direct contact with the target weed)

Herbicides may be applied preplant, preemergent, or post-emergent—before planting, before the weed grows, or after the weed grows (Figure 18-6). For annual, biennial, and simple perennial weeds, almost any control method is satisfactory before the plant forms seeds. Perennials that spread by rhizomes, stolons, or bulbs require special methods. Weed seeds remain viable for varying periods of time, depending on species. Many common weed seeds will remain viable in the soil for 20 years or longer.

History of Weed Control

For centuries, people have struggled in the battle against weeds. The early settlers used very crude hand tools made from wood and rocks to control weeds. More advanced mechanical means of weed control were gradually developed. Horse-drawn plows and cultivators were a big improvement over earlier methods. The development of the tractor gave the farmer still another more advanced weapon to use against weeds.

Crop rotations and good cropping practices are excellent methods of weed control. For example, cutting pasture and hay crops several times during the year prevents seed formation and weakens the weed plants present. Planting soybeans on narrow rows (20 inches or less) will produce shading earlier in the season.

During the last 30 years, chemicals known as herbicides have been developed and are now widely used to control weeds in crops. Many of

Figure 18-6 Many crops are sprayed from the air to prevent diseases or destroy insect pests. (Photo courtesy of USDA)

these herbicides are selective and specific for certain weeds. For example, 2,4-dichlorophenoxyacetic acid (2,4-D), one of the first herbicides used extensively in crop production, can be used to kill broadleaf weeds such as cocklebur, ragweed, jimsonweed and bitterweed without damage to corn or other grasses.

Basagran or Blazer can be sprayed overtop soybeans to control cocklebur, morning glory, smartweed, and other broadleaf weeds without harming the soybean crop. Such herbicides are referred to as postemergence herbicides.

Herbicides that will kill the weeds as they germinate and do not injure the crop can be sprayed on the soil surface when the crop is planted. These chemicals are called **pre-emergence herbicides** because they are applied before emergence of the crops and weeds. Examples are atrazine or Lasso for corn and Canopy or Scepter for soybeans.

Herbicides that must be mixed or incorporated into the soil before planting are referred to as **preplant incorporated herbicides**. Examples are Treflan for soybeans and Sutan for corn.

Herbicides and modern farm machinery make the battle against weeds easier each year. Hand labor, the big weapon against weeds a century ago, has been almost eliminated from crop production today. Herbicides play an important role in enabling the American farmer to produce enough food for himself or herself and 130 other people.

Noxious Weeds

Some weeds are more undesirable, troublesome, and difficult to control than others. There may be several reasons why a particular species is hard to control, but usually this is related to the reproduction methods. Usually, weeds that reproduce by more than one method are more difficult to control than those that reproduce by seeds only. For example, johnsongrass reproduces both from seeds and rhizomes. For this reason, control measures must be applied that will destroy the rhizomes as well as the seeds in the soil.

The department of agriculture in each state recognizes the importance of controlling weeds. Laws have been passed that restrict the sale of crop seeds containing more than a designated amount of certain weed seeds. These weeds are designated **noxious weeds**. Noxious weeds are those arbitrarily defined by law as being especially undesirable, troublesome, and/or difficult to control.

Weeds can be designated as—

+ Prohibited Noxious—illegal to sell, offer for sale, or transport for sale crop seed containing weed seeds
+ Noxious—crop seed for sale can only contain small specified numbers of weed seeds

Table 18-1 lists some examples of weeds, with their common and scientific name.

Table 18-1 Some Common Weeds and Their Scientific Names

Common Name	Scientific Name	Common Name	Scientific Name
Barnyard grass	*Echinochloa crusgalli*	Henbit	*Lamium amplexicaule*
Bermuda grass	*Cynodon dactylon*	Horse nettle	*Solanum carolinense*
Bindweed	*Conuolvulus sp.*	Jimsonweed	*Datura stramonium*
Bitterweed (sneezeweed)	*Helenium tenuifolium*	Johnsongrass	*Sorghum halepense*
Black-eyed Susan	*Rudbeckia huta*	Lamb's-quarter	*Chenopodium album*
Broadleaf plantain	*Plantago major*	Little barley	*Hordeum sp.*
Broomsedge	*Andropogon virginicus*	Morning glory	*Ipomoea sp.*
Buckhorn plantain	*Plantago lanceolata*	Mouse-ear chickweed	*Cerastium vulgatum*
Buttercup	*Ranunculus sp.*	Nutsedge (nut grass)	*Cyperus sp.*
Cocklebur	*Xanthium strumariun*	Peppergrass	*Lepidium virginicum*
Common chickweed	*Stellaria media*	Pigweed (redroot)	*Amaranthus retroflexu*
Common milkweed	*Asclepias syriaca*	Pigweed (spiny)	*Amaranthus spinosus*
Common ragweed	*Ambrosia artemisiifoli*	Purslane	*Portulaca oleracea*
Crabgrass	*Digitaria sp*	Red sorrel	*Rumex acetosella*
Curly dock	*Rumex crispus*	Shepherd's purse	*Capsella bursa-pastoris*
Daisy fleabane	*Erigeron strigosus*	Smartweed	*Polygonum sp.*
Dandelion	*Taraxacum officinale*	Thistle	*Cirsium sp.*
Foxtail	*Setaria sp.*	Three-seeded mercury	*Acalypha virginica*
Goldenrod	*Solidago sp.*	Wild carrot	*Daucus carota*
Goose grass	*Eleusine indica*	Wild garlic	*Allium vineale*

Allelopathy

When the chemicals produced by one plant affect another, it is known as **allelopathy.** The effect is usually harmful but sometimes the released chemicals make another plant more productive. Harmful allelopathic interactions are sort of a biological chemical warfare between plants. Several horticultural field crops, tree crops, and many weeds are known to be allelopathic.

Cucumbers and other members of the cucumber family produce chemicals that inhibit the growth of some plants in their vicinity. This cucumber-produced weed control suppresses many weeds that would otherwise compete for water, nutrients, and light. The stubble from wheat and rye reduce the growth of several of the broadleaf weeds.

Black walnut trees release chemicals that are inhibitory to many other species. The allelopathic chemicals in the leaves are washed into the soil by rain or are released after the leaves decompose. Black walnut leaves should never be used as mulch or for composting.

Some weeds release substances that are toxic to crop plants. Examples of weeds that exhibit allelopathy include crabgrass, giant foxtail, johnsongrass, and common sunflower.

PLANT GROWTH REGULATORS

Why do seeds germinate and grow into plants? Why do some plants produce fruit and seeds and others do not? Why does one shoot on a tree or plant grow taller than the others? If this branch or shoot is cut off, why does another take its place and become the main branch or stem? The answers to these questions are found in plant growth regulators.

Plant growth regulators are defined as organic compounds, other than nutrients, that affect the physiological processes of plants when applied in small quantities. The compound could be either a promoter or an inhibitor of plant growth. Plant growth regulators can be further defined as either natural or synthetic (human-made) compounds that alter a plant's life process or structure in some way so as to increase yields, improve quality, or aid in harvesting.

The term plant hormone is restricted to naturally occurring plant growth regulators. The term plant growth regulator includes synthetic compounds as well as natural hormones. The general term regulator covers a wide area and applies to any material that modifies a plant's physiological processes. Plant growth regulator, not hormone, should be the term used when referring to agricultural chemicals used for crop modification.

Four Classes of Regulators

The four classes of regulators are auxins, gibberellins, cytokinins, and inhibitors. Auxins is a term generally used for a group of compounds characterized by their capacity to induce lengthening in shoot cells, apical dominance, root formation in cuttings, leaf abscission, seedless fruits, wound tissue production, and other uses. Figure 18-7 illustrates the chemical structure of some of the common plant growth regulators.

Gibberellins are compounds that stimulate cell division, cell lengthening, or both. They promote seed germination, break dormancy, promote flowering, and so forth.

Cytokinins stimulate cell division in plants and break seed dormancy.

Inhibitors comprise a diverse group of plant regulators that inhibit or retard a physiological or biochemical process in plants. The naturally occurring inhibitory effect may be upon any of the processes stimulated by other hormones.

People have little control over hormonal actions. However, synthetic plant growth regulators can be applied directly for a specific use. Growth regulators are widely used for their specific action on food and nonfood crops. The response of plants or plant parts to a given plant growth regulator varies with the kind of plant. One type of plant may respond differently depending on its age, environmental conditions, physiological stage of development, and stage of nutrition. So synthetic plant growth regulators may act as an auxin one time at a certain rate and as an inhibitor when applied at a different rate to the same plant.

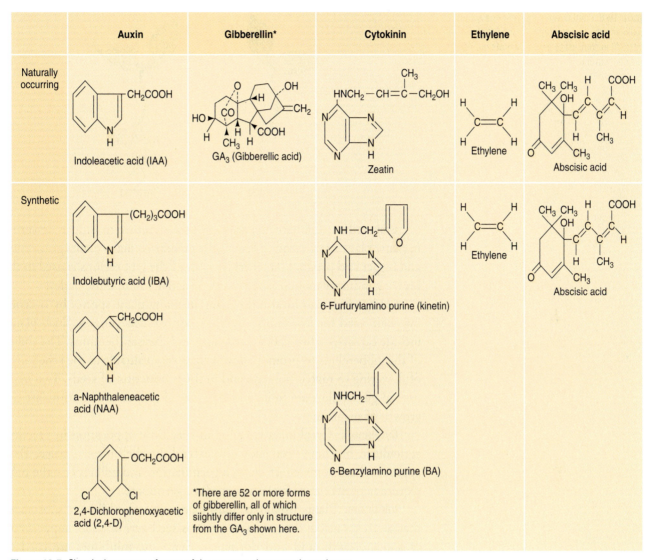

	Auxin	Gibberellin*	Cytokinin	Ethylene	Abscisic acid
Naturally occurring	Indoleacetic acid (IAA)	GA₃ (Gibberellic acid)	Zeatin	Ethylene	Abscisic acid
Synthetic	Indolebutyric acid (IBA); a-Naphthaleneacetic acid (NAA); 2,4-Dichlorophenoxyacetic acid (2,4-D)	*There are 52 or more forms of gibberellin, all of which siightly differ only in structure from the GA₃ shown here.	6-Furfurylamino purine (kinetin); 6-Benzylamino purine (BA)	Ethylene	Abscisic acid

Figure 18-7 Chemical structure of some of the common plant growth regulators.

Practical Uses of Plant Growth Regulators in Agriculture

Auxins, gibberellins, and inhibitors have all found some use in production agriculture. Some of these uses have been in the field or in the greenhouse. Growers use regulators to increase production, improve the products, or increase the efficiency of a natural process.

Auxins. The use of auxins in commercial agriculture began in the 1940s when naphthalene acetic acid was applied to prevent the preharvest drop of apples. Another auxin, indolebutyric acid (IBA), is used for initiating and/or accelerating the rooting of cuttings. This is very important in agriculture because of the commercial nurseries that sell rooted cuttings. The faster and better they root, the more valuable they are to both the nursery operator and the buyer. Easy-to-root plants are carnations, chrysanthemum, coleus, English ivy, forsythia, impatiens, philodendron, poinsettia, rose, and willow. Difficult-to-root plants are apple, fir, and maple. Figure 18-8 illustrates the effect of IBA on a cutting.

Figure 18-8 Plant cutting showing heavy root development when treated with indolebutyric acid (IBA) in contrast to very few roots on the untreated cutting.

Untreated   Treated

Gibberellins. Gibberellins are used in agriculture to induce flowering in many plant species and as growth stimulants to make plants grow taller and bigger and produce larger fruits. One of the gibberellins is used in the brewing industry to increase the enzyme activity in malting barley.

Gibberellins can induce flowering in many plant species by overriding their need for low temperatures to initiate this process. Such plants include carrots, endive, cabbage, turnips, and chrysanthemums. The ability of the gibberellins to promote flowering is very valuable in the scheduling of flowering to match holidays and in the production of seeds. Two main uses of gibberellins are to increase the size of grapes and to stimulate the growth of sugarcane.

Cytokinins. Cytokinins have found very little application in practical agriculture, but have been shown to extend the shelf life of lettuce. They appear to have some effect on seed germination and will replace the light requirement necessary for lettuce seeds to germinate.

Inhibitors. Plants usually contain many natural inhibitory substances that are responsible for processes such as seed germination, suppression of shoot growth, and bud dormancy. Synthetic inhibitors are used to control many of these processes in agriculture today. One of the first growth inhibitors used in commercial agriculture was maleic hydrazide. This material is used to prevent sprouting of potato tubers and onions. It is also used to control suckering of tobacco and excessive growth in grasses. Inhibitors can be sprayed on wheat to reduce stem length and thereby reduce lodging, or can be used to shorten the height of the poinsettia to produce a more compact, shorter plant for the Christmas season. Inhibitors are used on other floral crops and by commercial vegetable producers to increase yields of cucumbers, to stimulate uniform ripening of tomatoes, and to promote the development of red color in peppers.

A growth inhibitor applied as a sugarcane ripener causes a 10–20 percent increase in sugar content, depending on the variety of sugarcane. It is a common treatment in sugarcane-producing areas of the United States.

Foliar growth inhibitor applied to cotton may produce a darker leaf color, reduce internodal length, produce an open canopy, and induce better boll retention and earlier maturity. The reduction in plant size is beneficial to cotton growing on land that produces large cotton plants. The size reduction also allows for more air circulation within the plant canopy,

Figure 18-9 Foliar growth inhibitors applied to a cotton plant creates a compact and shorter plant with many bolls (right). The untreated cotton plant (left) is larger with fewer bolls.

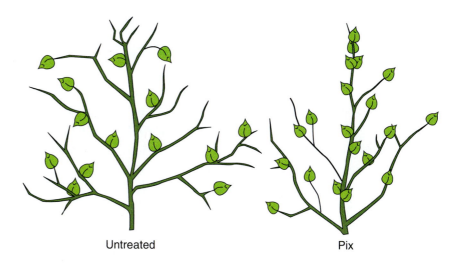

Untreated   Pix

reducing boll rot. The smaller size makes machine picking easier. Another inhibitor used in cotton production is sprayed on cotton to stimulate the opening of bolls. Figure 18-9 illustrates the effect of a foliar growth inhibitor on the cotton plant.

Growth inhibitors are also used on golf courses to retard the growth of the grass so it does not have to be mowed as often.

DISEASES

Plant disease is the rule rather than the exception. Every plant has disease problems of one sort or another. Fortunately, plants either tolerate these maladies or they are not very serious in most years. Plant pathologists consider almost any abnormal growth pattern by a plant to be evidence of disease.

Plant disease is caused by a large array of **biotic** (living) agents such as—

+ Fungi
+ Nematodes
+ Bacteria
+ Viruses
+ Mycoplasma
+ Viroid
+ Parasitic seed plant

Plant diseases can also be caused by a large array of **abiotic** (nonliving) factors such as—

+ Nutrient deficiencies
+ Water stress
+ Temperature stress
+ Combination of water and temperature stress

These agents cause abnormal and harmful physiological processes in the host plant.

All Because of a Plant Disease

Potatoes were first widely grown in Ireland. The soil and climate conditions in Ireland made potato production a natural for the Irish common folk. Before long, potatoes had become the most important food there. Then tragedy struck. Because pieces of the potato are used to grow the new crop, the Irish farmers were saving their own "seed" potatoes from year to year. Gradually, all the potatoes grown in Ireland came to have the same genetic makeup. Along with the potato, a disease of potato was also introduced into Ireland. It is called late blight of potato, and we now know that it is caused by the fungus *Phytophthora infestans*.

For several consecutive years, the environmental conditions favored the rapid development and spread of this destructive disease. After two consecutive crop failures, a famine resulted in 1847. The Irish potato famine changed the history of both the Irish and American nations, and of agriculture in general. One million people died because of the famine, and another 1.6 million emigrated, mostly to the United States. Many who were forced to emigrate were the brightest and most ambitious people of Ireland. They were quickly assimilated into American society and have made a large contribution to the development and success of our nation. The Irish people lost a third of their population between 1847 and 1860.

Late blight of potato was the first disease known with certainty to be caused by a microorganism. Heinrich Anton de Bary described the nature and life history of the causal fungus and proved unequivocally that the disease was indeed caused by a living organism and not by spontaneous generation. This work supported the "germ theory" and gave rise to the scientific disciplines of plant and animal pathology.

Because the epidemic that affected Ireland also affected other European countries, there was widespread interest in being able to predict when this dreaded disease might attack. As a consequence, the first models that described plant and pathogen growth were constructed for this disease. Modeling of plant responses has reached much higher levels now, but it has its beginnings in the Irish potato famine.

The grower needs to understand the difference between infectious disease, caused by biotic agents, and noninfectious disease, caused by abiotic agents.

Plant pathogens (disease-causing organisms) can cause various symptoms to appear on affected plants—

+ Dwarfing of growth
+ Yellowing of foliage
+ Leaf spotting
+ Blasting of grain heads
+ Stem cankers
+ Fruit rot
+ Seed decay
+ Damping off (destruction of seedlings near the soil line)
+ Wilt
+ Defoliation
+ Root rot
+ Galls

Although biotic infectious organisms can cause disease, they are all very different from each other.

Fungus

A fungus is an organism (plant) with no chlorophyll that reproduces by means of structures called spores and usually has filamentous growth; for example, molds, yeasts, and mushrooms (Figure 18-10). Fungi produce diseases like stem rust, corn smut, powdery mildews, brown rot,

Figure 18-10 Lighter, discolored barley heads infected with *Fusarium* head scab fungi. (Photo courtesy of USDA ARS)

Figure 18-11 *Cercospora* leaf spot disease symptoms on a sugar beet leaf. (Photo courtesy of USDA ARS)

and damping off. Fungi reproduce mainly by means of spores. Fungi are particularly damaging to plant propagation operations.

Bacterium

A **bacterium** is a single-celled, microscopic organism with cell walls and no chlorophyll that reproduces by fission (Figure 18-11). These bacteria cause diseases such as galls, leaf spots, soft rots, scabs, and systemic disorders. Bacteria are a significant cause of plant disease because they can multiply very rapidly when proper environmental conditions are present. Bacteria are as damaging to plant propagation operations, as are fungi.

Virus

Viruses are submicroscopic, subcellular particles that require a host cell in which to multiply. It is difficult to say whether a virus is a living or nonliving agent (Figure 18-12). In plants they cause such symptoms as stunting, leaves with yellow mosaic patterns, flower break, and vein

Figure 18-12 A leaf showing the effects of a mosaic virus. (Photo courtesy of USDA ARS)

clearing. Veins are chlorotic—lacking green color; without chlorophyll. Viruses can multiply only in living cells.

Mycoplasma

Mycoplasma is a microscopic bacteria-like organism that lacks a cell wall and appears filamentous.

Viroid

A viroid is a virus-like particle that lacks the outer protein coat of a virus particle.

Nematode

Nematodes are microscopic roundworms, usually living in soil, many of which feed on plant roots. They cause galls on roots, root lesions, injure root tips, and sometimes cause excessive root branching. Nematodes reproduce by eggs.

Parasitic Seed Plant

A parasitic seed plant is a higher plant with chlorophyll that lives parasitically on other plants. The best known example is mistletoe.

Fungi and bacteria cause such plant diseases as leaf spots and fruit, stem, or root rots. Plant viruses, viroids, and mycoplasmas often cause growth distortion, stunting, and abnormal coloration. Nematodes can cause stunting and root distortion. Parasitic seed plants cause a general weakening of the host plant.

DISEASE DEVELOPMENT

Understanding how plant diseases develop helps the grower control disease. By the time it becomes obvious that a plant has a disease, it is generally too late to do anything about it in that growing season. Plants cannot be cured in the way people expect their own ills to be cured.

The process by which diseases develop can be summarized into five distinct phases—

1. Inoculation
2. Incubation
3. Penetration
4. Infection
5. Disease

Inoculation

Inoculation is the introduction of the pathogen to the host plant tissue. Wind or rain or running water can move pathogens—as can birds, insects, people, or equipment—and introduce pathogens to a host plant. Some pathogens move themselves short distances, but most rely on other means of transport.

Incubation

Incubation is a period of development during which the pathogen undergoes changes to a form that can penetrate or infect the new host plant. Some fungi, for instance, grow a structure called a penetration peg that can grow through the cell walls of the plant.

Penetration

Penetration is the process of getting inside the plant. It may be an active or passive process. Some pathogens produce enzymes to dissolve the cutin and cellulose layers of plant material between themselves and the cell contents. Some pathogens can swim through water on a plant's surface and into the plant through natural openings (such as stomata, lenticels, or hydathodes) or through wounds. Some pathogens are put inside the plant by insects or by pruning tools or by driving rain.

Infection

When the pathogen invades the plant tissue and establishes a parasitic relationship between itself and the host, infection has occurred.

Disease

When the host plant responds to the presence of the pathogen, a disease exists. The host's response results in symptoms of the disease, such as chlorosis or necrosis.

OTHER PESTS

Invertebrate pests include insects and their relatives, nematodes, snails, and slugs. The term invertebrate signifies animals without backbones (no vertebrae). Insects have three body parts—head, thorax, and abdomen—and six legs. Ticks, mites, and spiders have only two body parts and eight legs.

Nematodes are a large group of unsegmented worms that can be plant parasites (Figure 18-13).

Snails and slugs are mollusks that prefer cool, moist surroundings.

Vertebrates are animals with backbones. They include fish, amphibians, reptiles, birds, and mammals. Most concern for crops is with birds and mammals. Pest birds harbor pathogens, eat or damage crops, cause damage to buildings, or make too much noise. Rodents are mammals that interfere with people or cause harm to crops. Rats, mice, squirrels, and voles cause most of our vertebrate problems. Animal pests are similar to weeds. Any animals that are out of place—such as deer in a hayfield, stray dogs with sheep, and livestock that break through a fence—are pests.

Insects

Most injuries to plants by insects result directly or indirectly from their attempts to secure food. They are people's chief rival for the available food

Figure 18-13 Whiplike larva of root-knot nematode (*Meloidogyne incognita*) magnified 500 times, shown here penetrating a tomato root. Once inside, the larva establishes a feeding site, which causes a nutrient-robbing gall. (Photo courtesy of USDA ARS)

supply of the world. When insects desire food that people also desire, they become our enemy and we call them harmful. Because of their diversity, there are insects that desire apparently every kind of organic material found on earth. Some prefer living tissue, some prefer dead tissue, and others prefer raw, sweet, sour, hard, or soft foods. Insects that help people by pollinating, providing food and other helpful materials, are called beneficial. Some insects are both beneficial (pollination) and harmful (competing with humans for food).

Insects can be harmful in many different ways (Figure 18-14). Some harmful effects of insects include:

+ Chewing of plant parts—leaves, roots, stems, bark, flowers, or fruit
+ Sucking sap from plant parts
+ Boring holes between the surfaces of leaves (leaf miners) or other plant parts
+ Laying eggs in some parts of the plant

Figure 18-14 Grasshoppers of the class Insecta can cause great damage to crops if not controlled. (Photo courtesy of University of Idaho)

+ Using parts of plants for their shelter or nests
+ Carrying other insects to the plant and establishing them there
+ Disseminating organisms of plant disease and making wounds through which other harmful organisms may enter

Insect control is often divided into four types: physical, cultural, biological, and chemical. Physical control would include the direct removal of insects, by interrupting their physiological processes, prevention of their entry into an area, or physically destroying them with machinery. Examples include:

+ Light at night interrupts insect behavior
+ High temperatures can kill insects in stored grain
+ Low temperatures prevent insect attack on furs and fabric
+ Aluminum foil, screens, trenches, sticky bands, and traps used as barriers to keep out insects

Cultural control can prevent pest damage. Some examples include:

+ Use of crop rotation interrupts an insect's food supply (Figure 18-15)
+ Soil tillage and removal of crop residues to reduce insects' food supply
+ Early or delayed planting can lessen the amount of food available when the insect is in its larval stages
+ Use of resistant (to insect pests) varieties and strains of plants
+ Destruction of weeds that may act as a host plant or shelter for the insect

Biological control is the use of other insects or pathogens to control economic pests. Some parasites deposit eggs on their victim. The larvae then consume the pest. Then the adult parasite emerges from the insect "mummy." Examples include:

+ Spotted alfalfa controlled by a parasitic wasp
+ Purple scale of oranges controlled by wasps

Figure 18-15 Preparing the ground for crop rotation; leaving crop residues on the soil surface at harvest reduces soil erosion significantly. (Photo courtesy of USDA NRCS)

In another type of biological control, predators kill and consume pests. Examples include:

+ Assassin bugs suck life fluids from pink boll worms
+ Ladybird beetles and their larvae eat aphids
+ Vedalia beetles have been imported from Australia to control cottony cushion scale in California and Florida citrus

Sometimes pathogens can be used to control insect pests. Examples include:

+ The bacteria *Bacillus thuringiensis* kills butterfly and moth larvae
+ Use of water suspensions of spores of brown, red, and yellow fungi for whitefly and scale control on Florida citrus has been successful

With 250 viruses known to be pathogenic to insects, investigations continue on new and innovative applications for biological control.

Chemical control usually implies the use of liquids, gases, powders, or granules to control insects. Stomach poisons are used to control chewing insects. The poison can be put on the leaves of the plants or in the soil (or water) and taken into the plant system. Contact poisons are used on both chewing and sucking insects. This method implies contacting the insect directly with the chemical. Fumigants are actually poison gasses that enter the insect's respiratory system. The nervous system is most affected by these control measures.

INTEGRATED PEST MANAGEMENT (IPM)

Emphasis is now being placed on the Integrated Pest Management (IPM) approach to pest control. IPM combines crop monitoring with many techniques including chemical, farming (cultural), biological, and mechanical controls to reduce pest populations in ways that are both profitable and environmentally sound (Figure 18-16). Overall, IPM reduces dependence on pesticides. In some cases, however, more chemicals are used under an IPM program than had been previously.

IPM programs carefully analyze pest situations before any control tactics are used. The first step in an IPM program is identification of the pests and beneficial organisms present in the environment being managed. Often, beneficial organisms keep pest populations so low that pesticide use will not save money. As a result, the need for pesticide application is prevented. Trained workers or scouts make periodic reports on the presence of pests and beneficial organisms in the field, lawn, or area being managed. Based on data gathered by the scouts, an IPM specialist makes recommendations about whether pest populations are high enough to warrant action.

Only when the pests threaten to lower the economic value of the crop or livestock, or when they create health problems, are controls used; otherwise, pests are left to be controlled by natural enemies in their environment and minor damage is tolerated. In many cases, constant monitoring of pests has been found to prevent wasteful pesticide applications.

Figure 18-16 Integrated Pest Management (IPM) makes use of farming practices, chemical control, mechanical control, biological control, and monitoring for safe, effective pest control.

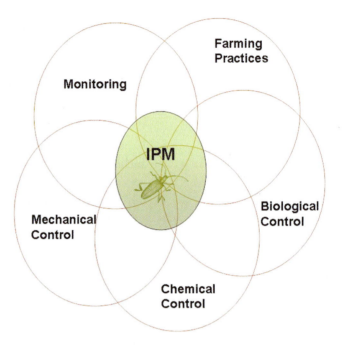

As Table 18-2 shows, the level of IPM used varies from none to a high level.

Pesticide Versus Nonpesticide Control

If action is necessary, the IPM specialist chooses a control tactic that will be most effective and least detrimental to beneficial organisms and the environment. Although chemical controls predominate, dependence on biological controls is increasing as research identifies natural predators, parasites, or diseases of pests. Other nonpesticidal controls for insects are irradiation to produce sterile offspring and the release of pheromones (sexually attracting odors that lure insects away from crops).

If pesticides are used, they are applied in the proper amounts and at the proper time. If possible, pest-specific chemicals are chosen instead of broad-spectrum pesticides, which could kill many nontarget organisms in addition to the target pest. One new insecticide available now is the *Bacillus thuringiensis*. This bacterium kills caterpillars without harming beneficial organisms.

Table 18-2 Characteristics of four levels of IPM practices

Level of IPM	Characteristics
None	No preventive practices, monitoring or alternative, nonchemical control methods; reliance on chemical controls
Low	Scouting, crop rotation, pest-free seed/plant material, pest-resistant varieties, cultivation, attractant baits/crops, selective pesticides, edge treatment, sprayer calibration
Medium	Weather-based forecasting, nutrient and water monitoring, green manures/compost, precision agriculture, induced resistance activators, elimination of alternate host, pest biotype monitoring
High	Release of beneficial organisms, biocontrols, pheromones, trap crops, soil solarization, interactive pest/weather/crop models, primarily a nonchemical preventive approach

SUMMARY

A weed is any plant growing where it is not wanted. Weeds compete with crops for water, nutrients, and light, reducing the yield of the crop. Growers use mechanical, cultural, biological, and chemical methods to control weeds.

Controlling plant growth by the use of plant growth regulators can modify crops, making them more productive and more efficient, and can change the timing of normal events such as flowering or germinating. The four general classes of plant growth regulators include the auxins, gibberellins, cytokinins, and inhibitors.

Plant diseases are cause by living organisms such as fungi, bacteria, viruses, nematodes, mycoplasma, and viroids. Disease can also be caused by such non-living factors as nutrient deficiencies, water stress, and temperature stress, or a combination of these. Pathogens can cause a variety of symptoms to appear on plants. Some of these include dwarfing, yellowing, stem cankers, damping off, wilt, defoliation, root rot, and galls. Infectious disease follows a five-step phase in plants—inoculation, incubation, penetration, infection, and disease.

Insects, snails, spiders, mites, rats, mice, nematodes, voles, and many others, depending on the location and the crop, can damage crops. Insects cause damage to crops mostly through their attempts to secure food. Insect control is physical, cultural, biological, or chemical.

Integrated Pest Management combines all the methods of pest control to reduce pests in a profitable and environmentally sound method.

Review

True or False

1. A weed that germinates in the spring and dies in the fall after the first frost is called a winter annual.

2. The quality of many farm products is reduced because of weeds.

3. Inhibitors comprise a diverse group of plant regulators that inhibit or retard a physiological or biochemical process in plants.

4. Bulbous perennials propagate by rhizomes.

5. Crop rotations and good cropping practices are excellent methods of weed control.

Short Answer

6. A _____ herbicide is applied after the crop and weeds emerge.

7. List three ways in which weeds may be spread.

8. List three functions of auxins in plants.

9. How do weeds lower the yield of many crops?

10. Name the five phases of disease development.

11. List four general categories of weed control.

12. Name three abiotic and three biotic causes of disease.

13. List the four groups of plant growth regulators.

14. Give four symptoms of a plant disease.

Discussion

15. What is allelopathy?

16. Describe what it means if a weed is designated as "prohibited noxious."

17. Discuss four uses of plant growth regulators.

18. Why do most producers of poinsettias use plant growth regulators?

19. Describe biological weed control.

20. Define a weed.

21. Give an example of mechanical, cultural, biological, and chemical controls of insects.

22. Describe three types of life cycles that are exhibited by weeds.

23. Discuss four ways insects harm plants.

Knowledge Applied

1. Develop a computerized presentation that examines and explains the adage: "One year of seeds equals seven years of weeds."

2. Set up demonstration plots to compare the growth and yield rates of unweeded plots of plants grown in chemically and culturally weed-controlled beds. Also calculate the cost in time and money for the various methods demonstrated. Examples could include:

   ✦ Cultural drip irrigated versus sprinkler irrigated
   ✦ Use of tall crop to shade weeds; for example, corn
   ✦ Pre- and postemergents, systemic, or contact herbicides

3. Collect and examine insect-damaged plant specimens. Also, collect and mount the insects causing damage to plants.

4. Examine a variety of chemical formulations (insecticides) used for insect control and discuss their uses, expense, possible threat to the environment if misused, and potential danger to the applicators of the insecticides.

5. Collect the seeds from one mature plant of several of these weeds: broadleaf plantain, johnsongrass, morning glory, ragweed, pigweed, Dallis grass, jimsonweed, and cocklebur. Count the number of seeds and record them in a chart you create. (You may wish to count the seeds of one spike of pigweed or ragweed and estimate the total production by multiplying the total length of spikelets by the number of seeds per inch of spikelet length.) A magnifying glass will be of great value when counting small seeds such as pigweed seed. Compare the weed seed production with a few crop plants such as soybean, beans, peas, wheat, barley, and corn. How do you think this comparison of seed production influences the capacity of weeds to compete with crops?

6. Select two of the following places and visit each.

   a. Visit an apple orchard and ask if any plant regulators were used and what specific action they were used for. Ask about timing, concentration, types of regulators, and results. On which varieties of apples can the material be used? On which can they not be used? What benefit would a chemical thinner be to an apple grower?
   b. Visit a nursery and see cuttings being rooted and determine which plant growth regulator is being used or how many are used. How were they applied? When? Concentration? Results obtained? Plants treated?
   c. Visit a commercial greenhouse in early fall and check on potted plants grown for holiday seasons. Determine if plant growth regulators are being used to reduce growth, enhance flowering, or stimulate rooting. Determine the time and concentration applied, and plants to which regulators were applied.
   d. Visit a cotton farm and see if a growth inhibitor was used on some cotton fields and not on other fields. Note the difference in leaf size, plant height, and number and size of bolls of treated and untreated plants. If you were a cotton farmer, would you use a growth inhibitor?

Resources

Coleman, E. (1995). *The new organic grower* (Revised ed.) Chelsea, VT: Chelsea Green Publishing Co.

Norris R. F., Caswell-Chen, E. P., & Kogan, M. (2002). *Concepts in integrated pest management.* Englewood Cliffs, NJ: Prentice Hall.

Simpson, B. B., & Ogorzaly, M. C. (2000). *Economic botany: Plants in our world* (3rd ed.). New York: McGraw-Hill.

United States Department of Agriculture. (1961). *Seeds: The yearbook of agriculture.* Washington, DC: Author.

United States Department of Agriculture. (1993). *IPM and biological control of plant pests: Field crops.* Beltsville, MD: Author.

United States Department of Agriculture. (1994). *Allelopathy: The effects of chemicals produced by plants.* Beltsville, MD: Author.

United States Department of Agriculture. (1994). *Cultural and mechanical weed control.* Beltsville, MD: Author.

Walters, D., Newton, A., & Lyon, G. (2007). *Induced resistance for plant defense: A sustainable approach to crop protection.* Ames, IA: Blackwell Publishing.

Zimmer, G. F. (2000). *The biological farmer.* Austin, TX: Acres U.S.A., Publishers.

Internet

Internet sites represent a vast resource of information. The URLs (uniform resource locators) for World Wide Web sites can change. Using one of the search engines on the Internet such as Google or Yahoo!, find more information by searching for these words or phrases: insect pests, crop protection, weeds, weed control, noxious weeds, plant diseases, plant pathogens, weed control, herbicides, biological control of pests, plant growth regulators, and integrated pest management or IPM.

Chapter 19
Genetic Engineering and Biotechnology

TODAY, AGRICULTURE *needs a new infusion of science and technology and new capabilities that will restore and enhance the competitiveness of United States agriculture in the world marketplace. The products of biotechnology offer some of the most exciting opportunities to meet these urgent needs.*

After completing this chapter, you should be able to—

- Define biotechnology, genetic engineering, and related terms

- Understand the basic processes of biotechnology research

- Recognize the degree of progress made in biotechnology research up to this point

- Identify the latest developments or applications resulting from biotechnology research in plant science

- Describe future impacts of biotechnology research and genetic engineering

- Discuss environmental, ethical, control, and conflict of interest concerns brought about by biotechnology research

- Name five plants altered by genetic engineering

- List four goals of genetic engineering in plants

- Describe the process of genetic engineering

- Explain how DNA controls formation of proteins

- Describe a transgenic plant

- Diagram how the genetic code is translated

- Give three advantages of genetic engineering over traditional selective breeding

KEY TERMS

Amino Acid
Biocontrol
Bioinformatics
Biopharmaceuticals
Biotechnology
Cross-Protection
Genetic Code
Genetic Engineering
Genomics
Helix
Messenger RNA (mRNA)
Micropropagation
Nucelotide
Plasmids
Proteomics
Recombinant DNA
Ribosome
Transfer RNA (tRNA)
Transgenic

BIOTECHNOLOGY DEFINED

The term **biotechnology** refers to an array of related basic sciences that have as their centerpiece the use of new methods for the manipulation of the fundamental building blocks of genetic information to create life forms that might never emerge in nature—life forms that can expand and enhance the well-being of humans.

The American Association for the Advancement of Science called genetic engineering one of the four major scientific revolutions of this century, on a level with unlocking the atom, escaping the earth's gravity, and the computer revolution. The newfound ability to manipulate cellular machinery has been termed a biotechnology revolution. It could have as profound an effect on our society as has the information revolution occurring alongside it. Many believe that the impact of biotechnology will be as great or greater in agriculture as in medicine.

The biotechnology revolution in agriculture is part of an overall increasing sophistication of biological techniques for improving the production, processing, and marketing of food and fiber. Biotechnology, for instance, allows an acceleration of the process of selection and breeding that has been under way for over 100 years.

In the case of agriculture, biotechnology is built on a broad base of existing and ongoing scientific research that supports and enhances the use of the new methods in genetic engineering and related techniques (Figure 19-1). This science base helps define what should be genetically engineered and enables the products of fundamental research in the laboratory to be practically applied in the field.

Figure 19-1 Genetic engineering (the process of manipulating or inserting genes in an organism) is being used on more and more crops. The crops or products represented here have either already been genetically engineered or are involved in ongoing or planned transgenic studies. (Photo courtesy of USDA ARS)

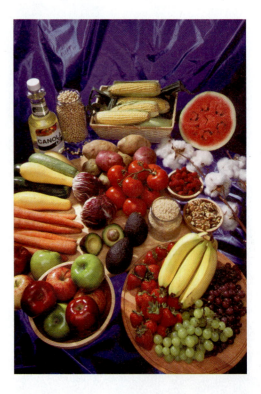

GENETIC ENGINEERING

Every cell in plants, animals, and microbes contains the genetic information to allow perpetuation of that cell or organism. The study of the structure, chemistry, and function of this genetic material has been the basis of understanding that has enabled the biotechnology revolution to come of age.

Genetic engineering, or manipulation, involves taking genes from their normal location in one organism and either transferring them elsewhere or putting them back into the original organism in different combinations. Its value to biotechnology is twofold.

First, scientists can take useful genes from plant and animal cells and transfer them to microorganisms such as yeasts and bacteria that are easy to grow in large quantities. Products that once were available only in small amounts from an animal or plant are then available in large quantities from rapidly growing microbes. One example is the use of genetically engineered bacteria to produce human insulin for treating diabetes.

The second benefit holds particular promise for plant and animal breeders. Genetic engineering allows desirable genes from one plant, animal, or microorganism to be incorporated into an unrelated species, thus avoiding the constraints of normal cross-breeding. A wider range of traits is available to the breeder, and these traits can be incorporated more quickly and more reliably into target species than is possible with conventional methods.

Needed Knowledge

Genetic material is arranged in helical strands containing the code for triggering the characteristic functions of that organism in succeeding generations. The discovery of the structure of the DNA helix in the 1950s (Table 19-1), the unraveling of genetic code in the 1960s, and the development and refinement of the tools of genetic engineering in the 1970s has caused something fundamentally different to happen in biotechnology science. Using enzymes as "genetic scissors," the genetic structure of cells can be snipped apart and reconstructed in combinations impossible to achieve by natural reproduction (Figure 19-2). Scientists

Table 19-1 Brief History of Genetic Engineering

Year	Event
1944	DNA identified as genetic material
1953	Double-strand DNA structure identified
1973	First transgenic bacteria prepared
1976	First genetic engineering company (Genetech) established
1980	First patent for genetically engineered microbe
1982	Approval of first genetically engineered drug
1986	First field test of genetically engineered plant
1987	Genetic engineering patent extended to higher life-forms

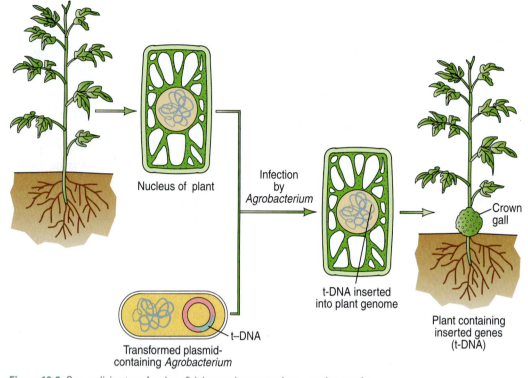

Figure 19-2 Gene splicing transfers beneficial genes in one species to another species.

not only can alter existing genes, but construct totally synthetic genes to cause the organism to perform desired functions.

To make genetic engineering work, methods are being developed to transfer genetic material into plants and animals and to make sure that the function that has been engineered is expressed at the right place and at the right time. But genetic engineering is only a part of biotechnology. The total picture includes understanding the physiology and biochemistry of the function of interest and knowledge of the existing genetic codes that regulate the process. This allows scientists to understand what to genetically engineer to produce a more desirable organism. Once such a product has been created in the laboratory, a variety of techniques such as tissue culture (Figure 19-3) are often needed to recreate an organism that can compete in a practical ecosystem. The techniques of plant breeding and development are used to take the final product back to the field.

The Genetic Code

Before scientists could undertake such genetic manipulation, they had to unravel the secrets of the genetic code. They discovered that DNA is a long, double-stranded molecule wound in a spiral called a **helix** (Figure 19-4). Each gene is a segment of the DNA strand that usually codes for a particular protein. Proteins, like DNA, are also long, chainlike molecules. They are constructed from 20 different **amino acid** building blocks. They are extremely versatile molecules, ranging from a few dozen to several hundred amino acids long. Unlike the regular spiral formed by

Figure 19-3 A rose plant that began as cells grown in a tissue culture. (Photo courtesy of USDA ARS)

Figure 19-4 DNA strands divide and bases attach themselves to the new strands to form identical genes for new cells.

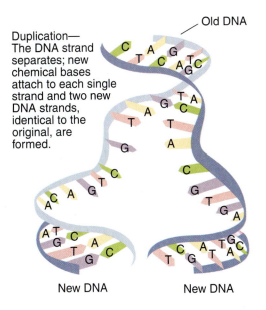

Duplication—The DNA strand separates; new chemical bases attach to each single strand and two new DNA strands, identical to the original, are formed.

Old DNA

New DNA   New DNA

DNA strands, proteins fold and twist into an enormous variety of three-dimensional shapes. The bodies of plants and animals possess thousands of different kinds of protein and each plays a specific role in life. That role can be structural or physiological; for example, there are proteins involved in the photosynthetic processes.

The DNA code is translated into amino acid sequences in proteins through an intermediary called **messenger RNA** (mRNA)—a single-stranded molecule similar to one side of the double-stranded DNA. To be able to control the protein-manufacturing process, scientists needed to understand the detail of the DNA coding system.

The DNA Jigsaw

The DNA molecule contains subunits called **nucleotides**. Each nucleotide comprises a sugar component (deoxyribose), a phosphate component, and one of four different bases—adenine (A), guanine (G), thymine (T), and cytosine (C). Scientists discovered that DNA was formed from two strands of nucleotides, held together by the bonds between the bases on opposite strands. The entire structure is like a ladder. The sides are formed by the sugar and phosphate groups and the rungs are the bases. The bases in the "rungs" are matched in pairs like pieces in a jigsaw. The two strands forming the ladder are then wound around one another to form the helix (Figure 19-4).

These DNA molecules contain the blueprint for all proteins made in a cell. Each sequence of three bases along the DNA strand is a chemical code for one of the 20 amino acids—the building blocks of proteins.

Translating the Code

To make the proteins, the DNA molecule is unravelled, the strands separate, and the cell makes a copy of the relevant part in the form of single-stranded mRNA. The mRNA then moves to the cell's "factories" called **ribosomes**, where it acts as a template for protein manufacture. The code for the protein is read off the base sequence on the mRNA, and the appropriate amino acids are added to the protein one by one, aligned against the mRNA code by small segments of RNA called **transfer RNA** (tRNA; Figure 19-2).

The coding system is universal. It is basically the same in all animals, plants, and microorganisms. A piece of DNA from a plant inserted into the chromosome of a bacterium makes perfect sense to the bacterial cell.

Scientists have known the detailed amino acid sequence of many key proteins for a long time. Once they also understood which base sequences in DNA were represented by which amino acids, they could identify the genes in the chromosome that coded for particular proteins.

Recombinant DNA Technology

Identifying the genes is not enough. The next step is to be able to copy the gene and insert it into other cells—cells that can be grown easily using existing microbiological techniques, or cells of other plants or animals where the protein is required (Figure 19-5). To do this, scientists used new biochemical techniques, involving special enzymes, to break the DNA strand at chosen points, insert new segments, and "stitch" the strand back together again. The result, known as **recombinant DNA**, is DNA that incorporates extra segments bearing genes it had not previously contained.

Insertion of genes into different organisms is made much easier by the existence of bacterial **plasmids**—small circles of DNA that are much smaller than the bacterial chromosome. Some of these plasmids can pass readily from one cell to another, even when the cells are far apart on the

Figure 19-5 Research geneticist Ann Blechl and colleagues were the first to insert modified *Fusarium graminearum* genes into wheat plants, which may lead to wheats that are more resistant to Fusarium Head Blight. (Photo courtesy of USDA ARS)

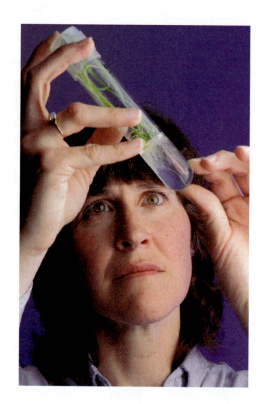

evolutionary scale. Using the special "cut-and-paste" enzymes mentioned earlier, scientists can insert genes from one organism into a bacterial plasmid, then insert the recombinant plasmid into a living microorganism, where it will direct the synthesis of the desired proteins. Human insulin for treating diabetes can now be produced in this way.

So far, scientists have used genetic engineering to produce, for example:

+ Improved vaccines against animal diseases such as foot rot and pig scours
+ Pure human products such as insulin and human growth hormone in commercial quantities
+ Existing antibiotics by more economical methods
+ New kinds of antibiotics not otherwise available
+ Plants with resistance to some pesticides, insects, and diseases
+ Plants with improved nutritional qualities to enhance livestock productivity

TARGETING AGRICULTURE

In the past, agriculture has been an energy- and labor-intensive industry. Biotechnology offers the opportunity to reduce the costs related to both these factors in future operations. Inherent resistance to pests and disease can reduce the use of chemical pesticides, thus reducing the cost of production and the potentially harmful environmental effects of such practices. The possible uses of biotechnology for agriculture are limited only by imagination and initiative (Figure 19-6).

Figure 19-6 Since World War II, American agriscience has progressed at a breathtaking pace. Here a technician checks on futuristic peach and apple "orchards." Each dish holds tiny experimental trees grown from laboratory-cultured cells to which researchers have given a new gene. (Photo courtesy of USDA ARS)

The total system for food and fiber production is extremely diverse and multifaceted, providing a broad range of potential applications of biotechnology. Biotechnology is not only enhancing the traditional enterprises in food and fiber production, it also is producing new high-technology industries that are in themselves providing new jobs and producing new goods and services. Thinking of biotechnological applications is sometimes limited to production agriculture, where an exciting new array of scientific breakthroughs is being developed. Other exciting areas related to production agriculture include the new application of biotechnology to food processing and manufacturing, to new methods for ecologically sound disposal of wastes, and to biochemical engineering, in which totally new products are being produced from agricultural residues using biotechnological tools.

Biological Factories

One of the early uses of biotechnology has been to use simple organisms such as bacteria and yeast as so-called biological factories to produce biologically active compounds. Genetic codes for these compounds are inserted into the genetic makeup of these simple organisms along with genetic instructions that cause the production of the desired material.

Through these techniques, for instance, human insulin is now produced and is replacing insulin from animal sources for treatment of diabetes in humans (Figure 19-7).

Interferon, a biological anticancer treatment and antiviral material previously available in only minute quantities, can now be produced inexpensively and in large quantities using biological factories.

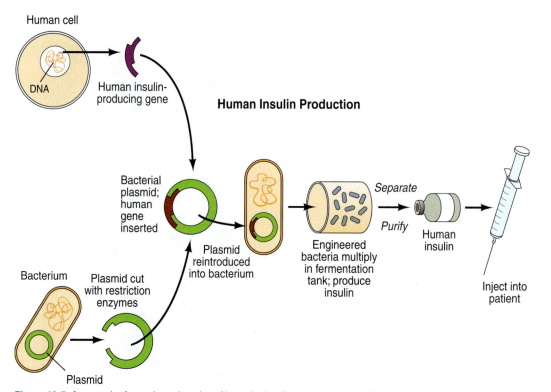

Figure 19-7 As a result of genetic engineering of bacteria, insulin is now readily available and relatively inexpensive.

Hormones, such as bovine growth hormone, have been manufactured using this method. The product is being injected in dairy cattle to enhance milk production by some 20–30 percent.

Diagnostic reagents and improved vaccines for animal and human disease are being produced using these techniques.

Early progress has been rapid in this area because the genetics of these organisms is relatively simple.

Plant-Water Relationships

In many parts of the world, water is the limiting factor in food production. Biotechnology is being used to greatly enhance the production of plants with high drought-stress tolerance. These plants will maintain yields in environments with much less water, and there is the promise of developing plants that can use brackish water. These developments could have a profound effect on stretching water resources and will be crucial as water for irrigation becomes less available and more expensive.

Plant Productivity

Plant growth and development has been the subject of investigation for decades, but until recently it has remained poorly understood. Biotechnology makes it possible to isolate, characterize, and manipulate specific genes. This new technology provides a powerful tool to understand plant growth and development and a way to directly manipulate the process. Opportunities in this field include altering chemical composition of the plant product, improving processing quality, producing plants that are resistant to stress or herbicides, altering plant size, and changing the ratio of grain to stalk (Figure 19-8).

Figure 19-8 Perhaps genes could be introduced to make wheat a perennial crop. (Photo courtesy of USDA ARS)

Nitrogen Fixation. The transfer of nitrogen-fixing genes to plants is possible with the techniques of biotechnology. Nitrogen-fixation genes might be transferred to plants that do not now have this nitrogen-fixation capacity. These genes would have to be in a form that could be incorporated into the plant genome, replicated, and expressed. The genes would have to be expressed in an environment amenable to nitrogen fixation where the enzyme, nitrogenase, could be protected from oxygen and where the enzyme system could tap into the sources of reductant and energy from the plant.

The full complement of nitrogen-fixation genes has been cloned from *Klebsiella pneumoniae* and transferred to and expressed in another bacterium, *Escherichia coli*.

Plant Disease Resistance

Genetic engineering offers an exciting and environmentally sound way of reducing the cost and increasing the effectiveness of plant pest control through the development of genetic resistance to disease. Because disease resistance is controlled by relatively few genes, this area is among the most favorable candidates for early application of biotechnology to plants.

The search for resistant genes involves screening cultivars or species of plants to identify individuals that exhibit resistance to infection, replication, or spread of a pathogen. If the resistance trait is the result of expression of a single gene or genetic locus, plant breeders begin the task of introducing the resistance trait into a cultivar having desirable agroeconomic traits that will ultimately be released to the farmer. The plant-breeding process (Figure 19-9) usually requires more than five years and considerable evaluation of progeny to eliminate plants that contain undesirable traits in addition to the disease-resistance trait.

Tobacco Mosaic Virus. Tobacco mosaic virus causes the leaves of some important crop plants, including the tomato, to wither and die. Scientists have incorporated into the tomato plant a gene that protects it from

Figure 19-9 A plant geneticist evaluates different genetic sources of alfalfa to identify plant traits that would increase growth and enhance the conversion of plant tissues into biofuel. (Photo courtesy of USDA ARS)

infection. It has the same effect as a vaccine for humans. This approach is now being applied to other viral diseases in crops.

Manipulating Microbes. Plant microbes affect agriculture both detrimentally and beneficially. Economic losses and human suffering result from plant diseases caused by microbes; therefore, much ecological research deals with plant pathogens (disease-producing agents). Unfortunately, with so many important crops, diseases, and agroecosystems, only a small proportion can be investigated intensively.

Scientists focus on microbial influences that maintain plant health. One of the successful applications involves cross-protection, in which the infection of plant tissues by one virus suppresses the disease caused by another closely related strain of the virus. The protecting strain must have negligible impact on the host. Such strains have been found naturally, but also can be created in the laboratory via biotechnology. Cross-protection has been used successfully in protecting citrus trees from severe strains of *Citrus tristeza* virus.

Another successful example involves the bacterium that causes crown gall of stone fruits and other plants: a nonpathogenic (or "friendly") strain produces an antibiotic that inhibits the pathogenic strain. Because the two strains are closely related, the nonpathogen survives in the same niches as the pathogen, and responds similarly to environmental fluctuations. Consequently, upon deliberate release of the biocontrol agent, close association of the two bacteria is assured.

For crown gall, the biocontrol agent was naturally occurring, but with biotechnology it will be possible to engineer normal resident nonpathogenic microbes into biocontrol agents.

Some microbes can be used as biocontrol agents for weeds. For example, a fungus is used to control northern jointvetch in rice and soybean fields. Knowledge of the ecology and epidemiology of this fungus contributed to the development of a rational, effective biological control (biocontrol) approach. Even though the genetic and biochemical bases of pathogenicity are unknown for this pathogen, it is so specific in its actions and is relatively unable to be dispersed widely that it makes a desirable biocontrol agent. With biotechnology, other pathogens of weeds can be altered for use in biocontrol. The ecology of specific candidates will have to be well described to assure selection of those with the greatest potential for safe and effective use.

Controlling Plants. Mobile or transposable elements provide an opportunity to isolate and identify genes that would enhance crop quality and productivity. These mobile elements provide a direct link between plant characteristics; for example, disease resistance, plant height, organ shapes, and the DNA molecules that control the particular traits.

Transposable elements provide one way of physically isolating genes that control complex plant traits because they provide a direct connection between the observable trait and the DNA molecule that controls it. Even though all the intermediate steps in the process (the mRNA, the proteins produced by the genes, and the pathways) may be unknown,

Technology for Plant Genome Mapping and Genetic Diagnostics

Genetic mapping requires the processing of thousands of samples. The application of genetic markers in large-scale quantitative trait-mapping experiments and in plant breeding was limited by the lack of diagnostic technology for processing thousands of individual samples. This is changing with the development of automated technology for marker-based genetic analyses. The sample throughput and cost requirements for plant diagnostics are very different from those of medical DNA diagnostics. An automated DNA extractor has been developed that is capable of processing one sample per minute in continuous operation, using disposable, bar-coded vessels. These vessels are also used for sample collection and assure sample identity preservation throughout the process.

Five to ten μg of DNA are obtained per sample. Automatically extracted DNA samples are quantitated by measuring fluorescence. The concentration data are used to instruct a pipetting robot to automatically dilute DNA to a standard concentration for suitable analyses. Pipetting workstations are also used for the preparation of amplification reactions.

The amplifications are performed in a custom-made thermocycler, consisting of three water baths and a robotic arm. The thermocycler is capable of performing up to 3,000 reactions, in 200-reaction heat-sealed vessels. DNA amplification products are automatically loaded on agarose gels, using a system based on agarose-solidified gel plugs. High-resolution agarose permits the resolution of alleles differing by two to four base pairs. Gels are imaged using a digital camera and stored in a database for semiautomated scoring. The genotypic information, along with the phenotypic data from the field and biochemical measurements, is made accessible to the breeder through a "breeders' workstation," a custom software suite that allows data display and analysis suitable for breeding applications.

Automation of the DNA isolation process, allele identification, and data handling meets the needs of marker-assisted plant breeding and large-scale genetic experimentation.

the gene responsible for the trait can be identified and physically isolated using a tagging procedure.

Nutritional Quality of Plants

Many plant foods are deficient in nutrients or lose nutritional value during storage. Some plants have other features that are not optimum for human or animal health. Genetic engineering can be used to both improve and retain nutritive value as well as to modify undesirable properties of plant products. For instance, through genetic engineering, the composition of dietary fats can be modified to reduce their possible contribution to cardiovascular disease.

+ An example: Scientists are working to develop a sulphur-rich feed plant for sheep. Research has shown that sulphur supplements in the diet help sheep produce better-quality wool fiber. Scientists believe it would be more cost-effective to feed the sheep on pastureland that was naturally sulphur-rich. Using biotechnology, scientists have developed alfalfa strains that produce a sulphur-rich protein in their leaves. They now plan to develop pasture grasses with the same characteristics.

Biological Control of Pests

Biological control exploits natural factors in the life cycle of harmful insects as a means of control. Some possibilities include use of highly specific insect pathogens (bacteria, viruses, fungi) to produce insect disease or death or to use unique viruses that interfere with the insect immune system, making it more vulnerable to disease. Also, insect chromosomes or genes with some ability to control population growth are under study, as well as methods that interfere with normal growth and maturation. All these processes of biological control are potentially capable of being enhanced through the use of genetic engineering to improve the effectiveness of the crop insect's natural enemy. These processes are highly specific to single insect species and thus are highly desirable environmentally as alternatives to chemical pesticides (Figure 19-10).

Figure 19-10 Biotechnology seeks to reduce or eliminate the use of chemicals. (Photo courtesy USDA ARS)

+ An example: An insecticidal protein has been successfully incorporated into tomato plants to provide protection from some leaf-eating insects. The protein comes originally from *Bacillus thuringiensis*, a naturally occurring bacterium that lives in the ground. Using genetic engineering techniques, scientists have inserted the gene for this protein into the plant's genetic material. When an insect eats the modified plant, the protein is released and the insect dies.

+ An example: Glyphosate is an environmentally friendly, widely used broad-spectrum herbicide. It is easily degraded in the agricultural environment and works by interfering with an enzyme system that is present only in plants. Unfortunately, the herbicide kills crop plants as well as weeds, but scientists have now used genetic engineering methods to breed crop plants that are glyphosate-resistant. By planting these modified crops, farmers can control weeds by spraying with glyphosate alone.

TRANSGENIC PLANTS

Transgenic plants and animals result from genetic engineering experiments in which genetic material is moved from one organism to another, so that the latter will exhibit a desired characteristic. Scientists, farmers, and business corporations hope that transgenic techniques will allow more precise and cost-effective animal- and plant-breeding programs. They also hope to use these new methods to produce animals and plants with desirable characteristics that are not available using current breeding technology. Table 19-2 list genetically modified plants.

In traditional breeding programs, only closely related species can be cross-bred, but transgenic techniques allow genetic material to be transferred between completely unrelated organisms, so that breeders can incorporate characteristics that are not normally available to them. The

Table 19-2 Genetically Modified Plants

Crops	Vegetables	Flowers	Trees
Alfalfa	Asparagus	Arabidopsis	Apple
Canola	Cabbage	Petunia	Pear
Corn	Carrot		Poplar
Cotton	Cauliflower		Walnut
Flax	Celery		
Potato	Horseradish		
Rice	Lettuce		
Rye	Peas		
Soybean	Tomato		
Sugar beet			
Sunflower			
Tobacco			

modified organisms exhibit properties that would be impossible to obtain by conventional breeding techniques.

Although the basic coding system is the same in all organisms, the fine details of gene control often differ. A gene from a bacterium will often not work correctly if it is introduced unmodified into a plant or animal cell. The genetic engineer must first construct a transgene, the gene to be introduced. This is a segment of DNA containing the gene of interest and some extra material that correctly controls the gene's function in its new organism. The transgene must then be inserted into the second organism.

Making a Transgene

All genes are controlled by a special segment of DNA found on the chromosome next to the gene and called a promoter sequence. When making a transgene, scientists generally substitute the organism's own promoter sequence with a specially designed one that ensures that the gene will function in the correct tissues of the animal or plant and also allows them to turn the gene on or off as needed. Figure 19-11 shows a common transgenic plant. For example, a promoter sequence that requires a light "trigger" can be used to turn on genes for important growth regulators (hormones) in plants. The plant would not produce the new hormone unless provided the appropriate trigger.

Inserting the Transgene

Unlike animals, plants do not have a separate germ line (eggs and sperm) and all cells of a plant retain the capacity to develop into a whole plant. This makes inserting the transgene much simpler. The transgene can be

Figure 19-11 Tomatoes—a transgenic plant. (Photo courtesy of USDA ARS)

introduced into a single cell by a variety of physical or biological techniques, including using viruses or derivatives to carry the new gene into the plant cells.

Tissue culture techniques can then be used to propagate that cell and encourage its development into a transgenic plant, all of whose cells contain the transgene. Once the plant is produced, nature takes over and increases plant numbers by normal seed production.

Uses of Transgenic Techniques

Transgenic methods have now been developed for a number of important crop plants such as rice, cotton, soybean, oilseed rape, and a variety of vegetable crops like tomato, potato, cabbage, and lettuce. New plant varieties have been produced using bacterial or viral genes that confer tolerance to insect or disease pests and allow plants to tolerate herbicides, making the herbicide more selective in its action against weeds and allowing farmers to use less herbicide.

For example, a new variety of cotton has been developed that uses a gene from the bacterium *Bacillus thuringiensis* to produce a protein that is specifically toxic to certain insect pests, including bollworm, but not to animals or humans. (This protein has been used as a pesticide spray for many years.) These transgenic plants should help reduce the use of chemical pesticides in cotton production, as well as in the production of many other crops that could be engineered to contain the *Bacillus thuringiensis* gene. In another case, a gene from the potato leaf roll virus has been introduced into a potato plant, giving the plant resistance to this serious potato disease.

Transgenic technologies are now being used to modify other important characteristics of plants such as the nutritional value of pasture crops or the oil quality of oilseed plants like linseed or sunflower.

Novel Uses

With techniques similar to those used to make insulin-producing bacteria, it may be possible to develop animals that produce other useful **biopharmaceuticals**—drugs produced by living tissues. For example, researchers have developed transgenic animals such as cows and sheep that secrete economic quantities of medically important chemicals in their milk. The cost of these drugs may be much less than for those produced using conventional techniques.

Eventually, it also may be possible to develop crops for nonfood uses by modifying the starches and oils they produce to make them more suitable for industrial purposes, or to use plants rather than animals to make antibodies for medical and agricultural diagnostic purposes. In the cut flower industry, transgenic research may yield products such as blue carnations as novelty items.

Advantages over Selective Breeding

Transgenic technology is an extension of agricultural practices that have been used for centuries—selective breeding and special feeding or fertilizing programs. It may reduce or even replace the large-scale use of pesticides and long-lasting herbicides. Transgenic technology is still experimental and is still very expensive. If it could be made commercially viable, it would offer a number of advantages over traditional methods.

Compared with traditional methods, transgenic breeding is:

+ More specific—Scientists can choose with greater accuracy the trait they want to establish. The number of additional unwanted traits can be kept to a minimum.
+ Faster—Establishing the trait takes only one generation compared with the many generations often needed for traditional selective breeding, where much is left to chance.
+ More flexible—Traits that would otherwise be unavailable in some animals or plants may be achievable using transgenic methods.
+ Less costly—Much of the cost and labor involved in administering feed supplements and chemical treatments to animals and crops could be avoided.

Micropropagation (Tissue Culture)

Plant breeders already use **micropropagation** techniques in which whole plants are grown from single cells or from small plant parts for rapid multiplication of identical, disease-free plants (Figure 19-12). If necessary, genetic engineering can be used to incorporate desired characteristics from other species into the cell prior to propagation.

Figure 19-12 Papaya plantlets raised in a petri dish by a process called micropropagation. The plantlets will undergo genetic testing to see whether they will yield so-called perfect flowers, which will develop into fruit. (Photo courtesy of USDA ARS)

THE FUTURE

Current research will see the improvement and development of crops for specific purposes. Plants that require less water could be developed for countries with arid climates. Crop plants engineered to be tolerant to

Table 19-3 Long-Term Goals of Plant Biotechnology

Agronomic Traits	Quality Traits	Specialty Chemicals
Herbicide tolerance	Oil composition	Plastics
Insect control	Solids content	Detergents
Virus resistance	Nutritional value	Pharmaceuticals
Fungal resistance	Consumer appeal	Food additives

salt could be farmed in salt-damaged farmland or could be irrigated with salty water. Crops with higher yields and higher protein values are also possible. Biotechnology will help by—

* Improving farming productivity
* Protecting our environment by allowing reduced and more effective use of chemical pesticides and herbicides
* Reducing food costs

Table 19-3 summarizes the long-term goals of plant biotechnology.

"OMICS" REVOLUTION

Following biotechnology and genetic engineering has been the development of three new fields of study: genomics, bioinformatics, and proteomics. This "omics" revolution in science has been very fast and, in many cases, borders on overwhelming.

Genomics

The term "genome" originated in 1930. It was used to denote the totality of genes on all chromosomes in the nucleus of a cell. Incredibly, DNA was not identified as the genetic material of all living organisms until 1944. The genetic code was elucidated in 1961, and with these fundamental insights in hand it was possible to contemplate the concept that biological organisms had a blueprint consisting of finite numbers of genes. The sequence of these genes encoded all of the information required to specify the reproduction, development, and adult function of an individual organism.

Genomics is operationally defined as investigations into the structure and function of very large numbers of genes undertaken in a simultaneous fashion. Because all modern genomics have arisen from common ancestral genomes, the relationships between genomes can be studied with this fact in mind. The study of all of the nucleotide sequences, including structural genes, regulatory sequences, and noncoding DNA segments, in the chromosomes of an organism plant or animal is called genomics. The massive interest and commitment to genomics flows from the generally held perception that genomics will be the single most fruitful approach to the acquisition of new information in basic and applied biology in the next several decades

Despite the importance of plants, much basic research on plant functions still needs to be done. For example, scientists need to learn more about how plants grow, how they protect themselves from disease or insects, and how they respond to their environment. One of the best ways to do this is to study the information encoded in a plant's DNA, the material that makes up its genes.

An understanding of genomics means that information gained in one organism can have applications in other, even distantly related organisms. Comparative genomics enables the application of information gained from facile model systems to agricultural or nonmodel taxa. The nature and significance of differences between genomes also provides a powerful tool for determining the relationship between genotype and phenotype. Plant genomics holds the promise of describing the entire genetic repertoire of plants. Ultimately, plant genomics may lead to the genetic modification of plants for optimal performance in different biological, ecological, and cultural environments for the benefit of humans and environment.

Recent scientific advances have been made through private and public investments in studying DNA structure and function not only in humans but in other organisms such as plants. A new biological paradigm can be extended to improving the useful properties of plants that are important to humanity. Solutions to many of the world's greatest challenges can be met through the application of plant-based technologies. For example, the revitalization of rural America will come from a more robust agriculture sector; reductions in greenhouses gases can be achieved from the production of plant biofuels for energy; chemically contaminated sites can be rehabilitated economically using selected plants; and worldwide malnutrition can be greatly reduced through the development of higher yielding and more nutritious crops that can be grown on marginal soil.

Molecular plant-breeding efforts have received a boost in the past decade from massive amounts of gene and genome sequence information (genomics), which has been used to generate molecular markers for marker-assisted breeding in species such as alfalfa—the number-one forage crop—white clover, and tall fescue.

Functional genomics can produce massive amounts of data on the majority of genes of a sequenced organism. Functional genomics includes transcriptomics, proteomics, and metabolomics, which identify and quantify thousands to tens of thousands of gene transcripts (RNA), proteins, and metabolites, respectively, in cells, tissues, or organs. By revealing concerted changes in RNA, proteins, and metabolites during plant development, under optimal or stressed genes scientists determine important plant characteristics or traits.

Other disciplines involved in plant genomics are genetics, biochemistry, biophysics, molecular and cell biology, and physiology. They are also used to elucidate the precise biological function(s) of specific gene products. Ultimately, combining genomics, functional genomics, and other approaches aimed to identify genes with key roles in plant metabolism, growth, development, and response to the environment, will be of value to plant-breeding efforts to improve plant performance, yield, and quality.

Genomics and functional genomics approaches are used to identify genes and processes that enable plants to respond and adapt to environmental challenges such as pathogen attack (bacterial, fungal, or viral), intensive grazing, or abiotic stress such as drought and soil acidity/aluminum toxicity. Another target of this type of research is plant secondary metabolism, which yields an amazing array of compounds of industrial and medicinal value. Secondary metabolism is crucial for the production of structural compounds in plants such as lignin, which affects digestibility of plant material.

Bioinformatics

The part of the "omics" revolution that has made it possible to analyze and interpret all of the genomics data is bioinformatics; for example, genomics generates data and bioinformatics provides the analytical tools enabling those data to be interpreted.

Bioinformatics is the study of the inherent structure of biological information and biological systems. It puts together the ever-increasing avalanche of systematic biological data with the analytic theory and practical tools of computer science and mathematics.

Genomics data can be viewed as the greatest encoding problem of all times. These bioinformatics problems open the way for biologists to collaborate with mathematicians and computer scientists because it aims to translate "biology problems" into new challenges that are interesting to theoreticians: problems of information content, structure, and encoding. By contrast, the common view and practice of bioinformatics is simply an application of existing mathematical and computer science to biological problems; this translates biological questions into the language of information content, structure, and encoding, so that mathematicians and computer scientists are needed to help solve this problem.

Bioinformatics is a rapidly developing branch of biology and is highly interdisciplinary, using techniques and concepts from informatics, statistics, mathematics, chemistry, biochemistry, physics, and linguistics. It has many practical applications in different areas of biology and medicine. Research in bioinformatics includes method development for storage, retrieval, and analysis of the data. Bioinformatics is a rapidly developing branch of biology.

Proteomics

Included in the "omics" revolution are a number of powerful tools including variations on the theme of proteomics. Proteomics aims to identify and characterize the expression pattern, cellular location, activity, regulation, posttranslational modifications, molecular interactions, three-dimensional structures, and functions of each protein in a biological system.

The potential of proteomics for plant improvement lies in the outstanding developments in robotics and nanotechnology that have been achieved in the last 10 years, allowing the genome sequencing of many organisms. The new genomics science highlights our ignorance of the newly discovered sequenced genes.

In plant science, the number of proteome studies is rapidly expanding after the completion of the *Arabidopsis thaliana* genome sequence and proteome analysis of other important or emerging model systems. This analysis of plants is subject to many obstacles and limitations, as in other organisms, but the nature of plant tissue, with its ridged cell walls and complex variety of secondary metabolites, means that extra challenges are involved that may not be faced when analyzing other organisms.

Proteomics is a complementary approach to solving many problems: from mutant characterization to genetic variability estimates, from the establishment of genetic relationship to the identification of the genes involved in the response to biotic or abiotic stresses, from the definition of genes coregulated or coaffected by a given molecule to the quantification of their amounts. Proteomics, with its recent development in mass spectrometry and database management, is today a very powerful new tool to deal with the global approaches of modern biology.

Proteomics holds a promise of a great revolution. If genomes and microarrays gave us a glimpse into the blueprint of life, proteomics was going to unravel the working end of the cell: the protein machinery. But, as has become rapidly apparent, proteins were not going to give up their secrets without a fight. Proteins are much more diverse in their properties than nucleic acids, so a single protocol—for example, preparation or analysis—is unlikely. There is no polymerase chain reaction for proteins, so the amount of starting tissue and detection sensitivity are critical limitations. Protein concentrations extend over a far greater dynamic range than do nucleic acids. Proteomics must deal with differences in abundance of six to eight orders of magnitude, meaning that the few proteins that are most abundant often interfere with detection of low-level proteins. None of these problems are insurmountable, but they have slowed the appearance of the expected biological results, as each problem needed to be solved individually.

With the sequencing of the *Arabidopsis* genome, there is a mature proteomics technology platform. Now is the time to bring the resources and tools together. Proteomics gives us the molecular mechanisms that will fill in the gaps that have existed in the defined genetic pathways. Proteomics will blend perfectly and powerfully with genetics. The revolution has begun and it continues at a rapid pace.

BIOTECHNOLOGY POLICY—PUBLIC PERCEPTION AND THE LAW

Although the opportunities for using biotechnology in agriculture are truly fantastic, they have triggered public concern and a corresponding need to develop methods to assure that biotechnology is used in an environmentally sound manner. Some people fear that widespread use of plants and animals with altered genetic characteristics may threaten the environment by disturbing the existing balance between organisms.

This balance is a dynamic one. Because gene mutation and changes in gene position within chromosomes are normal events in all living organisms, organisms with new properties are constantly emerging. However, transgenic technology does expand the scope of these events. Careful examination of the properties of the transgenic organism is essential before it is studied outside the closed environment of the laboratory.

Agricultural researchers are enhancing traditional methods of manipulating plant and animal germplasm used for over 100 years, as well as using the guidelines in recombinant DNA studies directed by human applications. In both cases, there is a sound track record of safety associated with research and its products. Recombinant DNA techniques have been safely employed since the mid-1970s through an essentially self-imposed series of safety guidelines and reviews within the scientific community. Formerly under the control of the National Institutes of Health, these safety procedures are now the responsibility of the U.S. Department of Agriculture. Methods and procedures are being completed to assure continued safety in applying recombinant DNA techniques for agricultural research and in producing biotechnology products.

For more readings on the safety of biotechnology check out the readings at the FDA website (http://www.cfsan.fda.gov/~lrd/biotechm.html) or the USDA websites (http://www.aphis.usda.gov/biotechnology/index.shtml).

SUMMARY

Success in biotechnology and genetic engineering is based on our increasing understanding of the genetic code. Biotechnology research benefits agriculture, forestry, and industrial processing by diversifying crops and crop products, with increasing concern and care for the environment. Some plants have already benefited from genetic engineering.

Much current research focuses on understanding and developing useful promoter sequences to control transgenes and establishing more precise ways to insert and place the transgene in the recipient. There is still much that needs to be done to improve our knowledge of specific genes and their actions, and of the potential side effects of adding foreign DNA and of manipulating genes within an organism.

A formal regulatory system exists to examine areas of risk or uncertainty before field testing is approved. The government, industry, and other interested groups are also considering more general questions on the uses of this new technology, including ethical questions and sociological consequences.

Following biotechnology and genetic engineering has been the development of three new fields of study: genomics, bioinformatics, and proteomics. The "omics" revolution in science continues at a rapid pace.

Review

True or False

1. The structure of DNA was discovered in the 1980s.

2. Lysine is one of the four bases in DNA.

3. DNA is a double helix.

4. Genetic engineering is used to produce transgenic plants.

5. To a cell, DNA is DNA whether it came from a plant or animal, or from a different species.

Short Answer

6. Name the four bases found in DNA.

7. List five crops that have been genetically engineered.

8. What are four of the goals of biotechnology?

9. Identify DNA, recombinant DNA, tRNA, and mRNA.

10. _____ are small circles of DNA used to pass DNA from one cell to another.

Critical Thinking/Discussion

11. Explain how DNA controls the formation of proteins.

12. Why are people concerned about biotechnology and genetic engineering?

13. What future impacts could genetic engineering have on agriculture?

14. Define genetic engineering, biotechnology, and transgenic.

15. Describe the advantages that genetic engineering has over traditional selective breeding.

16. Do the benefits of genetic engineering and biotechnology outweigh the risks?

17. Discuss the "omics" as they relate to biotechnology.

Knowledge Applied

1. Using ordinary supplies, construct a three-dimensional model of DNA.

2. Find sites on the Internet where scientists are involved in the genetic engineering of plants. Report on your findings.

3. Develop a computerized presentation that explains the relationship between, DNA, recombinant DNA, RNA, mRNA, tRNA, ribosomes, endoplasmic reticulum, and protein synthesis.

4. Investigate and report on the patents issued for transgenic plants and animals during the past five years.

5. Report on the bacteria and eukaryotes that are used in recombinant DNA research.

6. Make a poster presentation on genetic mapping in plants. Show what a genetic map looks like. Find out what type of progress is being made in genetic mapping and in which plants.

Resources

Brown, J., & Caligari, P. (2007). *An introduction to plant breeding.* Ames, IA: Blackwell Publishing.

United States Department of Agriculture. (1986). *Research for tomorrow: The yearbook of agriculture.* Washington, DC: Author.

United States Department of Agriculture. (1992). *New crops, new uses, new markets: The yearbook of agriculture.* Washington, DC: Author.

United States Department of Agriculture. (1994). *Biotechnology and sustainable agriculture: A bibliography.* Beltsville, MD: Author.

Walters, D., Newton, A., & Lyon, G. (2007). *Induced resistance for plant defence: A sustainable approach to crop protection.* Ames, IA: Blackwell Publishing.

Internet

Internet sites represent a vast resource of information. The URLs (uniform resource locators) for World Wide Web sites can change. Using one of the search engines on the internet such as Google or Yahoo!, find more information by searching for these words or phrases: biotechnology, agroeconomic, agroecosystems, biocontrol, bioinformatics, biopharmaceuticals, genetic code, genetic engineering, genomics, DNA, RNA, micropropagation, proteomics, and transgenic.

PART 4

Crops—Applied Plant and Soil Science

Chapter 20
Major Agronomic Crops

THIS CHAPTER COVERS crops that could be considered major for a couple of reasons. They could be considered major because of their economic impact to the United States. They could be called major because of their importance in the diet. Nevertheless, the crop grown in a specific area is probably considered major to that area, so obviously all major crops are not covered in this chapter. For the purposes of this book, these crops are considered as major agronomic crops: corn, cotton, potatoes, soybeans, sugarcane, sugar beets, and wheat.

After completing this chapter, you should be able to—

- Describe the culture of corn, cotton, potatoes, soybeans, sugarcane, sugar beets, and wheat

- Explain the genetic and the environmental factors influencing the production of corn, cotton, potatoes, soybeans, sugarcane, sugar beets, and wheat

- Compare the growing requirements for corn, cotton, potatoes, soybeans, sugarcane, sugar beets, and wheat

- Know the botanical name for corn, cotton, potatoes, soybeans, sugarcane, sugar beets, and wheat

- Recognize the importance of crop rotations

- Identify important pests and their control for the production of corn, cotton, potatoes, soybeans, sugarcane, sugar beets, and wheat

- Discuss the fertilizer needs of corn, cotton, potatoes, soybeans, sugarcane, sugar beets, and wheat

- Identify four elements of a comprehensive disease-management program

CORN (MAIZE)

The botanical name for corn is *Zea mays*. Worldwide, corn is better known as **maize**. Major corn growing areas of the United States are shown in Figure 20-1 and the historical trend for corn production is shown in Figure 20-2.

Several varieties of corn are grown including dent corn, flint corn, flour corn, sweet corn, popcorn, and pod corn (Table 20-1).

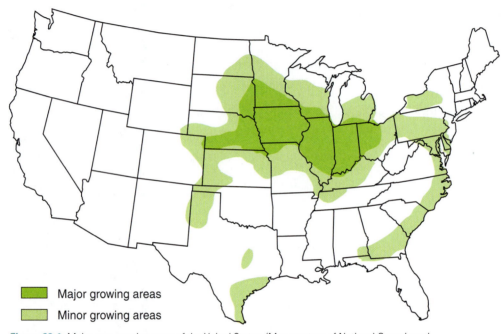

■ Major growing areas
■ Minor growing areas

Figure 20-1 Major corn growing areas of the United States. (Map courtesy of National Oceanic and Atmospheric Administration/USDA)

Figure 20-2 Historical trend for corn production. (Chart courtesy of USDA)

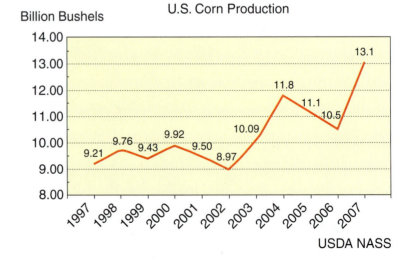

U.S. Corn Production

Billion Bushels

USDA NASS

Table 20-1 Botanical Varieties of Corn

Type	Variety	Features
Dent corn	indentata	Primary commercial feed corn; grain yellow or white; kernels form dent on top upon drying
Flint corn	indurata	Produced primarily for starch
Flower corn	amylacea	Easily ground and is used for production of cornmeal or corn flour
Sweet corn	saccharata	Sugary kernels when slightly immature; the horticultural vegetable crop
Popcorn	praecox	Subform of flint corn with a hard, starchy endosperm; moderately high moisture content; fibrous endosperm explodes when heated
Pod corn	tunicata	Thought to be an ancestor or relative to corn; individual kernels enclosed in husks or pods; not produced commercially

Successful corn production requires an understanding of the various management practices and environmental conditions affecting crop performance. Planting date, **seeding rates,** hybrid selection, **tillage,** fertilization, and pest control all influence corn **yield.** A crop's response to a given cultural practice is often influenced by one or more other practices. The keys to developing a successful production system are—

* To recognize and understand the types of interactions that occur among production factors, as well as various yield limiting factors
* To develop management systems that maximize the beneficial aspect of each interaction

Knowledge of corn growth and development is also essential to use cultural practices more efficiently to obtain higher yields and profits.

Temperature

Corn can survive brief exposures to adverse temperatures, with low-end adverse temperatures being around 32°F and high-end ones being around 112°F. Growth decreases once temperatures dip to 41°F or exceed 95°F. Optimal temperatures for growth vary between day and night, as well as over the entire growing season. The optimal average temperatures for the entire crop growing season, however, range between 68 and 73°F.

Planting Date

The recommended time for planting ranges from mid-April to mid-May. Approximately 100 to 150 growing degree days (heat units) are required for corn to emerge (see Chapter 11). Improved seed vigor and seed treatments allow corn seed to survive up to 3 weeks before emerging if soil conditions are not excessively wet. An early-morning soil temperature of 50°F at the 1/2- to 2-inch depth usually indicates that the soil is warm enough for planting. Corn germinates very slowly at soil temperatures below 50°F. The latest practical date to plant corn ranges from about June 15 to July 1. Plantings after these dates yield no more than 50 percent of normal yields.

Corn should be planted only when soils are dry enough to support traffic without causing soil compaction. The yield reductions resulting from "mudding the seed in" may be much greater than those resulting from a slight planting delay. No-tillage corn can be planted at the same time as conventional types, if soil conditions permit. Planting a full-season hybrid first, then alternately planting early-season and mid-season hybrids, allows the grower to take full advantage of maturity ranges and gives the late-season hybrids the benefit of maximum heat unit accumulation. Full-season hybrids generally show greater yield reduction when planting is delayed compared with short- to mid-season hybrids.

Planting hybrids of different maturities reduces damage from diseases and environmental stress at different growth stages (improving the odds of successful pollination) and spreads out harvest time and workload.

When planting is delayed past the optimum dates or if a crop needs to be replanted, it may be necessary to switch hybrid maturities. Hybrids planted in late May or early June mature at a faster thermal rate (require fewer heat units) than the same hybrid planted in late April or early May.

Other factors concerning hybrid maturity need to be considered when planting is delayed. Although a full-season hybrid may still have some yield advantage over shorter-season hybrids planted in late May, it could have significantly higher grain moisture at maturity than earlier maturing hybrids if it dries down (loses moisture as seeds mature) slowly. Moreover, many short- to mid-season hybrids have excellent yield potential. For plantings in late May or later, the dry-down characteristics of hybrids should be considered.

Seeding Depth

The appropriate planting depth varies with soil and weather conditions. For normal conditions, planting corn 1½ to 2 inches deep provides frost protection and allows for adequate root development. Shallower planting often results in poor root development and should be avoided in all tillage systems. In April, when the soil is usually moist and evaporation rate is low, seed should be planted shallower, no deeper than 1½ inches. As the season progresses and evaporation rates increase, deeper planting may be advisable.

Row Width

Most hybrids that are currently available perform well in 30-inch row spacing. Research from across the United States indicates average yields may increase 3–5 percent by planting in narrow rows (15–22 inches) compared with 30-inch rows. These yield increases occur at both moderate and high **plant populations** and high and moderate yield levels. Row spacing less than 30 inches may be the most effective option for growers to consider, particularly if the narrow corn spacings are matched with equipment used to produce other crops, such as soybeans and sugar beets. Key changes for narrowing rows include tractor and combine rims and tires, combine heads, and planter modifications.

Plant Populations and Seeding Rates

When corn is produced for grain, recommended plant populations at harvest range from 20,000 to more than 30,000 plants per acre, depending on the hybrid and production environment. Populations for corn silage may exceed those for grain by 2,000–4,000 plants per acre. Advances in corn hybrid genetics include improvements in stalk quality and resistance to barrenness. As a result, many newer corn hybrids may require harvest populations greater than 24,000 plants per acre to achieve their yield potential.

Most seed companies specify a range in seeding rates for the various hybrids they market. These seeding rate guidelines should be followed closely. Higher seeding rates are recommended for sites with high yield potential with high soil fertility levels and water-holding capacity.

Uneven plant spacing and emergence may reduce yield potential. Seed should be spaced as uniformly as possible within the row to ensure maximum yields and optimal crop performance, regardless of plant population and planting date. Corn plants next to a gap in the row may produce a larger ear or additional ears (if the hybrid has a prolific tendency), compensating for missing plants. These plants, however, cannot make up for plants spaced so closely together in the row that they compete for sunlight, water, and nutrients. Crowding often results in barren plants or ears that are too small to be harvested (**nubbins**), as well as stalk lodging and ear disease problems.

Uneven emergence affects crop performance because competition from larger, early emerging plants decreases the yield from smaller, later emerging plants. The primary causes of delayed seedling emergence in corn include:

1. Soil moisture variability within the seed depth zone
2. Poor seed-to-soil contact resulting from cloddy soils
3. Improper adjustment or functioning of the planter
4. Soil temperature variability within the seed zone
5. Soil crusting prior to emergence
6. Occurrence of certain types of herbicide injury
7. Variable insect and/or soil-borne disease pressure

Making Replant Decisions

Often, 10–15 percent of planted seeds fail to establish healthy plants because of stand losses (plant death) resulting from insects, frost, hail, flooding, or poor **seedbed** conditions. A poor stand may call for a decision on whether or not to replant a field. The first rule in such a case is not to make a hasty decision. Corn plants can and often do outgrow leaf damage, especially when the growing point is protected beneath the soil surface (up until about the six-leaf stage). If new leaf growth appears

Truly American Potato

O Creator! Thou who givest life to all things and hast made men that they may live, and multiply. Multiply also the fruits of the earth, the potatoes and other food that thou hast made, that men may not suffer from hunger and misery.

—Inca Prayer

The white or Irish potato actually originated in the Andes highlands from Chile to Colombia. There are at least 150 wild species of this plant ranging as far north as Colorado. Archaeologists commonly find pottery in the Andean highlands that predates the rise of the Inca Empire, made in the shape of the potato. Potatoes were cultivated by the Incas for 2,000 years before contact with the Spanish. The potato was so important to the Incas that they included reference to it in their prayers.

Besides eating the potato fresh, the Incas made a dried product from it called chuño. Chuño may be the world's first freeze-dried product. On nights when the Incas knew there would be a freeze, they would lay potatoes in a single layer on large cloth sheets on the ground. The potatoes would freeze; in the morning before the potatoes thawed, the Incas would walk on them, which partially crushed the potatoes. Then as the temperature warmed during the day, most of the moisture in the potatoes would escape. In this way, the Incas produced a product that could be stored without the specialized facilities used in more technological societies like our own. By being able to store these dried potatoes from growing season to growing season, the Incas helped assure an adequate supply of food throughout the year. Chuño is still made the same way today in many areas of the Andes, and it is often offered for sale in local markets.

The potato was first taken to Europe in 1537, but people there did not immediately accept it as a food. At first the pope of the Roman Catholic church declared that potatoes were the food of the devil because there was no reference to potatoes being used as food in either the Old or New Testament of the Bible. For those people not under the religious direction of the pope, there were other reasons for not rapidly accepting the potato as a food. The first introductions were short-day conditions as regard to potato set. This means that the potato plants required less than, or nearly equal to, 12 hours of daylight to produce the maximum number of potatoes. Northern Europe has long-day conditions during summer, when temperatures will allow the growing of potatoes, so yields were low. Not until large potatoes that could produce under long-day conditions were introduced from Chile did production in Europe began in earnest.

within a few days after the injury, then the plant is likely to survive and produce normal yields.

When deciding whether to replant a field, growers use the following information:

1. Original planting date and plant stand
2. Earliest possible replanting date and plant stand
3. Cost of seed and pest control for replanting

Pests

Insects of corn include corn earworm, European corn borer, and aphids. Insects are generally controlled by extensive use of **insecticides** from tassel emergence through kernel drying. Integrated Pest Management (IPM) using scouting, biological models of pest populations, targeted insecticide use, and "soft" or narrow-target pesticides have reduced the use of insecticides, improved production profitability, and reduced environmental hazard.

Diseases affecting corn are Southern leaf **blight**, Northern leaf blight, and diplodia rot. Diseases are controlled primarily through selection of disease-resistant cultivars, good management techniques, and applying appropriate targeted **fungicides** based upon biological-meteorological models.

Fertilizer Requirements

Fertilizer requirements vary according to soil tests. A corn crop removes nitrogen, phosphate, potassium, and various micronutrients from the soil. These must be replenished by a fertilization program.

Crop Rotations

The corn-soybean rotation is by far the most common cropping sequence used in the Midwest. This crop rotation offers several advantages over growing either crop continuously. Benefits to growing corn in rotation with soybeans include more weed-control options, fewer difficult weed problems, less disease and insect buildup, and less nitrogen fertilizer use. Corn that is grown following soybeans typically yields about 10 percent more than continuous corn.

No-till cropping systems, which leave most of the prior crop residue on the surface, are more likely to succeed on poorly drained soils if corn follows soybeans or meadow rather than corn or a small grain, such as wheat. This yield advantage to growing corn following soybeans is often much more pronounced when drought occurs during the growing season.

Genetically Engineered Corn

Bt corn has been enhanced through biotechnology (genetically modified) to resist damage by pests, namely corn borers (Figure 20-3). The corn borer is one of the most destructive insects to damage or reduce corn production and one of the most economically significant crop pests in the world.

Figure 20-3 Healthy European corn borer. (Photo courtesy of USDA ARS)

The mechanism that Bt corn uses to protect itself is based on naturally occurring soil bacteria, *Bacillus thuringiensis*, from which the name Bt is derived. Bt bacteria produce proteins deadly to specific target insects while being safe for nontarget animals. One of these proteins, produced from the gene *Cry1Ab*, targets corn borers. The tools of biotechnology were used to insert the *Cry1Ab* gene into Bt corn, providing a novel solution for protection of corn from corn borer attack.

Planting Bt corn is more effective than conventional treatments because once the corn borers enter the corn stalk they attack the plant from the inside. Conventional treatments are not always effective, or even possible. Because conventional pest sprays cannot reach the attacking corn borers, the pests continue to eat away at the inside of the corn stalk.

"Roundup-Ready" Corn

The chemical name for Roundup, a Monsanto product, is glyphosate. "Roundup-ready" corn is genetically engineered to permit direct, "over the top" application of the herbicide glyphosate, allowing farmers to spray both their crops and crop land with the herbicide so as to be able to kill nearby weeds without killing the crops

Glyphosate kills plants by blocking an enzyme, called EPSP synthase, which carries out a key step in the production of one class of amino acids. This causes the treated plant to stop growing, turn yellow, and die a few days after treatment because it cannot make proteins. Roundup-ready plants have been provided with the genetic instructions to make a different kind of EPSP synthase, one that is not blocked by glyphosate.

Harvesting

Corn is harvested in the fall with a combine (Figure 20-4). Frequently, moisture in corn must be removed before storage. The process is called drying down.

Figure 20-4 Corn is harvested with a combine and has many uses.

BIOFUEL

According to the National Corn Growers Association (www.ncga.com/), ethanol is a significant market for U.S. corn, consuming more than 1.8 billion bushels to produce 4.9 billion gallons of renewable fuel. In 2007, ethanol plants helped rejuvenate rural communities across the country by creating high-paying jobs, boosting local tax revenues, and creating partnership opportunities for local businesses. Additionally, ethanol helps the environment by reducing greenhouse gas emissions and displacing the harmful additive MTBE (methyl tertiary butyl ether) from reformulated gasoline. Ethanol plants also produce a coproduct called distiller's dried grains with solubles (DDGS), a high-protein, high-energy livestock feed.

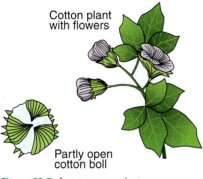

Cotton plant
with flowers

Partly open
cotton boll

Figure 20-5 A mature corn plant.

COTTON

The botanical name for cotton is *Gossypium* spp. Of the more than 30 species, only 3 are of commercial importance. Cotton is a dicot, an annual of almost tropical and/or subtropical origin. Its growth range is limited by length of frost-free growing season. It is classified as an agronomic, fiber crop and is grown primarily for its fiber but also for its seed meal and seed oil. Figure 20-5 illustrates a mature cotton plant.

Major Production Areas

The major production areas for cotton worldwide are China and the United States. In the United States, production extends from central California to the southeastern U.S. coast (South Carolina), and south to the southern extreme of Texas. Figures 20-6 and 20-7 show the production areas for cotton and the historical trends in the United States

Estimated annual world production of 80 million bales is produced by 70 countries. The U.S. production is approximately 10 million bales per year. Each bale weighs 500 pounds.

Production

Cotton is produced primarily for lint (fibers) from the mature cotton **boll**. However, the seeds left after lint removal (**ginning**) are used for seed oil production and as animal feed or feed supplement (cotton meal). Fibers are protuberances on epidermal cells of the ovule seed coat.

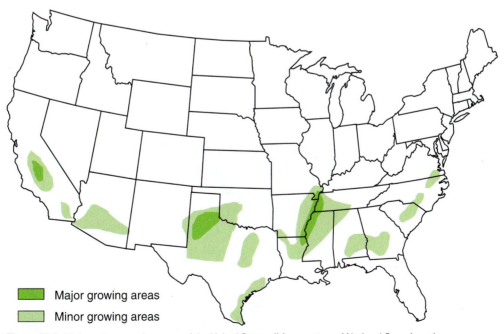

Major growing areas

Minor growing areas

Figure 20-6 Major cotton growing areas of the United States. (Map courtesy of National Oceanic and Atmospheric Administration/USDA)

Figure 20-7 Historical trends for cotton production. (Graph courtesy of USDA)

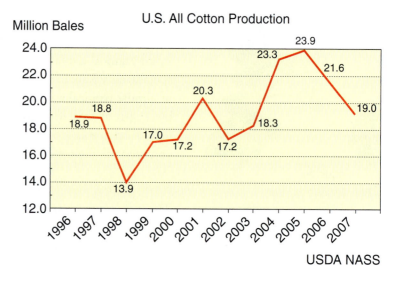

U.S. All Cotton Production

Million Bales

USDA NASS

Cotton is a highly mechanized crop, from planting through harvest through ginning. Originally, the crop required extensive hand labor for weed control, harvesting, and separating lint and seeds.

Cotton is a complete, perfect flower (Figure 20-8). It is self-fertile and self-pollinating with the aid of wind and insects. Cotton flowers have multiple anthers but a single, fused pistil.

Many cultivars of cotton are available. Selection of a cotton variety/cultivar should be made on yielding ability, quality, and resistance to wilt.

Resistance to the complex association of Fusarium wilt with nematodes is important in fields where the disease is known or suspected to occur. Cotton cultivars are evaluated under field conditions for their reaction to this combination of problems. The cotton root knot nematode commonly is associated with the wilt disease, but other nematodes may also contribute to the problem. Only cultivars resistant to the complex should be planted in fields known to have nematode or Fusarium wilt problems.

Figure 20-8 Cotton flower with beneficial insect, ladybird beetle. (Photo courtesy of USDA ARS)

Weed Control

Successful weed control is essential for economical cotton production. Weeds compete with cotton for moisture, nutrients, and light. The greatest competition usually occurs early in the growing season. Late-season weeds, although not as competitive as early-season weeds, may interfere with insecticide applications and may cause harvesting difficulties.

Herbicides. Herbicides are the most effective means for controlling weeds in cotton. To be effective, however, herbicides need to be matched with the weed problem. Preplant and/or pre-emergence applications are important for ensuring that the cotton has the initial competitive advantage over the weeds. Once this is achieved, then postemergence-directed applications can be utilized to extend the weed control through the season.

Cultivation. Cultivation can be used if effective weed control is not achieved with herbicides. However, if weeds have been controlled with herbicides, little benefit exists from cultivation. If cultivation is needed, growers should avoid throwing soil around small cotton plants to minimize disease problems.

Important Pests

Problem insects include—
+ Cotton boll weevil (Figure 20-9)
+ Thrips plant bugs
+ Pink bollworm, budworm

Two major diseases concern cotton growers—damping off from fungi and bacterial blight.

Insects are controlled primarily through extensive use of insecticides. Use of IPM technologies of scouting, introduction of biological controls (predators), targeted "soft" or narrow-spectrum pesticides, and use of biological insecticides is reducing the total chemical load and also production costs. The market is increasing for "organic" cotton produced without insecticides.

Diseases are controlled through use of resistant cultivars and proper management practices; for example, sanitation, drainage, and fungicides. Seeds for planting are typically treated with fungicide to prevent seedling diseases.

Chemicals can be used just before or at planting to reduce nematode population densities. However, their use is not always justified. The grower must compare the cost of treatment with the dollar value of the anticipated yield improvement.

Nematodes. Cotton root knot, reniform, and sting nematodes can be important cotton pests. The cotton root knot nematode is found in all soil types. Reniform nematodes are found most often in fine-textured soils and wet areas. Sting nematodes are limited to very sandy soils (generally greater than 85 percent sand) with low organic matter.

Figure 20-9 Cotton boll weevil. (Photo courtesy of USDA ARS)

Nematodes reduce cotton yields noticeably by themselves and are also important for increasing incidence and severity of Fusarium wilt. Nematode management is critical to managing Fusarium wilt.

Crop rotation can help keep preplant nematode population densities from becoming too high and is important in managing a soil-borne fungal disease such as Fusarium wilt. Grass family crops such as small grains, corn, sorghum, millet, and forage grasses are good crops to rotate with cotton because they support few root knot and reniform nematodes. They are less beneficial if sting nematodes are found in the field because that nematode reproduces well on nearly all grasses. Peanut is not susceptible to any of the three important nematode pests of cotton. It may safely precede cotton in most fields. Soybean should not precede cotton because it has the potential to increase several nematode pathogens of cotton.

Bt Cotton

As in corn, genetic engineering was used to insert the Bt gene into cotton, providing a new solution for pest control. Many pest problems existed before the introduction of Bt cotton in 1996. Pest such as tobacco budworms, cotton bollworms, and pink bollworms developed resistance to the synthetic pesticides and farmers were losing a significant amount of their cotton. Bt cotton reduces the cost per acre of pesticide application and is more effective at reduction in pests. Bt cotton yields about 5 percent more on average than traditional cotton.

Production Requirements

Cotton is planted after all danger of frost is past and soil temperatures are warm 68–77°F. Plants do well with high temperatures during the growing season. This crop requires high light intensity and ample soil moisture. Cotton needs well-drained, sandy loam soil and requires high fertilization. The major production limitation is the length of the growing season.

Crop Rotations

Crop rotations are an important part of a good cotton weed control program. Certain weeds, especially broadleaves, may be less difficult to control in a preceding crop such as corn. Other benefits of crop rotation may include reduction in insect, disease, and nematode problems both in cotton and succeeding crops.

Land Management Practices

Crop destruction after harvest reduces the levels of pests and pathogens available to attack the following crop. Under some conditions, cotton roots can survive a long time after harvest. Fields should be tilled as soon as possible after harvest to stop reproduction of nematodes and other pests that can live in (on) cotton roots or stubble, and to allow natural population decline to begin. Nematodes that can damage cotton can build up on weeds and other crops that may precede it, and many other crops can be affected by the nematodes that will build up on cotton. Thus, killing the roots after harvest by plowing them up and exposing them to sunlight is important to prepare for either cotton or a different crop to follow cotton.

Winter cover crops can be important for holding land against erosion after harrowing or plowing in the fall to destroy roots of the preceding crop. Most small grains will fit this purpose. The cover crop should be killed or incorporated into the soil early enough that it will not interfere with cotton planting.

Harvesting

Cotton is harvested with a combine (Figure 20-10) and hauled to a gin. At the gin the cotton is separated from the seed. The cotton is baled into 500-pound bales for further processing. The seed is both sold for livestock feed directly or the oil may be extracted from it and then sold as cotton seed meal.

Figure 20-10 Harvesting cotton. (Photo courtesy of USDA ARS)

POTATOES

The potato is a member of the Solanaceae, or nightshade, family. It is related to such vegetables as tomato, pepper, and eggplant. It is also related to important drug plants such as datura and belladonna. The potato of world commerce is a tetraploid, which means it has four sets of chromosomes in *Solanum tuberosum*. Wild potatoes are usually diploid (two sets of chromosomes). This is important for plant breeders who may want to move favorable characters, such as disease resistance, from wild populations into commercial cultivars. Other important species of potato are *S. andigena, S. phureja, and S. stenotonum*.

Figure 20-11 shows the top potato-producing states, and Figure 20-12 shows the historical trend for potato production.

Figure 20-11 Potatoes will grow in all states. This chart shows the top potato-producing states.

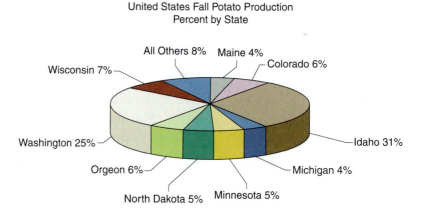

United States Fall Potato Production
Percent by State

All Others 8%
Maine 4%
Colorado 6%
Wisconsin 7%
Washington 25%
Oregon 6%
North Dakota 5%
Minnesota 5%
Michigan 4%
Idaho 31%

Chart may not add 100% due to rounding.

Figure 20-12 Historical perspective of potato production in the United States, all types. (Graph courtesy of USDA NASS)

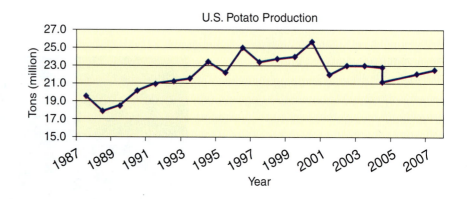

U.S. Potato Production

Morphology

The potato is a herbaceous dicot that reproduces asexually from tubers; hence, the species name *tuberosum*. The tubers form at the end of underground stems called stolons. The **tubers** are the edible portion of the plant; botanically they are stems, not roots. They are stems because they contain all the morphological features of stems. They have buds (the eyes), leaf scars (the eyebrows), and lenticels (Figure 20-13). Lenticels are small pores that allow stems to exchange gases with the air. The tuber is a storage organ. The plant produces sugars in the leaves. These sugars are converted to starch, which is stored in the tubers. The potato then uses the starch to produce new plants the following growing season.

Potato flowers range in color from white to purple, and produce small berries that are very poisonous. The berries contain viable seed, but in the past these were not used to produce new potato plants because the offspring were highly variable in such traits as yield and quality—size, taste, color, firmness and so on. The leaves and aboveground stems of the potato plant are also toxic to humans.

Figure 20-13 A potato plant and a potato with sprouts.

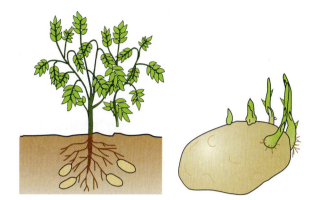

Areas of Potato Production

The potato is a very versatile crop. It is produced commercially on every continent on Earth except Antarctica. On a worldwide basis, the leading producer of potatoes is the Soviet Union, followed by Poland and the United States. Potatoes are produced commercially throughout Western Europe, but only sporadically in their native range. In the native areas, potatoes are produced mainly for direct consumption. In the United States, potatoes are produced commercially in every state (Figure 20-14). The leading state in potato production is Idaho. Idaho produces more potatoes for both the fresh and processing markets. Overall, Washington ranks as the second leading state in potato production.

Climatic Requirements

The potato is a cool-season crop. Mean temperatures in the range of 60–65°F are optimal for high yields. The optimum temperature for emergence is around 72°F. Higher temperatures seem to retard emergence.

Soil temperature is very important in potato growth. Because tuber pieces are planted, this is one of the main determining factors in the growth pattern of the plant. In cool soils the tubers sprout and emerge slowly. At 52°F it takes 30–35 days for complete emergence. The formation of tubers decreases at soil temperatures above 68°F and is almost completely inhibited at temperatures above 84°F. This presents a problem in tropical areas where soil temperatures are usually in this range. A goal of plant breeders is to find potato strains adapted to tropical conditions. As a rule of thumb, for every 1°F above the optimum temperature, yield is reduced 4 percent.

Soil temperature is also important in determining yield. The actual optimum day/night temperatures depend on the cultivar being grown. Soil temperatures of 61°F nights and 64°F days give the best yield and the highest amount of starch in Russett Burbank potatoes. If soil temperatures get too high, tuber abnormalities, such as knobby tubers, result.

The temperate region of the world is ideal for potato production because the plants are started and the tops are established in spring, when temperatures are low. The warm weather of summer causes high

Figure 20-14 Harvested Idaho potatoes on a conveyor belt on their way to storage.

rates of photosynthesis, which produces plenty of starch for transport to the tubers. The length of the growing season should be from 90 to 120 frost-free days. In parts of the world where there are shorter growing seasons, potatoes can be grown because the extremely long days of summer compensate.

Tuber Initiation

Both photoperiod and temperature affect tuber initiation (tuberization). In general, short days initiate tubers. Under long days, tuberization occurs if night temperatures are well under 68°F. Maximum tuber set occurs when nights are around 54°F. Interestingly, the temperature-sensitive part of the plant is the top, and not the stolon or root.

Low nitrogen levels in the plant favor tuber initiation. This is an important consideration when determining when to make **side-dress** applications of fertilizer to a potato crop. High light intensity also seems to enhance the tuberization process.

Soil, Water, and Nutrition

A loose-textured, well-drained soil with a pH of 5.0 to 6.5 is best, but potatoes will grow on almost any soil. Diseases can cause problems on soil that has high clay content. Generally speaking, the soil needs to be about four feet deep as this is the depth of rooting of the potato plant. If the soil texture is too "tight" (high in clay) the tubers that form can be misshapen. A total of 18–30 inches of water is needed throughout the entire life of the crop. This can be in the form of rainfall and/or irrigation. Potatoes respond better if the supply of water is uniform throughout the part of the season when the plant is actively growing. Potatoes require about 0.75 inches of water per week.

Potatoes usually require the application of nitrogen (N), phosphorus (P), and potassium (K) fertilizers. The normal method of fertilizer application is to apply half the needed fertilizer prior to planting and half as a side-dress application. If side-dressing is too late, the plants tend to have high levels of nitrogen, which inhibits tuber production.

Planting

Potatoes are propagated vegetatively in developed countries, including the United States. Seed pieces weighing 1.5–2 ounces are used. Large potatoes can be cut by hand with a sharp knife, although commercial growers use automated equipment for this procedure. When the potatoes are cut, they need to be kept at 64–70°F and 85–90 percent relative humidity for several days. This allows the wound to heal and protects the tuber from soil-borne decay. A good dusting with a fungicide is often substituted for this practice. Each piece of the cut tuber must contain an "eye." Because the eye of the potato is actually a bud, this is where the potato sprout will emerge. The pieces are planted 2–4 inches deep in beds 30–36 inches apart with 9–12 inches between the seed pieces.

Because potatoes are propagated vegetatively, the use of disease-free planting stock is very important for good potato production. Many virus diseases are passed from insects feeding on the foliage of potato plants. The disease moves through the stem into the tuber. If infected tubers are used as seed, the new plants are also infected with the disease. Certified seed potatoes are grown in cool areas where the symptoms of various diseases can easily be seen. If diseased plants are found, they are removed from the field, as well as are all the tubers from those plants. This is a very expensive process. In many developing countries, farmers either cannot afford to buy certified seed stock, or the methods for growing them are unavailable. This has led to the development of true botanical seed for potato production.

True Potato Seeds

The benefit of using pieces of tuber as seeds is that all individuals planted from that tuber have the same genetic makeup. In essence, they are clones. This allows growers to know what to expect when a certain cultivar is planted, as far as yield components and other quality factors. The seed that is produced in potato berries will germinate and grow into a plant with tubers. In the past, however, the yield from these plants has been very variable. Over the last 10–15 years, new breeding techniques have been developed that allow for the introduction of high-yield genes into all the progeny that grow from these seeds. This has led to the production of several commercial cultivars that are grown from true botanical potato seeds. These cultivars are presently used mostly by home gardeners and professional market gardeners. The development of this technology should aid in the spread of potato production into developing countries, where the nutritional benefits of potatoes are greatly needed.

Pests

The major insect pests of potato include wireworms, white grubs, Colorado potato beetle, aphids, and leafhoppers. Wireworms and white grubs are important because they feed directly on the tubers, causing injury that can lead to sites where diseases can enter. Aphids are important in the spread of virus diseases. They pierce the foliage with their stylet (similar to what a mosquito uses) and draw out the plant sap. While they are drawing the sap, they inject saliva into the plant to begin digestion. This saliva also contains viruses, which are injected into the plant. A great deal of the effort of commercial growers is aimed at controlling aphids.

The diseases of major importance are early and late blight, Fusarium and verticillium wilt, scab, bacterial ring rot, blackleg, and at least 10 viruses. More and more viruses are found every year in different parts of the world.

Weeds are a major problem in potato production. If the potato crop can be kept entirely weed-free for the first six weeks after the emergence of the crop, there is very little competition from weeds. The potato plants get large enough that the ground is shaded and very little additional weed growth occurs. Research has shown that there is no benefit, in terms of yield, from controlling weeds longer than the first six weeks after crop emergence.

Harvesting

As the potato tubers mature, the tops of the plants begin to turn yellow. In areas of early frost in the fall, the potato vines are killed. Chemical vine killers are used in most of North America. After the vines die, tubers are allowed to cure in the ground from several days to one week. This allows the potato skin to "set" by getting a little thicker and more firm. Large commercial growers dig and handle the tubers by machine (Figure 20-15). Damage to the tubers opens infection ports through which disease can enter, especially bacterial soft rot in storage. Bruising lowers the quality and the price received by the grower.

Curing. After harvest, the tubers must be hardened before they can be stored for long periods. This is usually done in a period of 4–5 days at 61–70°F in high humidity. This is usually done by packing houses. Some growers own their own packing houses, but many are owned by brokers. Once the potatoes are cured, they can be packed for immediate consumption or stored for use later.

For those potatoes that are going to the processor, it is not necessary to cure the tubers after harvest, and the vines are not killed prior to harvest.

Storage

Potatoes can be stored for almost a year under the proper conditions. Very cool temperatures, 40–50°F, at 90 percent relative humidity are ideal. The potatoes must be protected from freezing as this destroys tubers. If temperatures are too low, the starch contained in the tuber is converted to sugars. Then, when the potatoes are cooked, the sugar turns brown. This is especially bad for the processing industry.

Figure 20-15 Potato harvesting occurs in the fall after the vines have been knocked down/killed.

Figure 20-16 High-tech storage allows potatoes to be stored until processing or packaging.

Potato tubers must be stored in the dark to prevent greening (Figure 20-16). The green color is caused by chlorophyll, which is what plants use to trap the sun's energy. But sunlight also causes the production of the alkaloid solanine. Green potatoes contain solanine. Solanine is a toxic substance and can cause illness in most cases; it can cause death if too much is consumed.

SOYBEANS

The botanical name for the soybean is *Glycine max*. It is a native of Asia, where it was cultivated long before written history. Major soybean production areas are shown in Figure 20-17. Figure 20-18 shows the historical trend for soybean production in the United States.

Planting Date

The date of planting has more effect on soybean grain yield than any other production practice. Yield loss resulting from delayed planting ranges from one-fourth bushel to more than one bushel per acre per day, depending on the row width, date of planting, and plant type. Regardless of planting date, row width, or plant type, the soybean crop should develop a closed canopy (row middles filled in) prior to flowering or by the end of June, whichever comes first. Generally, when planting in early May, rows must be less than 15 inches apart to form a canopy by late June. An early canopy results in high yields because more sunlight is intercepted and converted into yield than when row middles do not fill in until late in the growing season. Assuming a half bushel per acre per day yield loss with delayed planting, a 10-day delay in planting 300 acres would decrease total production by 1,500 bushels.

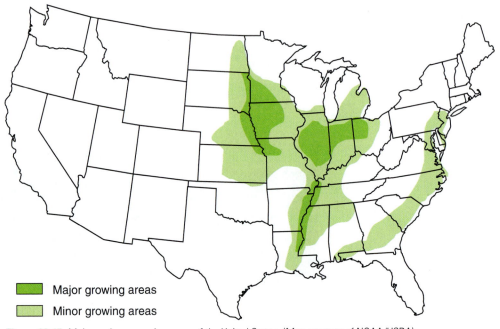

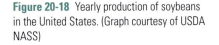

Figure 20-17 Major soybean growing areas of the United States. (Map courtesy of NOAA/USDA)

Figure 20-18 Yearly production of soybeans in the United States. (Graph courtesy of USDA NASS)

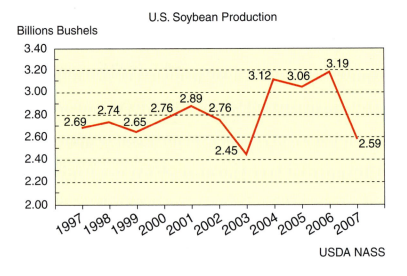

Adequate, vigorous stands are more difficult to obtain with early planting. Seed treatments, good seed-soil contact, and reduced seeding depths help establish vigorous stands. Herbicide programs must provide weed control for a longer time until the crop is large enough to suppress weed growth through competition. Narrow rows provide the needed competition for weeds sooner than wide rows.

Row Spacing

The average row width is less than 12 inches. For any planting date, variety, or seeding rate, yields increase as row widths decrease because of earlier canopy formation and increased sunlight interception. Yields can be increased by one-third bushel per acre for each inch the row width is decreased from 30 inches.

Currently, most grain drills offer good depth control and space seeds uniformly in the row. These new drills produce yields similar to those obtained with corn-bean planters designed to plant in 15-inch rows. However, those yields can be obtained only if a good seedbed has been prepared and if the tractor tire tracks are removed ahead of the drill where tillage is used.

Seeding Rate

Soybeans are not very responsive to changes in seeding rates. Growers select seeding rates by the number of viable seeds per foot of row. Seeding rates expressed in pounds per acre are not reliable because seed size varies significantly from year to year and variety to variety. Tall plants with small, weak stalks will be produced with excessive (250,000 seeds per acre) seeding rates and lodging will likely occur. If seeding rates are too low (60,000 seeds per acre), the plants will be short, have many branches, and the pods will be close to the soil surface, making harvest losses excessive.

Seeding rates that produce from 80,000 to 140,000 plants per acre at harvest are adequate for soybeans planted before May 15. If plantings are delayed until early June, the seeding rate should be increased sufficiently to obtain a harvest population of 100,000–160,000 plants per acre.

When seeding rates are reduced or when row widths are narrowed (less than 15 inches), the importance of uniform spacing of plants in the row becomes greater. For all row spacings, skips, or blank spacing greater than the distance between the rows, yields decrease. Poor distribution of an adequate number of plants in the row may result in a yield loss of up to 5 percent, depending on the variety's growth habit.

A number of determinate semidwarf variety types are available. Generally, these plant types are shorter than the indeterminate varieties and do not spread to fill wide row middles. So the determinate semidwarf types generally respond to higher seeding rates and to narrow rows. Row widths should not exceed 15 inches.

Although some of these varieties have higher yield potentials than some taller-growing varieties, they require a high level of management. The most productive soils, narrow rows, good fertility, increased seeding rates, and early planting are all required for best yields. Because plant height does not increase after the onset of flowering, rapid growth early in the season is critical for good yields. Determinate semidwarf varieties do not perform well when grown using no-till culture.

Planting Depth

One inch to 1½ inches is the ideal planting depth where tillage is used. Three-fourths inch to one inch is ideal for no-till seeding. The soil should be free of large "clods" to ensure good seed-soil contact and good seed coverage. Shallow planting—¾ to 1 inch—in late April promotes more rapid emergence than deeper planting, but this may increase exposure to herbicides, which may damage young seedlings. Soil temperatures at one-inch depth are three to eight degrees warmer than at two-inch depth. After May 15, the air temperatures are higher and the probability of crusting increases. It is a poor practice to plant deeper than 1 to 1½ inches because a crust may form above the seed and reduce emergence. It takes the combined pressure of many plants to break through the crust. In the process, many of the hypocotyls are broken and the seedlings do not emerge. When planted at a one-inch depth, the seed is more likely to be inside the crust layer. As the seed "swells" in the germination process, the soil crust is broken and a higher percentage of emergence occurs.

Variety Selection

Most soybean varieties have genetic yield potentials well over 100 bushels per acre. A variety's adaptability to the environment and production system where it will be used sets the production system yield potential. Varieties should be selected with characteristics that will help them perform well in the cultural system and environment to be used rather than on their yield record alone. Where excessive growth and lodging are problems, varieties that are medium to short in height with good standability should be selected.

Maturity information should be used to select varieties that mature at different times to allow for timely harvest in the fall. Generally, each ten-day delay of planting in May delays maturity three to five days in the fall. Full-season varieties with a bushy growth habit give the best yields in wide rows (Figure 20-19).

Diseases

Disease concerns of soybeans include root rot, stem rot, seed rot, soybean cyst nematode, and soybean rust.

Root Rot. Phytophthora root rot is a serious soybean disease everywhere soybeans are grown. Varieties are most susceptible in the seedling stage. Saturated soil with temperatures above 60°F are ideal conditions for

Figure 20-19 A soybean plant.

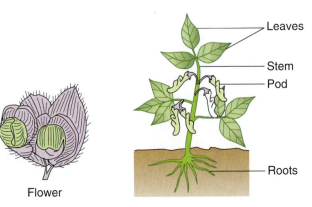

Flower

Leaves

Stem

Pod

Roots

infection. Susceptible varieties should not be grown in poorly drained soils or on soils known to have a history of the disease. Seeds of varieties with good field tolerance should be treated with a fungicide that aids in the control of phytophthora damping off, or the soil may be treated. Resistant varieties should be treated with a seed treatment fungicide to control phytophthora damping off if they are to be planted in poorly drained soil using reduced or no tillage.

Pythium and rhizoctonia root rots are also common and varieties are susceptible. Damage to plant stands is greatest on poorly drained soils and during seasons of high rainfall. Pythium is controlled by fungicide seed treatments, and seedling infections of *Rhizoctinia* are controlled by seed treatments containing fungicide.

Stem Rot. Sclerotinia stem rot may be severe when wet weather occurs during flowering. Some varieties are less susceptible than others, but there is no known resistance. Stem symptoms appear as water-soaked lesions followed by cottony growth and eventually large, black, irregular-shaped sclerotia resembling mouse droppings. Wide rows (20–30 inches) and reduced seeding rates aid in control by permitting air to move through the canopy to dry plant leaves and the soil surface.

Brown stem rot can severely reduce yield. The fungus enters the plants through the roots and slowly grows upwards into the xylem, where it interferes with the flow of water. The disease symptoms develop after flowering and are identified by an internal browning of the stem in August. Foliar symptoms are rare, but the leaves of infected plants may suddenly wilt and dry 20–30 days before maturity and remain attached to the plant.

Seed Rot. Phomopsis seed rot can be severe when rainfall occurs intermittently during the dry down and harvest. The longer soybeans are in the field after ripening, the greater the incidence of seed rot. Harvesting soon after the soybeans mature (15–20 percent moisture) decreases the amount of seed damage. Use varieties with a range of maturities to allow for a more timely harvest. Yield and grain quality losses are greater when soybeans are not rotated with other crops.

Soybean Cyst Nematode. Soybean cyst nematode (SCN) was first found in the southeastern states in the early 1950s. The disease has progressed north and is now a problem in Minnesota, Wisconsin, Iowa, Nebraska, Illinois, Indiana, Missouri, Michigan, and Ohio. In some of the fields surveyed, the population of the nematode was quite high, indicating a long-term residence, as 8–10 years can elapse between introduction of SCN and its appearance as a damaging parasite.

Soybean cyst nematode injury is easily confused with other crop production problems, such as nutrient deficiencies, injury from herbicides, soil compaction, or other diseases. The first field symptoms are usually detected in circular to oval patches of stunted, yellowed plants. Symptoms are most evident in late July or August when plants are under drought stress or in fields with low fertility. When populations of nematodes are high, the symptoms may even occur under normal-to-optimal growing conditions. Affected areas may increase in size each year in the direction of tillage.

Many times, growers are not able to detect any differences between affected and nonaffected plants other than lower yields. Aboveground symptoms do not allow positive diagnosis of the problem because the root system is the infection site. Laboratory examination is necessary for positive diagnosis and for determining population levels. Roots from heavily diseased plants may be stunted and generally have few rhizobium nodules. Before implementing control measures, growers sample soil from suspect fields analyzing for SCN.

When SCN is detected in an area, institute control procedures immediately. No single method will completely control SCN once it is introduced into the field. Methods of disease control are aimed at reducing nematode populations sufficiently to reduce yield losses. A combination of disease-control practices will be required including prevention of spreading, crop rotation, resistant varieties, nematicides, soil fertility, and planting dates.

Soybean Rust. Asian soybean rust is a serious foliage disease that has the potential to cause significant soybean yield losses. Two fungal species, *Phakopsora pachyrhizi* and *Phakopsora meibomiae*, cause soybean rust and are spread primarily by wind-borne spores that can be transported over long distances. Asian soybean rust, *P. pachyrhizi*, the more aggressive of the two species, was first reported in Japan in 1903 and was confined to the Eastern Hemisphere until its presence was documented in Hawaii in 1994. Currently, distribution of *P. pachyrhizi* includes Africa, Asia, Australia, Hawaii, and South America. *P. pachyrhizi's* rapid spread and severe damage, with yield losses from 10–80 percent, have been reported in Argentina, Asia, Brazil, Paraguay, South Africa, and Zimbabwe.

Asian soybean rust was not reported in the continental United States until the fall of 2004. The U.S. Department of Agriculture released an official notice of the confirmation of soybean rust on soybean leaf samples collected in Louisiana. Subsequently, the fungus was detected on plants from a number of additional states. Now that Asian soybean rust has been found in the continental United States, it is critical that anyone involved in soybean production be familiar with the disease and its identification and management.

The most common symptom of soybean rust is a foliar lesion. On the upper leaf surface, initial symptoms may be small, yellow flecks or specks in the leaf tissue. These lesions darken and may range from dark brown or reddish brown to tan or gray-green in color. Extended periods of cool, wet weather during the growing season favor soybean rust epidemics. Rust pustules appear on the leaf surface 9–10 days after infection, and spores are usually evident soon after their appearance.

Ideally, resistant varieties are the more practical, economical means of managing soybean rust. Commercial soybean varieties currently grown in the United States have little or no resistance to soybean rust. Soybean breeders are working to identify sources of resistance and to incorporate resistance into soybean varieties suitable for U.S. production. For the immediate future, the use of foliar fungicides may be one of the main tools for managing soybean rust.

Tillage

A desirable seedbed for soybeans should be smooth enough to permit the planting equipment to place the seed at a uniform depth. The soil particles should be fine enough to assure good seed-soil contact for rapid germination and emergence. The greater the time required for emergence, the greater the time for disease infection and loss of stand.

The freezing and thawing action on clay, silty clay, and silty clay loam soils tilled in the fall or winter to produce stale seedbeds usually have excellent seedbeds for early no-till spring planting. Tilling these soils when too wet in the spring results in a rough, cloddy seedbed. Also, spring tillage often causes compaction of the soil below five-inch depths, which restricts root growth and reduces availability of water and nutrients.

Silt loam soils and those soils with less than 2 percent organic matter, but with good drainage, tend to have the most desirable seedbed with a late winter or spring tillage system if the soil moisture level is satisfactory for tillage. Where tillage is used, the type and amount has little effect on yield, provided it is adequate to permit the establishment of a uniform stand and phytophthora root rot is not a problem. This is true for both heavy- and light-textured soils where organic matter contents are 2 to 6 percent.

Rhizobium Inoculation

Inoculation is usually not necessary if soybeans have been grown in the field within the past three years. If inoculation is necessary, the seed should be inoculated at the time of planting. Commercial inoculants usually contain strains of bacteria that fix nitrogen more rapidly than native strains, but they usually do not survive well from year to year. In situations in which soybeans have not been grown on the field for the past five years, the soybeans should be inoculated immediately prior to planting. Uniform inoculum application in the seed box is difficult because the seeds must be moistened if dry formulations are to adhere in adequate amounts to be effective.

When soybeans are planted in a field for the first time, the inoculation procedures fail to produce enough nitrogen for a good crop. When the number of nodules is insufficient to supply adequate nitrogen, some nitrogen will be supplied to the crop in the form of urea or ammonium nitrate. Applying 60–80 pounds of actual nitrogen can increase yields by 8–12 bushels per acre. This supplemental nitrogen should not be applied within 40 days of emergence, but rather should be applied prior to mid-flower, which is usually the second or third week of July, depending on variety maturity, planting date, and weather.

Harvesting

Soybeans are harvested in the fall with a combine (Figure 20-20) when the plants are mature and the beans have approximately 14 percent moisture. Harvest may be started at 17–18 percent moisture when air drying is available. Harvesting the crop when beans are above 12 percent moisture avoids cracking seed coats and "splits." When soybean seed is extremely dry (8–10 percent moisture), harvesting will cause more shattering and seed injury.

Following harvest, soybeans are used in a wide variety of products (Figure 20-21).

Figure 20-20 Harvesting soybeans.

SUGAR

Sugar, or table sugar, that favorite sweetener, refers to sucrose, a white crystalline solid disaccharide. Sugar is extracted from two main crops—sugarcane and sugar beets. Interestingly, sugarcane is a tropical plant and sugar beets are a cool-weather crop.

Sugarcane

The botanical name for sugarcane is *Saccharum officinarum*. Sugarcane is a tropical grass native to Asia, where it has been grown in gardens for over 4,000 years. It is the product of interbreeding four species of the *Saccharum* genus, and is a giant, robust, sugary plant. Methods for manufacturing sugar from sugarcane were developed in India about 400 BC. Christopher Columbus brought the plant to the West Indies, and today sugarcane is cultivated in tropical and subtropical regions throughout the world. Over 62 percent of the world's sugar comes from sugarcane. Figure 20-22 shows a typical sugarcane plant.

Florida is the largest producer of sugarcane in the United States, followed by Louisiana, Hawaii, and Texas in order of production. Sugar beets

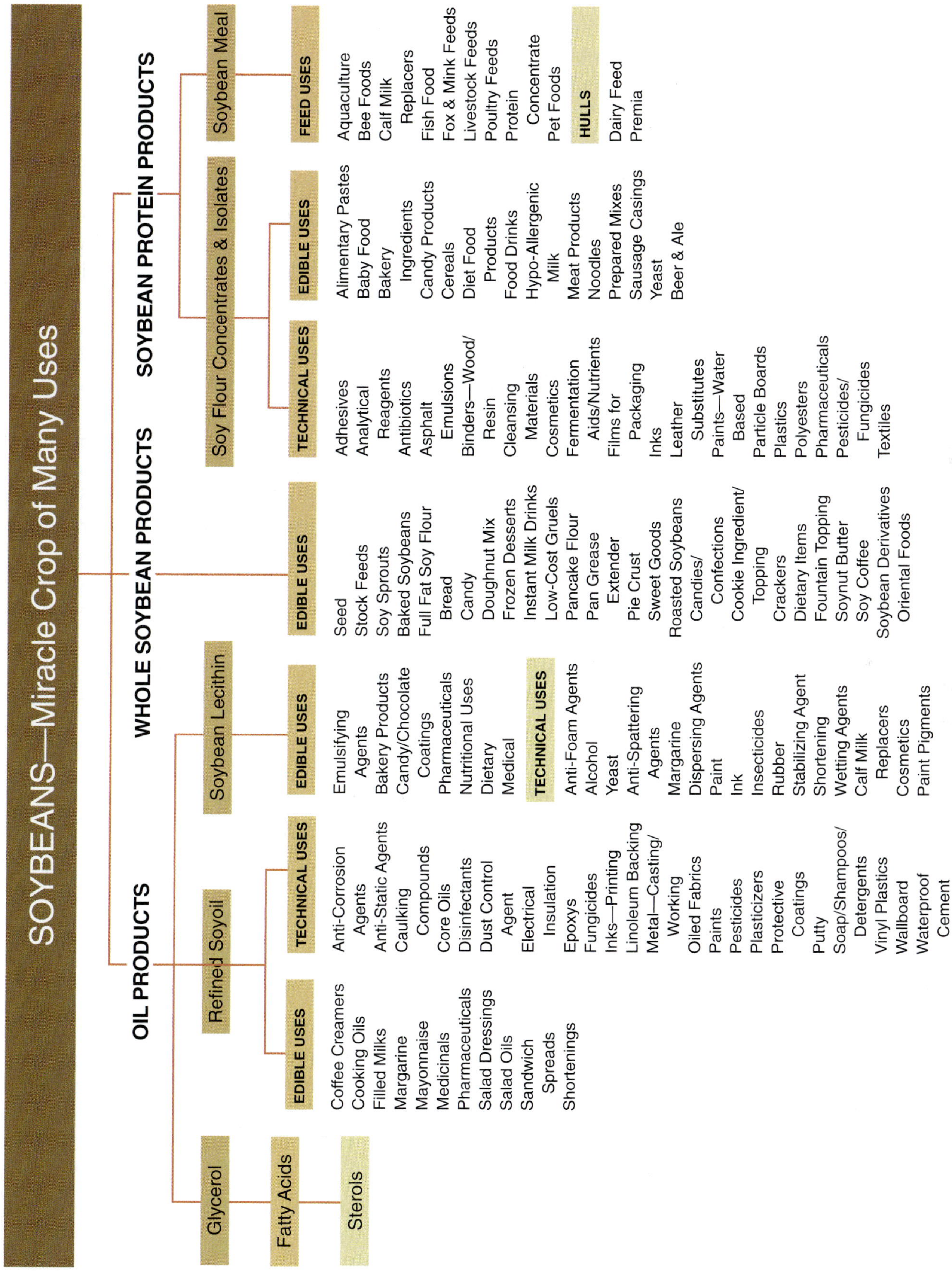

SOYBEANS—Miracle Crop of Many Uses

OIL PRODUCTS   **WHOLE SOYBEAN PRODUCTS**   **SOYBEAN PROTEIN PRODUCTS**

Glycerol

Fatty Acids

Sterols

Refined Soyoil

EDIBLE USES
Coffee Creamers
Cooking Oils
Filled Milks
Margarine
Mayonnaise
Medicinals
Pharmaceuticals
Salad Dressings
Salad Oils
Sandwich
 Spreads
Shortenings

TECHNICAL USES
Anti-Corrosion
 Agents
Anti-Static Agents
Caulking
 Compounds
Core Oils
Disinfectants
Dust Control
 Agent
Electrical
 Insulation
Epoxys
Fungicides
Inks—Printing
Linoleum Backing
Metal—Casting/
 Working
Oiled Fabrics
Paints
Pesticides
Plasticizers
Protective
 Coatings
Putty
Soap/Shampoos/
 Detergents
Vinyl Plastics
Wallboard
Waterproof
 Cement

Soybean Lecithin

EDIBLE USES
Emulsifying
 Agents
Bakery Products
Candy/Chocolate
 Coatings
Pharmaceuticals
Nutritional Uses
Dietary
Medical

TECHNICAL USES
Anti-Foam Agents
Alcohol
Yeast
Anti-Spattering
 Agents
Margarine
Dispersing Agents
Paint
Ink
Insecticides
Rubber
Stabilizing Agent
Shortening
Wetting Agents
Calf Milk
 Replacers
Cosmetics
Paint Pigments

EDIBLE USES
Seed
Stock Feeds
Soy Sprouts
Baked Soybeans
Full Fat Soy Flour
Bread
Candy
Doughnut Mix
Frozen Desserts
Instant Milk Drinks
Low-Cost Gruels
Pancake Flour
Pan Grease
 Extender
Pie Crust
Sweet Goods
Roasted Soybeans
Candies/
 Confections
Cookie Ingredient/
 Topping
Crackers
Dietary Items
Fountain Topping
Soynut Butter
Soy Coffee
Soybean Derivatives
Oriental Foods

Soy Flour Concentrates & Isolates

TECHNICAL USES
Adhesives
Analytical
 Reagents
Antibiotics
Asphalt
 Emulsions
Binders—Wood/
 Resin
Cleansing
 Materials
Cosmetics
Fermentation
 Aids/Nutrients
Films for
 Packaging
Inks
Leather
 Substitutes
Paints—Water
 Based
Particle Boards
Plastics
Polyesters
Pharmaceuticals
Pesticides/
 Fungicides
Textiles

EDIBLE USES
Alimentary Pastes
Baby Food
Bakery
 Ingredients
Candy Products
Cereals
Diet Food
 Products
Food Drinks
Hypo-Allergenic
 Milk
Meat Products
Noodles
Prepared Mixes
Sausage Casings
Yeast
Beer & Ale

Soybean Meal

FEED USES
Aquaculture
Bee Foods
Calf Milk
 Replacers
Fish Food
Fox & Mink Feeds
Livestock Feeds
Poultry Feeds
Protein
 Concentrate
Pet Foods

HULLS
Dairy Feed
Premia

Figure 20-21 The soybean has become famous for its many products and is growing in importance in world markets. (Chart courtesy of American Soybean Association)

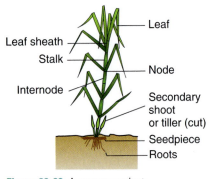

Figure 20-22 A sugarcane plant.

are grown in North Dakota, Minnesota, Wyoming, California, and 11 other states, and these provide approximately the same amount of sugar as the domestic sugarcane industry (Figure 20-23).

Planting. Sugarcane planting takes place from September through January. Because sugarcane is a multispecies hybrid, the seeds will usually produce plants different from the parents. As with any other plant in which this happens, parts of the mother plant must be planted in order to produce the desired daughter plant (also called a clone). Stalks, which ordinarily would be milled for sugar, are harvested from mature fields, cut into short 20-inch segments, laid in furrow rows 5 feet apart, and then covered with soil. Cane stalks have buds every 2–6 inches, and each of these buds has the capability to sprout rapidly when buried in moist soil. Within two to three weeks shorts emerge and, under favorable conditions, produce secondary shoots to give a dense stand of cane. The development of secondary shoots is called **tillering.**

Typically, a sugarcane field is replanted every three to four years. After a field has been harvested, it is maintained free of weeds and a second crop of stalks is produced from the old plant stubble. The second crop is harvested about one year after the first harvest. Between three and four annual crops are usually taken from one field before replanting. When production declines to an unacceptable level because of insects disease, or mechanical damage, the old cane plant is plowed under after harvest, and the land is prepared for replanting.

A sugarcane plant is capable of producing seeds, but seeds are not used for commercial planting. Seeds are so small (1,000 per gram or

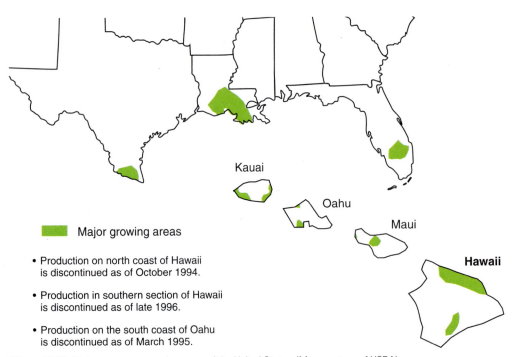

Figure 20-23 Major sugarcane growing areas of the United States. (Map courtesy of USDA)

500,000 per pound) that they cannot be planted directly in the field. Furthermore, sugarcane does not breed true, and every seedling is genetically a new variety. Varietal uniformity within fields is important for commercial production; therefore, only stalks are used as planting material.

During winter and spring, sugarcane plumes are the flowers and seed heads of the cane plant. Each plume consists of several thousand tiny flowers, each capable of producing one seed. The cool winter weather in Florida ordinarily prevents development of the seeds. During breeding for variety development, special precautions are taken in order to produce viable seed.

Culture. After planting, weeds are controlled with cultivation and herbicides. Water must be pumped out of the field when rainfall is excessive. When the soil gets too dry, the crop is irrigated by allowing the water to flow back into the ditches that are normally used for drainage. Water quickly seeps from the ditches throughout the entire field because the muck soil is so porous.

Growth. The growth of the cane plant does not proceed at a uniform rate. Development starts slowly in the germinating bud and increases gradually until a maximum is reached. This is followed by a gradual decrease (Figure 20-24). The period in which the plant is growing rapidly is called the grand growth period.

Cane growth is governed by a complex interaction of internal and external factors. The external factors include moisture, temperature, light, soil condition, and nutrition.

Nutrition. The essential elements for a healthy crop include carbon, hydrogen, oxygen, nitrogen, phosphorus, potassium, calcium, magnesium, boron, chlorine, copper, iron, manganese, molybdenum, sulfur, zinc, and possibly silicon. Nitrogen has the greatest influence on cane ripening. Nitrogen deficiency is common in sandy soil and is uncommon on organic soils.

Ripening. Storage of sucrose in the stalk is known a **ripening**. Ripening is a joint-to-joint process, and the degree of maturity of the individual joints depends on their age. As the plant matures, a uniform content of sucrose is found throughout the stalk except for the top few internodes and

Figure 20-24 A growth curve for sugarcane. (Graph courtesy of University of Florida, Cooperative Extension Service)

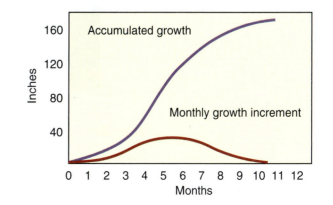

the belowground stool. Certain plant growth regulators can accelerate the accumulation of sucrose in the stalk.

Harvesting. Sugarcane is harvested from October through March. If there are no freezes, sugar yields are highest after January 1, but some fields must be harvested before they have reached maximum yield to allow time for processing the whole crop through sugar mills where sugar is extracted from the cane juice. Fields are burned the day before harvest. The fires are rather spectacular but of short duration (a 40-acre field burns in 15–20 minutes). The fires burn off dead leaves that would otherwise hinder harvest, interfere with milling machinery, and absorb sugar during the extraction process that cannot be recovered. Burning is done only in the daytime, when dispersal of the smoke by air currents causes minimum nuisance. In the past, most of the sugarcane was cut by hand using cane knives. Improvements in mechanical harvesters have resulted in a strong movement away from hand harvesting. Acreage that is machine-harvested has increased to over 50 percent (Figure 20-25).

Hand-cut and certain machine-cut cane is piled in rows and then loaded into tractor-drawn wagons by machines called continuous loaders. The other machine-cut cane is deposited directly into wagons by the harvester. Four-wheel drive tractors haul 16 tons of cane out of the field with each four-wagon load. At special ramps near the field, the cane is dumped from the wagon into highway trailers or rail cars for transport to the mills (Figure 20-26). Rail cars carry 25 to 30 tons each and highway trailers carry 20 tons per load.

Figure 20-25 Sugarcane harvested and loaded into high dump trailers will be transferred into trucks for shipment to a sugar mill. (Photo courtesy of USDA ARS)

Sugar Beet

Sugar beet companies have developed excellent sugar beet varieties through breeding and field trials. Present varieties have yield potentials of more than 30 tons per acre when grown on the best soils using the

Figure 20-26 Sugar mill yard at the Cora Texas Manufacturing Company, White Castle, Louisiana. (Photo courtesy of USDA ARS)

Figure 20-27 The sugar beet plant.

best possible cultural practices under favorable climatic conditions. Figure 20-27 shows a diagram of the sugar beet plant.

The botanical name for sugar beet is *Beta vulgaris*. Figure 20-28 shows where sugar beets are grown in the United States.

Seeding. Because the sugar beet seed is small, precision in depth of seed placement is important for the best emergence. Recommended depth of seeding ranges from ¼ inch to ¾ inches. Regardless of the depth of planting, the soil must be firm enough to ensure good seed-to-soil contact, which allows moisture to move from the soil into the seed.

Date of Planting. Planting before the end of April is ideal. May plantings have been successful. Growers plant sugar beets any time after the middle

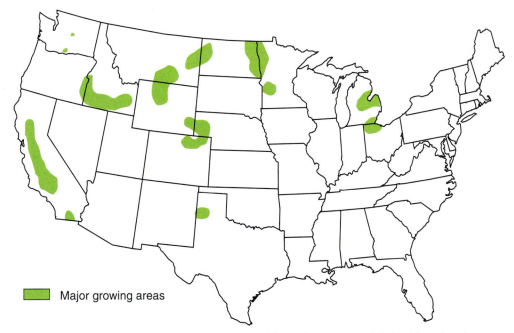

■ Major growing areas

Figure 20-28 Major sugar beet growing areas of the United States. (Map courtesy of National Oceanic and Atmospheric Administration/USDA)

of March as soil conditions permit to complete planting before the end of April. Some years there are a few days in late March when planting can be completed. Growers test the soil in late summer and they apply fertilizer and plow in the fall for fine-textured, nearly level glacial till or lakebed soils.

Spring tillage on fall-plowed soil should be no more than three inches deep. This allows for band placement of row fertilizer on wet soil with sufficient dry, loose soil to cover the fertilizer band. Covering knives and press wheels ahead of the seeding units firm loose soil over the fertilizer band. A slow planting speed is important to cover the fertilizer band and place seeds properly.

Plant Population and Row Width. The expression "100-percent stand" is defined as 100 beets per 100 feet of row. A 100-percent stand, by definition, would result in 17,400 plants per acre in a 30-inch row. A population of 25,000 beets per acre at harvest has consistently given higher sugar content and yield than the lower plant populations. Beets planted in row widths of 22 and 30 inches, with beets spaced about 9 inches apart, result in 31,600 and 23,000 plants per acre, respectively.

The best way to maintain adequate stands is through narrow rows. Available planting and harvesting equipment (Figure 20-29) can plant in row widths as narrow as 18 inches. In sugar beet production, the widest row width recommended is 30 inches. This is a compromise for growers who prefer a common row width for the production of sugar beets, soybeans, and corn.

Planting to a Stand. Growers are reducing the cost of hand labor (thinning) by planting to a stand (the correct number of plants per acre). Because sugar beets tend to compensate in growth over a wide population range, the practice of planting to a stand is used on most farms. Although quality peaks at about 25,000 plants per acre, tonnage is essentially the same over a population range of 20,000–35,000 plants per acre.

Figure 20-29 Sugar beets harvested from the field are first topped, then pulled from the ground and loaded into a truck.

Growers who know the emergence potential of their soil can calculate a seed spacing that will provide a stand in the 20,000–35,000 range over a wide range of emergence conditions.

The practice of planting to a stand can be made successful by—

+ Providing adequate drainage
+ Planting in April
+ Using a precision planter
+ Accurately controlling depth of planting
+ Placing fertilizer so it does not interfere with accurate seed depth
+ Having good seed-soil contact
+ Using recommended herbicides for weed control
+ Controlling diseases

Cultivation. Cultivation needs to begin soon after beet emergence. This practice is useful to control weeds and aerate the soil, which helps deter seedling diseases. If row guides are established during the planting operation, accuracy and speed of cultivation improve. As cultivation progresses throughout the season, soil should be moved toward the beet row. This facilitates the harvest operation. Spreading excess soil on beet crowns encourages disease.

Fertility Program. Row fertilizer should be placed to the side and two inches below the seed to prevent manganese deficiency. Growers apply 30–40 pounds of nitrogen per acre in row fertilizer. This helps prevent manganese deficiency and increases early leaf development. Row nitrogen is especially important when preplant nitrogen was not applied. For maximum crop safety, diammonium phosphate (18-46-0) and urea (45-0-0) should be avoided in a sugar beet row fertilizer. About 50 pounds per acre of phosphorus (P_2O_5), 20 pounds per acre of potassium (K_2O), 30–40 pounds per acre of nitrogen (N), and 6 pounds of manganese (Mn) should be applied in the row fertilizer. Sometimes side dress nitrogen is applied in late June. In some years this late application of nitrogen may reduce sugar yield per acre.

Nitrogen. Applying too much nitrogen reduces quality without increasing tonnage. Growers apply nitrogen at planting, if possible. If side-dressing is necessary, producers side-dress when beets are very small. A survey of grower practices showed those growers with the highest sugar production per acre applied most of their nitrogen ahead of planting, and those with the lowest yields side-dressed most of their nitrogen.

Phosphorus and Potassium. Growers apply phosphorus and potassium according to a soil test made the previous fall.

Weed Control. Weeds can be a serious problem in sugar beets because they can severely reduce yield and interfere with harvesting. Until recent years, hand-hoeing and cultivation were the accepted methods for controlling weeds. With the development of monogerm seeds and more effective herbicides, hand-labor costs for sugar beet production has decreased greatly.

Figure 20-30 After sugar beets are harvested, they are stockpiled until processed.

In many instances, growers have good early-season weed control only to have annual grasses and broadleaves come in late. For season-long control, a pre-emergence herbicide must be followed with poste-mergence ones.

Harvesting. Sugar beets are topped and pulled from the ground in the fall. They are hauled to a processing plant where they are stockpiled (Figure 20-30) for processing during the fall and winter months, or they are immediately processed.

WHEAT

Figures 20-31 through 20-34 show the major growing areas for wheat, and Figure 20-35 illustrates the historical production trends of wheat in the United States.

The botanical name for wheat is *Triticum* spp. The wheat plant is illustrated in Figure 20-36.

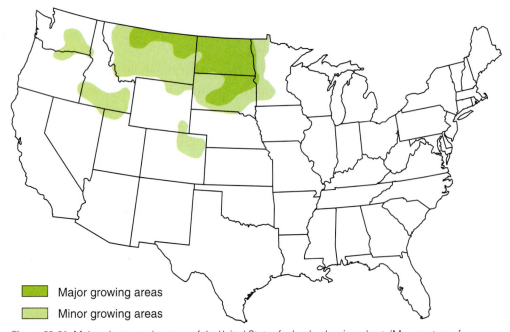

■ Major growing areas

■ Minor growing areas

Figure 20-31 Major wheat growing areas of the United States for hard red spring wheat. (Map courtesy of National Oceanic and Atmospheric Administration/USDA)

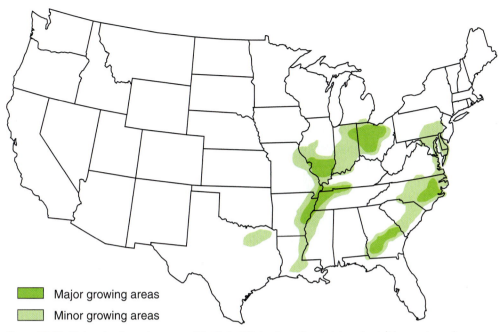

Figure 20-32 Major wheat growing areas of the United States for soft red winter wheat. (Map courtesy of National Oceanic and Atmospheric Administration/USDA)

Variety Selection

In many areas, a wheat variety performance trial is conducted annually to measure yield and other agronomic characteristics that are important to producers of the crop. Results of this trial are published and are available from county extension offices.

Figure 20-33 Major wheat growing areas of the United States for white winter wheat. (Map courtesy of National Oceanic and Atmospheric Administration/USDA)

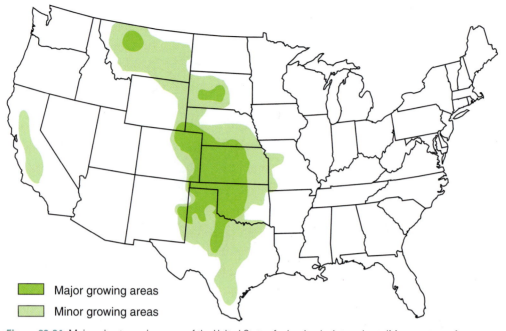

Figure 20-34 Major wheat growing areas of the United States for hard red winter wheat. (Map courtesy of National Oceanic and Atmospheric Administration/USDA)

Figure 20-35 Historical perspective of all wheat production in the United States, 1997–2007. (Graph courtesy of USDA NASS)

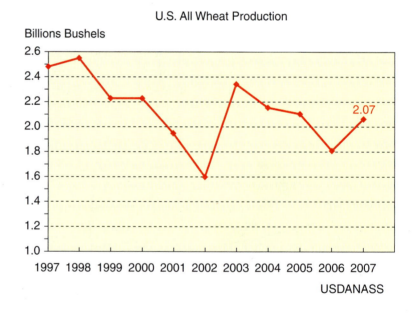

Variety selection should be based on winter hardiness, standability, disease resistance, and yield potential. Although differences in winter hardiness exist among varieties, planting date has the greatest effect on winter survival. The yield potential of available varieties is generally in excess of 150 bushels per acre. This yield is not approached, however, primarily because of a short grain fill period caused by high air temperatures in late June. The ideal air temperature during grain fill is 68–76°F.

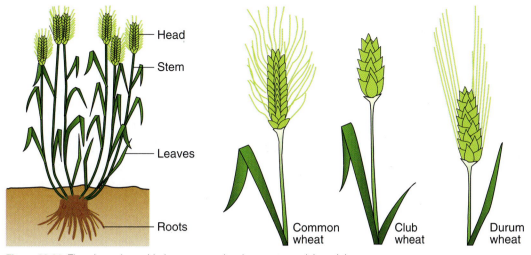

Figure 20-36 The wheat plant with three common heads—common, club, and durum.

Disease must be controlled if high yields are to be obtained. Both varietal resistance and fungicides are available and may be combined to provide a wide spectrum of protection. Although most available varieties have excellent standability, excessive seeding and nitrogen rates, or their combination, cause lodging, which results in reduced yield.

Seeding Date

Wheat should never be seeded prior to the "fly-safe date" because of the possibility of severe damage by diseases and hessian flies (Figure 20-37). The best time for seeding is a 14-day period starting 7 days after the fly-safe date. Long-term average yields are highest from seeding made during that time. Seeding during that time usually produces ample growth for winter survival and reduces the likelihood of fall disease infections and attack by potentially damaging insects.

Occasionally, when freezing weather is delayed until late November or early December, wheat seeded more than three weeks after the fly-safe date is equal in yield to that seeded during normal planting time. Because of reduced fall growth, late-seeded wheat is less winter hardy and more susceptible to spring heaving.

Figure 20-37 The mature hessian fly, an insect pest of wheat. (Photo courtesy of USDA ARS)

Seeding Rate

Often, wheat is seeded at excessive rates, which increases the potential for lodging and disease, both of which severely reduce yields. Yield loss from higher seeding rates is primarily a result of increased lodging and disease and decreased seed size, which leads to increased harvest loss.

Because seed size varies from variety to variety and year to year, seeding rates should be based on the number of seeds per foot of row rather than pounds per acre. When seeded at the normal time, a seeding rate of 13–21 seeds per foot of 7-inch row (75–120 pounds per acre) is recommended. Table 20-2 shows the pounds of seeds needed per acre for various combinations of seed size and seeding rate. One million to 1.5 million seeds per acre is the ideal seeding rate.

Table 20-2 Pounds of Wheat Seed Needed for Combinations of Seeding Rate and Seed Size

Seeds/Foot of Row (7 in.)	Seeds/Pound in 1000s				
	13	14	15	16	17
13	75	70	65	61	57
16	92	85	80	75	70
19	109	102	95	89	83
21	121	113	105	99	92
24	138	129	120	112	106

Normal seeding rates are 13–21 seeds per foot of a 7 inch row.

Row Width

Most grain drills are available with an optional 10-inch row spacing that aids in residue movement through the drill when seeding no-till, making the effect of 10-inch row spacings on wheat yields of interest to producers. Yields decrease as row spacings become greater than 10 inches.

Lodging Control

Lodging is a serious deterrent to high yields. Cultural practices that tend to increase grain yield also increase the likelihood of lodging (when grain falls over to the ground). Using recommended seeding rates (13–21 seeds per foot of a 7-inch row), applying proper rates of nitrogen, and selecting lodging-resistant varieties prevents lodging in high-yield environments where yields of 100 bushels per acre are anticipated.

When lodging occurs, the severity of foliar disease increases, resulting in reduced grain yield and quality. Additional effects of lodging are reduced straw quality and slowed harvest. The prevention of lodging increases dividends through a combination of reduced input costs and improved grain and straw quality.

Disease Control

Disease is often the major factor limiting yield of wheat in midwestern states. Yield losses as high as 30–50 percent are common where no disease control is practiced. Effective disease management requires knowledge of the diseases most likely to occur in a production area. Producers should fine-tune their disease-control strategies for those few diseases encountered each year. Correct diagnosis is the cornerstone of effective control, and producers with little experience identifying diseases should seek help from competent sources such as university extension service or an agricultural consulting service.

A comprehensive wheat disease-management program consists of the following practices:

1. Select varieties with resistance to the important diseases in the area. Monitoring wheat diseases aids a producer in selecting varieties with resistance to the common diseases of his or her community.

2. Plant well-cleaned, disease-free seeds, treated with a fungicide that controls seeding blights, bunt, and loose smut.

3. Plant in a well-prepared seedbed (Figure 20-38) after the fly-safe date. Waiting five to ten days after the fly-safe date ensures escape from fall infections of barley yellow dwarf and lessens the potential of root rot and early-season foliar diseases.

4. Rotate crops; never plant wheat where the previous crop was wheat or spelt. A two- to three-year rotation from wheat prevents most pathogens from surviving in fields.

5. Plow down residues from heavily diseased fields, especially those affected by Cephalosporium stripe or take-all. Plowing enhances decomposition of residue and death of the disease-causing fungi.

6. Use a well-balanced fertility program based on a soil test. Apply sufficient amounts of phosphorus, nitrogen, and potassium in the fall for vigorous root and seedling growth. Spring **top-dress** with nitrogen at the rate recommended to achieve the yield goal. Excessive nitrogen increases the severity of foliar disease and lodging.

Figure 20-38 Planting wheat with a no-till planter.

7. Control grass weeds. Destroying volunteer wheat, quack grass, and other grass weeds in and around potential wheat fields reduces the amount of disease inoculum available to infect the crop.

8. Apply fungicides only if warranted. Scout fields from flag leaf emergence through flowering. Foliar fungicides are able to control the following diseases: powdery mildew, leaf rust, *Septoria tritici* leaf blotch, *Septoria nodorum* leaf, and glume blotch. Leaf rust and powdery mildew are the most severe diseases. Know symptoms, severity ratings, and disease thresholds before scouting fields. Fungicide application following flowering is usually not economical.

Severe head scab may develop when planting wheat no-till into corn stubble. The potential for disease is great because *Gibberella zeae*, the fungus that causes stalk rot in corn, also causes head scab in wheat. Thus, wet weather during flowering can result from the high level of inoculum coming from corn residues. Table 20-3 lists some of the common wheat diseases.

Seed treatments are an important part of any wheat production system. Seed-borne diseases include loose smut, common bunt, and seed-borne *Septoria* and scab. When wheat is planted into a well-prepared seedbed, with good moisture for quick emergence, controlling these diseases and establishing good stands is easy. However, seed treatment cannot compensate for planting in soil that is too wet, too dry, poorly prepared, or for planting at the wrong depth. No-till seeding increases the likelihood of several diseases developing.

Seed treatments protect young seedlings as they establish themselves and obtain nutrients from the soil for further growth. For imperfectly prepared seedbeds, such as no-till, or when wet or dry weather reduces the chances of timely emergence, seed treatments are added insurance that stands will be adequate for optimum yield. *Fusarium* and *Pythium* are two soil-inhabiting fungi that aggressively attack young seedlings when the soil environment is less than optimum for germination and growth. *Fusarium* tends to be more of a problem in dry soil during the warmer periods, whereas *Pythium* is a problem in cold, wet soils.

Growers should buy **treated seeds** or have seeds properly cleaned and treated with a broad-spectrum fungicide or a combination of fungicides to control a range of diseases. Many of the seed-treatment products on the market are combinations of fungicides that control a number of seedling problems. Seed treatments are effective in controlling diseases only. They do not compensate for poor soil conditions or improper adjustment and operation of the drill during seeding operations.

Fertilization

Fertilization programs for wheat include consideration of nitrogen, phosphorus, and potassium needs.

Table 20-3 Some Common Wheat Diseases and Disorders

Disease or Disorder	Symptoms	Environment	Control
Head scab	Spikelets of head turn straw colored, glume edges with pink spore masses, kernels shriveled white to pink in color	Warm, wet weather during flowering period	Seed treatment for infected seed; crop rotation; plow down corn residues
Powdery mildew	Powdery white mold growth on leaf surfaces	High humidity, 60°–75°F, high nitrogen fertility, and dense stands	Resistant varieties; crop rotation; delayed planting; fungicides
Leaf rust	Rusty red pustules scattered over leaf surface	Light rain, heavy dew, 60°–77°F, 6–8 hour leaf wetness for germination and infection	Resistant varieties; balanced fertility; fungicides
Septoria tritici leaf blotch	Leaf blotches with dark brown borders, gray centers speckled with black fungal bodies	Wet weather from mid-April to mid-May, 60°–68°F, rain three to four days a week	Seed treatment; plant less susceptible varieties; crop rotation; balanced fertility; fungicides
Septoria nodorum leaf and glume blotch	Lens-shaped chocolate brown leaf lesions with yellow borders, brown to tan blotches on upper half of glumes on heads	Wet weather from mid-May through June, 68°–80°F, rain three to four days each week	Seed treatment; plant less-susceptible varieties; crop rotation; balanced fertility; fungicides
Tan spot	Lens-shaped, light brown leaf lesions, yellow borders	Moist, cool weather during late May and early June	Plow down infested residues; crop rotation; balanced fertility; fungicides
Cephalosporium stripe	Chlorotic and necrotic interveinal strips extending length of leaf	Cold, wet fall and winter with freezing and thawing causing root damage	Crop rotation; bury infested residues; control grassy weeds; lime soil top H 6.0–6.5
Take-all	Black, scruffy mold on lower stems and roots, early death of plants	Cool, moist soil through October-November and again in April-May	Crop rotation; control weed grasses; balanced fertility; use ammonium forms of N for spring top-dress; avoid early planting
Fusarium root rot	Seedling blight (pre- and post-emergence) wilted, yellow plants, roots and lower stems with whitish to pinkish mold; root rot plants have brown crowns and lower stems	Dry, cool soils, drought stress during seed filling	Seed treatments for seedling blight; delayed planting; balanced fertility; avoid planting after corn
Barley yellow dwarf	Stunted, yellow plants; leaves with yellowed or reddened leaf tips	Cool, moist seasons	Delay planting until after the Hessian fly-safe date; balanced fertility
Wheat spindle streak mosaic	Discontinuous yellow streaks oriented parallel with tapered ends forming chlorotic spindle shapes	Cool, wet fall followed by cool spring weather extending through May	Resistant varieties

Nitrogen. Providing adequate nitrogen for the wheat crop is an important step toward high yields. However, as the nitrogen rate increases, the potential for lodging and disease also increases. Depending on the level of soil organic matter, carryover nitrogen from previous crops, and the yield goal, the amount of fertilizer nitrogen required varies greatly.

In general each 1 percent of organic matter supplies 8–12 pounds of nitrogen per acre through organic matter oxidation. Nitrogen available from a previous soybean crop ranges from 20–40 pounds per acre.

Starter nitrogen at planting is usually not necessary when wheat follows soybeans because of the nitrogen left by the soybeans. If starter nitrogen is used, limit the rate to 20 pounds per acre; higher rates may cause excessive vegetative growth and delay the hardening-off process for winter.

Spring nitrogen top-dress for winter wheat should be applied between March 1 and April 15. If top-dressed too early and a freeze occurs after dormancy is broken, stand may be reduced. The nitrogen should be applied before spring growth starts to stimulate tiller growth and promote larger heads. In high-nitrogen areas, growers select a variety with short, stiff straw that resists lodging.

Phosphorus. Small grains respond well to phosphorus fertilizer on soils testing up to 90 pounds of phosphorus per acre (45 ppm). To maximize yields, the soil phosphorus level should be 90 or higher. A 1:4:2 ratio provides the proper balance of starter fertilizer.

Potassium. The small grain response to potassium application is less than that of phosphorus.

Considerations for No-Till Wheat

The successful production of no-till wheat requires the proper management of a different set of inputs than for other crops because wheat must survive the winter while maintaining vigor and fighting disease organisms. Key factors for success include having a smooth seedbed; proper seeding depth, rate, and date; the absence of carryover herbicides; and proper seed treatments and residue management.

Wheat normally follows soybeans in the crop rotation because many soybean varieties can be harvested early enough for timely wheat seeding. Historically, seedbed preparation has been limited to one or two diskings following soybean harvest. No-till drills have eliminated the need for tillage, and wheat is the easiest of the major crops to produce successfully without tillage. However, adjustments to some cultural practices are necessary and increase the profitability of no-till wheat.

Wheat grows well on a range of soils but does not grow well on poorly drained soil, especially during wet periods. The major causes of stand loss are standing water and the formation of ice sheets where water accumulates. Adequate surface and subsurface drainage is absolutely necessary, and this is more important for wheat than for other crops. Soil depressions caused by combines, grain carts, and tractors are major problems some years. Wheat should not be no-tilled in fields that were wet at the time of soybean harvest or where soil compaction is present.

Soybean straw and chaff should be spread evenly across the entire harvest swath to eliminate interference with proper functioning of seeding equipment. Uniform distribution of residue helps keep soil moisture and temperature constant across the field, which improves uniformity of growth and staging.

Grain drills should be adjusted to penetrate crop residue and place the seed one inch deep and in good contact with the soil. Adequate down pressure, along with a sturdy depth gauge mechanism and press wheels, are necessities for proper seed placement. Seeds must be covered and the seed slit closed. When emerging seedlings begin to dry because of exposure, emergence is retarded, stands become erratic, and seedlings are vulnerable to cold injury.

Excessive drilling speed often results in shallow seeding, which can also result in poor stands and reduced winter survival. Drilling speed should be reduced as soil roughness increases, but should never exceed five miles per hour because faster speeds result in erratic stands. Wherever a uniform depth of planting is difficult to maintain, a harrow following the press wheels sometimes improves stand establishment and uniformity.

The severity of several wheat diseases increases when tillage is removed from the production system. Typically, no-till seeding reduces the amount of vegetation produced in the fall. For this reason, and to assure fall tillering, no-till seedings should be made as soon as possible after the fly-safe date. Twenty pounds of starter nitrogen should be applied preplanting, along with the other recommended fertilizers, to accelerate root system development and vegetative growth.

The benefits of no-till wheat include reduced production costs and much less stand loss resulting from heaving in spring. Eliminating tillage helps retain soil moisture needed for germination and emergence, and can result in more rapid emergence and better stands when soil moisture is low at seeding.

Harvesting

Growers harvest wheat with a combine when it reaches maturity and begins to dry down to market moisture contents; about 9–14 percent depending on variety. Some growers will store wheat in bins on the farm while others transport the grain directly from the field to a local elevator. Wheat should be harvested without delay after it begins to dry in the field. Repeated precipitation and drying begins to decrease quality and test weight.

INTENSIVE WHEAT MANAGEMENT

Wheat yields are determined by three major factors:
1. Genetic potential of each variety
2. Management practices
3. Environment or weather

Two of the three factors (genetic potential and management practices) are under the producer's control, but he or she can do little about the environment other than select good soils or sites to grow the crop and possibly irrigate in times of inadequate rainfall. However, all three factors must be

Figure 20-39 Modern combines harvest wheat.

optimized to produce yields in excess of 100 bushels per acre, requiring large modern equipment for harvesting (Figure 20-39).

A complete management system must be used. Changing only one or two management practices without adjusting others to produce a balanced program increases the cost of production and reduces yield. Some of the steps to better wheat yields are:

1. Select highly productive and fertile soils.
2. Select varieties with high yield potentials.
3. Seed at the proper time.
4. Use proper seeding rates and stiff-strawed varieties.
5. Supply adequate levels of fertilizer nutrients.
6. Control disease and weeds by applying controls at the proper stages of growth.
7. Visit fields weekly from late April through June to spot any potential problems so they can be dealt with before they reduce yields.

SUMMARY

Major agronomic crops such as corn, cotton, potatoes, soybeans, sugarcane, sugar beets, and wheat have some very specific growing requirements. Because of these growing requirements, each of these crops grow in unique areas of the United States. Depending on cultural practices, final product, pest problems, and other environmental factors, growers must select the appropriate variety or cultivar. Based on this selection, growers prepare the seedbed, the fertilization, irrigation, and pest management to maximize the yield from the crop. Much of successful production involves proper timing; for example, timing of planting, timing of irrigation, timing of herbicides and pesticides, and timing of harvesting. Obviously, the weather patterns can change the timing of many crop-production activities. Crop rotation is often helpful in controlling disease, insect pests, and nematodes. Integrated Pest Management (IPM) at some level is always an option.

Review

True or False

1. Different hybrid varieties of corn can be used for different planting dates.

2. Corn should be planted at a depth of 1/4 inch.

3. Boll weevils destroy soybean plants.

4. Soybeans are extracted for sugar.

5. Lodging is a problem associated with wheat.

Short Answer

6. A good _____ plan can help control disease, insects, or nematodes.

7. Give the common botanical name for corn, cotton, potatoes, soybeans, sugarcane, sugar beets, and wheat.

8. What is a common crop rotation for corn?

9. List two ways insect pests are often controlled.

10. What is the relationship between the nitrogen, phosphate, and potassium removed by a crop and the fertilization program?

Critical Thinking/Discussion

11. Compare the growing of sugarcane with the growing of sugar beets.

12. Compare the growing of cotton with the growing of potatoes.

13. When selecting seeds, why does a grower study the features of different varieties or cultivars?

14. Most crops have a planting date or season; why is this important?

15. Compare the major growing areas for corn and soybeans with that of wheat and potatoes.

Knowledge Applied

1. Visit an equipment dealer and collect information and pictures of the equipment used in one of these crops: corn, cotton, potatoes, soybeans, sugarcane, sugar beets, and wheat.

2. Make a seed collection of the seeds or other methods of propagation used for corn, cotton, potatoes, soybeans, sugarcane, sugar beets, and wheat.

3. Develop a presentation (computer) showing the insect pests of corn, cotton, potatoes, soybeans, sugarcane, sugar beets, or wheat. Show the type of damage the pests cause to the plants and indicate how the pests are controlled.

continues

Knowledge Applied, *continued*

4. Make a presentation on nematodes. Describe their life cycle and some of the damage they cause. Suggest ways they are controlled. Determine if their damage is limited just to plants.

5. Compile a list of the agronomic crops grown in your area. Select one of these and provide a complete profile of how the crop is grown in your area. Indicate the seedbed preparation, seeding rate, fertilization, pests control, harvesting, and marketing. Find out how growers purchase their seed stock.

6. Use a soil sample report to develop a fertilization plan for corn, cotton, potatoes, soybeans, sugarcane, sugar beets, or wheat.

7. Create a presentation of the advantages and disadvantages of using a crop with the Bt gene or a Roundup-ready crop.

Resources

Acquaah, G. (2004). *Principles of crop production: Theory, techniques, and technology* (2nd ed.). Englewood Cliffs, NJ: Prentice Hall Career & Technology.

Connors, J. J., & Cordell, S. (Eds.). 2003. *Soil fertility manual.* Tucson, AZ: Potash & Phosphate Institute.

Hay, R., & Porter, J. (2006). *The physiology of crop yield* (2nd ed.). Ames, IA: Blackwell Publishing.

Jones, J. B., Jr. (2002). *Agronomic handbook: Management of crops, soils, and their fertility.* Boca Raton, FL: CRC Press.

McMahon, M., Kofranek, A. M. & Rubatzky V. E. (2006). *Hartmann's plant science: Growth, development, and utilization of cultivated plants* (4th ed.). Englewood Cliffs, NJ: Prentice Hall Career & Technology.

Rieger, M. (2006). *Introduction to fruit crops.* Binghamton, NY: Food Products Press.

Simpson, B. B., & Ogorzaly, M. C. (2000). *Economic botany: Plants in our world* (3rd ed.). New York: McGraw-Hill.

Vorst, J. J. (1998). *Crop production* (4th ed.). Champaign, IL: Stipes Publishing Company.

Internet

Internet sites represent a vast resource of information. The URLs (uniform resource locators) for World Wide Web sites can change. Using one of the search engines on the Internet such as Google or Yahoo!, find more information by searching for these words or phrases: agronomic crops, crop rotation, blight, fungicide, inoculation, insecticide, seeding rates, lodging, no-till, yield, corn (maize), cotton, potatoes, soybeans, sugarcane, sugar beets, wheat, Bt crops, and Roundup-ready.

Chapter 21
Cereal Grains, Forage Grasses, and Oil Seeds

THE MAJOR OBJECTIVE *of a grain production system is the capture, fixation, and storage of sunlight energy. The most important components of such a system are timely planting, disease control, proper fertilization, and variety selection. Small grains specifically discussed here include barley, oats, rye, sorghum, spelt, and triticale. Rice, wild rice, and buckwheat are also discussed in this chapter. Wheat was covered in Chapter 20.*

Forage grasses capture the energy of the sun and then they are harvested directly by livestock or by mechanical methods and fed to livestock later. Forage grasses include Kentucky bluegrass, Canada bluegrass, orchard grass, perennial ryegrass, reed canary grass, smooth bromegrass, tall fescue, timothy, prairie grass, garrison grass, and Bermuda grass.

Oil seed crops capture sunlight energy and store oil in their seeds; the oil is extracted for human use. The major oil seed crop, soybeans, was covered in Chapter 17. Other oil seed crops include, but are not limited to, canola, safflower, and sunflower.

After completing this chapter, you should be able to—

- Describe the cultural practices for growing barley, oats, rye, sorghum, spelt, triticale, rice, wild rice, and buckwheat

- Discuss the cultural practices for growing forage grasses

- Give the common and scientific name for five forage grasses

- Name six insect pests and tell which crops they affect

- Identify four criteria for selecting forage grasses

- List important pests and their control for the production of small grains

- Identify how triticale and buckwheat differ from other small grains

- Discuss the fertilizer needs of small grains

- Compare the culture of small grains with forage grasses

- Give the botanical name and common name for three crops grown for their oil seed

- List warm-season and cool-season small grains, forage grasses, and oil seeds

- List the important pests of oil seed production and describe their control

- Identify the five growth stages of small grains

- Name four factors to consider when selecting a small grain variety

- Describe a fungal disease

- Compare rice with wild rice

- Explain the importance of seed quality

KEY TERMS

Boot
Broadcast Planting
Broadleaf
Bunch Grass
Combine
Companion Crop
Drill
Endophyte
Forage
Grass
Heading
Heads
Jointing
Legume
Levee
Oil Seed
Palatability
Percent Purity
Ratoon
Resistant
Sod-Forming
Tillers
Volunteer Plants
Winter Hardy

GROWING SMALL GRAIN

Growers recognize the management factors that will help them reach the best small grain yields. They also need to know which small grains will grow best on their farms.

Small grains are important to the economy. They are used both as cash grain crops and as livestock feed. Small grains are well adapted to most of the soils, and many are adapted to a cool climate. This same climate, however, often results in difficult harvest conditions. All of the small grains are produced for their grain, but oat is also produced for its straw and is often grown as a **companion crop** to establish **forage** crops.

Seed Development

Small grain seeds contain carbohydrates, protein, fat, vitamins, and minerals. Grain at harvest has 13–15 percent moisture. If kept just below 13 percent moisture, it can be stored for up to two years with a minimum of spoilage. The seed has three main parts that consist of a seed coat (pericarp), an embryo, and an endosperm. The seed coat is the outside covering of the seed. It protects the embryo and the endosperm from injury. The embryo, or germ, is the living part of the seed that becomes the new plant when the seed is planted. The endosperm contains the starch food supply for the embryo during germination (Figure 21-1).

Germination and Early Growth

When the seed of a small grain crop is planted, it germinates or sprouts if the temperature is favorable and there is enough moisture. The seed absorbs water and uses its own food reserves. When the root breaks through the seed coat it quickly grows downward and anchors the seedling in the soil, absorbing water and nutrients. These roots may eventually go 3–6 feet deep and 3–4 feet horizontally. During the early stage of development, young plants are very susceptible to injury from drought and from sunny, hot, dry weather. Next, the young shoot (protected by a sheath called coleoptile) breaks through the seed coat opposite the root and grows upward. The sharp-pointed sheath breaks through the soil surface and stops growing.

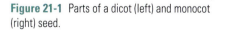

Figure 21-1 Parts of a dicot (left) and monocot (right) seed.

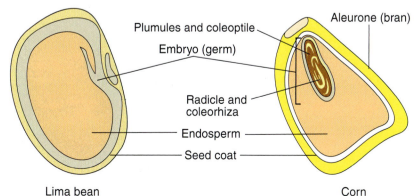

The growing point of small grain plants remains below the soil line until the first visible node appears above the soil line. By then, small grain plants have produced several additional shoots (called **tillers**) from the crown, followed by growth (elongation) of the main stem. Because the growing point remains protected below the soil line during early development, small grain plants can withstand the cold temperatures of early spring.

Growth Stages

The five basic growth stages of small grains include:
1. Seedling
2. Tillering
3. Jointing
4. Boot
5. Heading

The seedling stage is from germination to the stage at which tillers appear. The tillering stage is when the first side shoots (tillers) appear from the crown. At **jointing** stage the nodes begin to separate and can be felt in the lower part of the stem. During the **boot** stage the head can be felt inside the upper leaf sheath and the flag (last) leaf has developed. Spikes of barley, rye, and wheat or panicles of oat emerge during the **heading** stage. During the later part of this stage the florets are fertilized and the kernels develop.

Seedbed Preparation

The purpose of primary and secondary tillage is to prepare a firm seedbed with adequate moisture for good germination and seedling development (Figure 21-2). Good seed-soil contact is essential when the grain is seeded. Dry, loose soil makes an unsatisfactory seedbed. Secondary tillage to remove any weeds that are growing also is important. Too much

Figure 21-2 Tillage prepares field for seeding in a firm seedbed at times of adequate moisture for good germination and seedling development. (Photo courtesy of USDA NRCS)

tillage can pulverize the soil, which can lead to soil erosion due to wind and water or crusting after rainfall. Working plant residues into the soil near the surface helps to control erosion and protect the seed. Disking and harrowing the land before seeding is a common method of preparing a seedbed. Field cultivators can be used instead of a disk. Initial preparation of the seedbed in the fall helps the soil to dry and warm up faster in the spring and makes earlier seeding possible.

Selecting a Variety

Selecting high-quality seeds of an adapted variety is important for profitable small grain production. Other cultural practices cannot make up for lack of seed quality and freedom from weeds. Factors that are important in selecting a variety include:

+ Yield
+ Maturity
+ Disease resistance
+ Straw strength
+ Shatter resistance
+ Plant height
+ Grain quality

In most cases, relative yields may be the most important factor, but susceptibility to a disease—even one that occurs only one year out of five—can affect any yield advantage.

Quality Factors. Important quality factors include protein content, milling, and baking of wheat; and malting characteristics, protein, kernel plumpness, and test weight in barley. In oats, quality factors are groat protein, groat percentage, and test weight.

The importance of characteristics such as straw strength and plant height may vary from field to field and from year to year. Straw strength may not be as important on fields with low soil moisture reserves, but it is very important in preventing lodging on fields with good moisture and high fertility. Straw strength is particularly important when oat is used as a companion crop. Lodging may damage forage seedlings.

Disease Resistance. Disease resistance also is important in selecting a variety, especially if a particular disease has been a problem in fields. The important diseases are leaf (crown) rust and smut in oat; stem and leaf rust, tan spot, and scab in spring wheat; take-all in winter wheat; stem rust, loose smut, and spot blotch in barley; and ergot in rye. Winter hardiness is an important consideration in selecting a variety of winter wheat or rye.

Planting

Small grains should be seeded as early as possible after the frost is out of the ground. Although germination of these cereal crops begins at 24 to 36°F, soil temperatures should be 40°F before planting. Early seedings generally produce higher yields than later seedings because small grain plants develop best during cool, moist growing conditions. If temperatures go above 90°F during pollination, yields can be drastically reduced because of poor pollination and seed set. By seeding early, small grain crops normally complete their development before the hot weather in July. Oat, barley, and spring wheat should be seeded from late April to early May. Winter rye and wheat should be planted early enough in the fall so the plants can become well established before the first killing frost.

Winter rye should be planted in early to mid-September; winter wheat should be planted in late August and early September. Small grains usually are planted about 1–2 inches deep, depending on soil moisture and soil texture. Wheat, barley, and rye cannot be planted as deep as oat. Semidwarf wheat varieties cannot be planted as deep as tall varieties. A grain drill with press wheels is the best machine for seeding small grains because it distributes the seed evenly at a uniform depth and gives good soil-seed contact.

Fertility Needs

Although small grains require lesser amounts of the major nutrients than do forage crops or corn, adequate amounts of nitrogen, phosphorus, and potassium must be present for good yields. Generally these three elements are added according to soil tests, past cropping history, and expected yield goals. The major portion of nutrients is taken up by the small grain plant between the tillering and heading stages. Much of the nitrogen and phosphorus in the whole plant is removed with the grain.

To avoid nitrogen and phosphorus deficiencies, small grain producers usually apply phosphorus and nitrogen at seeding time with the grain drill. This is especially important in cool, moist soils in which little soil nitrogen and phosphorus is readily available to plants when the soil temperature is below 50°F. Using a fertilizer attachment on the drill also is important because of the higher efficiency of uptake of the plant nutrients, especially nitrogen and phosphorus. When winter rye and winter wheat are planted, phosphate and potassium fertilizers should be applied in the fall (Figure 21-3).

Nitrogen should be top-dressed on these crops the following spring. Most nitrogen fertilizer sources can be used for small grain production. The nitrogen should be applied just prior to seeding time. The amount of nitrogen needed depends primarily on the previous crop and on yield potential or yield goal. Nitrogen, for example, may not be needed if the previous crop was a good legume. Soil testing offers the only effective means of determining phosphorus and potassium levels in a soil. This information can be used to determine the additional quantity needed to attain good yields.

Weeds

Many weeds, both broadleaf and grass types, can be troublesome in small grain fields. These weeds must be controlled either by tillage before the small grain is planted or by using selective herbicides. A good stand of

Figure 21-3 No-till planting and fall fertilizing of wheat. (Photo courtesy of USDA NRCS)

vigorous small grain plants will compete with weeds fairly well. Early seeding of high-quality seeds in a good seedbed with adequate fertility will get the crop off to a good start before "warm-season" weeds have a chance to get established. Warm-season weeds include annual grasses such as foxtail, wild oats, and quack grass, and annual broadleaf weeds such as mustard, pigweed, lamb's-quarters, and wild buckwheat. If the seeding date of the small grain is delayed, these weeds must be controlled with selective herbicides. Annual broadleaf weeds will overgrow small grain even if the crop is planted early. In such cases, a broadleaf herbicide must be used.

Diseases

Fungi, along with bacteria and certain viruses, are responsible for several diseases that affect small grain production. Many of these diseases are spread by spores that are carried by wind or water. Fungal diseases can survive and overwinter on dead plant material, in stored seeds, and in the soil, and can infect the crop the following growing season. Fungal diseases may attack the leaves or the developing **heads** of small grain plants.

Among the leaf diseases, leaf rust, Septoria leaf blotch, and Helminthosporium leaf spot are the most common. Smuts and ergot damage the heads of small grain plants and can be carried over into next year's crop by the infected seeds. Many of these problems can be controlled or reduced by treating seeds, by using certified seeds, by planting **resistant** varieties, and by applying fungicides. Many disease-resistant varieties of small grains have been developed.

BARLEY

The botanical name for barley is *Hordeum vulgare* (Figure 21-4). Two cultural practices are used to grow barley—winter and spring. Figure 21-5 shows the major areas where spring barley is grown in the United States.

Figure 21-4 Most of the barley grown in the United States is used for livestock feed.

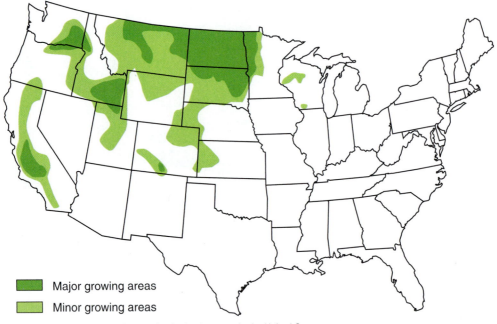

Figure 21-5 Major areas where spring barley is grown in the United States.

Barley is used mainly for livestock feed. Studies show that ton for ton, ground barley may be equal to corn in feeding value for dairy cows when used as 40–60 percent of the grain mixture. Barley consistently exceeds other spring grains in pounds of total digestible nutrients per acre. Barley often outproduces oats by more than 200 pounds of grain per acre.

It also is sold to the malting industry for use in making beer and other alcoholic beverages. Barley does particularly well when the ripening season is long and cool. Although it can withstand much dry heat, it does not do well in hot humid weather because of the prevalence of diseases under such conditions. It grows better with moderate rather than excessive rainfall. Some spring barley varieties mature earlier than oat, rye, and wheat. Barley can be grown further north and at higher altitudes than any other cereal.

Barley is one of the most dependable crops in areas where drought, frost, and salt problems occur. It is best adapted to well-drained medium and fine-textured soils with a pH of 7.0–8.0. It generally does not produce satisfactory yields on sandy soils. Barley often lodges when grown on soils high in nitrogen, resulting in low grain yields. Nitrogen management is therefore very important. Barley has the highest tolerance for salts of any of the small grains.

Winter Barley

Although barley is an excellent animal feed and easily replaces corn in rations, very little winter barley is grown in colder areas because of a general lack of winter hardiness. If grown, barley should be seeded between September 15 and 30 to increase its chances of surviving the winter. Plant 15–20 seeds per foot of 7-inch row (90–120 pounds of high-quality seeds per acre). Soil pH between 6.5 and 7.0 and recommended levels of phosphorus will improve winter survival. Twenty pounds of nitrogen per acre and the recommended phosphate and potash should be applied before planting.

Spring Barley

Yield performance of spring barley in some areas has been erratic because of late spring planting due to cool, wet growing conditions, which delayed maturity. When seeding is accomplished early, fertilization is adequate, and good growing conditions follow, high yields (60 to 90 bushels per acre) of spring barley have been obtained.

Bowers. Bowers is an awnless six-rowed barley developed by the Michigan Agricultural Experiment Station and released in 1979. Bowers barley has exhibited medium-to-early maturity and medium height. The variety tillers well and usually has large heads. Bowers has exhibited some resistance to a number of barley diseases.

OATS

Figure 21-6 shows the oat plant. The botanical name for oats is *Avena sativa*.

When used as a companion crop, oat often is removed during early flowering for forage. Oat grain is mostly used as livestock feed, with a small amount sold for use in high-protein cereals. Oat grows well on a wide range of soil types, especially if the soil is well drained and reasonably fertile. Oat is less sensitive than wheat or barley to soil conditions, especially to acidity. Oat generally grows best on medium- and fine-textured soils, but it also will produce fair yields on light, sandy soils if there is sufficient moisture.

The oat plant requires more water for its development than does any other small grain. It also will yield better than the other small grains with less sunshine. Generally, however, oat yields are inferior to those of barley when moisture is a limiting factor. High temperatures during flowering can increase the proportion of empty spikelets, a condition called blast. Hot, dry weather during grain fill causes early ripening and reduced yield; hot, humid weather during this period favors diseases.

Nitrogen management is important in obtaining best oat yields. If too much nitrogen is present, lodging can be a problem and yields will suffer as a result.

Several varieties of spring oats with high yield potentials, good test weight, and stiff medium-to-short straw are available. Some varieties are resistant to the common diseases of oats.

Spring oats is the first crop to be planted in the spring. The selection of fields with well-drained soil is essential to permit timely planting. Spring oats should be seeded as early in the spring as soil conditions permit,

Figure 21-6 Oats provide bulk and protein to the diets of animals as well as food for humans.

preferably between March 1 and April 15. Grain yields decrease rapidly as seeding is delayed past mid-April.

The proper seeding rate is 15–20 seeds per foot of 7-inch row (75–100 pounds of high-quality seeds per acre). The seeds should be planted no more than 1 inch deep to assure rapid emergence. Although oats can be established using no-till seeding techniques, little if any crop residue should be present to allow the soil to warm rapidly and the seeds to emerge. Fall preparation of the seedbed eliminates the need for tillage in the spring and sometimes permits earlier planting than when tillage is needed. This technique also eliminates the soil compaction associated with soil preparation.

Soil pH should be above 6.0 unless a legume is also seeded, in which case the soil pH should be 6.5–7.0. Phosphate and potash can be applied in the fall or spring. Nitrogen should be applied in the spring anytime before emergence.

RICE

Rice is a member of the grass family. It is an annual monocot. The botanical name for rice is *Oryza sativa*. Common rice is lowland or paddy rice (requires continuous irrigation or flooded ponding). Upland rice is nonirrigated or not grown in flooded conditions. Wild rice is not from the same genus and is not related to paddy or upland rice species. Figure 21-7 shows a rice plant.

Archeological evidence indicates a sophisticated rice cultivation system existed in China over 7,000 years ago. Worldwide, 42 countries produce rice, ranging from mountainous Himalayan to lowland delta areas. Rice is the staple food in Asia, Latin America, parts of Africa, and the Middle East.

Within the United States, the rice-producing states include Florida, Arkansas, Louisiana, California, Texas, Mississippi, and Missouri (Figure 21-8). The United States ranks 10th in total world rice production but only 15th in land area devoted to rice. The U.S. rice yields are surpassed only by Korea.

Rice thrives under the hot and humid conditions that characterize areas of the southern United States during the summer months. Fields that are nearly level and can be diked and flooded easily are readily adapted to rice production.

Figure 21-7 U.S. long-grain rice. (Photo courtesy of USDA ARS)

Varieties

Rice is divided into short-, medium-, and long-grain varieties; certain varieties have distinct aromas and flavors. Rice can also be classified according to its cooking characteristics, such as stickiness. Rice is classified into three broad groups by starch and grain characteristics:

1. Waxy or glutinous rice
2. Common, translucent, or nonwaxy rice
3. Aromatic rice

Major growing areas

Figure 21-8 Major rice growing areas in the United States. (Image courtesy of USDA NDAA)

Rice cultivars are characterized by maturity date or length of growing season—

+ Very early cultivars—90 days from germination to harvest
+ Early cultivars—90–97 days
+ Intermediate to late cultivars—98–105 days

Rice is further classified by cultural adaptation. Japonica is usually short-grained kernels and grown in temperate zones. Indica is usually found in tropical climates and is typically long-grained.

Some common cultivars include Katy, Newbonnet, Lemont, Texmont, Della, Dellmont, Gulfmont, Labelle, Mars, Rexmont, Skybonnet, Starbonnet, and Tebonnet.

Seeding

Rice is planted in a level, well-prepared field. A grain drill, similar to what is used to plant other small grains, sows about 90 pounds of rice seeds per acre. Some rice is seeded with an airplane. Because of Florida's long growing season, rice plants can be harvested and then allowed to regrow for a second harvest or **ratoon** crop. Approximately half of the rice acreage regrows as a ratoon crop—it is regrown from young shoots or tillers.

The length of time from planting to harvest is determined by the rice variety and weather conditions. It takes an average of 120 days for the first crop to mature and another 85 days for the ratoon crop to mature.

Weed Control

Weeds are controlled by integrating cultural practices and herbicides. Field preparation includes complete weed removal through disking.

Immediately after planting, an herbicide is applied that inhibits weed seed germination or kills growing weeds. As soon as the rice plants reach five to six inches in height, the flood is applied, which further inhibits weed growth. In some areas rice is not planted on the same field two years in a row, so problem weeds can be controlled with normal crop rotation. When weeds are a problem, they often include barnyard grass and sprangletop.

Fertilizer

Soil pH needs to be 5.0–7.5. Fertilizer needs are determined by a soil test.

Culture

In some places, rice is used as a seasonal rotation crop with sugarcane and vegetables. Although rice does not compete with other crops for space directly, some accommodation is often necessary. Early harvested rice fields are usually the first fields replanted to sugarcane or vegetables. Sometimes only one rice harvest can be made as the additional time needed for ratoon crop growth and harvest would interfere with the subsequent crop planting.

Rice fields are flooded when the rice plant is about six inches high. The flood is maintained two to four inches deep until the rice grain has matured and begins to dry out. The flood is drained prior to harvest to ensure firm ground for harvest equipment. About one month after the harvest operation is complete, the flood is re-established to allow the ratoon crop to grow.

Rice is an aquatic crop and is flooded during its growing season. This period corresponds with the rainy season in some places like south Florida. Rice fields actually serve as temporary storage for this rainfall and allow for the gradual addition of this water back into the environment.

Paddy leveling and **levee** building are very critical to maintain optimum and uniform paddy water depth. Soils must hold water well after flooding. Flooding is used to control some weeds and insect pests but may lead to water-borne disease spread, some water weeds, and some water insect pests. Typically, flooding begins at tillering, although flooding at planting occurs in some instances.

Pests

The fungal disease, blast, is a serious problem for rice growers. The best control available is to plant blast-tolerant varieties, although some fungicide is used in case of high disease incidence. Other diseases of rice include seedling blight, sheath blight, red rice, leaf spot, stem rot, and root rot.

When rice is grown as a rotation crop with sugarcane or vegetables, certain insects such as the rice water weevil have not become a problem. Stinkbugs are a problem and must be controlled. Fields are scouted and, when population thresholds are exceeded, control measures with approved insecticides are started. Stinkbugs pierce the immature kernels

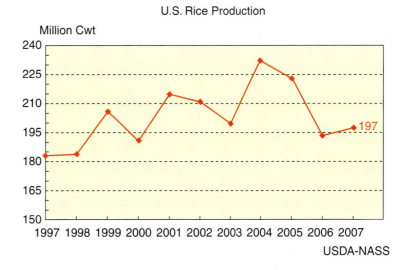

with their proboscis and, by sucking the juices, reduce the quality of the seed. Other insect pests include rice water weevil, leafhoppers, thrips, and grasshoppers.

Insect-feeding and wading birds are welcome additions to the rice fields. Blackbirds and bobolinks (small, migratory, sparrowlike birds) are not. Blackbirds not only pull up the new seedlings and eat the seeds but also eat the mature grain. Bobolinks eat large mounts of mature grain just before harvest. Both of these birds can seriously reduce yields.

Harvesting

Once the rice is mature and the fields have been drained, the rice field is harvested with a conventional grain combine. The moisture content of the rice kernel is about 19 percent at harvest, much too high for quality storage. The first stop for the rice after harvest is the drying bins. There, heated air is forced through the rice until the moisture content has been reduced to 12–13 percent. The rice is then stored in silos until it is milled and sold as white rice.

Figure 21-9 provides a historical perspective of rice production in the United States and Figure 21-10 shows how the yield per acre has increased over time.

RYE

Figure 21-11 illustrates a rye plant. The botanical name for rye is *Secale cereale*.

Rye is used as livestock feed and is sold for making bread. It often is grown as a green manure crop as well. Rye is not as palatable to livestock as other small grains but does have a feeding value about 80 percent that of corn. When used in livestock feed, rye generally makes up less than a third of the grain mixture. Ergot can be a problem in rye. There are two types of

Figure 21-10 Historical perspective of rice yields (pounds per acre) in the United States. (Image courtesy of USDA NASS)

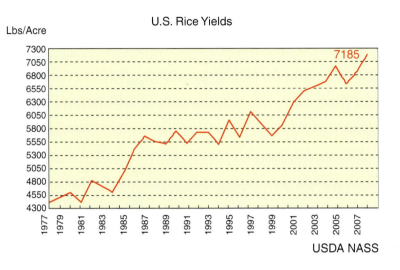

U.S. Rice Yields

Lbs/Acre

7185

USDA NASS

Figure 21-11 The rye plant, providing feed, livestock grain, and flour. Reproduced by permission from A. S. Hitchcock [revised by A. Chase], *Manual of the Grasses of the United States.* (Washington, DC: U.S. Department of Agriculture, 1950. Miscellaneous Publication No. 200.)

rye: fall rye, a winter type that is seeded in the fall, and spring rye, which is seeded in the spring.

Rye can withstand any kind of adverse weather conditions except heat. Rye sprouts more quickly and grows more vigorously than wheat at low temperatures. Its earliness frequently enables rye to escape injury from drought, although the heads are likely to be blasted if hot, dry winds occur when the plants are in blossom. Late-spring freezing can cause sterility in flowering plants. Winter rye is the most hardy of all cereals. It can survive temperatures too low for winter wheat production.

Winter hardy types can be seeded even where temperatures frequently fall to −40°F or where mean winter temperatures are about 0°F. Winter rye grows faster under cool temperatures than either winter or spring wheat. Under normal conditions, winter rye tillers appear mainly in the fall. Most growth ceases in the fall when mean temperatures drop below 40°F and begins again in the spring when temperatures rise above 40°F. Ripening takes place at temperatures of 50–68°F. Temperatures above 77°F may injure the crop. Highest rye yields usually are obtained on fertile, well-drained, medium- and fine-textured soils. Rye is adaptive, however, and is more productive than the other grains on infertile, sandy, acid soils.

Rye is an especially good crop for drained marshlands or newly cleared timberland. Under conditions favorable for winter wheat, rye usually yields less grain because of its shorter growth period and heavier straw yield. Rye often is seeded on less productive land and grows quite well with poorer seedbed preparation than is customary for wheat. Winter rye can utilize spring moisture more effectively than the spring-seeded small grains. As a result, winter rye generally will yield more in soils that become droughty in early summer. Rye volunteers freely because the grain shatters so readily. Because rye matures earlier than other grains, some of the seed is again shattered from **volunteer plants** before the regular planted crop is harvested. Consequently, it is difficult to eradicate rye in a system of continuous small grain.

Winter Rye

Winter rye is usually used as a winter cover crop because of its tolerance to adverse growing environments. Winter rye is the most winter hardy and earliest maturing cereal grain grown in some areas. It is more productive than other cereals on infertile, sandy, or acidic soils, and on poorly prepared land. When used as winter cover, winter rye is usually seeded at a rate of 60–90 pounds per acre.

When grown for grain production, winter rye should be seeded between September 20 and October 20 at the rate of 84–112 pounds of high-quality seeds per acre. The seeds should be planted 1 inch deep. When used as a winter cover crop or a green manure crop, it should be seeded in early September. Fertilization is similar to wheat, but nitrogen application should be limited to 40 pounds per acre. Rye competes well with weeds, and herbicides are generally not needed.

Few diseases attack rye, but ergot can be a serious problem. Rye is attacked by most of the insects that attack other small grains.

SORGHUM

The botanical name for sorghum is *Sorghum bicolor* Moench or *Sorghum vulgare* Pers (Figure 21-12).

Grain sorghum has the potential of becoming an important crop, particularly in areas where corn cannot be grown profitably because of spring flooding, droughty soil conditions, or late planting because of unfavorable weather.

In the past, grain sorghum has not been recommended in the Midwest because it has not been able to compete economically with corn. However, new grain sorghum single-cross hybrids have a yield potential similar to corn under good conditions, and a higher yield potential under certain adverse conditions. Farmers can grow these hybrids with available equipment, and the grain can replace corn in most animal rations. So far, good cash markets are limited. Figure 21-13 shows the major growing areas for sorghum.

Hybrids

A recent advance in grain sorghum for the eastern United States has been the release of several bird-resistant hybrids. The immature grain of these hybrids is high in tannin and quite bitter. Birds have not caused significant damage to such hybrids in tests, even though corn fields adjacent to some test plots were nearly destroyed. Sorghum hybrids that did not contain the genes necessary for tannin production were destroyed in bird-infested areas. Non-bird-resistant varieties developed in the western and southern United States can be grown where bird pressures are not present. Some 75- to 80-day maturity varieties are available for emergency crop situations.

Figure 21-12 Sorghum plants with grain heads.

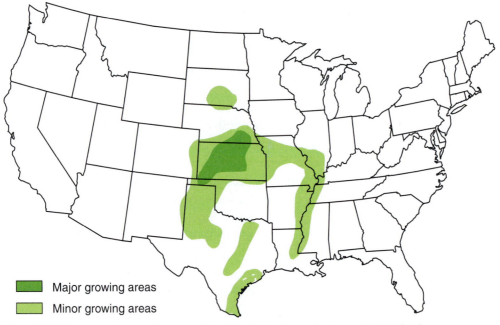

Figure 21-13 Major sorghum growing areas of the United States. (Image courtesy of USDA NOAA)

Cultural Practices

Some suggested practices for growing grain sorghum include—

1. Plant bird-resistant hybrids if birds are a problem. Non-bird-resistant sorghum is extremely susceptible to bird damage.
2. Plant one week later than the date recommended for corn. Risk of frost damage in the fall is greatly increased when grain sorghum is planted after June 10.
3. Plant in 30-inch or narrower rows, using a corn planter with special plates. A grain drill may also be used for 14- or 21-inch rows.
4. Plant from 8–12 pounds of seed per acre.
5. Fertilize according to soil test.
6. Consider using chemical weed control if weed problems are expected.
7. Harvest with a grain combine when seed moisture is 26 percent or less. See combine manufacturer's manual for proper machine adjustment.
8. Plan to dry the grain most years.
9. Check on marketing arrangements before planning to use this crop in a farming situation.

Present bird-resistant grain sorghums produce brown seed. This may affect the choice of market for this crop.

SPELT

Spelt, sometimes referred to as speltz, is a type of wheat that is not considered in the official grain standards. Spelt is primarily used as part of the grain ration for livestock and is usually ground or milled before

Figure 21-14 Spelt, a type of wheat.

feeding. It is an excellent replacement for oats in rations needing bulkiness. Pastry made from spelt flour does not trigger the allergic reaction that common wheat flour causes in people with celiac disease (5–6 percent of the population). Spelt can be grown in areas where winter wheat safely survives the winter. Common spelt is very susceptible to leaf rust, powdery mildew, and loose smut.

Spelt is tall, with moderately weak straw, and is late maturing compared with most wheat varieties. Figure 21-14 shows a spelt plant. Because spelt does not thrash completely free of its surrounding chaff, there is no official bushel weight. Usually, 20–30 percent of its total weight is the surrounding chaff. This should be considered when comparing yields with wheat. Common spelt is not resistant to loose or stinking smut. Treatment of the seed with a fungicide prior to planting helps prevent a smut problem, but the fungicide label should be consulted for clearance before use on spelt.

Hessian fly, greenbug, and wheat stem sawfly are the primary insects that attack wheat fields. Spelt also has similar susceptibility to these insects.

Spelt should be managed like wheat except for the seeding rate (15–20 seeds per foot of 7-inch row or 8–12 pecks per acre) and nitrogen application rate (50–70 percent of wheat). Spelt should not be planted following wheat, and the fly-safe date should be adhered to. Mature grain standing in the field dries more rapidly after rain than the other small grains because of the water shedding characteristics of the grain chaff. Combine settings include a slow cylinder speed (similar to soybeans) and very little air to the screens.

TRITICALE

Triticale is a human-made species resulting from a cross of wheat (*Triticum*) and rye (*Secale*). Hybrids of wheat and rye crosses date back to 1875, but intensive work on such hybrids was started only about 30 years ago at the University of Manitoba. Ideally, triticale combines the high yield potential and good grain quality of wheat with the disease and environmental tolerance (including soil conditions) of rye.

The protein content of triticale is normally higher than that of wheat, but the amino acid composition of the protein is similar to that of wheat. Feeding trials have found triticale to be unsatisfactory for hogs, and it produced less gain and feed efficiency in beef cattle than barley. Forage yields have been similar to those of wheat and rye. The cultural practices and fertilization practices recommended for wheat are satisfactory for triticale.

The primary producers of triticale are Germany, France, Poland, Australia, China, and Belarus. According to the Food and Agriculture Organization, about 13.5 million tons are harvested in 28 countries across the world.

WEED CONTROL FOR SMALL GRAINS

Small grains usually compete quite well with most weeds, especially when sound management practices are used to produce a good stand. Most broadleaf weeds that do become a problem can be controlled with herbicides. These herbicides reduce small grain yields if applied at the incorrect stage of growth. Grasses usually do not become an economic problem in small grains, and at the present time, no herbicides are available to control grasses if a problem with grasses occurs.

Small grains are very sensitive to triazine residues. To avoid an atrazine or simazine residue problem, growers do not use either of these products for weed control in corn when wheat will be seeded that fall.

WILD RICE

Wild rice is native to North America. Natural stands are found alongside the tidewater rivers (above the saline zone) in the Atlantic states and in shallow lakes and streams in northern Minnesota, northern Wisconsin, and southern Ontario and Manitoba, Canada. The grain is consumed by wildlife, especially ducks. It was used for human food 10,000 years ago when man moved into central North America. The American Indians harvested the grain from natural stands for food.

Wild Rice Plant

Wild rice is in the grass family of plants. It belongs to the genus *Zizania*. Rice belongs to the genus *Oryza*. *Zizania palustris*, the large-seeded type, grows in the Great Lakes region, is harvested for food, and is the species grown as a field crop. North American species have a chromosome number of $2n = 30$.

Wild rice caryopses are the fruits of the plant (Figure 21-15). The caryopsis is similar to kernels of the cereal grains. A large endosperm is surrounded by a thin layer of pericarp and aleurone. The embryo at one end of the caryopsis contains a cotyledon that can extend to the length of the caryopsis, a coleoptile that surrounds the terminal bud, and a coleorhiza that surrounds the primary root. The lemma and palea (hulls) remain on the seed during harvest but are removed during processing.

Wild rice seeds require three to four months of storage in cold (35°F) water before they will germinate. This dormancy is caused by an imperme-able and tough pericarp that is covered by a layer of wax and by an imbalance of growth promoters and inhibitors. Freshly harvested seeds can be forced to germinate by carefully removing (scraping) the pericarp from directly above the embryo. Scraped seeds cannot be planted directly, but they can be germinated in water and the seedlings transplanted.

The adventitious root system of the mature plant is shallow, with a lateral spread of 8–12 inches. The roots are straight, spongy, and generally white, but often rust-tinged by iron deposits. Root hairs are lacking.

Figure 21-15 Northern wild rice, *Zizania palustris* L. var. *interior.* Reproduced by permission from A. S. Hitchcock [revised by A. Chase], *Manual of the Grasses of the United States.* (Washington, DC: U.S. Department of Agriculture, 1950. Miscellaneous Publication No. 200.)

The stamens produce a large amount of pollen, which is wind-borne. The pistillate (female) inflorescence has 150–200 florets on branches that are oppressed and vary in length from 1 to 5 inches, with the upper ones being the shortest. Wild rice is usually cross-pollinated because the stigmas are receptive for three to four days and generally are pollinated from other plants before the stamens of the same stem shed pollen.

The mature caryopsis is purplish-black and is surrounded by the lemma and palea before and after harvest.

Wild rice requires from 106 to 130 days to mature in north central Minnesota, depending upon temperatures during the growing season and variety. Germination starts when water temperatures reach 42°F. Floating leaves are visible a month later, and upright (aerial) leaves 10 days later. Tillering begins after an additional 10 days. Jointing (elongation of internodes) starts 67 days after germination and the boot stage occurs eight days later. Flowering begins in late July and grain formation in August. Fields are harvested during late August and early September. Wild rice requires approximately 2,900 growing degree days to mature.

Site

Most of the wild rice in Minnesota is grown on low, wet land previously not farmed. Before considering an area for wild rice, a preliminary topographic survey should be made to determine if the site is flat enough to avoid expensive grading; too much grading will expose subsoil. The site also should have impervious subsoil so that fields can be flooded during most of the growing season. The site should be above the floodplain so the fields can be drained during late summer for harvest and fall tillage.

Seed Source and Handling

Growers plant new fields with the most shatter-resistant varieties available. New growers can obtain seeds from seed growers if arrangements are made before harvest in the fall. Some certified seeds of new varieties are available. Growers can save their own seeds but they should be from weed-free fields.

Seedbed Preparation

If large amounts of vegetation are tilled under in new fields being developed for wild rice, the area should be tilled a year or more before planting. It may also be desirable, particularly on peat soils, to grow small grains such as oats a year or two before planting wild rice. The initial cropping without flooding allows the vegetation to decompose and results in fewer problems with floating peat when the fields are flooded. A rotary tiller is commonly used to prepare the seedbed. Rototilling to a depth of 6 inches is satisfactory. A disk can be used to prepare a seedbed but it is not effective in destroying existing vegetation. A moldboard plow is not satisfactory in peat soil covered with vegetation because the turned-over soil tends to float when fields are flooded. Land-breaking plows cause less floating of soil but they should not be used in shallow peat where underlying clay is brought to the surface. It is difficult to establish wild rice in clay subsoil. The final seedbed should be devoid of ridges and hollows to assure good drainage.

Time of Seeding

Wild rice can be seeded either in the fall or spring. Fall seeding has the advantage of eliminating the need to store wild rice seeds over the winter. If seeds are not properly stored during the colder months, poor germination can result. In addition, seeding in the fall is desirable because fields generally are dry and ground equipment can be used. However, if the soil is too dry, flooding the fields immediately after fall seeding may be necessary so the seeds will not dry out.

Method of Seeding

Use of a grain drill assures seeding to a uniform depth. Wild rice will not emerge from depths greater than 3 inches. However, the seeds may be broadcast over the soil surface using a fertilizer spreader and then incorporated to a depth of 1–2 inches with a disk or harrow. Spring seeding usually is done over the water with an airplane.

Flooding the soil affects the behavior of fertilizer nitrogen as well as native soil nitrogen. The unique reactions that nitrogen undergoes in flooded soils result in considerable loss of applied nitrogen fertilizer. The use of added nitrogen generally is poorer in flooded soils than in well-drained soils.

BUCKWHEAT

Buckwheat, a broadleaf plant, is sometimes used as a crop suitable for either late spring planting or on land with low yield potential where production costs must be kept low. It is suitable as a honey crop, smother crop, or green manure crop. Most buckwheat grain is used as food for humans. It is a satisfactory partial substitute for cereal grains in livestock rations, but has a lower feeding value than cereal grains.

The botanical name for buckwheat is *Fagopyrum esculentum*; Figure 21-16 shows a buckwheat plant. It is more closely related to rhubarb than to the cereal grains. Unlike cereal grains, buckwheat has a strong tap root rather than fibrous roots like grain.

Several varieties (Mancan, Manor, Pennquad, Tempest, Tokyo, Winsor Royal) are available. Suitable planting dates range from May through July 20, or 10–12 weeks before frost. The crop flowers about four weeks after emergence, has an indeterminate growth habit, and is mature about 10 weeks after emergence. It grows vegetatively and flowers until terminated by frost. The crop grows best where the climate is cool and moist and it performs well over a wide range of soil types and fertility levels. Serious diseases affecting other dicot field crops have not been important in buckwheat.

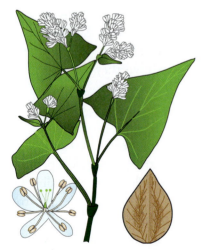

Figure 21-16 Buckwheat, *Fagopyrum esculentum*. Reproduced with permission from N. L. Britton and A. Brown, *An Illustrated Flora of the Northern United States, Canada and the British Possessions* (Vol. 1: 672.1913.) Courtesy of Kentucky Native Plant Society.

A Baker's Dozen of Best Management Practices

Crop production alters the physical and chemical makeup of vast areas of land, affecting soil and water on and off the farm. Some effects are undesirable, including soil erosion, sediment deposition, increased levels of phosphorus and nitrogen in bodies of water fed by agricultural watersheds, and the presence of nitrate and pesticides in drinking water. Agriculture will always affect soil and water resources, but is should attempt to minimize impacts whenever possible.

Best management practices (BMPs) minimize offsite damage while maintaining profitability and stability in farming operations. Some BMPs that farmers should consider to reduce the pollution potential of their operations include:

1. Loss of soil, fertilizers, and pesticides can be reduced by installing buffers and filter strips on field edges that border waterways, grassed waterways in areas of concentrated runoff, and terraces and sediment retention basins at appropriate locations in the field.

2. Conservation tillage and growing more acres of hay protect fields from erosion and help reduce the movement of fertilizer and pesticides into surface waterways.

3. In some areas, tile drainage allows earlier planting and promotes early crop development, increasing the amount of time the soil is protected from potential wind and water erosion.

4. Choosing pesticides and following labels carefully saves money and reduces potential contamination of soil and water. When in doubt, avoid the use of highly water-soluble, long-lived compounds on coarse-textured soils, particularly soils that lie over shallow aquifers or near streams.

5. Integrated pest management and pest scouting may reduce costs by reducing or preventing pesticide use. Crop rotations provide much more economical pest control than a pesticide application in continuous cropping systems.

6. Soil testing results indicate the nutrient reserves of the soil. Fertilizer rates should be adjusted to reflect these reserves. A comprehensive soil-testing plan increases profits.

7. Establishing realistic yield goals and managing to achieve them go hand-in-hand with soil testing when designing rational fertilizer programs.

8. Soil erosion and runoff remove most phosphorus from fields. Incorporation of phosphorus by tillage, or subsurface placement in conservation tillage systems, reduces phosphorus accumulation at the soil surface, thus reducing the amount lost to erosion and runoff.

9. Proper management of livestock waste also reduces loss of nutrients from fields. Manure analysis indicates nutrient concentrations in these materials and helps determine rates of application.

10. Adjust nitrogen fertilizer rates to compensate for previous crop and manure application. Legumes often supply significant quantities of nitrogen to a succeeding grass crop.

11. Follow recommended cultural practices. A well-designed cropping system protects the soil better and allows for more complete utilization of nutrients and pesticides than a poor one. Plant on time. Choose the most appropriate varieties or hybrids for each production system.

12. Grow every crop on the soil that gives it the comparative advantage for sustained high yields.

13. Discontinued use of poorly producing land improves the overall net return of an operation if that land does not generate enough income to cover out-of-pocket costs. Establishing permanent vegetative cover on such sites greatly reduces the loss of soil into waterways.

Volunteer plants can be a problem in crop sequences, but are easily controlled with herbicides. The normal seeding rate is 36–72 pounds per acre or 8–10 seeds per foot of 7-inch row. The depth of planting is 1–1½ inches deep. One hundred pounds of a 19-19-19 fertilizer is adequate to produce good yields on medium fertility soil. Normal yields are 20–40 bushels (48 pounds per bushel) per acre. The grain is usually worth 6–10 cents per pound if a market is available.

FORAGE GRASSES

This discussion of forage grasses includes Kentucky bluegrass, Canada bluegrass, orchard grass, ryegrass, reed canary grass, smooth bromegrass, tall fescue, timothy, prairie grass, garrison grass, and Bermuda grass. Management of all the forage grasses is similar and is covered in a separate section. Tables 21-1 and 21-2 summarize some of the features of the forage grasses.

Kentucky Bluegrass

Kentucky bluegrass (*Poa pratensis*) is one of the most predominant pasture grasses. It is adapted to well-drained loams and heavier soil types. Bluegrass will be dominant in pastures only if the soil has a salt pH of 5.3 or higher and at least a medium level of phosphorus. Although the grass often becomes dormant during dry weather, it will survive severe droughts. Figure 21-17 shows a Kentucky bluegrass plant.

Kentucky bluegrass is a rhizomatous plant that produces a dense sod. Optimum temperatures for growth range from 60 to 90°F. Serious injury will occur if the grass is subjected to continuous soil and air temperatures

Figure 21-17 Kentucky bluegrass is a grass found in the pastures and lawns of the Northeast.

Table 21-1 Agronomic Adaptation and Characteristics of Perennial Forage Grasses

Forage Species	Minimum Adequate Drainage[1]	Tolerance to pH < 6.0	Minimum Adequate Soil Fertility	Drought Tolerance	Persistence	Seedling Aggressiveness	Growth Habit
Kentucky bluegrass	SPD	Medium	Medium	Low	High	Low	Dense sod
Orchard grass	SPD	Medium	Medium	Medium	Medium	High	Bunch
Perennial ryegrass	SPD	Medium	Medium to high	Low	Low	Very high	Bunch
Reed canary grass	VPD	High	Medium to high	High	High	Low	Open sod
Smooth bromegrass	MWD	Medium	High	High	High	Medium	Open sod
Tall fescue	SPD	High	Medium	Medium	High	High	Variable
Timothy	MWD	Medium	Medium	Low	High	Low	Bunch
Garrison grass	VPD	Medium	Medium	High	High	Low	Open sod
Switchgrass	SPD	High	Low to medium	Excellent	High	Very low	Bunch
Big bluestem	MWD	High	Low to medium	Excellent	High	Very low	Bunch
Indiangrass	MWD	High	Low to medium	Excellent	High	Very low	Bunch
Eastern gamagrass	PD	High	Medium to high	Good	High	Very low	Bunch

[1]. Minimum drainage required for acceptable growth: WD = well-drained; MWD = moderately well-drained; SPD = somewhat poorly drained; PD = poorly drained; VPD = very poorly drained.

Table 21-2 Suitability of Perennial Forage Grass Species to Different Types of Management and Growth Characteristics

Species	Frequent, Close Grazing	Rotational Grazing	Stored Feed	Periods of Primary Production	Relative Maturity[4]
Kentucky bluegrass	HS[1]	HS	NR[2]	Early spring and late fall	Early
Orchard grass	NR	HS	HS	Early spring, summer, fall	Early-medium
Perennial ryegrass	NR	HS	HS	Early spring and late fall	Early-medium
Reed canary grass	NR	HS	HS	Early spring, summer, fall	Medium-late
Smooth bromegrass	NR	S[3]	HS	Spring, summer, fall	Medium-late
Tall fescue	NR	HS	HS	Early spring, summer, fall	Medium-late
Timothy	NR	S	NR	Late spring and fall	Late
Garrison grass	NR	S	NR	Spring, early summer	Medium-late
Switchgrass	NR	HS	HS	Summer	Very late
Big bluestem	NR	HS	HS	Summer	Very late
Indiangrass	NR	HS	HS	Summer	Very late
Eastern gamagrass	NR	HS	S	Summer	Very late

1. Highly suitable
2. Not recommended
3. Suitable
4. Relative time of flower or seedhead appearance in the spring; depends on species and variety. Warm-season grasses mature in midsummer, exact time varies by species.

of 100°F and above. Bluegrass is relatively unproductive in midsummer, but yields can be increased and sustained by favorable moisture and nitrogen fertilization.

At comparable growth stages, bluegrass contains more energy per pound of dry matter than smooth bromegrass. It is also extremely palatable during periods of rapid growth. Because of its prostrate growth habit, in early clipping studies it did not compare in yields with taller growing species such as bromegrass, orchard grass, or reed canary grass. However, later studies with grazing animals found it to be nearly equal with these grasses in productivity (Figure 21-18). The carrying capacity of these fields can be increased by weed control, addition of legumes, and fertilization.

Figure 21-18 Cows grazing on a grass pasture. (Photo courtesy of USDA ARS)

Pastures of bluegrass overgrown with weeds generally result from overgrazing and underfertilization. Bluegrass should not be grazed lower than two or three inches. Overgrazing opens the sod to weeds and reduces the vigor and growth rate of bluegrass. Overgrazing causes very poor root and rhizome development.

Controlling Pasture Weeds. Pasture weeds are annuals, biennials, and perennials. Annual species such as ragweed, fleabane, and sunflower originate from seeds each year and die after the seeds have matured in the summer or fall. Perennials originate from well-established root systems that survive for many years. Seeds of perennials also produce new infestations. Ironweed, goldenrod, dock, thoroughwort, and vervain are examples of this most difficult group to eradicate.

Repeated mowing or clipping may gradually reduce a stand of weeds. Mowing must be timed properly for good results. The early bud stage, when root reserves are low and weeds are under stress, is most favorable for effective control of biennials or perennials. Weeds mowed at this time have a slow recovery rate. To prevent seed production, annual weeds should be mowed before seed heads mature. The optimum stage for mowing annual and perennial weeds usually does not occur at the same time.

In many cases, spraying with an herbicide is the most effective way to control broadleaf weeds growing in bluegrass pastures. The best time to spray for weed control is when the weeds are growing rapidly and are in the vegetative stage. Using an herbicide and mowing or clipping is an excellent way to control weeds while maintaining lespedeza in a bluegrass sod.

Bluegrass and Other Forages. Legumes are often established in existing bluegrass fields to improve production and strengthen the production curve during the summer months. Bluegrass with a legume will produce nearly as much beef as orchard grass, timothy, or fescue with the same legume.

Birdsfoot trefoil, ladino clover, lespedeza, red clover, or alfalfa may be seeded in bluegrass sods. Trefoil is probably the best-suited companion crop with bluegrass because of its heavy summer production when bluegrass is semidormant. Trefoil–bluegrass pastures can be expected to produce twice as much as straight bluegrass fertilized with 60 pounds of nitrogen. Production with trefoil will also be much more uniform throughout the growing season than on heavily fertilized straight bluegrass pastures.

To add legumes to existing bluegrass fields, producers till the grass sod during late fall, winter, or very early spring enough to kill approximately one-half the sod. Growers will band-seed the trefoil or other legumes using a high-analysis phosphate fertilizer such as 0-45-0.

The legume seeding should be done during January or February if it is broadcast. If it is band-seeded with a drill, it may be done as late as March. Seedings made after April 1 must compete with the grass.

Fertilizing Bluegrass. Pure bluegrass sods should be fertilized annually with at least a 60-20-20 fertilizer. In fields with medium-to-high potash levels, the potash may be omitted. If legumes are to be maintained in a stand, the nitrogen should be omitted and a higher level of potash and phosphate used. For example, a 0-30-60 fertilizer is well suited to bluegrass–lespedeza or trefoil pastures.

Although Kentucky bluegrass usually appears spontaneously, it is occasionally seeded as a pasture grass. When bluegrass is seeded for this purpose, common or commercial seeds are used. It is usually sown in a mixture with other grass and legume seeds. Usually two to four pounds per acre of bluegrass seeds are added to a normal forage mixture. When seeded alone, use 10 pounds per acre of certified quality seeds. Seeding in the fall before October 15 is best, depending on location.

Canada Bluegrass

Canada bluegrass (*Poa compressa*) is also found in many Kentucky bluegrass fields in Missouri. This is usually a signal that the field is lacking in lime or phosphate, or both, because Canada bluegrass is more tolerant of acidic, low-phosphate soils than is Kentucky bluegrass.

Canada bluegrass has bluer foliage than Kentucky bluegrass, matures later, is less productive, and makes very slow or little recovery after it is grazed. It is an inferior species; fields infested with it should have a soil test taken and the soil nutrient deficiencies should be corrected.

Orchard Grass

Orchard grass is a versatile perennial bunch-type grass (no rhizomes) that establishes rapidly and is suitable for hay, silage, or pasture. It has rapid regrowth, produces well under intensive cutting or grazing, and attains more summer growth than most of the other cool-season grasses. Orchard grass tolerates drought better than several other grasses (Table 21-1). Orchard grass grows best in deep, well-drained, loamy soils. Its flooding tolerance is fair in the summer but poor in the winter. Orchard grass is especially well suited for mixtures with tall legumes, such as alfalfa and red clover. The rapid decline in palatability and quality with maturity is a limitation with this grass. Timely harvest management is essential for obtaining good quality forage.

The botanical name for orchard grass is *Dactylis glomerata*; Figure 21-19 shows an orchard grass plant. Improved varieties of orchard grass, with high yield potential and improved resistance to leaf diseases, are available. When seeding orchard grass-legume mixtures, growers usually select varieties that mature at about the same time. The later-maturing varieties are best suited for growing with alfalfa because they match the maturity development of alfalfa and are easier to manage for timely harvest to obtain good quality forage.

Perennial Ryegrass

Perennial ryegrass is a bunch grass suitable for hay, silage, or pasture. Perennial ryegrass produces excellent quality and palatable forage, is a vigorous establisher, has a long growing season, and is high yielding under good fertility when moisture is not lacking (Tables 21-1 and 21-2).

Figure 21-19 Orchard grass, *Dactylis glomerata*. Reproduced by permission from A. S. Hitchcock [revised by A. Chase], *Manual of the Grasses of the United States*. (Washington, DC: U.S. Department of Agriculture, 1950. Miscellaneous Publication No. 200.)

Figure 21-20 Perennial ryegrass, *Lolium perenne.*

Figure 21-21 Reed canary grass, *Phalaris arundinacea.*

Because it is less winter hardy than other grasses, perennial ryegrass is best seeded in combination with other grasses and legumes. Perennial ryegrass can be grown on occasionally wet soils. It is less competitive with legumes than orchard grass, and is usually later to mature than orchard grass. Like orchard grass, perennial ryegrass can withstand frequent cutting or grazing. It is more difficult to cut with a sickle bar mower and is slower to dry than other grasses.

The botanical name for perennial ryegrass is *Lolium perenne.* Figure 21-20 illustrates a ryegrass plant.

Reed Canary Grass

Reed canary grass (*Phalaris arundinacea*) is a tall, leafy, coarse, high-yielding perennial grass (Figure 21-21), tolerant of a wide range of soil and climatic conditions (Table 21-2). It can be used for hay, silage, and pasture. It has a reputation for poor palatability and low forage quality. This reputation was warranted in the past, but new varieties are available that make this forage an acceptable animal feed, even for lactating dairy cows. Reed canary grass grows well in very poorly drained soils, but is also productive on well-drained upland soils. It is winter hardy, drought-tolerant (deep-rooted), resistant to leaf diseases, persistent, responds to high fertility, and tolerates spring flooding, low pH, and frequent cutting or grazing. Reed canary grass forms a dense sod. Limitations of this grass include slow establishment, expensive seeds, and forage quality that declines rapidly after heading.

Only low-alkaloid varieties (Palaton, Venture, Rival) are recommended if the crop is to be used as an animal feed. These varieties are palatable and equal in quality to other cool-season grasses when harvested at similar stages of maturity. Common reed canary grass seed should be considered to contain high levels of alkaloids, and is undesirable for animal feed.

Smooth Bromegrass

Smooth bromegrass is a leafy, sod-forming perennial grass that is best suited for hay, silage, and early spring pasture. It spreads by underground rhizomes and through seed dispersal (Tables 21-1 and 21-2). Smooth bromegrass is best adapted to well-drained silt-loam or clay-loam soils. It is a good companion with cool-season legumes. Smooth bromegrass matures somewhat later than orchard grass in the spring and makes less summer growth than orchard grass. It is very winter hardy and, because of its deep root system, will survive periods of drought. Smooth bromegrass produces excellent quality forage, especially if harvested in the early heading stage. It is adversely affected by cutting or grazing when the stems are elongating rapidly (jointing stage). It should be harvested for hay in the early heading stage for best recovery growth.

The botanical name for smooth bromegrass is *Bromus inermis* and Figure 21-22 shows the plant. Improved high-yielding and persistent varieties are available. Some varieties are more resistant to brown leaf spot,

Figure 21-22 Smooth bromegrass, *Bromus inermis.* Reproduced with permission from N. L. Britton and A. Brown, *An Illustrated Flora of the Northern United States, Canada and the British Possessions* (Vol. 1: 672.1913.) Courtesy of Kentucky Native Plant Society.

Figure 21-23 Fescue (*Schedonorus phoenix*), is a widely adapted grass, but palatability as a livestock feed is a problem. Reproduced with permission from Wetland Flora: Field Office Illustrated Guide to Plant Species (Washington, DC: U.S. Department of Agriculture Natural Resources Conservation Service. Provided by NRCS National Wetland Team, Fort Worth, TX.)

which may occur on smooth bromegrass. These improved varieties start growing earlier in the spring and stay green longer than common bromegrass, which has uncertain genetic makeup.

Tall Fescue

Tall fescue is a deep-rooted, long-lived, sod-forming grass that spreads by short rhizomes. It is suitable for hay, silage, or pasture for beef cattle and sheep (Tables 21-1 and 21-2). Tall fescue is the best cool-season grass for stockpiled pasture or field-stored hay for winter feeding. It is widely adapted and persists on acidic, wet soils of shale origin. Tall fescue is drought-resistant and survives under low fertility conditions and abusive management. It is ideal for waterways, ditch and pond banks, farm lots, and lanes. It is the best grass for areas of heavy livestock and machinery traffic. Figure 21-23 illustrates a tall fescue plant.

Many older tall fescue pastures contain a fungus (**endophyte**) growing in the plant. The fungal endophyte is associated with poor palatability in the summer and poor animal performance. Several health problems may develop in animals grazing on endophyte-infected tall fescue; this is especially so in breeding animals. Deep-rooted legumes should be included with tall fescue if it is to be used in the summer. Legumes improve animal performance, increase forage production during the summer, and dilute the toxic effect of the endophyte when it is present.

The botanical name for tall fescue is *Schedonorus phoenix*. Newer endophyte-free varieties or varieties with very low endophyte levels (less than 5 percent) are recommended if stands are to be used for animal feed. Kentucky 31 is the most widely grown variety, but most seed sources of this variety are highly infected with the endophyte fungus, and should not be planted. When buying seeds, growers should make sure the tag states that the seeds are endophyte-free or have a very low percentage of infected seeds. Because endophyte-free varieties are less stress-tolerant than endophyte-infected varieties, they should be managed more carefully.

Timothy

Timothy (*Phleum pratense*) is a hardy perennial bunch grass that grows best in cool climates. Its shallow root system makes it unsuitable for droughty soils. It produces most of its annual yield in the first crop. Summer regrowth is often limited because of timothy's intolerance to hot and dry conditions. Timothy is used primarily for hay and is especially popular for horses. It requires fairly well-drained soils (Tables 21-1 and 21-2). Timothy is less competitive with legumes than most other cool-season grasses and is adversely affected by harvesting or grazing during the jointing stage (stem elongation). Because it is easily weakened by frequent cutting, a sufficient recovery period is necessary for accumulation of energy reserves for regrowth. Figure 21-24 shows the timothy plant.

Prairie Grass

Prairie grass (Matua) is a tall-growing perennial bunch grass introduced from New Zealand and the only variety of prairie grass currently sold in the United States. It is adapted to well-drained soils with medium-to-high fertility and pH of 6.0 or higher. It is a type of bromegrass, but differs from smooth bromegrass in that it does not spread by rhizomes, and it produces seed heads throughout the growing season. This grass is not as winter hardy as the other cool-season grasses. It winterkills easily in colder states.

Although its forage quality compares well with that of other cool-season grasses, prairie grass may contain lower levels of trace elements. Prairie grass grows well during the fall and can be managed to provide early spring grazing. It should be harvested or grazed on a monthly schedule during the fall. Stockpiling growth dramatically reduces winter survival of this grass.

Garrison Grass

Garrison grass, also called creeping meadow foxtail, is a vigorous, sod-forming, cool-season grass. It is especially well suited for low meadows and pastures, and survives extended periods of standing or running water. Garrison grass is adapted to a broad range of soils, but sufficient moisture is required for good growth. It is very cold hardy, and has good drought tolerance. Specially designed grassland drills are the best method of seeding this grass because the seeds are light and fuzzy. A cultipacker seeder can be used in the fully opened position for planting.

Figure 21-24 Timothy (*Phleum pratense*) grows best when temperatures are 65–72°F. Reproduced by permission from A. S. Hitchcock [revised by A. Chase], *Manual of the Grasses of the United States.* (Washington, DC: U.S. Department of Agriculture, 1950. Miscellaneous Publication No. 200.)

Perennial Warm-Season Grasses

The native, perennial, warm-season grasses have potential to produce good hay and pasture growth during the warm and dry midsummer months. These grasses initiate growth in late April or early May, and produce 65–75 percent of their growth from mid-June to mid-August. Warm-season grasses produce well on soils with low moisture holding capacity, low pH, and low phosphorus levels. Warm-season grasses complement cool-season grasses by providing forage when the cool-season grasses are less productive. The management requirements for warm-season grasses are different than those normally followed for cool-season grasses, particularly in establishment procedure, fertilization, and grazing management.

Switchgrass, Indian grass, and big bluestem are winter hardy and grow in many areas. They do best on deep, fertile, well-drained soils with good water-holding capacity. Gama grass and Caucasian bluestem also have potential in some areas (Tables 21-1 and 21-2).

Switchgrass. Switchgrass (*Panicum virgatum*) is noted for its heavy growth during late spring and early summer. It provides good warm-season pasture and high-quality hay for livestock. For erosion control, switchgrass is one of the most valuable native grasses on a wide range of sites. It is a valuable soil-stabilization plant on strip mine spoils, sand dunes, dikes, and other critical areas. Additionally, it is suitable for low windbreak plantings in truck crop fields.

Switchgrass (Figure 21-25) provides excellent nesting and fall and winter cover for pheasants, quail, and rabbits. It holds up in heavy snow, particularly the Shelter and Kanlow cultivars. The seeds provide food for pheasants, quail, turkeys, doves, and songbirds.

Interest in switchgrass as a renewable biofuel resource has been increasing in recent years, primarily in the southern United States. Several organizations are cooperating to evaluate several upland types of switchgrass for use as a biomass energy resource. The development of hybrid cultivars

Figure 21-25 Switchgrass (*Panicum virgatum*) is being studied as a possible renewable biofuel. (Photo courtesy of USDA ARS)

Figure 21-26 Improved varieties of Bermuda grass (*Cynodon dactylon*) make good pastures and turf in the South. Reproduced by permission from A. S. Hitchcock [revised by A. Chase], *Manual of the Grasses of the United States.* (Washington, DC: U.S. Department of Agriculture, 1950. Miscellaneous Publication No. 200.)

for increased biomass yield will ultimately result in cultivars for the central and southern United States. Results of a variety of studies may contribute to producers cashing in on a growing demand for renewable fuels and a decrease on our dependency on fossil fuels.

Bermuda Grass. Bermuda grass (*Cynodon dactylon*) is grown extensively throughout all the tropical and subtropical areas of the world and in the southern areas of the United States. Bermuda grass is an aggressive plant that spreads rapidly by seeds, stolons, and rhizomes. It is used for hay crops, pasture, and lawns and turf in warmer areas. Bermuda grass grows best on fertile clay soils in full sun.

Improved hybrid cultivars are more productive, more frost-tolerant, and more drought-resistant than common Bermuda grass. Figure 21-26 shows the Bermuda grass plant.

MANAGING FORAGE CROPS

All forage crops respond dramatically to good management practices. Higher yields and improved persistence result from paying attention to the basics of good forage management as outlined here.

Seeding Year Management

Establishing a good stand is critical for profitable forage production (Figure 21-27). Begin by selecting species adapted to soils where they will be grown. Crop rotation is an important management tool for improving forage productivity. Crop rotation reduces disease and insect problems. Disease pathogens accumulate and can cause stand establishment

Figure 21-27 Fine grass and legume seeds must be slightly covered and left in a compact seedbed. The cultipacker presses the seeds into the seedbed.

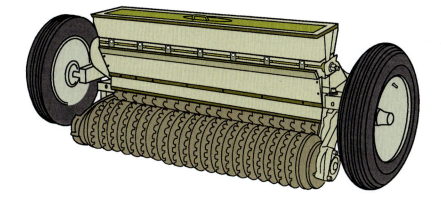

failures when seeding into a field that was not rotated out of a forage crop for at least one year. Rotating to another crop for at least a year is the preferred method.

Growers ensure that fields are free of any herbicide carryover that can harm forage seedlings. Current labels on herbicides provide more information on crop rotation restrictions.

Many producers are successfully using reduced tillage and no-tillage seeding methods (see Figure 21-29).

Fertilization and Liming

Proper soil pH and fertility are essential for optimum economic forage production. Take a soil test to determine soil pH and nutrient status at least six months before seeding. This allows time to correct deficiencies in the topsoil zone. The topsoil in fields with acidic subsoils should be maintained at higher pHs than fields with neutral or alkaline subsoils. Producers maintain topsoil zone pH at the following levels:

+ pH 6.8 for forage legumes on mineral soils with subsoil pH less than 6.0
+ pH 6.5 for forage grasses on mineral soils with subsoil pH less than 6.0
+ pH 6.5 for alfalfa on mineral soils with subsoil pH greater than 6.0
+ pH 6.0 for other forages on mineral soils with subsoil pH greater than 6.0
+ pH 6.0 for Kentucky bluegrass pastures

Soil pH should be corrected by application of lime when topsoil pH falls 0.2–0.3 pH units below the recommended levels. With conventional tillage plantings, lime should be incorporated and mixed well in the plow layer at least six months before seeding. If more than four tons per acre are required, half the amount should be plowed down and the other half incorporated lightly into the top two inches. If low rates of lime are recommended or if a split application is not possible, the lime should be worked into the surface rather than plowed down. This assures a proper pH in the surface soil where seedling roots develop and where nodulation begins in legumes.

```
MEADOW PASTURE MIX
LOT #J574   NET 50#
PURITY   GERM
22.00    80    BROME
21.00    88    ORCHARD
09.75    90    TIMOTHY
10.25    88    FESCUE
18.50    90    PERENNIAL RYE
06.00    90    BLUE GRASS
05.95    90    CLOVER

3/96 TESTED
0.75 CROP
5.05 INERT
0.75 WEED

GLOBE SEED AND FEED CO., INC.
TWIN FALLS, ID 83301
```

Figure 21-28 A grass seed label gives important information.

Corrective applications of phosphorus and potassium should be applied prior to seeding, regardless of the seeding method used. Fertilizer recommendations for forage legumes and tall grasses are found in extension bulletins and through soil tests. Starter applications of 0-40-40 may be applied at seeding. Seedling vigor of cool-season forage grasses is enhanced on many soils by nitrogen applied at seeding time. Growers may apply 10–20 pounds of nitrogen per acre when seeding grass-legume mixtures, and 30 pounds per acre when seeding pure grass stands. Starter nitrogen applications of 10 pounds per acre may be beneficial with pure legume seedings, especially under cool conditions and on soils low in nitrogen. Manure applications incorporated ahead of seeding may also be beneficial.

Seed Quality

High-quality seeds of adapted species and varieties should be used. Seed lots should be free of weed seeds and other crop seeds, and contain only minimal amounts of inert matter. Certified seed is the best assurance of securing high-quality seeds of the variety of choice (Figure 21-28). Purchased seeds account for only 20 percent or less of the total cost of stand establishment. Buying cheap seeds and seeds of older varieties is false economy. The producer should always compare seed price on the basis of cost per pound of pure live seeds, calculated as follows:

+ **Percent purity** = 100 percent – (percentage inert matter + percentage other crop seed + percentage weed seed)
+ Percent pure live seed (PLS) = percentage germination × percentage purity
+ Pounds of PLS = pounds of bulk seed × percentage PLS

Weed Management in Forage Crops

The best way to minimize weed problems in forage crops is to establish and maintain a vigorous stand of the forage crop. Most of the weed control in healthy forage crops is from competition provided by the forage crop. Growers fertilize properly, follow good cutting management practices, use disease-resistant varieties, control problem insects, and use herbicides selectively and judiciously to keep the forage stand productive and competitive with weeds.

Weeds reduce forage yields, but more importantly they often reduce forage quality and palatability. Because the feed value of weeds differs among species, the relative importance of controlling them varies by species.

Perennial and biennial weeds should be controlled with herbicides in the previous crop or prior to planting a forage crop.

Herbicides are usually needed the most during establishment, especially in pure legume seedings. Grassy weeds are usually more competitive against legumes than broadleaf weeds. To ensure good stand establishment, growers should provide good weed control during the first 60 days after planting. Preplant or postemergence herbicides are available. With a postemergence herbicide program, the grower can take the "wait and see" approach to weed control, and base herbicide use on need. Because timing of application is critical with a postemergence program, growers should check the labels for application details and restrictions.

Late summer seedings are usually less subject to serious infestations of annual grasses and most summer annual broadleaf weeds. Preplant herbicides are not recommended. Winter annual weeds may germinate in late summer seedings and severe infestations can cause stand-establishment problems. Herbicides are available for control of these weeds, and fall applications are more effective than spring applications.

TILL AND MINIMUM-TILL SEEDING

Many producers are successfully adopting minimum and no-tillage practices for establishing forage crops. Advantages include soil conservation, reduced moisture losses, lower fuel and labor requirements, and seeding on a firm seedbed. All forage species can be seeded no-till. No-till forage seedings are most successful on silt loam soils with good drainage. Consistent results are more difficult on clay soils or poorly drained soils. Weed control and/or sod suppression is absolutely critical for successful no-till forage establishment because most forage crops are not competitive in the seedling stage.

Fertilization

Growers soil test well ahead of seeding to determine lime and fertilizer needs. For no-till seedings, growers should take soil samples to a 2-inch depth to determine pH and lime needs, and to a normal 8-inch depth to determine phosphorus and potassium needs. If possible, producers should make corrective applications of lime, phosphorus, and potassium earlier in the crop rotation when tillage can be used to incorporate and thoroughly mix these nutrients throughout the soil. When this is not feasible, lime, phosphorus, and potassium applications should be made at least eight months or more ahead of seeding to obtain the desired soil test levels in the upper rooting zone.

Lime and phosphorus move slowly through the soil profile. Once soil pH, phosphorus, and potassium are at optimum levels, surface applications of lime and fertilizers maintain these levels. Attempts to establish productive forages often fail where pH, phosphorus, or potassium soil test values are below recommended levels, even when corrective applications of these nutrients are surface applied or partially incorporated just before seeding.

Seeding Equipment and Methods

Because most forage crops have small seeds, careful attention must be given to seed placement. A relatively level seedbed improves seed placement. A light disking may be necessary before attempting to seed. Seeds are planted shallow (1/4–1/2 inch, in most cases), in firm contact with the soil (Figure 21-29).

Figure 21-29 Grass drill to interseed native grasses into a cool-season field of grasses. (Photo courtesy of USDA NRCS)

Crop residue must be managed to obtain good seed-soil contact. Chisel plowing or disking usually chops residue finely enough for conventional drills to be effective. When residue levels are greater than 35 percent, no-till drills are recommended.

Establishment

When following corn, growers should plant as soon as the soil surface is dry enough for good soil flow around the drill openers and good closure of the furrow. Perennial weeds should be eradicated in the previous corn crop. If perennial weeds are still present, growers should apply Roundup before seeding. If any grassy weeds or winter annual broadleaf weeds are present in the field, growers can use an herbicide before seeding.

Most drills can handle corn grain residue, but removal of some of the residue (e.g., for bedding) often increases the uniformity of stand establishment. Most drills do not perform as well when corn stalks are chopped and left on the soil surface.

No-till seeding in August following small grain harvest permits the producer to optimize small grain and straw yields without also being concerned with trying to establish an interseeded forage crop in the spring. No-till seeding of the forage crop helps conserve valuable moisture.

Weeds should be effectively controlled in the small grain crop, and herbicides applied at the proper stage of small grain development. Growers should remove straw after small grain harvest. It is not necessary to clip and remove stubble. It may be removed if additional straw is desired. Stubble should not be clipped and left in the field, as it may interfere with good seed-soil contact when seeding forages.

Insect Control

Insects can be a serious problem in no-till seedings, especially those seeded into old sods. Slugs can be especially troublesome where excessive residue is present from heavy rates of manure applied in previous years. Chemical control measures for slugs are limited.

OIL SEED CROPS

Although many other plants worldwide are used as an oil seed crop, only canola, safflower, and sunflower are discussed here. Soybeans were discussed in Chapter 20.

Canola

Canola refers to rapeseed (*Brassica* spp.), low in erucic acid (less than 2 percent) and glucosinolates (less than 30 mcgmol/g of oil-free meal). Canola has the lowest percentage concentration of saturated fatty acids of eight commonly used vegetable oils. "Canola" is a name registered by the Western Canadian Oilseed Crushers Association. Producers need to be precise about the type of rapeseed they are planting, as well as how and where it will be marketed.

Canola Types. Canola consists of two species of the mustard family, *Brassica campestris*, called Polish turnip rape, and *Brassica napus*, known as Argentine rape (Figure 21-30). Both fall- and spring-planted varieties are available in both types. Michigan research has shown fall-planted varieties outyield spring plantings by 30–40 percent and can be combine-harvested directly rather than being windrowed.

Cultural Practices. Canola is susceptible to Sclerotinia wilt and should not be planted in rotation with other susceptible crops (soybean, sunflower, and dry bean) if this disease becomes a problem. Canola plantings in a field should be at least four years apart. Canola is also susceptible to white rust, downy mildew, and black spot. When direct combining, some seed are likely to shatter and produce volunteer plants the following season.

Figure 21-30 A field of canola or rape—a relative of the radish—in bloom.

Phenoxy herbicides and many commonly used corn and soybean herbicides control volunteer canola.

Canola can be grown on most soil types, but is best suited to well-drained and noncrusting loam soils. Because the seed is small, canola responds to good seedbed preparation. The seedbed should be reasonably smooth, firm, and moist within 1 inch of the surface. Seeds should be planted 3/8–3/4 inch deep. Alfalfa seeding equipment works well. Some **broadcast plantings** have died over winter because of dry soil conditions and poor seed-soil contact. Other kinds of equipment can be used. They should be used with a cultipacker. Five to seven pounds of seeds should be used per acre, depending on seed size and soil conditions.

Fertilization practices are similar to those used in wheat production, with some additional nitrogen. Nitrogen and potassium should not be placed in direct contact with the seeds, but should be broadcast. Twenty to thirty pounds of nitrogen should be applied prior to planting, with an additional 80–120 pounds applied in the winter or early spring before growth starts. Do not use a floater to apply the spring nitrogen, as many plants under the tires may be killed if they are not in the correct growth stage or weather conditions are not ideal when the fertilizer is applied.

Because canola seedlings are very sensitive to weed competition, they should be seeded in clean fields at narrow row spacings. This results in an early leaf canopy that shades or smothers weed growth. Mixtures of canola with mustard and wild garlic reduce the market value of the crop.

Canola is ripe when plants turn a straw color and seeds become dark brown. This occurs approximately July 1 for winter varieties. Combine cylinder speed should be 1/2 to 3/4 of that used for wheat. Seed moisture should be near 11 percent for direct combining. Seed moisture should be lowered to 9 percent if seeds are to be stored for a long time. Because canola shatters easily (one to five bushels per acre at harvest), volunteer plants may grow following harvest and/or the next season. Newly harvested seed may "sweat," so air movement should be provided in all canola storage bins. Also, because canola seed is small enough to pass through many bin aeration floors, something such as nylon window screen may be needed on the bin floor and other ventilation channels.

Most canola varieties presently being planted mature at midwheat harvest. Longer maturity varieties are also available. Varietal identity throughout the marketing program may become important in the future as the oil components and characteristics of present varieties are being altered genetically. These varieties will have specific, unique uses in the world oil market. Double-cropping soybean after canola is possible.

Safflower

Safflower (Figure 21-31) is grown as an alternate crop with small grains in dry land areas of north central, south central, and eastern Montana, northwest South Dakota, and western North Dakota.

Figure 21-31 A field of safflower (*Carthamus tinctorius*) in bloom. (Photo courtesy of USDA ARS)

Growing safflower in conjunction with chemical weed control is an especially effective way to clean up land infested with grassy weeds. Inclusion of safflower in dry land cropping rotations enhances subsequent yields of small grains by controlling grassy weeds and several broadleaf species that build up in small grains. Safflower also disrupts the cycle of certain small grain diseases and insects.

Safflower production is particularly applicable to dry land areas where a deep-rooted, long-season, annual crop is needed to use surplus soil water from recharge areas that otherwise contribute to saline seeps.

The same equipment used in small grain production, with minor adjustments, can be used to plant and harvest safflower.

The botanical name for safflower is *Carthamus tinctorius*. Safflower varieties should be chosen for their resistance to disease and their ability to provide the expected yield under the management system being used.

Cropping Systems. Safflower is a deep-rooted crop. Once established, it withstands periods of drought longer than other annual crops. In terms of recrop yields as a percentage of fallow yields, safflower is more adaptable for recropping than other oil seed crops.

Soil Preparation and Weed Control. Safflower is a poor weed competitor during the three- to four-week period it remains in the rosette growth stage. After the rosette stage, plants elongate rapidly, branch extensively, and compete more successfully with weeds. Unless weeds are controlled during the rosette stage, they will drastically reduce yields and may interfere with harvesting. Preplant, soil-incorporated herbicides provide long-lasting control of most annual weeds affecting safflower. Preplant herbicides must be thoroughly incorporated into the soil to at least a two-inch depth. Thorough mixing is important because these herbicides do not move in the soil and kill only germinating weeds in the treated layer. Weed seeds can germinate and become established in untreated soil or clods where the herbicide was inadequately mixed, as well as in the untreated soil below the zone of incorporation.

Seeding Rate and Depth. Seeding rates vary depending on seeding date, seeding depth, seed size, and seed quality (germination percentage). The proper rate ranges from 15 to 25 pounds per acre of pure live seeds. Light seeding rates may result in lower branching, delayed maturity, and excessive weed competition. High seeding rates generally result in thick stands with higher incidence of disease and overcrowding, especially in years of low precipitation.

For solid drill plantings, rows may be 6–14 inches apart. Row spacing greater than 14 inches permits increased air movement and penetration of sunshine into the crop canopy. This may reduce the incidence of leaf diseases but also favors weed competition, excessive lower branching, and delayed maturity. A row spacing of 14 inches or less is recommended. A 20-pounds-per-acre seeding rate planted in 12-, 10-, 8-, and 6-inch rows would require six, five, four, and three seeds per linear foot of row, respectively.

Safflower should be planted one to two inches deep in a moist, firm seedbed. Planting too deep, especially in a soil susceptible to crusting, will usually result in poor emergence. Two-inch depth bands may be used on planters to control seeding depth.

Planting Date. Safflower is quite frost-tolerant in the seedling and rosette stage and withstands temperatures as low as 20°F. Thus, safflower can be planted in April or early May. With a minimum growing season of about 120 days, early planting allows safflower to take full advantage of the entire growing season for highest yields. Growers should generally not plant after May 20 because early-September frosts and disease injury often drastically reduce seed yields and quality of late plantings. Safflower yield performance trials have shown that late planting usually results in shorter plants with less branching and lower yields, even when frost and disease injury are not a problem.

Fertilization. Fertilizer needs of safflower depend on where safflower is inserted into dry land crop rotations. In general, the limiting nutrient restricting safflower yields on recrop land is nitrogen (N). On fallow, the limiting nutrient restricting safflower yields is phosphorus (P). With good weed control, safflower grown on recrop yields between 500 and 1,500 pounds per acre. Actual yield depends upon stored soil water, growing season precipitation, and the total amount of soil and fertilizer nitrogen available. On summer fallow, safflower yields between 1,200 and 2,200 pounds per acre with good weed control.

Insects. Some insect pests may attack safflower and damage the plants. However, yields are seldom affected appreciably, unless stands are drastically reduced. Because safflower plants have the ability to produce new branches and heads to compensate for insect damage, yields are not significantly reduced until high-level infestations of insects occur over a long period. In general, it is advisable to plant safflower early to minimize damage from a late season buildup of safflower insect pests.

Wireworms, army cutworms, thrips, and lygus bugs are the most common destructive insects encountered on safflower. Other destructive insects that occasionally invade safflower fields include seed corn maggots, grasshoppers, alfalfa loopers, leafhoppers, sunflower moths, clover stem borers, and armyworms.

The most damaging spring insect pests—such as wireworms, army cutworms, and seed corn maggots—are damaging because they reduce stands by destroying germinating seeds and seedlings. Wireworms and seed corn maggots may be controlled with insecticide seed treatments. Most cutworm species may be controlled with timely insecticide spray applications.

Thrips and lygus bugs are early to middle-late season pests that cause browning, bronzing, and blasting of buds. Damage is usually most severe in irrigated and late-planted safflower. Control is not recommended for thrips unless 25–30 percent of the early buds are bronzed and blasted before the onset of bloom. Insecticides are not recommended for lygus bug control unless one lygus per eight or nine buds is present.

Large numbers of honey bees and wild bees are attracted to safflower during the flowering period because of its abundant pollen and nectar. Bees are beneficial insects that aid in pollination of safflower and improve seed set. In using insecticides against safflower pests, consideration must be given to insect numbers, stage of plant growth, harmful effect on plants and residues, and the presence or absence of beneficial insects like bees and ladybird beetles.

If chemical control is necessary, insecticides should be applied before the flowering stage, or applied late in the evening or early morning, to minimize bee kill. Bee keepers in the area should be notified in advance of any insecticide application.

Diseases. Bacterial blight caused by the bacterium *Pseudomonas syringae* and Alternaria leaf spot caused by the fungus *Alternaria carthami* are the two most serious diseases of safflower in Montana and the Dakotas.

Bacterial blight occurs in spring months when cool, wet weather prevails. Wind-driven rain enhances infection. Symptoms are dark, water-soaked lesions on stems and leaf petioles and reddish-brown spots with pale margins on leaves and severe necrosis of the terminal buds. Older lesions are whitish in appearance. As the weather becomes drier and warmer, the plants recover and resume normal growth.

Alternaria leaf spot occurs when heavy dews or rainy periods prevail during the safflower flowering period (mid-July to August). Characteristic large, brown spots, resembling fingerprints, develop on leaves. Seeds may be discolored if infection is severe. Alternaria leaf spot may also cause seed rot and damping off of untreated seed.

Other diseases of safflower noted are rust, Sclerotinia head rot, and root rots. These diseases are usually avoided by planting treated and disease-free seeds and following proper crop rotations. Safflower should never be seeded on safflower stubble because of the buildup of rust, root rots,

Figure 21-32 A sunflower, *Helianthus annuus.* (Photo courtesy of USDA ARS)

and other diseases. At least two crops should intervene between safflower crops to avoid losses from these diseases. Risk of disease or insect losses may also increase if safflower is seeded on sunflower, field bean, mustard, or rapeseed stubble.

Sunflower

The top five sunflower-producing states are Kansas, North Dakota, South Dakota, Minnesota, and Colorado. "Bird seed or confectionery-type" sunflowers were grown in north central Ohio during the early 1970s. The oil type has been predominant since 1979. Several large bird feed formulators, as well as many country elevators, are purchasing oil-type sunflower for use in bird seed mixes. In other areas sunflower as a single crop should be considered for marginal land.

The botanical name for the sunflower is *Helianthus annuus.* Figure 21-32 shows a sunflower.

Cultural Practices. Growers plant 25,000 plants per acre from May 20 to June 15 for single-crop production. They apply 75–100 pounds of nitrogen, 50 pounds of phosphorus, and 50–75 pounds of potassium per acre to produce 1,800–3,000 pounds of seeds. When planting sunflower as a double crop, they may apply 50–75 pounds of nitrogen prior to or immediately following planting. Granular nitrogen may reach the soil surface more easily than liquid, especially if large amounts of wheat stubble or organic matter are present.

Herbicides are applied according to the weeds expected. If cultivation is used for weed control in a single crop, it must be done before plants are 1 foot tall. Only a few herbicides are labeled for weed control in sunflower. However, sunflower seedlings are strongly rooted and usually not injured by rotary hoeing or other similar implements that may be needed to supplement herbicidal weed control in single-crop plantings. Herbicides registered for sunflower are often used in other crops and are familiar to most growers.

If no-till sunflowers are to be planted, either full-season or double-crop, a nonionic surfactant should be applied to burn down any existing vegetation. Growers should apply the tank mix after sunflowers are planted, but before they emerge.

A special sunflower attachment for a grain combine head is necessary to prevent large harvest losses. Harvested grain needs to be aerated for a short time to prevent heating. Seeds can be harvested at 20-percent moisture, but the market grade is set at 10-percent moisture. The use of some type of harvest aid is necessary on most single-crop flowers to dry them down and even the maturity of individual heads. These harvest aids should be applied when the sunflower heads are yellow and bracts are turning brown (seed must be physiologically mature). Sunflowers may be harvested 7–21 days after application, but should not be grazed by livestock or fed to livestock. Birds like to eat sunflower seeds as they near maturity, and this may become a problem.

Table 21-3 Sunflower Seed Production

Year	Harvested (1,000 acres)	Yield (pounds)	Production (1,000 pounds)	Dollars per Cwt	Total Value of Production ($1,000)
2007	2,009.5	1,437	2,888,555	21.30	606,991
2006	1,770	1,211	2,143,613	14.50	308,832
2005	2,610	1,540	4,018,355	12.10	487,654
2004	1,711	1,198	2,049,613	13.70	272,732
2003	2,197	1,213	2,665,226	12.10	316,214
2002	2,167	1,131	2,451,247	12.10	294,595

Source: U.S. Department of Agriculture National Agricultural Statistics Service

A producer should have an acceptable market established prior to growing sunflower. Sunflower silage approaches corn silage nutrient-wise but has lower tonnage.

Table 21-3 provides sunflower seed production data for the past five years in the United States.

SUMMARY

The major objective of a grain production system is the interception, fixation, and storage of sunlight energy. The most important components of such a system are timely planting, disease control, proper fertilization, and variety selection. Small grains specifically discussed here include barley, oats, rye, sorghum, spelt, triticale, and wheat. Rice, wild rice, and buckwheat are also discussed in this chapter. Because of their growing requirements, many of the grains have major areas of production. Small grains are used as part of a crop rotation. They may be grown for human or animal food, or both. Grains come with their unique management problems to maximize yields and combat pests.

Forage grasses capture the energy of the sun and then they are harvested directly by livestock or by mechanical methods and fed to livestock later. Forage grasses include Kentucky bluegrass, Canada bluegrass, orchard grass, perennial ryegrass, reed canary grass, smooth bromegrass, tall fescue, timothy, prairie grass, garrison grass, and Bermuda grass. Depending on how and where they will be used, the grower must select the proper type. Often grasses are mixed with each other or with legumes to increase their usefulness to producers. Grasses too must be properly managed.

Oil seed crops capture sunlight energy and store oil in their seed, which is extracted for human use. Oil seed crops include soybeans, canola, safflower, and sunflower. Soybeans are the major oil seed of the United States. Canola, safflower, and sunflower are oil seeds that offer alternatives to growers in other areas of the United States. When raising an alternative crop like canola, safflower, or sunflower, growers rely on established markets for the crop. After extracting the oil from oil seeds, the by-product, oil seed meal, is used as a livestock feed.

Review

True or False

1. Buckwheat is a close relative to the wheat plant.

2. Wild rice grows best in a subtropical climate.

3. Bermuda grass grows best in a state like Montana.

4. Canola is a type of rapeseed.

5. Endophyte may be found in timothy.

Short Answer

6. What are four factors to consider when selecting a small grain variety for growing?

7. Name the five growth stages of small grains.

8. Give the common and the scientific name for three plants grown for their oil seed.

9. Name six insect pests and the crops they affect.

10. Name three cool-season forage grasses and three cool-season small grains.

11. How was triticale developed?

Critical Thinking/Discussion

12. Explain the importance of seed quality when planting a crop such as a small grain or forage grass.

13. Compare the production of rice to that of wild rice and the production of buckwheat to that of oats.

14. Discuss the reasons for growing small grains.

15. Name five forage grasses and give their botanical names, too. Indicate how each is used for livestock feed—pastured or harvested, or both.

16. Compare the culture of canola, safflower, and sunflower for oil seed.

Knowledge Applied

1. Visit an equipment dealer and collect information and pictures of the equipment used in one of these crops: barley, oats, rye, sorghum, spelt, triticale, rice, wild rice, buckwheat, forage grasses, canola, safflower, or sunflower.

2. Make a seed collection of the seeds or other methods of propagation used for barley, oats, rye, sorghum, spelt, triticale, rice, wild rice, buckwheat, forage grasses, canola, safflower, and sunflower.

3. Develop a presentation (computer) showing the insect pests of one of the small grains—barley, oats, rye, sorghum, spelt, triticale, or rice. Show the type of damage the pests cause to the plants and indicate how the pests are controlled.

continues

Knowledge Applied, *continued*

4. Search the Internet for production guidelines for one of these crops—barley, oats, rye, sorghum, spelt, triticale, rice, wild rice, buckwheat, forage grasses, canola, safflower, or sunflower. Print the results of your search.

5. Visit a grocery store. Read the labels on bottles of vegetable oil and make notes as to the oil seed source for the product. Put this information, the product name, and the botanical name for the oil seed in a chart. Also try to find oil seeds not specifically mentioned in this chapter. For example, where would you find cotton seed oil or linseed oil being used?

Resources

Acquaah, G. (2004). *Principles of crop production: Theory, techniques, and technology* (2nd ed.). Englewood Cliffs, NJ: Prentice Hall Career & Technology.

Barnes, R. F., Nelson, C. J., Moore, K. J., & Collins, M. (2007). *Forages: The science of grassland agriculture* (6th ed.). Ames, IA: Blackwell Publishing.

Connors, J. J., & Cordell, S., eds. (2003). *Soil fertility manual*. Tucson, AZ: Potash & Phosphate Institute.

Hay, R., & Porter, J. (2006). *The physiology of crop yield* (2nd ed.). Ames, IA: Blackwell Publishing.

Jones, J. B., Jr. (2002). *Agronomic handbook: Management of crops, soils, and their fertility*. Boca Raton, FL: CRC Press.

McMahon, M., Kofranek, A. M., & Rubatzky, V. E. (2006). *Hartmann's plant science: Growth, development, and utilization of cultivated plants* (4th ed.). Englewood Cliffs, NJ: Prentice Hall Career & Technology.

Simpson, B. B., & Ogorzaly, M. C. (2000). *Economic botany: Plants in our world* (3rd ed.). New York: McGraw-Hill.

United States Department of Agriculture. (1948). *Grass: The yearbook of agriculture*. Washington, DC: Author.

United States Department of Agriculture. (1950-51). *Crops in peace and war: The yearbook of agriculture*. Washington, DC: Author.

United States Department of Agriculture. (1953). *Plant diseases: The yearbook of agriculture*. Washington, DC: Author.

United States Department of Agriculture. (1961). *Seeds: The yearbook of agriculture*. Washington, DC: Author.

Internet

Internet sites represent a vast resource of information. The URLs (uniform resource locators) for World Wide Web sites can change. Using one of the search engines on the Internet such as Google or Yahoo!, find more information by searching for these words or phrases: forage grasses, barley, oats, rye, sorghum, spelt, triticale, rice, wild rice, buckwheat, oil seeds, fungal diseases, companion crop, endophyte, legume, ratoon, winter hardy, percent purity, percent live seed, small grain, and seed quality.

Additionally, the PLANTS Database (http://plants.usda.gov/index.html) provides standardized information about the vascular plants of the United States and its territories. It includes names, plant symbols, checklists, distributional data, species abstracts, characteristics, images, crop information, automated tools, onward Web links, and references. The PLANTS database is a great educational and general use resource.

Chapter 22

Food Legumes and Forage Legumes

SOME MEMBERS *of the legume family are grown for their edible seeds. This includes dry beans, peas, and peanuts. Beans grown in the United States include navy or pea bean, dark and light red kidney, cranberry, yellow eye, pinto, and black turtle. Soybeans are also legumes grown for their seed but are not used much by humans. Soybeans were covered in Chapter 20. Legumes grown for their seed are known as pulses worldwide.*

Forage legumes are grown for their use as livestock feed. They may be grazed, harvested and stored, or both. Important forage legumes include alfalfa, red clover, bird's-foot trefoil, white clover, alsike clover, and annual lespedeza.

After completing this chapter, you should be able to—

- Describe the cultural practices for growing dry beans, peas, and peanuts

- Discuss the cultural practices for growing forage legumes

- Give the common and scientific name for five forage legumes

- Identify the scientific names for beans, peas, and peanuts

- Name six insect pests and tell which crops they affect

- Identify four criteria for selecting forage legumes

- List important pests and their control for the production of beans, pea, and peanuts

- Name four types of dry beans

- Discuss the fertilizer needs of beans, peas, and peanuts

- Compare the culture of dry beans and peas to forage legumes

- Identify the temperature or climate needed to grow beans, peas, and peanuts

- List five important pests of peanut production and describe their control

- Describe how the pod forms on the peanut plant

- Discuss the importance of crop rotation in peanut production

- Describe a fungal disease

- Describe how to maintain an alfalfa stand

- Explain the importance of seed quality to bean, pea, and peanut production

- Discuss the importance of calcium to the peanut plant

- Show the relationship of maturity—stage of blooming—to protein content of a forage legume, specifically alfalfa

BEANS

Dry beans are produced in a variety of colors and sizes (Figure 22-1). The botanical name for beans is *Phaseolus* spp. The common red kidney bean is *Phaseolus vulgaris*. Figure 22-2 shows a typical bean plant.

Soil and Climatic Requirements

The best soil for dry bean production is a loamy soil with high organic matter and good drainage. Fine-textured soils tend to be poorly aerated and susceptible to compaction problems, while coarse-textured sandy soils tend to be droughty and susceptible to wind erosion. The best bean soils are nearly level. Steeper slopes are susceptible to water erosion.

Beans need a frost-free season of 100 to 120 days, with frequent rains or proper irrigation during the period of rapid growth and plant development.

Variety and Seed Selection

Growers select high-quality, disease-free seeds. Certified seed is a dependable source of high-quality seed that has passed rigid quality standards. For colored bean types, Idaho-grown seeds can usually be considered disease-free, although they may carry mosaic virus infestation.

Figure 22-1 Beans are produced in a variety of colors and sizes. (Photo courtesy of USDA ARS)

Figure 22-2 Typical bean plant, *Phaseolus* spp. (Photo courtesy of USDA ARS)

Regardless of the source, growers should have seeds tested or ask their seed suppliers for results of disease tests, including common bean mosaic virus and bacterial blight. Regardless of blight test results, all bean seeds should be treated with streptomycin sulfate (along with an insecticide and fungicide) to control external bacterial organisms on the seed coat surface.

Planting

Growers plant beans, if possible, following corn or small grain seeded to a clover green manure crop, or after alfalfa. Planting beans after beans or after beets is not recommended. Producers try to choose fields for beans that are level or only slightly sloping, well-drained, medium to fine-textured, with good water-holding capacity. If the field has low spots where water is likely to collect after heavy rains, use a land leveler or construct open shallow-surface ditches running to an outlet to lead off excess water.

Where the soil is subject to wind erosion, some growers may seed a small grain (rye is excellent) for winter and early spring cover. They keep it mowed to a four- to six-inch height to prevent excessive moisture loss until tillage and planting. In fine-textured soils, rye can be used to dry out the ground in the spring to allow tillage under optimum moisture conditions. Growers use tillage to make the land suitable for beans to avoid soil compaction.

In already compacted soils, deep tillage is recommended. This must be done when the soil moisture content is at an intermediate level.

Soil and Fertilizer

Growers should follow soil test recommendations in using fertilizers. Beans are very sensitive to fertilizer applied in contact with the seeds. Starter fertilizer is applied 1 inch to the side and 2 inches below the seed. As a general guide, the micronutrients manganese and zinc may increase yields, particularly under high soil pH. When submitting a soil sample, these micronutrients need to be tested. Fertilizers containing boron should not be used.

Beans generally respond to nitrogen up to 40 pounds per acre, depending upon the previous cropping pattern. If a good legume stand or 10 tons of manure is applied, one can adjust nitrogen rates to 10 pounds per acre.

Planting Practices

Traditionally, the first part of June has been the preferred planting period, if soil moisture is favorable. With the development and release of full-season, direct-harvest types, a planting period can begin in late May, provided soil temperature (65°F or higher) and moisture are favorable.

Seeds should be placed at a uniform depth in moist soil approximately 1½ inches deep. Planting in dry soil or planting deep to reach moisture is not recommended. If a soil crust forms at time of emergence, a rotary hoe or other suitable farm tool may be used to break the crust. Table 22-1 provides planting rates for different classes of dry bean.

Table 22-1 Suggested Planting Rates for Field Beans

Type	Row Width	Seeds/Foot of Row	Approximate Pounds/Acre
Navy	28	4–5	40
Cranberry	28–32	3–4	60
Kidney	28–32	3–4	60
Yellow eye	28–32	4	60
Pinto	28	4	50
Black turtle	28	4–5	40

Weed Control

Herbicide should be chosen for the type of weed problem that exists. Growers develop a program for perennial weed control that involves control in nonbean years and some cultivation to hold these weeds down in the year of growing beans.

Diseases

Several serious diseases affect dry edible beans. Some of these diseases, such as bacterial blight and common bean mosaic virus, are seed-borne and can be perpetuated by planting disease-infected seeds. Others, such as Fusarium root rot and bean rust, are not perpetuated by seeds. Table 22-2 lists bean diseases and their control.

Other disease control measures include—

1. Plant disease-resistant varieties, when available, or seeds that have been tested for seed-borne diseases.
2. Plow under all bean refuse, preferably in the fall, to reduce the disease inoculum potential from the previous season.
3. Avoid conditions that favor Fusarium root rot and other **soil-borne** diseases. These conditions include poor soil aeration resulting from soil compaction or poor soil drainage, low soil temperatures, and planting where beans were grown the year before.
4. Avoid working in bean fields when they are wet.
5. Be prepared to apply agricultural chemicals when halo blight or bean rust is observed in the field.

Table 22-2 Field Bean Diseases and Their Control

Disease	Spread	Control
Halo bacterial blight	Splashing water, insects, animals, seed	Copper sprays, disease-free seed, crop rotation, seed treatment
Common and fuscous blight	Splashing water, insects, animals, seed	Disease-free seed, crop rotation, seed treatment
Common bean mosaic virus	Aphids	Disease-free seed
Root rots	Plowing, cultivation, etc.	Tolerant varieties
Bean rust	Windblown spores	Copper and sulfur
Bean anthracnose	Infected seed	Disease-free seed
White mold	Splashing water, wind	Fungicide sprays

Insect Pests

Good management that yields a clean, vigorous stand of beans will assure a minimum of problems with insect control. No special equipment or operations are needed, only those normally used in producing high yields of quality beans. Applying insecticides will require a spray rig; a granule applicator will also be needed if granular insecticides are to be applied.

Harvesting

Beans are pulled when approximately 90 percent of the leaves have fallen and the stems and pods have lost all green color. Plants are pulled early in the day when the pods and stems are tough. The first beans threshed should be monitored to adjust cylinder speed and clearance as necessary to keep splits and checks to a minimum. Weather conditions at harvest time are all-important to determine the best procedure (Figures 22-3 and 22-4).

Figure 22-3 Cutting beans is normally a two-step process. This new equipment developed by Pickett Equipment cuts the beans and windrows them in one pass over the field. (Photo courtesy of Pickett Equipment, Burley, Idaho)

Figure 22-4 After beans have dried in the field, they are threshed and transported to a bean warehouse.

Drying and Storage

If beans are harvested with more than 20–22 percent moisture, they can be stored only a few days before spoiling. At moistures of 16–18 percent, beans will store safely for several months. For long-term storage, they should be below 15 percent moisture. Beans can be dried in most commercial grain driers. If they are to be used for seed, drying temperatures should not exceed 100°F.

Marketing

Bean markets, like all other commodities, are controlled by a few very simple rules. The most important is the law of supply and demand. When supply exceeds demand, the price goes down and when demand exceeds supply, the price goes up. Although this is an oversimplification, it is the skeleton upon which the market is built.

PEAS

There are many varieties of peas (*Pisum sativum*). The processing or market will determine the variety grown. Peas are a cool-season crop, best planted in late summer or early fall, but early spring crops can be successful where temperatures are not hot during late spring. Both bush and vine types are available for both edible pod-type and regular shelling-type peas. Bush peas can be grown in most areas but vine types need the cooler, moister climate found along the west coast (Figure 22-5). Vine peas produce more and for a longer period but they require a trellis or other climbing support.

Soil

Fields should have uniform fertility, soil type, slope, and drainage to get a uniform pea crop. The best soils are silt loams, sandy loams, or clay loams. Peas need a good supply of available soil moisture, but yields may

Figure 22-5 Pea plant (*Pisum sativum*), a cool-season crop.

be reduced by overirrigating as well as underirrigating. Peas grown on wet soils develop shallow root systems that cannot supply the plant's water requirements when the soil dries out later in the season. Root rot is often a problem in wet soils. Determine corrective lime and fertilizer needs by a soil test. Growers adjust pH to 6.5 or higher for maximum yields.

Soil Temperature and Planting

Good germination will occur at 39–57°F. The land should be plowed and harrowed, and a **cultipacker** should be used lightly to ensure a firm seedbed. The land should be level in order to make harvesting more efficient. Plantings may be made as soon as the soil can be worked in the spring. Enation-resistant varieties may be planted throughout the entire planting season. Processing peas are scheduled on the basis of heat units (see Chapter 11). Planting and harvesting schedules are established by the processing company. Fresh market peas and edible-pod peas may be scheduled on the basis of heat units and by picking requirements for given plantings. In general, April plantings will require about 70 days to harvest, May plantings about 60 days, and June plantings about 55 days. Plantings should be about two weeks apart in April and early May and about one week apart from mid-May on. Growers plant enation virus-susceptible pea varieties before April 1 to reduce infection.

Seeding

Pea seeds number approximately 90–175 per ounce. Producers drill dwarf types for processing at a uniform depth of one-and-one-half to two inches into moisture, dropping three to six seeds per foot of row, with rows six to eight inches apart. They aim for a plant population of 480,000 plants per acre, and they try to avoid excessive overlaps and double planting along the edges of the field. This may cause uneven-colored peas and lack of uniformity at harvest.

A light rolling may be advantageous if moisture is adequate and not excessive. Heavy rolling or packing is likely to reduce root growth, fertilizer uptake, and pea root **nodulation**, and to increase the number of plants affected by root rot. Growers inoculate with *Rhizobium* bacteria in a planter box treatment when planting on soils on which peas had not previously been planted.

Fertilizer

Good management practices are essential if optimum fertilizer responses are to be realized. These practices include—

+ Use of recommended pea varieties
+ Selection of adapted soils
+ Weed control
+ Disease and insect control
+ Good seedbed preparation
+ Proper seeding methods
+ Timely harvest

Because of the influence of soil type, climatic conditions, and cultural practices, crop response from fertilizer may not always be accurately predicted. Soil test results, field experience, and knowledge of specific crop requirements help determine the nutrients needed and the rate of application. The fertilizer application for vegetable crops should ensure adequate levels of all nutrients. Optimum fertilization is essential for top quality and yields. Recommended soil sampling procedures should be followed in order to estimate fertilizer needs.

Nitrogen (N). Nitrogen rates depend on the soil test. Nitrogen can be banded with phosphorus (P) and possibly potassium (K) at planting time. Pea seeds should be inoculated immediately before seeding to ensure an adequate supply of nitrogen-fixing bacteria. A fresh, effective, live culture of the correct strain of *Rhizobium* should be used.

Phosphorus (P). Phosphorus is essential for vigorous early growth of seedlings. Preferably P, N, and where required, K, should be applied in a band below the seeds at planting time. When banding equipment is not available, fertilizers can be drilled with the seeds. Additional P_2O_5 and K_2O, when required, can be broadcast and plowed down prior to planting. Soil testing determines the amounts to apply.

Potassium (K). Potassium should be applied and plowed down before planting or banded at planting time as described in the previous section on P. Potassium should not be included with N and P when fertilizer is drilled with the seeds. Additional K, where required, should be broadcast and plowed down prior to planting.

Seedling injury from banded fertilizers tends to be more serious in drier soils, in coarse-textured, sandy soils where fertilizer band is close to the seeds. Phosphorus fertilizers are less injurious to seedlings than N and K fertilizers.

Sulfur (S). Plants absorb S in the form of sulfate. Fertilizer materials supply S in the form of sulfate and elemental S. Elemental S must convert to sulfate in the soil before the S becomes available to plants. The conversion of elemental S to sulfate is usually rapid for fine ground (less than 40 mesh) material in warm, moist soil. Sulfur in the sulfate form can be applied at planting time. Some S fertilizer materials such as elemental S and ammonium sulfate have an acidifying effect on soil. Sulfur is sometimes contained in fertilizers used to supply other nutrients such as N, P, and K, but may not be present in sufficient quantity.

Magnesium (Mg). Based on a soil test, magnesium might be required. Magnesium can be supplied in dolomite, which is a liming material and reduces soil acidity to about the same degree as ground limestone. Dolomite should be mixed into the seedbed at least several weeks in advance of seeding and preferably during the preceding year. An application of dolomite is effective for several years.

Lime. Peas are fairly sensitive to soil acidity and are responsive to liming of acid soils. Lime application is suggested when the soil pH is 6.0 or below, or when calcium levels are low.

Weed Control

Growers cultivate as often as necessary when weeds are small. Proper cultivation, field selection, and rotations can reduce or eliminate the need for chemical weed control.

Insect Control

Proper rotations and field selection can minimize problems with insects. Insect pests of peas are described in Table 22-3.

Table 22-3 Insect Pests of Peas

Insect	Description
Loopers (also alfalfa looper) (*Autographa californica*)	Slender, dark, olive-green worms with pale heads and three distinct dark stripes. Move in a looping manner.
Celery looper (*Anagrapha falcifera*)	Pale green with no distinct marking. Adults are gray, three-quarter-inch long with a white teardrop on both front wings.
Cutworms and Armyworms	Large grown larvae that feed on seedlings, leaves, and pods.
Grasshoppers (several spp.)	May cause considerable damage in years of grasshopper abundance.

Disease Control

Proper rotations, field selection, sanitation, spacings, fertilizer, and irrigation practices can reduce the risk of many diseases. Fields can be tested for the presence of harmful nematodes. Using seeds from reputable seed sources reduces risk from seed-borne diseases. Some diseases of peas include bacterial blight, stem rot, downy mildew, enation mosaic virus, leaf roll, powdery mildew, root rot, seed rot, and mosaic virus. These diseases are described in Table 22-4.

Harvesting, Handling, and Storage

The processor determines time of harvest according to the tenderometer reading, the number of other fields ready for harvest, weather, soil conditions, and the processor's need for quality. Generally, yields of shelled peas increase with increasing maturity, but quality decreases with maturity. With mobile viners the crop is cut and swathed into windrows, threshed out by the mobile viners following swathers. Peas must be delivered to the processing plant soon after harvest, especially when the weather is hot, to avoid "off" flavors. With the new pod-stripping harvesters, no swathing is needed.

Storage.   Green peas tend to lose part of their sugar content, on which much of their flavor depends, unless they are promptly cooled to near 32°F, relative humidity 90–95 percent, after picking. Hydrocooling is the preferred method of precooling. Peas packed in baskets can be hydrocooled from 70–34°F in about 12 minutes when the water temperature is 32°F.

Table 22-4 Diseases of Peas

Disease	Causative Agent/Vector	Signs	Control
Bacterial blight	Bacterium—*Pseudomonas Syringae pv. pisi*	Small, water-soaked spots on leaves, pods, stems; lesions coalesce and leaves become brown and die.	Rotate crops; use disease-free seed.
Basal stem rot	Fungus—*Phoma medicaginis var. pinodella*	Black, sunken lesions from soil line up to about 6 inches; plants wilt and die at any stage of development.	Rotate crops no vetch or alfalfa in rotation; use disease-free seed; treat seed; use spring-planted dry peas.
Black stem rot	Fungus—*Mycospharella pinodes*	Develops during winter causing blackening and killing of stem tissue; spotting of leaves, stem, and flower parts; seeds may be killed completely or survive in weakened condition.	Rotate crops, no vetch in rotation; use disease-free seed; remove diseased pea straw after harvest by burning; treat seed; plant on well-drained soil.
Downey mildew	Fungus—*Peronospora pisi*	Downey white to violet-colored fungus on lower leaf surface; leaves turn yellow and die; stems distorted and stunted; brown blotches on distorted pods.	Crop rotation to non-legumes; use disease-free seed; remove diseased pea straw after harvest by burning; plant on well-drained soil; apron seed treatment.
Enation mosaic virus	Virus—Pea aphid, *Acyrthosiphon pisum* Green peach aphid, *Myzus persicae*	Mottling, crinkling, stunting, followed by proliferation of tissue outgrowths on leaves and pods; pods badly distorted.	Use disease-resistant seed varieties.
Pea leaf roll	Virus—Bean leaf roll	Chlorotic mottling in terminal foliage, which becomes bright yellow; stunting; death of plants; yields reduced.	Use disease-resistant seed; insecticide applications may help.
Powdery mildew	Fungus—*Erysiphe polygoni*	Powdery white mycelium and spores on surface of leaves and stems; some varieties of peas may die; dwarfing; some varieties develop small brown to black necrotic spots on infected pods.	Use sulfur dust or wettable sulfur; some varieties resistant to disease; no serious losses when planted in the spring.
Root rot	Fungus—*Fusarium* sp. *Pythium* sp.	Plants are yellow and cease to grow rapidly; gray, reddish, or black lesions on lower stem and roots; taproot may be badly rotted.	Rotation; plant in well-drained soil; do not overirrigate; seed in spring when soil temperature is below 65°F; can use a fungicide for *Pythium* sp.
Seed rot	Fungus	Seeds fail to emerge; emerged seedlings often rot at soil line.	Treat seed with Captan at one teaspoon per pound of seed or Arasan at one teaspoon per pound of seed; problem reduced by copper drench of one ounce of copper fungicide in three gallons of water.
Mosaic virus	Virus—Pea aphid, *Acyrthosiphon pisum* Green peach aphid, *Myzus persicae* Potato aphid, *Macrosiphum euphorbiae*	Slight stunting; slight rosetting of upper leaves; striking downward roll of leaves; pods may have mottled appearance; affected seed may show slight brownish staining and seed coat rupture.	Isolate introduced pea cultivars; monitor principal or all seed lots for virus; apply systemic aphicides; use pea germ plasm resistant to virus.

Vacuum cooling also is possible, but the peas must be prewetted to obtain cooling similar to that by hydrocooling.

After precooling, peas should be packed with crushed ice (top ice) to maintain freshness and turgidity. Adequate use of top ice provides the required high humidity (95 percent) to prevent wilting. The ideal holding temperature is 32°F. Peas cannot be expected to keep in salable condition for more than one to two weeks even at 32°F unless packed in crushed ice. With ice, the storage period may be extended perhaps a week. Peas keep better unshelled than shelled. Also, researchers have demonstrated that the edible quality of green peas was maintained better when the peas were held in a modified atmosphere of 5–7 percent carbon dioxide at 32°F than in air for 20 days.

Packaging

Fresh market peas are hand harvested and the pods are commonly packaged in 30- to 32-pound bushel wirebound crates, or 28- to 30-pound bushel baskets.

PEANUTS

The scientific name for the peanut is *Arachis hypogaea*. Figure 22-6 shows a typical peanut plant with the pods underground. Peanut yields and quality rise and fall each year based on weather conditions during the growing season. Moisture and temperature are the two weather factors that have the most impact on crop yields.

Major and minor peanut-growing areas of the United States are indicated in Figure 22-7.

Germination and Sprouting

The germination process begins when soil temperatures are above 60°F and viable and nondormant seeds absorb about 50 percent of their weight in water. For practical purposes, soil temperatures need to be above 65°F for germination to proceed at an acceptable rate. Large-seeded Virginia-type peanuts planted under favorable moisture and temperature conditions will show beginning radicle (root) growth in about 60 hours. If conditions are ideal, sprouting of young seedlings should be visible in seven days for smaller-seeded varieties and ten days for larger-seeded varieties.

Root Growth

The peanut root grows rapidly following germination. The root tip grows downward in the soil through cell division and enlargement. By the time the main shoot breaks through the soil, the taproot will be 5–6 inches deep. Lateral roots develop from the taproot and may be 1–2 inches long at seedling emergence. The cotyledons supply the food for early root growth. Once the vegetative tissues are producing a food

Figure 22-6 Peanut plant showing peanuts (*Arachis hypogaea)* at the ends of the pegs. (Photo courtesy of USDA ARS)

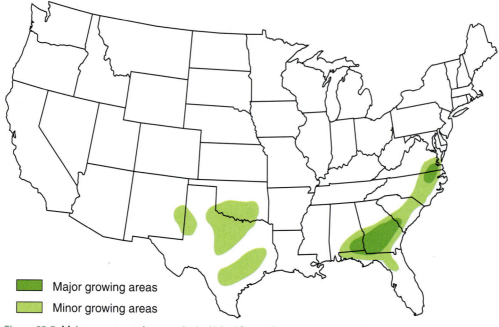

Figure 22-7 Major peanut-growing areas in the United States. (Image courtesy of USDA NOAA)

Major growing areas
Minor growing areas

supply (photosynthate) greater than their needs, some of the food is transported to the roots.

Any condition that subjects peanut roots to stress should be avoided so that adequate root capacity exists to supply the plant with moisture and nutrients. Lack of an adequate and efficient root system may limit yields.

Vegetative Growth

Vegetative growth is slow for the first three to four weeks, as leaf tissue is limited. As leaves develop and fully expand, the capacity for photosynthate production increases and vegetative growth is more rapid. In early growth stages, cool temperatures cause slow growth. As temperatures rise, vegetative growth increases rapidly. Scientists have determined that optimum peanut growth occurs at 86°F. Excessively high (above 95°F) or low (below 60°F) temperatures will slow growth.

High temperatures are often accompanied by drought conditions, and vegetative stress can be observed as visible wilting of the plants. Irrigation alleviates the stress of high temperatures or drought.

Other stress factors reduce the photosynthetic efficiency of the plant and limit yields. Leaf or stem diseases, insect infestations, and chemical phytotoxicity may reduce the vegetative surface area, slow photosynthate production, and reduce plant growth.

Reproductive Growth

Peanuts are indeterminate in growth habit. An indeterminate growth habit means that vegetative and reproductive growth occurs simultaneously. The plant must produce enough food to continue vegetative growth, as well as provide food for seed development.

The first flower develops about 30–40 days after emergence. Daily flower production is slow for the next two weeks. By mid-July, dozens of flowers may be visible each day on each plant. Temperature (about 86°F) and moisture must be favorable for flowering to continue at a steady rate. Any environmental stress may interrupt normal flowering.

Pollination and fertilization occur quickly in the flower. If fertilization of the ovule is successful, the **peg** begins to grow downward. It takes about ten days for the peg to penetrate into the soil. A week later, the peg tip enlarges and pod and seed development begins. With favorable temperatures, nutrient, and moisture conditions, mature fruit develops in nine to ten weeks (Figure 22-8).

Figure 22-8 Pollinating a peanut flower. (Photo courtesy of Texas Agricultural Extension Service)

Weather or biotic stresses may interrupt normal flowering and fruit development. High or low temperatures and big differences between day and night temperatures may stop flowering and fruit development. Drought conditions will slow growth. Diseases, insects, or weeds that effectively reduce leaf or root surface area may result in slower maturation.

Variety Selection

Yield and quality are two major factors that influence variety selection. Growers with significant disease or insect history may need to choose a variety with disease or insect resistance. Planting two or more varieties with different maturity dates permits efficient use of limited harvesting and curing equipment. Planting varieties with different genetic pedigrees reduces the risk of crop failure because of adverse weather or unexpected disease and insect epidemics.

Selecting and Managing Soil

Peanuts are best adapted to well-drained, light-colored, sandy loam soils. These soils are loose, friable, and easily tilled with a moderately deep rooting zone for easy penetration by air, water, and roots. A balanced supply of nutrients is needed because peanuts do not usually respond to direct fertilization.

Soil pH should be in the range of 5.8–6.2. Peanuts grown in favorable soil conditions are healthier and better able to withstand climatic and biotic stresses.

On most peanut farms, it is not possible to plant peanuts every year in the most suitable soil types. A grower plans the long-range use of fields considering rotations and disease, insects and weed problems, along with all crops grown on the farm. Maximum income normally results when the best balance is found between all crop factors.

Growers should maintain soil maps to understand the potential of each field for crop production. Records over several years will help determine the suitability of fields for peanuts and other crops. This information will help identify the limiting factors of all soil resources.

Crop Rotation

A long crop rotation program is essential for efficient peanut production. The peanut plant responds to both the harmful and beneficial effects of other crops grown in the fields. Research shows that long rotations are best for maintaining peanut yields and quality. A three-year rotation with two years of grass-type crops has been effective in reducing nematode and soil-borne disease problems and permiting better control of many weeds. These crops respond to heavy fertilization but leave adequate residual nutrients for healthy peanut growth.

Peanuts following tobacco, soybeans, or other legumes and some vegetable crops generally have higher disease losses than peanuts following corn, grain sorghum, or small grains. Cotton is a good rotational crop except that the taproot and other plant residue are difficult to manage. If peanuts must be planted directly behind cotton or tobacco, the taproots should be ripped up and shredded early in the fall to allow for maximum decomposition before land preparation.

Collar rot is frequently more severe in a rotation with cotton. A small grain cover crop can be planted to reduce the disease and nematode problems and to lessen the possibility of soil erosion during the winter. Growers with a short rotation program can expect a long-term buildup of diseases such as Southern stem rot, black root rot, and Sclerotinia blight.

Fertilizing

Peanuts respond better to residual soil fertility than to direct fertilizer applications. For this reason, the fertilization practices for the crop immediately preceding peanuts are extremely important. Grass-type crops generally respond well to direct application of fertilizer. Growers can fertilize these crops for maximum yields and, at the same time, build residual fertility for the following crop of peanuts. Peanuts have a deep root system and are able to utilize soil nutrients that reach below the shallower root zone of the grass-type crops.

The peanut crop is usually produced without applying any fertilizer materials during the production year. If peanut fields need fertilizers, they should be broadcast before land preparation. Fertilizing peanuts requires an understanding of the growth characteristics and nutrient needs of the plant and the ability of the soil to provide these needs.

Lime. Peanuts grow best on soils limed to a pH of 5.8–6.2, provided other essential elements are in balance and available to the plant. Dolomitic limestone is the desired liming material as it provides both calcium and magnesium.

Nitrogen. Roots of peanuts can be infected by *Rhizobium* bacteria. Nodules form on the roots at the infection sites. Within these nodules, the bacteria can convert atmospheric nitrogen into a nitrogen form that can be used by plants. This symbiotic relationship provides sufficient nitrogen for peanut production if the roots are properly nodulated. Direct applications of nitrogen to peanuts are not generally needed.

Growers should inoculate their peanut seed or fields to ensure that adequate levels of rhizobia are present in each field. Commercial inoculants are available that can be added to the seeds or put into the furrow with the seeds at planting. In-furrow inoculants are available in either granular or liquid form.

Potassium and Phosphorus. The most efficient and easiest way to apply potassium is to apply it to the crop preceding peanuts. This will usually increase the yield of the preceding crop and provides time for the potassium to leach downward into the area where the peanut root system obtains most of its nutrients. Generally, fields testing medium or higher in a soil test do not need additional potassium applied for peanut production.

Once a medium or higher level of phosphorus is achieved, it remains quite stable over a number of years. The addition of phosphorus-containing fertilizer to peanuts is not needed if it is applied to other crops in the rotation.

Calcium. Perhaps the most critical element in the production of large-seeded Virginia-type peanuts is calcium. Lack of calcium uptake by peanuts results in "pops" and darkened plumules in the seeds. Seeds with dark plumules usually fail to germinate.

Calcium must be available for both vegetative growth and pod growth. Calcium moves upward in the peanut plant but does not move downward. Thus, calcium does not move through the peg to the pod and developing kernel. The peg and developing pod absorb calcium directly, so it must be readily available in the soil solution.

Adequate soil calcium is usually available for good plant growth, but not for pod development or good-quality peanuts. It is important to provide calcium in the fruiting zone through gypsum applications. Gypsum should be applied to all peanuts, regardless of the soil type or soil nutrient levels. The calcium supplied through gypsum application is relatively water-soluble (compared with other calcium sources) and more readily available for uptake by peanut pegs and pods.

Each pod must absorb adequate calcium to develop normally. Gypsum is available in three forms—finely ground, granular, and phosphogypsum.

Several additional by-product gypsums are now on the market. The by-product materials vary in elemental calcium content. Studies show that all forms of gypsum are effective in supplying needed calcium when used at rates that provide equivalent calcium levels in the fruiting zone.

The use of gypsum on large-seeded peanuts is very effective in improving peanut seed quality and grades. Some research data indicate that high rates of gypsum may control or reduce the pod rot disease complex. Gypsum should not be broadcast before land preparation or before planting. This practice may result in the calcium being leached below the fruiting zone if there is high rainfall.

The best results are obtained when gypsum is applied in late June or early July. The availability of calcium supplied by gypsum application is also influenced by the amount of rainfall. Moisture is needed to solubilize gypsum and make the calcium available to the peanut fruit. In unusually dry years, peanuts may show symptoms of calcium deficiency, even when recommended rates of gypsum are applied.

Manganese and Boron. Two other elements often found to be deficient in peanuts are manganese and boron. Manganese deficiency usually occurs when soil is overlimed. Increasing the soil pH reduces the plant's uptake of manganese. Interveinal chlorosis is a symptom of manganese deficiency. A deficiency can be corrected by a foliar application of manganese sulfate. Boron plays an important role in kernel quality and flavor. Boron deficiency may occur in peanuts produced on deep, sandy soils. Deficient kernels are referred to as having hollow hearts. The inner surfaces of the cotyledons are depressed and darkened and are graded as damaged kernels.

Land Preparation

Historically, peanut growers have used the moldboard plow equipped with trash covers to prepare a smooth, uniform, and residue-free seedbed for planting. The burial of old crop residue and weed seeds has been effective in the long-term suppression of soil-borne diseases and short-term suppression of some weed problems.

Land preparation begins with the disposal or management of the crop residue from the previous crop. In order to promote decomposition during the winter, crop litter should be chopped or shredded and disked lightly. Seeding a cover crop can help reduce soil erosion during the winter months. Applications of lime, phosphorus, and potassium can be applied at this time if needed.

Most peanut soils should be broken in the spring with a moldboard plow. The plow should have the necessary attachments to bury litter. Chisel plowing may be acceptable in fields with light crop residue. Chisel plowing, as a rule, can be used for other crops in the peanut rotation.

Harrowing to break up clods and leave a smooth, firm, and clean seedbed usually follows plowing. Many peanut growers are bedding their peanut fields either in the fall or spring. Beds should be prepared several weeks before planting to allow them to warm up and gain uniform moisture levels. Many growers prefer planting on raised beds rather than

George Washington Carver

George Washington Carver was born of slave parents on a plantation in Missouri in 1864. When he was six weeks old, he and his mother were captured by a band of raiders and taken to Arkansas. His master, Moses Carver, bought him back from his captors for a racehorse valued at $300, but his mother was never heard of again.

The boy attended a one-room school about eight miles from the plantation. Later he worked and attended other schools when and where he could, and finally completed high school in Kansas. He worked, saved his money, and set out for an agricultural college that had accepted his application. But when he arrived and the school administrators saw his color, they refused to admit him. He worked for awhile in Kansas and then went to Simpson College in Iowa. After a year there, he entered Iowa State College, where he received his Bachelor of Science degree at the end of three years. He worked as assistant station botanist at the college and continued his studies leading to a Master of Science degree in 1896.

That year Booker T. Washington invited him to join the faculty of Tuskegee Institute as Director of Agriculture. He organized the agriculture department there and planned the first building devoted entirely to agriculture.

At Tuskegee he was able to carry on his research—research aimed at making life more abundant for people everywhere, but especially for those of his own South. Farmers appealed to Dr. Carver for help when the boll weevil threatened total destruction of the cotton crop in Alabama. He told them to raise peanuts. Then he set about to find new uses for the peanuts so that the farmers would be able to sell them. He developed about 300 products from peanuts, including peanut milk, plastics, stains and dyes, lard, flour, soap, and cooking oils. He developed more than 118 products from the sweet potato, including mucilage, starch, molasses, dyes, ink, and vinegar. He also developed new products from cotton, soybeans, wastes, and clay.

George Washington Carver died on January 5, 1943. He bequeathed his entire estate, amounting to more than $60,000, to the George Washington Carver Foundation, where it has been used as the beginning of an endowment fund.

flat planting. These beds often give faster germination and early growth, provide drainage, and may reduce pod losses during digging.

Planting

Planting dates for peanuts should be decided according to field conditions and expected weather patterns rather than calendar dates. As a general rule, peanuts should be planted as soon as the risk of a killing frost is over. Varieties require from 150 to 165 days to reach full maturity. Early plantings usually give higher yields, more mature pods, and permit earlier harvesting.

Peanuts should not be planted until the soil temperature at a four-inch depth is 65°F or above for three days when measured at noon. Favorable weather for peanut germination should also be forecast for the next 72 hours after planting. The soil should be moist enough for rapid water absorption by the seeds. The planter should firm the seedbed so there is good soil-to-seed contact.

Seeding rates vary considerably. Seeding rates may vary according to seed quality, but most growers seed at rates exceeding 100 pounds of seeds per acre. Growers should average planting three to four seeds per foot of row. Growers using air planters should be able to reduce seeding costs by using seed spacing of three to four inches.

Disease

A long rotation is the single most powerful disease management tool available to peanut growers. Resistant varieties are usually the least expensive procedure. Proper rotation and variety selection delays or prevents the onset of most serious disease problems. Once diseases become established, expensive chemicals often must be used.

The first step in controlling peanut diseases, which occur in spite of using long rotations and resistant plants, is to correctly identify diseases and their abundance. This can be accomplished by scouting the fields each week to monitor disease levels.

The second step is to determine the best control methods for the identified problems. Cultural and chemical controls are usually used in combination for maximum benefit. Cultural control methods such as rotation, resistant varieties, and crop residue destruction have the general effect of reducing the number of disease-causing organisms (pathogens) such as nematodes and fungi.

Pesticides are only useful if cultural practices have not sufficiently reduced pathogen levels below a certain threshold. Thresholds are levels of disease or weather favorable for disease that represent an economic threat to the crop.

Diseases of the peanut plant include peanut leaf spot, web blotch, and pepper spot.

Peanut Leaf Spot. Peanut leaf spot is caused by two different fungi—*Cercospora arachidicola* (early leaf spot pathogen) and *Cercosporidium personatum* (late leaf spot pathogen). It is often difficult to distinguish between these two diseases, particularly as symptoms vary depending on the variety grown. Early leaf spot has light-brown lesions, generally surrounded by a yellow halo, and can be found as early as June 1. Late leaf spot has darker spots, usually without a halo, and appears later in the season.

Web Blotch. Web blotch (*Phoma arachidicola*) is a sporadic problem that can be very serious. A large (½ inch), circular, dark area forms on the upper surface of the leaf, which dries and cracks as it ages. Severe defoliation often occurs in portions of a field but can spread over the entire field in a short time.

Pepper Spot. Pepper spot (*Leptosphaerulina crassiasca*) is present every year in all fields as very small dark lesions that "pepper" a leaf. Occasionally, lesions will fuse and kill large areas, resulting in a scorch symptom. Pepper spot has been associated with severe vine decline, which sometimes occurs after a heavy, late-season rain.

Phytotoxicity is caused by systemic insecticides applied at planting and is often confused with leaf spot. Chemical toxicity symptoms will usually be around the margins of the leaflets. These symptoms will not spread or form spores.

Fungicides for leaf spot diseases should be applied in one of two ways—

1. A set calendar 14-day schedule
2. A weather-based leaf spot advisory schedule

The weather advisory is just one component of making a spray decision. The decision to spray should be in the context of all available information. Considerations when using this program are as follows:

1. Spray as soon as possible after conditions become favorable, but not within 14 days of the last spray.
2. Try to spray before rain (if it has been ten days or more since the last spray).
3. Scout all fields at least once a week; revert to the 14-day spray schedule if any place in the field has over 20 percent leaf spot (one of five leaflets with a spot).
4. If spraying within three days of the advisory is impossible, then spray preventively (i.e., a 14-day schedule).
5. Chlorothalonil is the best fungicide to use on leaf spot.
6. If late leaf spot, web blotch, or heavy pepper spot is identified, revert to a 14-day spray schedule (with chlorothalonil).
7. Growers should begin listening to the advisory on June 20 and spray when the first advisories are issued.
8. If possible, try to avoid the use of chlorothalonil in fields with a history of Sclerotinia blight.
9. Use leaf spot-resistant varieties.
10. Rotate at least two years without peanuts.

Soil-Borne Pathogens and Diseases. Most soil-borne pathogens can easily survive one winter in sufficient numbers to infect a susceptible crop the next spring. Therefore, mapping the location and intensity of soil-borne diseases is an effective tool for deciding where certain cultural practices and/or chemical treatments will be applied the next time peanuts are grown. Numbers of these organisms eventually reach low levels when long rotations are used so that little or no plant disease or loss occurs. Some of these soil-borne pathogens include nematodes, fungi, and bacteria.

- *Nematodes.* Peanut nematodes include the northern root-knot (*Meloidogyne hapla*), peanut root-knot (*Meloidogyne arenaria*), lesion (*Pratylenchus brachyurus*), ring (*Criconemella ornata*), and sting (*Belonolaimus longicaudatus*) nematodes. Nematodes cause plants to be stunted, wilted, or off-colored. Nematode damage can also result in increased susceptibility to black root rot, which is discussed in detail later.
  - Root-knot nematode damage on the roots and pods may show up either as small, round lesions about the size of the head of a pin (*M. hapla*) or large swellings or galls (*M. arenaria*). Observing pods or roots for galls after digging may give a good indication of where a root-knot problem may be found.
  - Nematodes are often found in spots or small areas of fields. For the best use of nematicide dollars, fields should be treated for nematodes on a field-by-field or part of a field basis rather than treating the entire crop.
  - Planting crops that do not support the growth and reproduction of the nematode(s) found in a field reduces nematode numbers. Long rotations are the most effective method of controlling nematodes and can be used instead of nematicides. Table 22-5 shows some good and bad crop rotations for peanuts to prevent nematode problems.
- *Seed and seedling rots (many fungi).* Seed and seedling rots are caused by many different soil pathogens. Seeds will either not germinate (seed rot), germinate but not penetrate the soil surface (pre-emergence damping-off), or die shortly after emergence (postemergence damping-off). The result is a poor stand with skips. The primary problem often is either environmental (poor seedbed conditions) or poor seed quality, rather than disease. Seed and seedling diseases can usually be prevented by using high-quality seeds (greater than 70 percent germination) coated with a

Table 22-5 Rotational Crops for Nematode Control in Peanuts

Nematode	Root-knot	Ring	Lesion	Sting
Rating	Good	Good	Good	Good
Rotation	Cotton Fescue Small grains Corn	Cotton Tobacco Turfgrass	Corn Nursery stock	Small grains
Rating	Bad	Bad	Bad	Bad
Rotation	Legumes Tobacco Peanut	Corn Grass Peanut Soybeans	Peanut Soybeans Tobacco Legumes	Peanut Cotton Soybeans

good chemical seed treatment fungicide. No chemicals are available for postsymptom treatment.

+ *Southern stem rot (white mold).* Southern stem rot (caused by *Sclerotium rolfsii*) is found in all peanut counties of North Carolina. Limbs normally have white, stringy fungus growth and yellow-to-brown birdshot-sized balls (sclerotia) on the lower stems and leaf litter. These sclerotia distinguish southern stem rot from the other soil-borne diseases. Long rotations are preferable because the sclerotia, which can grow and infect healthy plants, live for a number of years in the absence of a host crop.

+ *Sclerotinia blight.* Sclerotinia blight (caused by *Sclerotinia minor*) is normally found in the most northerly counties of North Carolina and extensively throughout Virginia. This disease starts by killing individual limbs rather than causing an overall wilt.

  Fields with a history of serious problems should be treated when Sclerotinia blight is first observed or by the end of July, whichever comes first. Fields with a history of little or no Sclerotinia blight should not need to be chemically treated for this disease.

+ *Cylindrocladium black rot (CBR), or black root rot.* CBR (caused by *Cylindrocladium crotalariae*) is found in all peanut counties in North Carolina. Plants become light green and die as a whole, although some limbs may die before others. A blackened, rotting root system, which allows plants to be easily pulled from the ground, is characteristic of this disease. Long rotations help to reduce the amount of fungus in the soil. Peanut following peanut or soybean rotation is a formula for disaster.

  An important key in CBR management is the use of a resistant variety, which will have less disease and more yield than other varieties when planted in a field with a CBR problem. Root-knot nematode makes the disease worse on all peanuts. Ring nematode affects all but resistant varieties in the same way.

+ *Rhizoctonia limb and pod rot.* Caused by *Rhizoctonia solani*, this disease commonly infects peanut limbs, pegs, pods, and occasionally leaves. Young seedlings can also be affected by this disease on stem tissues just below the soil surface. Typically, small sunken lesions, light to dark brown with banded zones (target effect) form on stems at the soil surface. This disease is most common in moist fields or where the vines are thick. Irrigated fields are most severely attacked. Growing any crop other than peanut for at least two years between peanut crops will help.

+ *Botrytis blight* (Botrytis cinerea). With this blight stems have abundant gray mold and spores. Leaves may develop dark lesions. Botrytis blight often follows mechanical injury like tractor tire and freezing, when moisture is abundant.

+ *Pod rot (many fungi).* Pod rot can be a difficult disease to control because the causes are so diverse. Long rotations usually have a very positive effect. Some pod rot symptoms can be the result of poor calcium nutrition or excessive magnesium or potash levels, which weakens the hull, allowing various soil fungi to grow into and rot the pod.

+ *Tomato spotted wilt virus.* This disease has now spread to all peanut growing states. Once established in an area on wild hosts, outbreaks can occur in any season. Spread is accomplished primarily through infected thrips, which obtain the virus by feeding on infected plants (cultivated or wild plant species). Symptoms on peanut can be quite variable but usually include one or more of the following—stunting, pale yellow or white ring patterns on leaves, stunted small and malformed growth, undersized pods, and red seed coats. On some occasions, the whole plant may turn light green, resembling CBR disease.

Harvesting

The best-tasting peanut is a mature peanut. Growers must harvest and deliver mature peanuts to the market. Maturity affects flavor, grade, milling quality, and shelf life. Not only do mature peanuts have the quality characteristics that consumers desire, they are worth more to the producer. The indeterminate fruiting pattern of peanuts makes it difficult to determine when optimum maturity occurs. The fruiting pattern may vary considerably from year to year, mostly because of different weather conditions. Growers check each field before digging begins.

Heat units or growing degree days have been evaluated as a means of determining maturity. Research shows that 2,600 growing degree days are needed for the earliest varieties to mature (see Chapter 11).

FORAGE LEGUMES

The forage industry plays a major role in agriculture. Forage crops are environmentally friendly. They reduce soil erosion, improve soil tilth, help reduce pesticide use, and enhance agricultural profitability. The forage legumes include alfalfa, red clover, bird's-foot trefoil, white clover, alsike clover, and annual lespedeza.

Species Selection

The selection of forages for hay, pasture, and conservation is an important decision, requiring knowledge of both agronomic characteristics and potential feeding value of forage plants. The intended use of forages, dry matter, and nutritional requirements of livestock to be fed, seasonal feed needs, harvest and storage capabilities, and seasonal labor availability influence which species to grow. Tables 22-6 and 22-7 compare some characteristics of forage legumes.

Pure Stands versus Mixtures

The decision to establish a pure stand or a mixture of two or more species should be made before deciding which species to plant. Advantages of pure grass or legume stands are simpler management, more herbicide options, and greater forage quality potential. Pure legume stands decline in

Table 22-6 Suitability of Perennial Forage Legume Species to Different Types of Management and Growth Characteristics

Species	Frequent, Close Grazing	Rotational Grazing	Stored Feed	Periods of Primary Production	Relative Maturity[4]
Alfalfa	NR[1]	S[2]	HS[3]	Spring, summer, early fall	Early-medium
Red clover	NR	S	HS	Spring, summer, early fall	Medium-late
Bird's-foot trefoil	NR	HS	HS	Spring, summer, early fall	Medium-late
White clover, dutch	HS	HS	NR	Spring and fall	Early-medium
White clover, ladino	NR	HS	S	Spring, early summer, fall	Early-medium
Alsike clover	NR	S	S	Spring, early summer, fall	Late

1. Not recommended
2. Suitable
3. Highly suitable
4. Relative time of flower or seedhead appearance in the spring; depends on species and variety

Table 22-7 Agronomic Adaptation and Characteristics of Forage Legumes

Forage Species	Minimum Adequate Drainage[1]	Tolerance to pH < 6.0	Minimum Adequate Soil Fertility	Drought Tolerance	Persistence	Seedling Aggressiveness	Growth Habit
Alfalfa	WD	Low	High-medium	High	High	High	Bunch
Red clover	SPD	Medium	Medium	Medium	Low	High	Bunch
Bird's-foot trefoil	SPD	High	Medium	Medium	Medium	Low	Bunch
White clover, dutch	PD	Medium	Medium	Low	High	Low	Spreading
White clover, ladino	PD	Medium	High-medium	Low	High	Low	Spreading
Alsike clover	PD	High	Medium	Low	Low	Low	Speading

1. Minimum drainage required for acceptable growth: WD = well-drained; SPD = somewhat poorly drained; PD = poorly drained

forage quality more slowly with advancing maturity than grasses, providing a wider window of opportunity for harvesting good quality forage.

Legume-grass mixtures are common and have the potential to exploit the relative strengths of grasses and legumes. Grass-legume mixtures are often higher yielding and have more uniform seasonal production. Including legumes in a mixture reduces the need for nitrogen fertilizer, improves forage quality, and reduces the potential for nitrate poisoning and grass tetany compared with pure grass stands.

Including grasses in a mixture usually lengthens the usable life of a stand because they persist longer and are more tolerant of mismanagement than legumes. Grasses also reduce the incidence of bloat, improve hay drying, are usually more tolerant of lower fertility, and are stronger competitors

Figure 22-9 The alfalfa plant, *Medicago sativa.* Reproduced with permission from N. L. Britton and A. Brown, *An Illustrated Flora of the Northern United States, Canada and the British Possessions* (Vol. 1: 672.1913.) Courtesy of Kentucky Native Plant Society.

with weeds than legumes. The fibrous root system of grasses helps reduce erosion on steep slopes and reduces legume heaving. Growing grasses and legumes together often reduces the losses from insect and disease pests. Mixtures are generally more satisfactory for pastures than pure grass or pure legume stands.

ALFALFA

Unlike red or white clover, established alfalfa (*Medicago sativa*) is productive during midsummer except during extreme drought. Alfalfa is a taprooted crop and can last five years and longer under proper management. Whether grazed or fed as hay, alfalfa is excellent forage for cattle and horses. Figure 22-9 is a drawing of the alfalfa plant.

Once established, good management practices are necessary to ensure high yields and stand persistence. These practices include timely cutting at the proper growth stage; control of insects, diseases, and weeds; and replacement of nutrients removed in the forage. Alfalfa has superior forage quality when managed properly. The major problems are getting a stand and keeping it productive.

Site Selection and Soil Fertility

Alfalfa is best adapted to deep, fertile, well-drained soils with a salt pH of 6.0–6.5, but it can be grown with conservative management on more marginal soils. On sites that have more moderate drainage, growers should also seed a grass, such as orchard grass or bromegrass, with alfalfa to reduce winter heaving of the alfalfa. The grass acts as mulch during winter to reduce variations in soil temperature, which cause repeated freezing and thawing. Grasses also help prevent weed invasion by filling in spaces between alfalfa crowns.

Alfalfa requires high levels of fertility, especially phosphorus, for establishment. Soil should be tested 6–12 months ahead of planting to determine proper amounts of fertilizer and agricultural lime for successful establishment. Soil salt pH should be 6.0 or above, which allows for good nodulation by the plant roots. Growers should disk or plow down any needed limestone 6–12 months before seeding to give time for it to react in the soil and raise the salt pH.

In no-till seedings, growers should apply needed lime a year in advance because it cannot be incorporated. After two years of production, the producer should take another soil sample to determine if the soil needs additional limestone or fertilizer. Top-dressing limestone or fertilizer helps maintain production potential and ensures stand longevity.

Adequate available phosphorus is a key to establishing a vigorous stand of alfalfa. Phosphorus stimulates root growth for summer drought resistance, winter survival, and quick spring growth.

Nitrogen and potash are not as important as phosphorus for alfalfa establishment, but they are needed in small amounts. Soil test recommendations normally suggest 20–30 pounds of nitrogen at seeding along with

20–60 pounds of potash. The fertilizers should be worked into the soil to prevent direct contact with germinating seed.

Research suggests applying 20–30 pounds nitrogen for fall and early spring plantings to stimulate growth before development of nitrogen-fixing root nodules. Growers should not fertilize late spring seedings with nitrogen because of the potential for increased weed competition.

Variety Selection

Several varieties of alfalfa are available, but a limited number are adapted to certain areas. There is no single "best" variety for a particular location. The most recommended varieties are those that are consistently high yielding, moderately winter hardy, and have moderate or higher resistance to bacterial wilt, phytophthora root rot, and anthracnose.

Establishment

Alfalfa may be frost-seeded, broadcast, no-tilled, or drilled into a prepared seedbed. Growers frost-seed in January or February to allow freezing and thawing to work the seeds into the soil. Planting into killed vegetation using no-till techniques or into a prepared seedbed involves less risk of failure and produces denser, more uniform stands than frost-seeding.

Whether planting no-till or into a prepared seedbed, growers should place seeds no more than ¼ inch deep for maximum emergence. With a prepared seedbed, the soil should be very firm to ensure good soil-to-seed contact. When broadcasting, growers should firm the field with a culti-packer or roller before and after planting. Drills that are capable of precise seed depth control and have press wheels to firm the seedbed are also excellent.

In dry years, getting a firm seedbed is critical for seedling survival. In dry years, seedlings germinate and then die in a loose seedbed because water does not move up to the upper soil layer where the young roots are.

Companion Crop

Alfalfa is often fall-seeded with small grains such as wheat, oats, and barley. Otherwise, alfalfa is broadcast into these crops during winter. The companion crop prevents excessive soil erosion, decreases weed problems, protects young alfalfa seedlings, and provides some early spring forage before the alfalfa becomes productive. Although beneficial, the small grain companion crops also compete for light, water, and soil nutrients. The companion crop should be harvested for hay or silage no later than the boot stage to minimize competition. Alfalfa often provides one hay cutting in late August to early September when seeded with a companion crop.

Seeding Rates and Mixtures

When seeded alone, growers use 15 pounds per acre of certified seed, which is about the equivalent of 13 pounds per acre of pure, live seed (PLS). When seeded with a grass, 10 pounds per acre of bulk alfalfa seeds (equal to eight pounds per acre PLS) is sufficient. Seeding rates per acre for grasses in an alfalfa grass mixture are:

+ Bromegrass—ten pounds bulk (eight pounds PLS)
+ Orchard grass—six pounds bulk (four pounds PLS)
+ Tall fescue—ten pounds bulk (eight pounds PLS)
+ Reed canary grass—six pounds bulk (four pounds PLS)

Seeding a cool-season grass with alfalfa decreases the potential for heaving, reduces weed competition, lessens damage to soil structure by grazing animals, and reduces bloat potential when grazed. The grass will decrease forage quality but will be a major component in the first cutting only.

Growers make decisions about whether to include a grass based on the intended market or use of the alfalfa and on the winter-heaving potential of the site. If intended for dairy use or sale to a cubing plant, growers seed pure alfalfa. For grazing of beef or horses, growers use an alfalfa-grass mixture. On sites that have a high clay content subject to heaving, alfalfa-grass mixtures are recommended.

Maintaining Alfalfa Stands

Proper management can allow growers to maintain a productive stand of alfalfa for five or more years. An annual fertility program and proper harvesting management are major factors determining stand productivity and longevity. Insects, diseases, and weeds are problems that can reduce yields and length of stand (Figure 22-10).

Most alfalfa seedings initially have 15 or more plants per square foot. As the stand ages, some plants die and remaining plants spread to occupy the space. Alfalfa-grass mixtures can maintain productivity with only two alfalfa plants per square foot.

Figure 22-10 A good stand of alfalfa can last five years or longer with good management.

Annual Fertilization

Annual applications of phosphorus, potash, boron, and sometimes lime are necessary to maintain vigorous, productive stands. To avoid nutritional deficiencies, producers should apply fertilizer each year according to soil tests.

Phosphorus fertilization of established stands keeps plants vigorous so that high yields can be maintained over time.

Potash application improves winter survival of plants and lengthens the productive life of the stand. Alfalfa stands with fewer than three plants per square foot cannot maintain high yields and are often subject to increased weed invasion.

Annual fertilizer recommendations vary according to phosphorus and potassium levels in the soil, but will be close to 15 pounds of phosphate and 55 pounds of potash per ton of expected yield. Growers can apply fertilizer at any time. A single application following the first cutting or a split application following the first and third cuttings are both good options. Split applications are useful for irrigated alfalfa, for high-yielding alfalfa stands, or when applying high rates of potash.

Growers should include boron in the top-dress fertilizer at a rate of one pound of boron per acre per year. Boron is toxic to seedlings; it should not be applied at seeding.

Producers should soil test every two to three years to make sure that soil salt pH, phosphorus, and potassium levels are adequate. Also, where needed, growers should top-dress additional lime to keep the pH above 6.0.

Harvest Management

Stage of maturity at harvest determines hay quality and affects stand life. Forage quality (protein, energy value) declines rapidly as the plant begins to flower. Figure 22-11 illustrates what happens to protein content as the plant reaches different stages of flowering.

For spring-seeded established stands in the seeding year, growers take the first harvest at the mid- to full-bloom stage. Following harvests are made as flowers begin to appear.

For established stands, growers take the first and second cuttings when the plants are just beginning to bloom. For persistence of the stand, the grower may make one or two more harvests. Alfalfa should not be cut or

Figure 22-11 Graph of protein content versus flower stage of alfalfa.

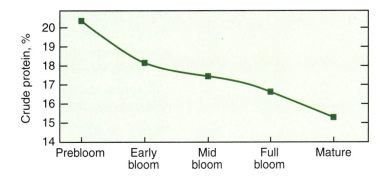

grazed from mid-September to the first of November. This allows the plant to store root reserves to overwinter. After November first, growers can take or graze a fifth cutting if the soil is well drained or a grass is used to help prevent winter heaving. With a four-cut system, a properly fertilized stand can last six or more years.

Harvesting alfalfa in the bud stage produces three to five cuttings of high-quality hay (Figure 22-12). This practice, however, reduces stand life to three or four years.

Figure 22-12 Much of the alfalfa in the West is harvested in large square bales for use as dairy feed.

Alfalfa can be grazed without a loss of stand using small pasture units and high stocking rates. Producers should use enough animals to remove most topgrowth in less than six to ten days. Animals are turned onto the alfalfa when alfalfa is in the bud stage. After grazing, alfalfa needs to regrow for 30–35 days. To reduce the chance of bloating, producers use poloxalene (bloat-inhibitor) blocks. Hungry animals should never be turned onto lush alfalfa pastures.

Insects

The alfalfa weevil and potato leafhopper are the two major pests of alfalfa. Regular monitoring of alfalfa fields is the best way to prevent economic injury from insects. Growers should spray or cut when insect populations reach economic thresholds, not after insect injury symptoms are apparent.

Alfalfa weevil adults lay eggs in the older alfalfa stems in late fall and early spring, and the larva damage mainly the first cutting. Chemical insecticides can be used when 25 percent of the tips are skeletonized and if there are three or more larvae per stem. Instead of spraying, some growers will cut the alfalfa when it is in the bloom stage, scouting regrowth for signs of damage.

The immature or nymph stage of potato leafhoppers stunts plants and yellows leaves. It also lowers yield and protein content by sucking juices from young upper stems. Leafhopper numbers can be large enough to warrant treatment before significant leaf yellowing occurs. Population thresholds for chemical control vary with plant height.

Weed Control

Weed control in alfalfa begins with establishing a uniform dense stand of alfalfa or alfalfa-grass mixture. Experts recommend a preplant-incorporated herbicide for conventional spring seedings of pure alfalfa. If growers want a grass in spring-seeded alfalfa, they plant the alfalfa alone using a preplant herbicide. The grass is drilled into the alfalfa stand the following spring. Numerous diskings during mid to late summer give adequate control for late summer seedings.

Control of weeds after alfalfa emergence depends on the individual weeds, the stage of growth of the alfalfa, and whether there is a grass with the alfalfa. Several herbicides control weeds in pure stands of alfalfa, but only a few are available for use in alfalfa-grass stands.

Diseases

Alfalfa is subject to several diseases, including phytophthora root rot, bacterial wilt, anthracnose, Sclerotinia root, and crown rot. The best control is prevention. Growers should choose a variety with a high level of resistance to phytophthora, bacterial wilt, and anthracnose. There is no varietal resistance to Sclerotinia.

Sclerotinia is particularly damaging to fall-seeded alfalfa stands south of the Missouri river. The disease has killed seedling stands, and older stands are also subject to damage. Cultural controls include deep tillage of alfalfa residue to bury the inoculum that is formed in the spring on infected alfalfa. A less practical control is to maintain a three- to four-year interval between forage legumes in a rotation. Red and white or ladino clovers are also hosts of Sclerotinia and can be a source of inoculum for subsequent alfalfa crops.

RED CLOVER

Red clover (*Trifolium pratense*) is a short-lived perennial legume grown for hay, silage, pasture, and as a green manure crop. Red clover is better adapted than alfalfa to soils that are somewhat poorly drained and slightly acidic. The greatest production occurs on well-drained soils with high water-holding capacity and pH above 6.0. Red clover is not as productive as alfalfa in the summer (Tables 22-6 and 22-7). Figure 22-13 shows a red clover plant.

Red clover is one of the easiest legumes to establish using no-till interseeding or frost-seeding techniques. Under high humidity, red clover is often difficult to dry quickly enough for safe hay baling. Harvesting it for silage or including a grass in the stand helps overcome this problem.

Figure 22-13 Red clover is one of the easiest forages to plant using the no-till interseeding or frost-seeding method.

Medium red clover varieties can be harvested three to four times per year. Mammoth red clover is late to flower and is considered a single-cut clover because the majority of its growth is produced in the spring. Most of the improved varieties are medium types and have good levels of disease resistance to northern and southern anthracnose and powdery mildew. These and other diseases can reduce stands quickly.

BIRD'S-FOOT TREFOIL

Bird's-foot trefoil (*Lotus corniculatus*) is a deep-rooted perennial legume that produces well in long-lay stands. More than other forage legumes, bird's-foot trefoil is tolerant of low-pH soils (as low as pH 5.0), moderate to somewhat poor soil drainage, marginal fertility, and fragipans. Bird's-foot trefoil can withstand several weeks of flooding, and tolerates periods of moderate drought and heat (Tables 22-6 and 22-7). Bird's-foot trefoil seedlings develop slowly, and early spring seedings are generally more successful than late summer seedings. Bird's-foot trefoil is subject to invasion by weeds when grown in pure stands, so it is best seeded with a grass companion. It produces excellent quality forage, has fair palatability, stockpiles well, and is nonbloating. Bird's-foot trefoil should be managed to allow for reseeding, which will help maintain its presence in forage stands.

Empire-type varieties have prostrate growth and fine stems, making them well suited to grazing. European-type varieties are more erect, establish faster, and regrow faster after harvest. Thus, they are well suited to hay production and rotational grazing. Most of the newer varieties are intermediate, with semierect to erect growth habit.

Bird's-foot trefoil is shown in Figure 22-14.

Figure 22-14 Bird's-foot trefoil (*Lotus corniculatus*) can withstand marginal soil conditions. Reproduced with permission from N. L. Britton and A. Brown, *An Illustrated Flora of the Northern United States, Canada and the British Possessions* (Vol. 1: 672.1913.) Courtesy of Kentucky Native Plant Society.

WHITE CLOVER

White clover (*Trifolium repens*) is a low-growing, short-lived perennial legume that is well suited for pastures. White clover improves forage quality of grass pastures and reduces the need for nitrogen fertilizer. White clover can be frost-seeded or no-till seeded into existing grass pastures. White clover has a shallow root system and does not tolerate prolonged dry spells. Although well-drained soils improve production, white clover tolerates periods of poor drainage (Tables 22-6 and 22-7).

Large white clover types, also known as ladino clovers, are more productive than the White Dutch or "wild white" clovers. Wild white clovers persist better under heavy, continuous grazing because they are prolific reseeders. Growers should purchase seeds of stated quality to be certain of obtaining pure seeds of the white clover desired.

ALSIKE CLOVER

Alsike clover (*Trifolium hybridum*) is a short-lived perennial legume that is tolerant of wet, acidic soils. Alsike clover tolerates soils with a pH as low as 5.0, which is too acidic for red clover and alfalfa. Alsike clover also grows better than red clover on alkaline (high pH) soils. Alsike clover tolerates flooding better than other legumes, making it well suited for low-lying fields with poor drainage. Alsike clover can withstand spring flooding for several weeks. A cool and moist environment is ideal for alsike clover growth. It has poor heat and drought tolerance. It usually produces only one crop of hay. It must be allowed to reseed itself to maintain its presence in pastures, otherwise it will last only about two years. Tables 22-6 and 22-7 compare the features of alsike clover with other legumes.

ANNUAL LESPEDEZA

Annual lespedezas (*Lespedeza* spp.) are spring-sown, warm-season legumes. They can be used for hay, pasture, and soil-erosion control. These species are relatively low yielding, but produce nonbloating forage of high nutritive value that furnishes excellent quality pasturage in late summer. They can be grown on acidic and low-phosphorus soils. They will respond to both lime and phosphorus fertilization. Annual lespedezas grow best on well-drained soils. They are dependable reseeders and can persist in pastures for years if allowed to reseed each year. Annual lespedeza can be used effectively in pasture renovation to improve animal performance and late-summer forage production, especially in endophyte-infected tall fescue pastures. Figure 22-15 shows a lespedeza plant.

Figure 22-15 The lespedeza plant produces a nonbloating forage of high nutritive value.

MANAGING FORAGES

All forage crops respond dramatically to good management practices. Higher yields and improved persistence result from paying attention to the basics of good forage management.

Seeding Year Management

Establishing a good stand is critical for profitable forage production. Begin by selecting species adapted to soils where they will be grown. Crop rotation is an important management tool for improving forage productivity, especially when seeding forage legumes. Crop rotation reduces disease and insect problems. Seeding alfalfa after alfalfa is especially risky because old stands of alfalfa release a toxin that reduces germination and growth of new alfalfa seedlings (called autotoxicity). This is especially true on heavy-textured soils. Disease pathogens accumulate and can cause stand establishment failures when seeding into a field that was not rotated out of alfalfa for at least one year.

Fertilization and Liming

Proper soil pH and fertility are essential for optimum economic forage production. Growers take a soil test to determine soil pH and nutrient status at least six months before seeding. This allows time to correct deficiencies in the topsoil zone. The topsoil in fields with acidic subsoils should be maintained at higher pHs than fields with neutral or alkaline subsoils. Growers should maintain topsoil zone pH at the following levels:

- pH 6.8 for forage legumes on mineral soils with subsoil pH less than 6.0
- pH 6.5 for alfalfa on mineral soils with subsoil pH greater than 6.0

Soil pH should be corrected by application of lime when topsoil pH falls 0.2–0.3 pH units below the recommended levels. With conventional tillage plantings, lime should be incorporated and mixed well in the plow layer at least six months before seeding. If more than four tons per acre are required, half the amount should be plowed down and the other half incorporated lightly into the top two inches.

Corrective applications of phosphorus and potassium should be applied prior to seeding regardless of the seeding method used. Growers should apply 10–20 pounds of nitrogen per acre when seeding grass-legume mixtures. Starter nitrogen applications of ten pounds per acre may be beneficial with pure legume seedings, especially under cool conditions and on soils low in nitrogen. Manure applications incorporated ahead of seeding may also be beneficial.

Seed Quality

High-quality seed of adapted species and varieties should be used. Seed lots should be free of weed seeds and other crop seeds, and contain only minimal amounts of inert matter. Certified seed is the best assurance of securing high-quality seeds of the variety of choice. Purchased seeds accounts for only 20 percent or less of the total cost of stand establishment. Buying cheap seeds and seeds of older varieties is false economy. Growers should always compare seed price on the basis of cost per pound of pure live seeds.

Seed Inoculation

Legume seeds must be inoculated with the proper nitrogen-fixing bacteria prior to seeding to assure good nodulation. Inoculation is especially important when seeding legumes in soils where they have not been grown for several years. Because not all legume species are colonized by the same strains of nitrogen-fixing bacteria, producers must be sure to purchase the proper type of inoculum for the forage legume to be planted. Because many seed suppliers distribute preinoculated seeds, growers should check the expiration date and reinoculate if necessary.

Seed Treatments

Apron-treated seeds are highly recommended for alfalfa and may be useful for red clover. Metalaxyl is a systemic fungicide for controlling seedling damping-off diseases caused by *Pythium* and *Phytophthora* during the first four weeks after seeding. These pathogens kill legume seedlings and cause establishment problems in wet soils. Many companies are marketing Apron-treated alfalfa seeds. Various other seed treatments and coatings are sometimes added to forage seed. Growers should always consider the cost versus the opportunities for benefit.

Spring Seedings

Depending on the area, forages should be planted as soon as the seedbed can be prepared after March 15. With early seeding, the plants become well established before the warm and dry summer months. Weed pressure increases with delayed seeding. Annual grassy weeds can be especially troublesome with delayed spring seedings. An herbicide program is usually essential when seeding pure legumes late in the spring.

Seeding without a companion crop in the spring allows growers to harvest two or three crops of high-quality forage in the seeding year, particularly when seeding alfalfa and red clover. Weed control is important during early establishment when direct seeding pure legume stands. Several preplant and postemerge herbicide options are available for pure legume seedings. Growers should select fields with little erosion potential when direct seeding into a tilled seedbed.

Small grain companion crops can be used successfully when seeding forages, if managed properly. Companion crops reduce erosion in conventional seedings and help minimize weed competition. Companion crops usually increase total forage tonnage in the seeding year, but forage quality will be lower than direct-seeded legumes. When seeding with a small grain companion crop, growers should reduce excessive competition, which may lead to establishment failures by observing the following—

+ Spring oats and triticale are the least competitive, while winter cereals are often too competitive.
+ Use varieties that are early-maturing, short, and stiff-strawed.
+ Sow companion small grains at 1.5 to 2.0 bushels per acre.
+ Remove small grain companions as early pasture or silage.
+ Do not apply additional nitrogen for the companion crop.

Late Summer Seedings

Late summer is an excellent time to establish many forages, provided sufficient soil moisture is available. August is the preferred time for most species, allowing enough time for plant establishment before winter. Growers should not use a companion crop with August seedings because they compete for soil moisture and can slow forage seedling growth to the point where the stand will not become established well enough to survive the winter.

Seeding Basics

The ideal seedbed for conventional seedings is smooth, firm, and weed-free. The soil should not be overworked thus conserving moisture and reducing the risk of surface crusting. Seedbeds should be firm before seeding to ensure good seed-soil contact and reduce the rate of drying in the seed zone. Cultipackers and cultimulchers are excellent implements for firming the soil. The lack of a firm seedbed is a major cause of establishment failures.

Seeding depth for most cool-season forages is 1/4–1/2 inch on clay and loam soils. On sandy soils, seed can be placed 1/2–3/4 inch deep. Seeding too deep is one of the most common reasons for seeding failures.

Seeding rate recommendations are related to seed size, germination, species vigor, and spreading characteristics. Increasing seeding rates above the recommended levels does not compensate for poor seedbed preparation or improper seeding methods. An established stand having about six grass and/or legume plants per square foot is generally adequate for good yields. About 20 to 25 seedling plants per square foot in the seeding year usually results in at least six established plants the following year.

Seeding Equipment

Forage stands can be established with many different types of drills and seeders, provided they are adjusted to plant seeds at an accurate depth and in firm contact with the soil. When seeding into a tilled seedbed, band seeders with press wheels are an excellent choice. If the seeder is not equipped with press wheels, cultipack before and after seeding in the same direction as the band seeding. This assures that seeds are covered and in firm contact with the soil.

Figure 22-16 Harvesting of forages is made easier with modern equipment such as this baler for large round bales. Additionally, having ground in forage is good protection against erosion. (Photo courtesy of USDA NRCS)

Seeding Year Harvest Management

Seeding year harvest (Figure 22-16) management of cool-season forages depends on time and method of seeding, species, fertility, weather conditions, and other factors. Forages seeded in August or early September should not be harvested or clipped until the following year. For spring seedings, it is best to mechanically harvest the first growth. This is especially true for tall-growing legumes. If stands are grazed, stock fields with enough livestock to consume the available forage in less than seven days. Grazing for a longer period increases the risk of stand loss. Soils should be firm to avoid trampling damage. The following are general harvest management guidelines for spring seedings, according to species.

+ *Bird's-Foot Trefoil.* Seedling growth of trefoil is much slower than alfalfa or red clover. Seeding year harvests should be delayed until the trefoil is in full bloom. When seeded with a companion crop, an additional harvest after removal of the small grain is generally not advisable.

+ *Red Clover.* When seeded without a companion crop, red clover can usually be harvested twice in the year of establishment. Under good conditions, up to three harvests are possible. It is important to harvest red clover before full bloom in the seeding year. If red clover is allowed to reach full bloom in the year of seeding, it often has reduced stands and yields the following year.

Fertilizing Established Stands

A current soil test is the best guide for a sound fertilization program. Forages are very responsive to good fertility. Adequate levels of phosphorus and potassium are important for high productivity and persistence of legumes. Forage fertilization should be based on soil test levels and realistic yield goals. Under hay-land management, forages should be top-dressed annually to maintain soil nutrient levels and achieve top production potential. Each ton of tall grass or legume forage removes approximately 13 pounds of P_2O_5 and 50 pounds of K_2O. These nutrients need to be replaced, preferably in the ratio of one part phosphate to four parts potassium.

Forages may be top-dressed with phosphorus and potassium at any time of the year. The timing of phosphorus and potassium applications is not critical when soil test levels are optimum. Growers avoid applications with heavy equipment when the soil is not firm. Soil conditions are frequently most conducive to fertilizer applications immediately following the first cutting or in late summer and early fall. Split applications, one-half after the first cutting and one-half in late summer or early fall, may result in more efficient use of fertilizer nutrients when high rates of fertilizer are recommended.

Legumes "fix" atmospheric nitrogen. Where the forage stand is more than 35 percent legumes, nitrogen should not be applied. Where grasses are the predominant forage, nitrogen fertilization is extremely important for production. Economic returns are usually obtained with 150–175 pounds of nitrogen per acre per year split three times during the year. In the summer, some forms of nitrogen are subject to surface volatilization, resulting in loss of available nitrogen. Ammonium nitrate is the best choice because surface volatilization losses are minimized.

SUMMARY

Some members of the legume family are grown for their edible seeds. This includes dry beans, peas, and peanuts. Beans grown in the United States include navy or pea bean, dark and light red kidney, cranberry, yellow eye, pinto, and black turtle. Soybeans are also a legume grown for their seeds but are not used much by humans. Legumes grown for their seed are known as **pulses** *worldwide. Because of their growing requirements many of the legumes have major areas of production. Legumes are used as part of a crop rotation. Legumes come with their unique management problems to maximize yields and combat pests.*

Forage legumes are grown for their use as livestock feed. They may be grazed, harvested and stored, or both. Important forage legumes include alfalfa, red clover, bird's-foot trefoil, white clover, alsike clover, and annual lespedeza. Forage legumes can be grown with grasses, depending on their planned use. Forages are environmentally friendly. They reduce soil erosion, improve soil tilth, help reduce pesticide use, and enhance agricultural profitability. Like other legumes, forage legumes have the ability to fix nitrogen.

Review

True or False

1. Alfalfa seeds should be planted 2–3 inches deep when establishing a stand.

2. None of the diseases of plants are seed-borne.

3. Viruses only cause disease in humans and animals.

4. Protein quantity rises in the alfalfa plant after it reaches full bloom.

5. All crop rotations help prevent or reduce the incidence of disease in a crop.

Short Answer

6. Give the common and scientific name of five forage legumes.

7. After pollination, the _____ forms in the ground on the peanut plant.

8. Are peas a warm- or cool-weather crop?

9. Name four common types of dry beans.

10. Give the scientific name for beans, peas, and peanuts.

Critical Thinking/Discussion

11. Describe how to maintain an alfalfa stand.

12. Why is calcium so important to the peanut plant?

13. Explain the importance of seed quality to bean, pea, and peanut production.

14. Describe the seedbed preparation for a forage legume.

15. Name five important pests in peanut production and describe how these are controlled.

Knowledge Applied

1. Visit an equipment dealer and collect information and pictures of the equipment used in one of these crops: dry beans, peas, peanuts, or forage legumes.

2. Make a seed collection of all the different types of dry beans. Try to identify the beans by scientific name, including the variety or cultivar.

3. Make a seed collection of all the forage legumes. Label the seeds with the common and scientific names.

4. Develop a presentation (computer) showing the insect pests of one of the dry beans, pea, or peanuts. Show the type of damage the pests cause to the plants and indicate how the pests are controlled.

continues

Knowledge Applied, *continued*

5. Search the Internet for production guidelines for beans, peas, peanuts, or one of the forage legumes. Print the results of your search.

6. Develop a management checklist for keeping disease out of a crop of peas, peanuts, or beans. Make the list include some practices that could be considered best management practices.

7. Collect and submit an alfalfa hay sample for analysis. Compare the results with some "table" values. Based on the analysis, decide if the alfalfa was cut too late or at just the right time.

Resources

Acquaah, G. (2004). *Principles of crop production: Theory, techniques, and technology* (2nd ed.). Englewood Cliffs, NJ: Prentice Hall Career & Technology.

Connors, J. J., & Cordell, S., eds. (2003). *Soil fertility manual.* Tucson, AZ: Potash & Phosphate Institute.

Duke, J. A. (2003). *Handbook of legumes of world economic importance.* New Delhi, India: Scientific Publishers.

Hardarson, G. G., & Broughton, W. J. (2003). *Maximising the use of biological nitrogen fixation in agriculture.* New York: Springer Publishing Company.

Hay, R., & Porter, J. (2006). *The physiology of crop yield* (2nd ed.). Ames, IA: Blackwell Publishing.

Jones, J. B., Jr. (2002). *Agronomic handbook: Management of crops, soils, and their fertility.* Boca Raton, FL: CRC Press.

McMahon, M., Kofranek, A. M., & Rubatzky, V. E. (2006). *Hartmann's plant science: Growth, development, and utilization of cultivated plants* (4th ed.). Englewood Cliffs, NJ: Prentice Hall Career & Technology.

Simpson, B. B., & Ogorzaly, M. D. (2000). *Economic botany: Plants in our world* (3rd ed.). New York: McGraw-Hill.

United States Department of Agriculture. (1950–51). *Crops in peace and war: The yearbook of agriculture.* Washington, DC: Author.

United States Department of Agriculture. (1953). *Plant diseases: The yearbook of agriculture.* Washington, DC: Author.

United States Department of Agriculture. (1961). *Seeds: The yearbook of agriculture.* Washington, DC: Author.

Internet

Internet sites represent a vast resource of information. The URLs (uniform resource locators) for World Wide Web sites can change. Using one of the search engines on the Internet such as Google or Yahoo!, find more information by searching for these words or phrases: food legumes, forage legumes, dry beans, peas, peanuts, disease-free seed, certified seed, protein content, or crop rotation.

Additionally, the PLANTS database (http://plants.usda.gov/index.html) provides standardized information about the vascular plants of the United States and its territories. It includes names, plant symbols, checklists, distributional data, species abstracts, characteristics, images, crop information, automated tools, onward Web links, and references. The PLANTS database is a great educational and general use resource.

Chapter 23
Vegetables

A VEGETABLE is the edible part of a plant that grows and dies in a single growing season. Many people enjoy a home garden, and it is part of a lifestyle. Vegetables can be used fresh or stored for later production. Many vegetables are grown on a commercial, large-scale basis.

After completing this chapter, you should be able to—

- Name ten different vegetable crops
- Describe three types of vegetable production
- Identify the basic principles of vegetable production
- Discuss the growing requirements of five vegetables
- Name five herbs and describe their growing conditions and preservation
- Describe cultural practices used in vegetable production

SCOPE OF VEGETABLE PRODUCTION

The study of vegetable production is olericulture. The production of vegetables can be classified into three categories:

1. Home gardening
2. Market gardening
3. Truck cropping

Home gardening usually involves vegetable production for one family, with little or no selling of the produce.

Market gardening refers to growing a wide variety of vegetables for local or roadside markets.

Truck cropping refers to commercial, large-scale production of selected vegetable crops for wholesale markets and shipping. Table 23-1 lists the average annual production of commercial vegetables in the United States that are tracked by the U.S. Department of Agriculture. Head lettuce, onions, watermelon, tomatoes, and cabbage are the top five commercially grown vegetables (Figures 23-1 and 23-2).

Table 23-1 Production of Commercially Produced Vegetables in the United States

Vegetable	Anual Production (1,000 CWT)[1]	Vegetable	Anual Production (1,000 CWT)[1]
Head lettuce	61,137	Broccoli	5,846
Onions	42,016	Cauliflower	3,921
Watermelon	28,856	Snap beans	3,317
Tomatoes	27,497	Honeydews	3,124
Cabbage	20,684	Garlic	2,006
Celery	17,820	Spinach	1,142
Carrots	16,243	Asparagus	1,069
Leaf/romaine	15,605	Escarole/endive	954
Sweet corn	15,522	Artichokes	789
Cantaloupe	13,656	Brussels sprouts	693
Bell peppers	7,389	Eggplant	651
Cucumbers	6,255		

[1] Based on a twenty-five-year average of data from the National Agricultural Statistics Service, USDA

Potatoes are a common vegetable in most home and market gardens. Potatoes are a major agronomic crop. They are grown on a large scale in several states; their production is covered in Chapter 20.

CATEGORIZING VEGETABLES

Vegetables can be categorized three ways:

1. Botanical classification
2. Edible parts
3. Growing season

Figure 23-1 Harvesting commercially grown lettuce near Yuma, Arizona.

Figure 23-2 Cultivating a field of commercial lettuce near Yuma, Arizona.

Botanical Classification

All vegetables belong to the angiosperm division of the plant kingdom. Next, vegetables are classified as either monocotyledons (one seed leaf) or dicotyledons (two seed leaves). From this point, vegetables are grouped by family, genus, species, and variety (Table 23-2).

Edible Parts

Vegetables are divided into three groups based on the part of the plant that is eaten:

+ Leaves, flower parts, or stems
+ Fruits or seeds
+ Underground parts

Table 23-2 Vegetable Varieties, Families, and Characteristics[1]

Common Name	Variety[2]	Plant Family[3]	Transplant-ability[4]	Pounds per 100 Feet	Days to Harvest[5]	Comments
Warm-Season Vegetables						
Beans, bush	Snap: Bush Blue Lake, Contender, Roma, Harvester, Provider, Cherokee Wax, Bush Baby, Tendercrop. Shell: Horticultural, Pinto, Red Kidney	Leguminosae (Pea)	III	45	50–60	Fertilize at one-fourth the rate used for other vegetables. Seed innoculation not essential in most soils. Flowers self-pollinate. Use shell beans green or dry.
Beans, pole	Dade, McCaslan, Kentucky Wonder 191, Blue Lake	Leguminosae (Pea)	III	80	55–70	See Beans, bush. Support vines. May be grown with corn for vine support.
Beans, lima	Fordhook 242, Henderson, Jackson Wonder, Dixie Butterpea, Florida Butter (Pole), Sieva (Pole)	Leguminosae (Pea)	III	80	65–75	See Beans, bush. Provide trellis support for pole varieties. Control stinkbugs, which injure seeds in pods.
Cantaloupes	Smith's Perfect, Ambrosia, Edisto 47, Planters Jumbo, Summet, Super Market, Primo, Luscious Plus	Cucurbitaceae (Melon)	III	150	75–90	Bees needed for pollination. Mulch to reduce fruit-rots and salmonella. Harvest at full-slip stage.
Corn, sweet	Silver Queen, Gold Cup, Guardian, Bonanza, Florida Staysweet, How Sweet It Is, Supersweet	Gramineae (Grass)	III	150	60–95	Separate super-sweets (last three varieties) from standard varieties by time and distance. Sucker removal not beneficial. Plant in two- to three-row blocks.
Cucumbers	Slicers: Poinsett, Ashley, Dasher, Sweet Success, Pot Luck, Slice Nice. Picklers: Galaxy, SMR 18, Explorer	Cucurbitaceae (Melon)	III	100	50–65	Bees required for pollination. Many new hybrids are gynoecious (female flowering). Monoecious varieties have M/F flowers. For greenhouse use parthenocarpic type.
Eggplant	Florida Market, Black Beauty, Dusky, Long Tom, Ichian, Tycoon, Dourga	Solanaceae (Nightshade)	I	200	90–110	Stake eggplants. Harvest into summer. Require warm weather. "Dourga" is white.
Okra	Clemson Spineless, Perkins, Dwarf Green, Emerald, Blondy	Malvaceae (Mallow)	III	70	50–75	Produces well in warm seasons. Okra is highly susceptible to root-knot nematodes.
Peas, southern	Blackeye, Mississippi Silver, Texas Cream 40, Snapea, Sadandy, Purplehull, Zipper Cream	Leguminosae (Pea)	III	80	60–90	See Beans, bush. The cowpea curculio is common pest. Tiny white grub infests seeds in pods. Good summer cover crop.
Peppers	Sweet: Early Calwonder, Yolo Wonder, Big Bertha, Sweet Banana, Jupiter. Hot: Hungarian Wax, Jalapeno, Habanero	Solanaceae (Nightshade)	I	50	80–100	Mulching especially beneficial. Continue care of peppers well into summer. Mosaic virus a common disease pest. Most small-fruited varieties are attractive but hot.
Potatoes, sweet	Porto Rico, Georgia Red, Jewel, Centennial, Coastal Sweet, Boniato, Sumor, Beauregard, Vardaman	Convolvulaceae (Morning Glory)	I	300	120–140	Sweet potato weevils are a serious problem. Vardaman bush type for small gardens.
Pumpkin	Big Max, Funny Face, Connecticut Field, Spirit, Calabaza, Cushaw	Cucurbitaceae (Melon)	III	300	90–120	Bees required for pollination. Foliage diseases and fruit-rot are common.
Squash	Summer: Early Prolific Straightneck, Dixie, Summer Crookneck, Cocozelle, Gold Bar, Zucchini, Peter Pan, Sunburst, Scallopini, Sundrops. Winter: Sweet Mama, Table Queen, Butternut, Spaghetti	Cucurbitaceae (Melon)	III	150	40–55	Summer types usually grow on a bush; winter squash have vining habit. Both male and female flowers on same plant. Common fruit-rot/drop caused by fungus and incomplete pollination. Bees required. Crossing occurs, but results not seen unless seeds are saved. Winter types store longest.

(continued on following page)

Table 23-2 Vegetable Varieties, Families, and Characteristics[1] (continued)

Common Name	Variety[2]	Plant Family[3]	Transplant-ability[4]	Pounds per 100 Feet	Days to Harvest[5]	Comments
Warm-Season Vegetables (continued)						
Tomatoes	Large fruit: Floradel, Solar Set, Manalucie, Better Boy, Celebrity, Bragger, Walter, Sun Coast, Floramerica, Flora-Dade, Duke. Small fruit: Florida Basket, Micro Tom, Patio, Cherry, Sweet 100, Chelsea	Solanaceae (Nightshade)	I	200	90–110	Staking, mulching beneficial. Flowers self-pollinated. May drop if temperatures too high or low, or if nitrogen fertilization excessive. Some serious problems are blossom-end rot, wilts, whitefly, and leafminers. Better Boy appears resistant to root-knot.
Watermelon	Large: Charleston Gray, Jubilee, Crimson Sweet, Dixielee. Small: Sugar Baby, Minilee, Mickylee. Seedless: Fummy	Cucurbitaceae (Melon)	III	400	85–95	Cue to spaced requirement, not suited to most gardens. Suggest small ice-box types. Plant fusarium wilt-resistant varieties. Bees require pollination.
Cool-Season Vegetables						
Beets	Early Wonder, Detroit Dark Red, Cylindra, Red Ace, Little Ball	Chenopodiaceae (Goosefoot)	I	75	50–65	Beets require ample moisture at seeding or poor emergence results. Leaves edible.
Broccoli	Early Green Sprouting, Waltham 29, Atlantic, Green Comet, Green Duke	Cruciferae (Mustard)	I	50	75–90	Harvest small multiple sideshoots that develop after main central head is cut.
Cabbage	Gourmet, Marion Market, King Cole, Market Prize, Red Acre, Chieftan Savoy, Rio Verde, Bravo	Cruciferae (Mustard)	I	125	90–110	Buy clean plants to avoid cabbage black-rot, a common bacterial disease that causes yellow patches on leaf margins. Watch for loopers.
Carrots	Imperator, Thumbelina, Nantes, Gold Pak, Waltham Hicolor, Orlando Gold	Umbelliferae (Parsley)	II	100	65–80	Grow carrots on a raised bed for best results. Sow seeds shallow and thin to proper stand.
Cauliflower	Snowball Strains, Snowdrift, Imperial 10-6, Snow Crown, White Rock	Cruciferae (Mustard)	I	80	75–90	Tie leaves around flowerhead at two- to three-inch diameter stage to prevent discoloration. For green heads, grow broccoflower.
Celery	Utah Strains, Florida Strains, Summer Pascal	Umbelliferae (Parsley)	II	150	115–125	Celery requires very high soil moisture during seeding/seedling stage.
Chinese cabbage	Micihili, Wong Bok, Bok Coy, Napa	Cruciferae (Mustard)	I	100	70–90	Bok Choy is open-leaf type; Michihili and Napa form round heads.
Collards	Georgia, Mates, Blue Max, Hicrop Hybrid	Cruciferae (Mustard)	I	150	70–80	Tolerates more heat than most other crucifers. Harvest lower leaves. Kale may also be grown.
Endive/ Escrole	Florida Deep Heart, Full Heart, Ruffec	Compositae (Sunflower)	I	75	80–95	Excellent ingredient in tossed salads. Well adapted to cooler months.
Kohlrabi	Early White Vienna, Grand Duke, Purple Vienna	Cruciferae (Mustard)	I	100	70–80	Both red and green varieties are easily grown. Use fresh or cooked. Leaves edible.
Lettuce	Crisp: Minetto, Ithaca, Fulton, Floricrisp. Butterhead: Bibb, White Boston, Tom Thumb. Leaf: Prize Head, Red Sails, Salad Bowl. Romaine: Parris Island Cox, Valmaine, Floricos	Compositae (Sunflower)	I	75	50–90	Grow crisphead type in coolest part of the season for firmer heads. Sow seeds very shallow, as they need light for germination. Intercrop lettuce with long-season vegetables.

Table 23-2 Vegetable Varieties, Families, and Characteristics[1] *(continued)*

Common Name	Variety[2]	Plant Family[3]	Transplant-ability[4]	Pounds per 100 Feet	Days to Harvest[5]	Comments
		Cool-Season Vegetables *(continued)*				
Mustard	Southern Giant Curled, Florida Broad Leaf, Tendergreen	Cruciferae (Mustard)	II	100	40–60	Consider planting in a wide-row system. Broadleaf type requires more space. Cooked as "greens."
Onions	Bulbing: Excel, Texas Grano, Granex, White Granex, Tropicana Red. Bunching: White Portugal, Evergreen, Beltsville Bunching, Perfecto Blanco. Multipliers: Shallots	Amaryllidaceae (Lily)	I	100	120–160	Plant short-day bulbing varieties. For bunching onions, insert sets upright for straight stems. For multipliers, divide and reset. Bulbing onions may be seeded in the fall, then trans-planted in early spring (Jan–Feb). Granex used for Vidalia and St. Augustine Sweets.
Parsley	Moss Curled, Perfection, Italian	Umbelliferae (Parsley)	II	40	70–90	Curley and plain types do well.
Peas, English	Wando, Green Arrow, Laxton's Progress, Sugar Snap, Oregon Sugar	Leguminoseae (Pea)	III	40	50–70	Edible podded type are Oregon (flat) and Sugar Snap (round)—be sure to trellis.
Potatoes	Sebago, Red Pontiac, Atlantic, Red LaSoda, LaRouge, Superior	Solanaceae (Nightshade)	II	150	85–110	Plant two-ounce seed pieces with eyes. Do not use table-stock for seed. Remove tops two weeks before digging to toughen skin. Varieties planted by seeds produce less than from tubers.
Radish	Cherry Belle, Comet, Early Scarlet Glove, White Icicle, Sparkler, Red Prince, Champion, Snowbelle	Cruciferae (Mustard)	III	40	20–30	Intercrop summer type with slow growing vegetables to save space.
Spinach	Virginia Savoy, Melody, Bloomsdale Longstanding, Tyee, Olympia	Chenopodiaceae (Goosefoot)	II	40	45–60	Grow during coolest months.
Turnips	Root/Tops: Purple-Top White Globe, Just Rite. Tops: All Top	Cruciferae (Mustard)	III	150	40–60	Grow for roots and tops. Broadcast seed in wide-row system or single file.

1. University of Florida Cooperative Extension Service.
2. Other varieties may produce well also. Suggestions are based on availability, performance, and pest resistance.
3. To practice crop rotation, group family members; avoid planting family members following each other.
4. Transplantability categories: I, Easily survives transplanting; II, Survives with care; III, Use seeds or containerized transplants only.
5. Days from seeding to harvest, values in parentheses are days from transplanting to first harvest.

For example, asparagus, chard, celery, lettuce, and cabbage represent edible leaves, flower parts, and stems. Sweet corn, okra, peas, peppers, tomatoes, and melons represent the vegetables of edible fruits or seeds. Onions, garlic, beets, radishes, potatoes, and turnips are vegetables categorized by their edible underground parts. Some vegetables may fit in more than one category. Turnips and beets are both edible underground vegetables and the leaves of both may also be eaten as a vegetable. Table 23-2 indicates the use of some of the vegetables.

Growing Season

Vegetables are either warm-season or cool-season crop according to their growing season. Table 23-2 also divides the vegetables into warm and cool seasons. In Chapter 11, Table 11-2 lists the actual temperature requirements of some vegetable crops.

STEPS IN VEGETABLE PRODUCTION

Planning goes before planting. Regardless of the size of the vegetable production enterprise, the individual must decide where, what, and how much to plant.

Site

Gardens need to be located for convenience on a site close to a source of water with at least eight to ten hours of direct sunlight. Where possible, the site should allow rotation for weed and other pest control.

Soil type at the site is important. Most vegetables do best in well-drained, loamy soil with a pH of 5.8 to 7 (Table 23-3).

Size

The size of the vegetable growing area will depend on the ground available and the use of the produce. What to plant also depends on the use and the market for the produce. How much to plant depends on the crop or crops selected and the amount of available space.

What to Plant

What to plant is determined by the type of vegetable enterprise. For example, a home garden and market garden will have a variety of vegetables determined by the likes of a family or the demands of a roadside or farmer's market. A large-scale, commercial operation will concentrate on large areas of single vegetable crops.

The days to harvest influences the type of vegetable grown. Some areas may not provide enough growing days to produce a crop (Table 23-2). Also, if the crop is not all consumed or sold fresh, some plans for storage and preservation should be made.

Diseases, insect pests, and crop rotation may further influence the vegetable crops grown.

Table 23-3 Nutrient Requirements and pH for Vegetables

Crops	Fertilizer Requirements[1]			
	Nitrogen	Phosphorus	Potash	pH
Asparagus	4	3	4	6–7
Bean, bush	1	2	2	6–7.5
lima	1	2	2	5.5–6.5
Beet, early	4	4	4	5.8–7
late	3	4	3	5.8–7
Broccoli	3	3	3	6–7
Cabbage, early	4	4	4	6–7
late	3	3	3	6–7
Carrot, early	3	3	3	5.5–6.5
late	2	2	2	5.5–6.5
Cauliflower, early	4	4	4	6–7
late	3	3	4	6–7
Corn, early	4	3	3	6–7
late	3	2	2	6–7
Cucumber	3	3	3	6–8
Eggplant	3	3	3	6–7
Lettuce, head	4	4	4	6–7
leaf	3	4	4	6–7
Onion	3	3	3	6–7
Parsley	3	3	3	5–7
Parsnip	2	2	2	6–8
Pea	2	3	3	6–8
Potato	4	4	4	4.8–6.5
Radish	3	4	4	6–8
Spinach	4	4	4	6.5–7
Squash, summer	3	3	3	6–8
winter	2	2	2	6–8
Tomato	2	3	3	6–7
Turnip	1	3	2	6–8

[1.] 1 = Light; 2 = Moderate; 3 = Heavy; 4 = Extra heavy

SITE PREPARATION

Site preparation includes primarily the soil. This includes the addition of organic matter, lime, or fertilizer. But first the area needs to be plowed, tilled, or spaded.

Plowing

Plowing, tilling or spading is the first step in preparing a seedbed. The soil is worked into a fine, firm seedbed for planting time. Commercial vegetable production may require special seedbed preparations (Figure 23-3). To maintain the physical structure of the soil, ground should never be plowed when it is too wet. Plowing should be done at least three weeks before planting.

Figure 23-3 Commercial spinach production requires a specialized seedbed to allow for mechanical harvest.

Organic Matter

Many soils benefit from applications of various forms of organic matter such as animal manure, rotted leaves, compost, and cover crops. These are applied by thoroughly mixing liberal amounts of organic matter into the soil well in advance of planting, preferably at least a month before seeding. For example, in some areas 25–100 pounds of compost or animal manure can be added per 100 square feet. When inorganic fertilizers are not used, well-composted organic matter may be applied at planting time (Figure 23-4). Because of inconsistent levels of nutrients in compost, accompanying applications of balanced inorganic fertilizer may be beneficial. Organic amendments low in nitrogen, such as composted yard trash, must be accompanied by fertilizer to avoid plant stunting.

Another type of organic matter used is green manure in the form of a cover crop. Off-season planting and plowing down of green-manure crops include legumes, ryegrass, lupine, and hairy vetch.

Figure 23-4 Community gardens grown on soil fertilized with dairy manure compost.

Liming

A pH test of the soil helps determine if liming is necessary for the type of crop being grown. If a soil test indicates a pH of 6.0 or less, lime should be applied (Table 23-4).

Table 23-4 Planting Guide for Vegetables

Crop	Seeds/Plant per 100 Feet	Spacing (Inches) Rows	Plants	Seed Depth (Inches)
Warm-Season Vegetables				
Beans, bush	1 lb.	18–30	2–3	1–2
Beans, pole	½ lb.	40–48	3–6	1–2
Beans, lima	2 lb.	24–36	3–4	1–2
Cantaloupes	½ oz.	60–72	24–36	1–2
Corn, sweet	2 oz.	24–36	12–18	1–2
Cucumbers	½ oz.	36–60	12–24	1–2
Eggplant	50 plts./1½ pkt.	36–42	24–36	½
Okra	1 oz.	24–40	6–12	1–2
Peas, southern	½ oz.	30–36	2–3	1–2
Peppers	100 plts./1 pkt.	20–36	12–24	½
Potatoes, sweet	100 plts.	48–54	12–24	—
Pumpkin	1 oz.	60–84	36–60	1–2
Squash, summer	1½ oz.	36–48	24–36	1–2
Squash, winter	1 oz.	60–90	36–48	1–2
Tomatoes, stake	70 plts./1 pkt.	36–48	18–24	½
Tomatoes, ground	35 plts./1 pkt.	40–60	36–40	½
Watermelon, large	⅛ oz.	84–104	48–60	1–2
Watermelon, small	⅛ oz.	48–60	15–30	1–2
Watermelon, seedless	70 plts.	48–60	15–30	1–2
Cool-Season Vegetables				
Beets	1 oz.	14–24	3–5	½–1
Broccoli	100 plts./⅛ oz.	30–36	12–18	½–1
Brussels sprouts	100 plts./⅛ oz.	30–36	18	½–1
Cabbage	100 plts./ ⅛ oz.	24–36	12–24	½–1
Carrots	⅛ oz.	16–24	1–3	½
Cauliflower	55 plts./⅛ oz.	24–30	18–24	½–1
Celery	150 pltS./⅛ oz.	24–36	6–10	¼–½
Chinese cabbage	125 plts./⅛ oz.	24–36	12–24	¼–¾
Collards	100 plts./⅛ oz.	24–30	10–18	½–1
Endive/Escarole	100 plts.	18–24	8–12	½
Kale	100 plts./⅛ oz.	24–30	12–18	½–1
Kohlrabi	⅛ oz.	24–30	3–5	½–1
Leek	½ oz.	12–24	2–4	½

(continued on next page)

Table 23-4 Planting Guide for Vegetables *(continued)*

Crop	Seeds/Plant per 100 Feet	Spacing (Inches) Rows	Spacing (Inches) Plants	Seed Depth (Inches)
Cool-Season Vegetables *(continued)*				
Lettuce: crisp, butterhead, leaf, and romaine	100 plts.	12–24	8–12	½
Mustard	¼ oz.	14–24	1–5	½–1
Onions, bulbing	300 plts./sets 1–1½ oz.	12–24	4–6	½–1
Onions, bunching	800 plts./sets 1–1½oz.	12–24	1–2	2–3
Onions, multipliers	800 plts./sets 1–1½ oz.	18–24	6–8	½–¾
Parsley	¼ oz.	12–20	8–12	¼
Peas, English	1 lb.	24–36	2–3	1–2
Potatoes	15 lbs.	36–42	8–12	3–4
Radish	1oz.	12–18	1–2	¾
Spinach	1oz.	14–18	3–5	¾
Strawberry	100 plts.	36–40	10–14	—
Turnips	¼ oz.	12–20	4–6	½–1

Source: Florida Cooperative Extension Service

Lime needs are best met two to three months before the crop is to be planted. The lime must be thoroughly mixed into the soil to a depth of six to eight inches and then watered to promote the chemical reaction.

Fertilizing

Unless very large quantities of organic fertilizer materials are applied, commercial fertilizer is usually needed for gardens. A soil test also helps determine the amount and type of fertilizer that should be added to the vegetable crop. Vegetable crops have different fertilizer needs (Table 23-3), and proper fertilization will increase yields.

PLANTING VEGETABLES

Vegetables all are started as seed. Some seeds can be sown where they will grow, and others need to be started in a special environment such as a cold frame, hotbed, or greenhouse and then transplanted. Table 23-2 indicates the transplantability for many vegetable crops.

Depending on the size of the vegetable enterprise, the type of seed and the quantity of vegetable seeds are planted by:

+ Hand, in rows or hills
+ Broadcasting, by hand or machine
+ Seeders, powered by hand or tractor

Figure 23-5 Transplanting tomatoes on a commercial operation.

If vegetables are to be transplanted as seedlings (Figure 23-5 and Table 23-2), they can be:

+ Hand set
+ Hand-machine set
+ Machine set

Seeding rate, row and plant spacing, and seeding depth are listed in Table 23-4.

CULTURAL PRACTICES

Common cultural practices include weed control, irrigation, and pest and disease control.

Weed Control

The primary purpose of cultivation is to control weeds. Weeds are easier to control when they are small. Depending on the size of the garden, weed control is best accomplished by hand-pulling, hoeing, mechanical cultivation, or **mulching**. Chemical herbicides may be used on larger operations.

Mulching. Mulching involves the artificial modification of the surface of the soil with straw, leaves, paper, polyethylene film, or a special layer of soil. The mulch is spread around the plants and between the rows. Mulching helps control weeds, regulates soil temperatures, conserves soil moisture, and provides clean vegetables. Plastic film is often used as mulch with cantaloupes, cucumbers, eggplants, summer squash, and tomatoes (Figure 23-6).

Irrigation

Commercial vegetable operations plan for irrigation to ensure a crop. Rainfall is seldom sufficient or uniform for a high-yielding vegetable

Safety First

With more poisonous (toxic) pesticides, the use of personal safety equipment is necessary to prevent accidental exposure to pesticides through your clothing, body openings, or skin. Some of this equipment will also protect your eyes and prevent inhalation of toxic chemicals. Personal safety equipment is effective only if it fits correctly, is cleaned and maintained, and is used properly. Always select equipment that gives maximum protection.

The greatest cause of pesticide poisoning is pesticides contacting the skin. Not all parts of your body will absorb pesticides at the same rate. Therefore, adequate equipment for one situation may not be satisfactory for another. The precautionary statement of the label will describe the hazards associated with a pesticide and the type of protection equipment that must be worn when handling, mixing, or applying the pesticide.

Recommended protective clothing and equipment will include some or all of the following:

+ Waterproof apron made from rubber or synthetic material
+ Waterproof boots or foot covering made of rubber or synthetic materials
+ A daily change of coveralls or clean outer clothing
+ Waterproof pants and jackets if there is any chance of becoming wet with spray (never wear cotton fabric without additional protective clothing when there is a chance of getting wet)
+ Face shield, goggles, or full-face respirator
+ Waterproof gloves without any type of absorption lining, and made from rubber or synthetic material
+ Waterproof, wide-brimmed hat with a nonabsorptive headband
+ Cartridge-type respirator, a fitted rubber face piece, and replaceable filters

Figure 23-6 Plastic mulch applied to a field of commercial vegetables.

crop. Depending on their location, some gardeners may get by with rainfall alone, but the dry regions of the West need irrigation.

The type of irrigation depends on the location. Vegetable operations use sprinkler systems, drip systems, surface irrigation, subirrigation (water permeates soil from below), and trickle irrigation.

Pest and Disease Control

Vegetable producers scout the crop twice weekly for insect damage and spray with the appropriate pesticide when needed. Soil-inhabiting insects including mole crickets, wireworms, cutworms, and ants can be controlled with a broadcast preplant application of diazinon. Baits containing Dylox or diazinon are effective for cutworms and mole crickets. Metaldehyde controls slugs.

Pesticides should be applied strictly according to manufacturers' precautions and recommendations. Use of pesticides should be limited to the controlling of insects and diseases only as necessary. Application should be stopped during the harvesting season. Finally, pesticides need to be applied in the early evening to avoid killing bees and thereby reducing pollination.

Some soils contain nematodes, microscopic worms that can seriously reduce growth and yield of most vegetables by feeding in or on their roots. Nematode damage is less likely in soils with high levels of organic matter and in which crops are rotated so that the same members of the same family are not planted repeatedly in the same soil. Excessive nematode populations may be reduced temporarily by soil **solarization**. To solarize soil, vegetation is removed, soil is broken up, and then it is wetted to activate the nematode population. Next, the soil is covered with a sturdy, clear plastic film during the warmest six weeks of summer. High temperatures (above 130°F) must be maintained during this time for the greatest reduction in nematodes.

Diseases can be controlled by exclusion—the purchase of disease-free plants only. Also, plants and growing areas should be examined for common symptoms of diseases to avoid gross movement of infested soil.

Eradication is another tool for disease control. Certain soil-borne diseases such as damping-off, root and stem rots, and wilts are especially troublesome on old crop sites. Site and crop rotation can slow or prevent the incidence of certain soil-borne diseases. Growers avoid growing vegetables of the same family repeatedly in one area, and they watch for early disease symptoms (Figure 23-7).

Choosing resistant varieties prevents diseases. Growers choose adapted varieties with resistance or tolerance to the diseases common in their area.

Planting fungicide-treated seeds prevents disease. Untreated seeds can be dusted with a captan or thiram fungicide. Many common diseases can be controlled with either chlorathalonil, maneb, or mancozeb fungicide. Powdery mildews can be controlled with triadimefon, sulfur, or benomyl, and rusts can be controlled with sulfur or ziram. Bacterial spots are controlled with basic copper sulfate plus maneb or mancozeb.

Figure 23-7 Cantaloupe plants showing the onset of yellow vine disease. Research continues to find disease-resistant varieties. (Photo courtesy of USDA ARS)

HARVESTING

Vegetables should be harvested when they are at the peak of maturity. At this point, the produce has reached its optimal size; the flavor is fully developed; texture is just right; it keeps best and produces a quality processed product. The best harvesting time varies with each vegetable crop. Some vegetables hold their quality for only a few days, and others can hold their quality over a period of several weeks (Figure 23-8). Table 23-5 provides guidelines for harvesting some of the vegetable crops.

Figure 23-8 Fresh fruits and vegetables require proper timing of harvest and care afterward to maintain flavor and nutritional value. (Photo courtesy of USDA ARS)

Table 23-5 Some Guidelines for Harvesting Vegetables

Vegetable	When to Harvest
Asparagus	Not until third year after planting when spears are six to ten inches above ground while head is still tight. Harvest only six to eight weeks to allow for sufficient top growth. The "fern" that then develops should be left to grow for the rest of the summer.
Snap beans	Before pods are full size and while seeds are about one quarter developed, or two to three weeks after first bloom.
Lima beans	When the seeds are green and tender, just before they reach full size and plumpness, and when pods first reach the stage when they open easily.
Beets	When 11/4 to 2 inches in diameter.
Broccoli	Before the individual flowers in the head begin to open their yellow flowers. The dark green head should be tight and flat-topped. Side heads will develop after the central head is removed, but will be smaller at maturity.
Carrots	When 1 to 1½ inches in diameter.
Cabbage	When heads are solid and before they split. Splitting can be prevented by cutting or breaking off roots on one side with a spade after a rain.
Cauliflower	Before heads are ricey, discolored, or blemished. Tie outer leaves together above the head when curds are 2 to 3 inches in diameter; heads will be ready about 12 days after tying.
Sweet corn	When kernels are fully filled out and in the milk stage as determined by the thumbnail test. Use before the kernels get doughy. Silks should be dry and brown, and tips of ears filled tight. Generally, corn is ready at 19 or 20 days after silking, unless weather is cool.
Cucumbers	When fruits are slender and dark green before color becomes lighter. Harvest daily at season's peak. If large cucumbers are allowed to develop and ripen, production will be reduced. Keep large fruit from forming; otherwise fewer cukes will be formed. For pickles, harvest when fruits have reached the desired size. Pick with a short piece of stem on each fruit.
Eggplant	When fruits are half grown, before color becomes dull.
Kohlrabi	When balls are two to three inches in diameter.
Muskmelons	When stem easily slips from the fruit, leaving a clean scar.

(continued on next page)

Table 23-5 Some Guidelines for Harvesting Vegetables *(continued)*

Vegetable	When to Harvest
Onions	For fresh table use, when they are one inch in diameter. For boiling, select when bulbs are about 1½ inches in diameter. For storage, when tops fall over, shrivel at the neck of the bulb, and turn brown. Allow to mature fully, but harvest before heavy frost.
Parsnips	Delay harvest until after a sharp frost. Roots may be left safely in ground over winter and used the following spring before growth starts. They are not poisonous if left in ground over winter.
Peas	When pods are firm and well-filled, but before the seeds reach their fullest size.
Peppers	When fruits are solid and have almost reached full size. For red peppers, allow fruits to become uniformly red.
Potatoes	When tubers are large enough. Tubers continue to grow until vines die. Skin on unripe tubers is thin and easily rubs off. Such tubers will not store well. For storage, potatoes should be mature and vines dead.
Pumpkins and squash	Summer squash are harvested in the early immature stage when skin is soft and before seeds develop. Winter squash and pumpkin should be well-matured on the vine. Skin should be hard and not easily punctured by the thumbnail. Cut fruit off vine with a portion of stem attached. Harvest before heavy frost.
Turnips	When two to three inches in diameter. Larger roots are coarse, textured, and bitter.
Tomatoes	When fruits are a uniform red, but before they become soft. High-quality fruit can be obtained by harvesting at any time after pink color is evident and leaving such fruits to sit indoors for a few days.
Watermelon	When the underside of the fruit turns yellow or when snapping the melon with the finger produces a dull, muffled sound instead of a metallic ring. Also, curly tendrils on fruit stem probably will be turning brown.

When harvesting produce from some vegetable plants like peas, beans, and cucumbers, growers take care not to damage plants. Injured plants may be killed and stop producing fruit. Also, growers should never harvest vegetables when foliage is wet, as this practice may spread plant diseases.

Depending on the type of operation and the type of crop. harvesting is done by hand or by mechanical harvesters. Mechanical harvesters for larger commercial operations are designed to prevent injury to the crop.

STORING

The most effective method ever devised to increase the storability of fresh horticultural commodities is cooling. As a general rule of thumb, every 18°F decrease in temperature increases storability two- to threefold. Reducing the temperature of a tomato, for instance, from a field temperature of 100°F to 50°F, will improve storability eight- to twenty fold. This shows the importance of avoiding even short holding periods at field temperatures. In this example of tomatoes, those held one to two hours at field temperatures before cooling to 50°F would store approximately one day less than those cooled immediately to 50°F. The remarkable effect of temperature on vegetable storage is not only from its influence on the living tissue of the commodity being stored, but also on decay organisms. The most crucial temperature range for decay control seems to be 32–40°F. At these temperatures many decay organisms become inactive, or nearly so. Optimal cooling strategies require rapid cooling to the lowest temperature the commodity can withstand without inducing chilling or freezing injury. The process of rapidly removing heat from vegetables before shipping is called precooling.

Hydrocooling is one method of precooling. In this method, vegetables are immersed in cold water long enough to lower the temperature to the desired level.

In addition to the effect of temperature, storage humidity also has a marked influence on storability. The rate of water loss is proportional to the gradient in the concentration of water between the commodity and its environment. For a given temperature, water loss is proportional to the difference between the relative humidity (RH) of the interior of the commodity and the relative humidity of the environment. The internal RH is usually considered to be near 100 percent, thus the gradient is usually 100 minus the RH of the air. Knowing this, one can calculate that a storage RH of 80 percent will cause a commodity to lose water five times faster than a storage RH of 96 percent, which is a typical storage humidity. For a tomato at 70°F, this would amount to about 0.5 percent weight loss per day.

Interestingly, the amount of water vapor the atmosphere can hold declines with temperature, so the gradient in water vapor concentration (the gradient equals the water vapor concentration inside the product minus that of the surrounding air) for a commodity at 80 percent RH would be smaller if it were held at lower temperatures. In fact, the gradient at 80 percent RH is roughly four times smaller at 50°F that at 80°F.

The combined effects of temperature and RH can be used to describe a common, but undesirable, field phenomenon that occurs when vegetable temperatures are higher than surrounding air temperatures. In this situation, the driving force behind water loss, the water vapor gradient, is higher for the warmer vegetables, and they lose moisture more rapidly than cooler vegetables even though both are exposed to air of the same humidity. For example, if the air temperature is 80°F and the commodity temperature is 100°F, the water vapor concentration in the product would be twice that if it were at 80°F, leading to markedly greater moisture loss. If the RH of the 80°F air were at 50 percent, the 100°F vegetables would lose moisture three times faster than those at 80°F. Even more to the point, the rate of water loss for the 100°F product would be approximately 30 to 50 times faster than the product stored properly at 50°F and 90–95 percent RH.

To minimize water loss and the resulting decline in yield, the temperature should be reduced to the minimum for the commodity, and the RH at that temperature needs to be elevated to the maximum for the commodity. Table 23-6 provides the storage temperature and RH for some common vegetables. Even when stored at the proper temperature and RH, storage time for vegetables varies (Table 23-6).

Table 23-6 Storage Conditions and Life for Common Vegetables

Vegetable	Temperature (°F)	Relative Humidity (%)	Storage Life
Peas	32	95–98	1–2 weeks
Peppers	45–55	90–95	2–3 weeks
Potatoes	38–40	85–95	4–5 months
Summer squash	41–50	95	1–2 weeks
Sweet corn	32	95–98	5–8 days
Tomatoes	46–50	90–95	4–7 days
Turnip	32	95	4–5 months

SOME COMMON HAND TOOLS

Some of the hand tools for horticulture/garden work include hand trowel, hoe, hand cultivator, shovels and spades, garden fork, and flathead rake. (Refer to Appendix D for additional information about hand tools.)

Hand Trowel

Hand trowels are useful for working in containers and small spaces, mixing potting media, planting seedlings, and for digging out weeds. Most have a wooden handle and broad or narrow blades. Blade widths come in various sizes and lengths depending upon the intended use of the tool. Narrow, long blades are good for planting bulbs and digging up weeds, and wider blades are useful for transplanting small potted plants to the garden and mixing potting soil.

Hoe

Long-handled hoes are used to remove sod, weed beds, edge beds, and dig planting holes. Hoes are often considered the most useful tool.

Hand Cultivator

A hand cultivator is a three-pronged fork with a short wooden handle. The tool is helpful to dislodge weeds and to cultivate in narrow spaces between plants. Hand cultivators with longer handles and four-pronged forks are useful for compost management and cultivating flower beds without stooping.

Hand Weeder

A hand weeder is a semipronged, sharp steel blade attached to a handle. Hand weeders are great for weeding, breaking up the soil, and cultivating. The fish tail design allows the gardener to cut the long taproots of weeds.

Shovels and Spades

Shovels come in various shapes and sizes. Shovels are used to move soil and compost, dig holes for transplanting, and cut through sod to break ground. Spades have a flatter blade than a shovel. Spades are used to dig planting holes, sever invasive roots, dig in unprepared ground to improve drainage, or to dig holes for larger shrubs and young trees.

Garden Fork

Garden forks have four flat tines. They are used to break up hard or compacted soil; work compost, peat, or manure into the soil; and to turn beds in preparation for the next season.

Flathead Rake or Soil Rake

Flathead rakes or soil rakes are metal rakes with a straight head and teeth that are a few inches long and about an inch apart. They are used to prepare and level planting beds.

Tool Care

The best way to protect the investment in tools is to clean them after each use. A stiff bristle brush will clean shovels, spades, forks, and hoes. A stiff putty knife will help to clean off caked-on soil. Tools should never be put away wet. Tools need to be dried with a towel before lightly oiling the metal portion of the tools with an oily rag. Then tools should always be stored in a dry place after use.

The blades of cutting tools should be sharpened frequently. Sharper tools make garden work easier. Blades are sharpened by stone or file, along the beveled side. Very dull or nicked blades require more work and might need to be taken to a professional for sharpening.

HERBS

Besides vegetables, many people consider growing some of the herbs on a large scale because there seems to be a market developing for these plants. Herbs are grown for seasoning, aroma, or medicine (Figures 23-9 and 23-10). Table 23-7 provides propagation, growing conditions, and preservation details for a number of common herbs.

Figure 23-9 Chives in bloom.

Figure 23-10 Mint plants from which oil is extracted and used in a variety of products such as chewing gum and toothpaste.

Table 23-7 Common Herbs

Name/*Latin Name*	Type of Plant	Method of Propagation	Growing Conditions	Preserving
Angelica *Angelica archangelica*	Biennial or perennial	Seed	Moist, rich, soil; partial shade	Crystallize stems in sugar syrup
Anise *Pimpinella anisum*	Annual	Seed, self sows	Full sun, rich soil	Store dried seed in closed container
Balm *Melissa officinalis*	Perennial	Seed or divide root clumps	Full sun or partial shade	Dry or freeze leaves
Basil *Ocimum basilicum*	Annual	Seed	Full sun, rich soil	Dry or freeze leaves
Bay, sweet *Laurus nobilis*	Shrub	Semi hard cuttings in fall	Sun or partial shade, well-drained soil	Dry
Borage *Borago officinalis*	Annual	Seed, self sows	Full sun, well-drained soil	Possibly dried, but difficult
Burnet *Poterium sanguisorba*	Perennial	Seed	Full sun, well-drained, light soil	None
Caraway *Carum carvi*	Biennial	Seed	Full sun	Dry seed heads
Chervil *Anthriscus cerfolium*	Annual	Seed, self sows	Partial shade, moist soil, well drained	Use fresh leaves or dried
Chives *Allium schoenoprasum*	Perennial	Seeds or divide clumps	Full sun or some shade, rich soil	Dry or freeze leaves, seeds fresh
Cicely, sweet *Myrrhis odorata*	Perennial	Seeds or divided roots	Shade or partial shade, moist soil	Dry or freeze leaves
Coriander *Coriandrum sativum*	Annual	Seed	Full sun	Dry seeds
Costmary *Chrysanthemum balsamita*	Perennial	Divided roots	Full sun or partial shade, rich soil	Dry or freeze leaves
Dill *Anethum graveolens*	Annual	Seed	Full sun, moist soil, well drained	Dry leaves and seeds
Fennel *Foeniculum vulgare*	Perennial	Seed	Full sun	Dry leaves and seeds

Table 23-7 Common Herbs *(continued)*

Name/*Latin Name*	Type of Plant	Method of Propagation	Growing Conditions	Preserving
Garlic *Allium sativum*	Perennial	Plant from separated cloves	Full sun, light soil	Dry bulbs
Horseradish *Armoracia rusticana*	Perennial	Root cuttings	Full sun or partial shade, moist soil	Store main root in sand in a cool, dark, dry place
Hyssop *Hyssopus officinalis*	Perennial	Seed, self-sows or divided roots	Full sun, light, well-drained soil	Dry or freeze leaves
Lovage *Levisticum officinale*	Perennial	Seeds or root division	Full sun or partial shade, rich, moist soil	Dry leaves
Marjoram *Origanum majorana*	Perennial	Seeds or divided plants, or tip cuttings	Full sun, sheltered, rich soil	Dry leaves
Mint *Mentha spp.*	Perennial	Cuttings, root division	Partial shade, rich, moist soil	Dry or freeze leaves
Oregano *Origanum vulgare*	Perennial	Seed, cuttings, root division	Full sun	Dry leaves
Parsley *Petroselinum crispum*	Biennial	Seed	Sun or partial shade, rich, moist soil	Freeze
Rosemary *Rosmarinus officinalis*	Perennial	Seed, cuttings, layerings	Full sun or partial shade, light, well-drained soil	Dry or freeze leaves
Sage *Salvia officinalis*	Perennial	Seed, cuttings, division	Full sun, sandy soil	Dry leaves
Savory *Satureja hortensis*	Annual	Seed	Full sun	Dry leaves
Sesame *Sesamum indicum*	Annual	Seed	Full sun	Dry seeds
Shallot *Allium ascalonicum*	Perennial	Plant from separated cloves	Full sun	Use fresh or dry bulbs
Sorrel *Rumex* species	Perennial	Seeds or root division	Full sun or partial shade, moist soil	Dry or freeze leaves
Tarragon *Artemisia dracunculus*	Perennial	Root division, cuttings	Full sun	Dry or freeze leaves
Thyme *Thymus vulgaris*	Perennial	Seed, cuttings, root division	Full sun, sandy soil	Dry leaves

SUMMARY

Vegetables are either monocotyledons or dicotyledons. The parts of vegetables that are eaten vary from the seeds or fruits, to the leaves, flower parts, or stems to some underground part. Vegetable-production enterprises vary in size from the home gardener to truck cropping. Site selection is important for a successful garden. Factors to consider in site selection include water supply, other vegetation, soil type, and size. Soil preparation for a garden means plowing or spading and can often mean adding organic matter, lime, and fertilizer. Depending on the type of garden crop, seeds can be planted directly into the garden, or the plants may be started in a greenhouse and transplanted.

Once the vegetables are planted, the crop requires cultivation, weed control, irrigation, mulching, and pest control for a successful crop. For the best quality, vegetables should be harvested at the peak of maturity. To maintain quality after harvest, vegetables need to be stored under the proper conditions. Vegetables vary in the length of time they can be stored. Precooling a vegetable at harvest increases the success of storage.

Review

True or False

1. Herbicides rapidly remove heat from vegetables and weeds.

2. Polyethylene film is a type of mulching material for growing vegetables.

3. Dicotyledons have one seed leaf.

4. For vegetable gardening, the soil should be plowed to a depth of one to two inches.

Short Answer

5. What name is given to a person involved in the study of vegetable production?

6. To correct the acidity level in the soil, _____ is commonly added.

7. When is the best time to transplant vegetable plants?

8. Name the process for the rapid removal of heat from a vegetable crop before storage or shipment.

Critical Thinking/Discussion

9. Describe factors to consider for the proper storage of vegetables.

10. Discuss how the pH of the soil can influence the type of vegetables grown.

11. What factors should be considered when selecting a site for vegetable production?

12. Describe the edible parts of plants often called vegetables. Give examples of each.

13. How should the soil be prepared and maintained for growing vegetables?

14. Discuss the care of the vegetable crop once it is planted.

Knowledge Applied

1. Use outdated seed or garden catalogs to prepare a display of warm- and cool-season vegetables.

2. Use the Internet or other library resources to discover any of the vegetables in this chapter that are commercially grown and for which the U.S. Department of Agriculture tracks their production. Report on your findings.

3. Develop a chart or a computer presentation with pictures illustrating the various insects that attack vegetable plants. Find pictures of insects in extension service bulletins or on the Internet.

4. Plant tomato seeds or some other warm-season vegetable in a greenhouse for later transplanting.

5. Develop an experiment to store a selected vegetable under different conditions; for example, light/dark, dry/humid, or warm/cool. Record what happens to the vegetable under each set of conditions. Which is the best storage condition for the vegetable? Compare other vegetables under the same storage conditions.

Resources

Hackett, C., & Carolane, J. (1982). *Edible horticultural crops: A compendium of information on fruit, vegetable, spice and nut species.* New York: Academic Press.

Hodges, D. M. (Ed.). (2003). *Postharvest oxidative stress in horticultural crops.* Boca Raton: FL, CRC Press.

Huang, K. C. (1998). *The pharmacology of Chinese herbs* (2nd ed.). Boca Raton: FL, CRC Press.

Kowalchik, C., & Hylton, W. H. (Eds.). (2000). *Rodale's illustrated encyclopedia of herbs.* Emmaus, PA: Rodale Press.

Newcomb, D., & Newcomb, K. (1989). *The complete vegetable gardener's sourcebook.* New York: Prentice Hall Press.

Reader's Digest. (2000). *Illustrated guide to gardening.* Pleasantville, NY: The Reader's Digest Association.

Reiley, H. E. (2006). *Introductory horticulture.* Albany, NY: Delmar Cengage Learning.

Ryder, E. J. (1999). *Lettuce, endive and chicory (crop production science in horticulture).* Cambridge, MA: CABI Publishing.

Wijesekera, R. O. B. (1991). *The medicinal plant industry.* Boca Raton: FL, CRC Press.

Internet

Internet sites represent a vast resource of information. The URLs (uniform resource locators) for World Wide Web sites can change. Using one of the search engines on the Internet such as Google or Yahoo!, find more information by searching for these words or phrases: hydrocooling, market gardening, commercial vegetable production, organic vegetable production, herbs, home gardening, olericulture, truck cropping, vapor concentration, solarization, vegetable production, and food preservation.

Additionally, the PLANTS database (http://plants.usda.gov/index.html) provides standardized information about the vascular plants of the United States and its territories. It includes names, plant symbols, checklists, distributional data, species abstracts, characteristics, images, crop information, automated tools, onward Web links, and references. The PLANTS database is a great educational and general use resource.

Chapter 24
Small Fruits

SMALL FRUITS *include grapes, blueberries, strawberries, and the brambles. Brambles are red, black, and purple raspberries, and the erect and trailing blackberries. Also included as brambles are loganberries, boysenberries, dewberries, and tayberries. Small fruits are grown in a variety of locations and climates, depending on the type. Many are able to grow in cooler climates.*

After completing this chapter, you should be able to—

- Describe a fertilizer program for a vineyard
- List the characteristics of ripe grapes
- Name the procedures for weed control in vineyards
- Explain how to select a site for a vineyard
- Describe a site and its preparation for strawberry production
- Explain the cultural practices of weed control, mulching, and fertilizing
- List eight steps for better strawberry production
- Diagram four planting systems for strawberries
- List the three main soil requirements for highbush blueberry production
- Identify the two types of commercially grown blueberries
- Outline a fertilizer and pruning program for bramble fruits
- List five key points to consider when selecting a site for brambles
- Identify seven measures for disease prevention in raspberry planting
- Describe currant and gooseberry production

KEY TERMS

Brambles
Canes
Heeling In
Hill System
Laterals
Plasticulture
Pruning
Runners
Training
Trellis

GRAPES

Grapes are a popular homegrown fruit as well as an important commercial crop. Grapes have many uses and are very nutritious. They are the best-known vine fruit. They are consumed fresh, as juices and wines, as raisins, jam and jelly, and as frozen products. Grapes are native to the United States. Many of the domesticated varieties have native wild grape ancestry in at least one parent. American fine wine grapes are native to central Asia. Grapes are easily grown commercially in most areas of the United States and have a wide range of flavors. They do not grow well in arid sections and in areas having extremely high temperatures and very humid conditions.

Basically, three kinds of grapes are grown in the United States:

1. European or Old World grapes (*Vitis vinifera*). These make up about 95 percent of the grapes grown in the world and they are used for table, raisin, and wine grapes.
2. North American grapes (*Vitis labrusca* or *Vitis rotundifolia*). These are hardy, native, disease-free grapes. Well-known examples include the Concord, Delaware, and Niagara grape.
3. Hybrids. These grapes are crosses between the European grape varieties and the native American species. They are selected for the quality of fruit from their European ancestors and for their disease and insect tolerance from their native American ancestors.

Grapes require three to four years for the vines to mature before producing fruit (Figure 24-1). They also require at least a growing season of 140 frost-free days.

Site

A standard variety like Concord needs 160–165 days for its growing season. Variety selection will depend on the growing season. The varieties listed in the variety section are early to midseason types.

The vines are healthier if good weather occurs after the crop is produced so that the vines recover from producing the crop and have time to get ready for winter.

The soil should be well drained, 30 inches deep, and with a pH of 5.0–8.0. An excessively rich soil, high in organic matter, produces heavy, late-maturing crops with low sugar content. Light soils produce light yields of early-maturing fruit with high sugar content, but vine growth is weak. There should be a minimum of cold-warm-cold cycles during the winter. Exposure to alternating cold and warm temperatures causes winter injury. The site should allow for good air drainage to reduce the amount of frost injury.

Varieties

When selecting varieties, growers consider the number of days required to mature the crop and the cold temperature hardiness. Varieties listed here mature fairly early and are hardy. Other varieties should be considered

Figure 24-1 Flame seedless grapes ready for harvest. (Photo courtesy of USDA ARS)

only if they mature in a time comparable with the growing season for a specific area. Other characteristics are the type and uses of the fruit.

Popular varieties are:

* Concord—a blue variety, hardy; fruit good for juice, jelly, jam, wine
* Niagara—a white grape, somewhat less hardy than Concord
* Delaware— a pink-red grape, somewhat less hardy than Concord, low vigor
* Fredonia—a blue variety, less productive than Concord
* Golden Muscat—a white variety, vigorous, productive
* Moore Early—a blue variety, fruit quality not as good as Concord; the fruit sometimes cracks badly
* Beta—very early, very hardy, productive, a blue grape, berries with high sugar and acid
* Seyval—early midseason, a white grape, moderately vigorous, productive

Depending the on variety and local conditions, the days required for grapes to mature ranges from 140 to 170 days.

Table Varieties

Fresh table grapes come in three basic colors: green (sometimes called white), red, and blue-black. More than 50 kinds of table grapes are currently in production, but the following sections describe 17 major varieties. Each variety possesses a distinct color, taste, texture, and history.

Bloom. Table grapes are often covered with natural bloom, which is a delicate white substance common on many soft fruits. The bloom protects the grape from moisture loss and decay.

The Greens.

* Perlette Seedless—The first grape of the season, the Perlette is light in color, almost frosty green with a translucent cast; the berries are nearly round. Perlette means "little pearl" in French.
* Thompson Seedless—Almost everyone is familiar with this grape's light green color, oblong berries, and sweet, juicy flavor. The variety may have originated in southern Iran.
* Sugraone—The Sugraone berry is bright green and elongated. The fruit offers a light, sweet flavor and a distinctive crunch.
* Calmeria—This grape carries the nickname "lady fingers," so called for its elongated, light green and delicately sculpted berries. A winter treat, this seeded grape has a mild, sweet flavor with an unforgettable tang.

The Reds.

* Flame Seedless—The result of a cross between Thompson Seedless, Cardinal, and several other varieties, the Flame Seedless is a round, crunchy, sweet grape with a deep-red color (Figure 24-1).
* Crimson Seedless—This blush-red variety has firm, crisp berries with a sweetly tart, almost spicy, flavor.

- Emperatriz—These red, medium-sized berries show slight yellow hues throughout their firm skins. The Emperatriz is sweet and fruity.
- Rouge—These seeded, dark-red berries are large and oval with a firmly crisp, thick-skinned texture and a mildly sweet, earthy flavor.
- Ruby Seedless—Grown commercially in the San Joaquin Valley since 1968, the Ruby Seedless is a deep-red, tender-skinned grape.
- Emperor—Large, deep-red clusters and a lasting flavor characterize this seeded variety that was first planted in California in 1863. In East Coast cities, where European traditions remain strong, the Emperor is very popular.
- Red Globe—The large, remarkable clusters of the Red Globe contain plum-sized, seeded berries. The Red Globe is popular for both eating and decorating during the holiday season
- Christmas Rose—Another relative newcomer, this light-red seeded variety ripens through December. Developed from four older varieties, the berries are large with a tart-sweet flavor.

The Blue-Blacks.

- Exotic—Developed in 1947 in Fresno, California, Exotic berries are plump and juicy and grow in long, beautiful clusters. A cross between the red Flame Tokay and the Ribier, this seeded grape is crisp and mild in flavor.
- Fantasy Seedless—These blue-black sweet berries are oval, thin-skinned, and firm. Fantasy's conical clusters have medium-sized berries with pale green flesh and a mellow flavor.
- Ribier—This dark blue-black seeded grape crossed the Channel from Orleans, France, in 1860 to become an English "hothouse" variety. The skins are firm and the taste is mild.
- Niabell—This Concord-type variety features thick-skinned, round berries ranging in color from purple to black with an earthy, rich flavor.

Planting

Growers plant grape vines in the spring. Nurseries offer either one- or two-year-old plants. The two-year-old plants cost more than one-year-old plants. At planting time, the dead or broken roots are cut off and the tops are pruned back to two or three buds. The grower should set vines at eight-foot spacings in rows eight to nine feet apart and plant the vines two inches deeper than they were growing in the nursery.

Propagation

Grapes are propagated commercially by hardwood cuttings. If a disease-resistant rootstock is needed for certain varieties, plants are grafted or budded to a disease-resistant seedling rootstock. Chapter 17 details hardwood cutting.

Trellis Construction

Only one type of trellis and training method will be discussed. The trellis is the two-wire trellis. Growers should use durable posts about three inches

Plant Breeders Produce New Berry Varieties

Producing better varieties of plants is an on-going challenge for plant breeders. Thanks to the work of scientists at the Fruit Laboratory in Beltsville, Maryland, growers will have four new selections of commercially available blueberries from which to choose. The Little Giant variety is bred for the cooler climates of Washington, Oregon, Michigan, New Jersey, and New York. It offers an alternative variety for frozen and processing markets, and can be planted with other northern highbush blueberry varieties for cross-pollination.

Pearl River, Magnolia, and Jubilee are new varieties and more suited to the warmer climates in the Gulf Coast and southeastern United States. They can be interplanted with other Southern highbush blueberry varieties to ensure fruit set, early ripening, and maximum yield.

Each of the four new varieties is productive and disease-resistant.

Also, two new June-bearing strawberries have been introduced by the Agricultural Research Service plant breeders for the Middle Atlantic and adjacent regions. Primetime is a midseason berry. It bears fancy, good-quality, large fruit. Latestar is a late-season variety. It produces large, attractive fruit. Both varieties are recommended for shipping and local markets and both varieties resist multiple fungal diseases. They produce well on either light or heavy soils, and in matted rows or in hill culture. Plants are available in season at nurseries.

in diameter. About two-and-one-half feet of the eight-foot long post should be buried in the ground. Posts are set at 16-foot intervals along the row. End posts may need to be longer to give good support and strength. If the end posts are weak or poorly braced the trellis will be weak and sag.

Wires fasten to the post so they slide through the fastening device. This allows the tension on the wires to be increased to keep them taut. The bottom wire should be three feet from the ground; the top wire about five feet. Number 10 galvanized wire is suitable for use on grape trellises (Figure 24-2).

Figure 24-2 New grapes and trellis system.

Training

The training system described is called the four-arm Kniffin. Plants trained to the four-arm Kniffin system consist of a vertical stem that reaches to the top wire plus an upper and lower branch on each side of the trunk. These branches are tied to the wires. During the first year, the strongest shoot is trained to grow upward and is fastened to the first wire. During the second growing season, it reaches and is tied to the top wire. During the winter, cut off the portion of the trunk that extends above the wire. During the second growing season, the vine will have sent out side branches near the bottom wire. One branch on each side of the trunk should be trained to grow on the bottom wire. During the second winter these side branches should be pruned back to five buds each at the time the excess trunk is cut off. During the third growing season, the upper branches form.

Pruning

Fruit is produced only on the previous year's wood. Vines allowed to grow for many years without pruning accumulate much old, unproductive wood. The object of **pruning** is to remove all the old wood and leave four **canes** that will produce next year's crop. The four canes are the four arms or branches described in the training section. In addition, two to four other canes, called renewal spurs, are left to produce the fruiting canes for the following year. Renewal spurs only have two to three buds and, like the four fruiting canes, originate from the trunk.

Grape pruning regulates the health and vigor of the vines. This is done by pruning the vines so the fruit load is about what the vines are able to mature. The system is based on the vigor of the vines in the previous growing season.

Grapes are pruned during the dormant season in late winter or early spring. Buds that have begun to swell are easily knocked off during the pruning operation. When pruning, growers select the four canes that will be the fruiting wood for the coming growing season. They also select the canes to serve as the renewal spurs. The renewal spurs should originate on the trunk near the point where a wire crosses. This positions canes properly. Once the renewal spurs and fruiting wood have been selected, all other growth is removed. The wood that is left should be dark brown and slightly larger than a pencil.

Weed Control and Fertilizing

Weeds are controlled by cultivating no deeper than three to four inches. The amount of fertilizer needed varies with the soil. Soil testing and foliar analysis will help monitor plant needs. Growers avoid giving too much nitrogen.

Harvesting

Grapes should be left on the vine until fully ripe because ripening stops once the grapes are harvested. Color is not a good indicator of ripeness as it changes two to three weeks before the grapes are ripe. Sugar content also increases as grapes become ripe. The stem of ripe grape clusters will be brownish and wrinkled. Ripe grapes are easily pulled from the cluster.

Pest Control

Grapes are sprayed for insects and fungal diseases. Growers should consult a local county agricultural agent or locate the recent U.S. Department of Agriculture bulletin on spraying grapes. Major diseases that require control are black rot, powdery mildew, and downy mildew. The major insect pests include the flea beetle, leaf hopper, berry moth, and Japanese beetle.

STRAWBERRY

The strawberry is moderately easy to grow when compared with other types of fruits. A strawberry planting lasts about three years before it needs replanting.

Site

A good site slopes about 2 feet in 100 feet. This allows good drainage of both water and air. Good air drainage allows the cold air to flow off, reducing susceptibility to frost injury. Plants in low areas may be covered by water during winter thaws. Then, when the temperature drops after the thaw, the freezing water may kill the plants.

The best growing conditions are full sun and sandy-to-gravely loam with a pH of 6.0–6.5 and a good supply of organic matter. Good drainage is essential for best growth. Strawberry **runners** root better on light soils. Everbearing varieties may give the best results on rich soil. Growers avoid sites infested with nutsedge, quack grass, or persistent problem weeds, and they spray problem weeds with herbicides before planting the strawberries. If the strawberries follow sod there could be problems with white grubs or wireworms.

Flower Bud Formation

June-bearing plants form flower buds in the short days of late summer and fall. This is why the strawberry patch should not be neglected after the berries have been picked. Poor care late in the season leads to a poor crop the following year. Low temperatures are needed for the flower buds to complete their development. Everbearers form flower buds in the longest days of summer and flower and fruit in summer and fall. The longer days of summer trigger runner formation (Figure 24-3).

Figure 24-3 Parts of the strawberry.

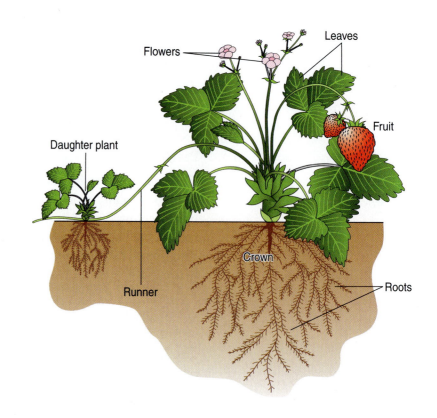

Figure 24-4 Strawberries (Chandler variety) ready for market. (Photo courtesy of USDA ARS)

Varieties

Many varieties of strawberries are available (Figure 24-4). Some varieties are listed here. Everbearing varieties may not be as productive or have as high a quality as June-bearing varieties. Some varieties include:

+ Raritan—midseason, good flavor and yields but susceptible to red stele and wilt diseases
+ Delite—late, resistant to red stele and wilt but forms too dense a matted row
+ Redchief—midseason, resistant to red stele and wilt but berries are hard to cap
+ Holiday—midseason, large, firm berries but plants are not very disease-resistant
+ Guardian—midseason, resistant to red stele and wilt but berries are light-fleshed, rough, and green tipped
+ Earliglow—early, resistant to red stele and wilt
+ Midway—midseason variety
+ Marlate—late, high-quality berries but low yields; berries have light flesh, are hard to cap
+ Gem—everbearer, hardy and productive but a poor runner producer, berries are soft, acidic
+ Ozark Beauty—everbearer, best everbearer for north-central United States
+ Scarlet—midseason variety for home gardens

Planting

Strawberries fit into a crop rotation system when grown in the vegetable garden. Strawberries should not follow strawberries, tomatoes, peppers, or eggplant. These crops are all susceptible to verticillium wilt. Where wilt has been a problem, growers should use only disease-resistant varieties. The soil should be tested before planting. The best yields of strawberries are obtained from new plants set new each year. Everbearers give the best crop the year they are planted. June-bearing plants give the best crop the year after they are planted.

Growers purchase virus-free plants. Healthy strawberry plants have medium-to-large crowns and large root systems consisting of light-colored roots. Plants with black roots are old and should not be planted.

Strawberries can be planted as soon as the soil can be worked in the spring. When the plants arrive, they should be unpacked and either planted or heeled in. **Heeling in** is the temporary planting of plants in a trench. Plants should not be left heeled in for longer than two to three weeks.

Strawberry plant crowns should be at soil level. If set too deep, the crown rots. The roots should not be allowed to dry out while the plants are waiting to be planted. Roots are spread out like a fan when planting, then firm down the soil around them. The spacing depends on the training system used. Early spring planting promotes the formation of highly productive runner plants. Planting systems for strawberries vary, depending on the environment and production goals. In many of these systems, the plants are grown on raised beds as annuals. This results in removal of the plants, plastic mulch, and irrigation system at the end of every season.

Some planting systems include raised bed **plasticulture**, matted row system, hedgerow system, spaced bed system, and **hill system**.

Raised Bed Plasticulture. Organic and conventional growers in California and Florida, where most of the nation's strawberries are produced, tend to favor the raised bed plasticulture system (Figure 24-5). They grow plants as annuals, transplanting strawberry crowns in the late summer or early fall. Production starts in the late winter and continues through the summer and into late fall, depending on the area and the varieties grown. Because methyl bromide is not allowed in organic production, crop rotation, green manure crops, and compost are critical to control soil-borne diseases and pests.

Beds can be narrow or wide. Plastic mulch is used in both narrow and wide beds and can vary from a single strip of plastic laid between the plants to full bed coverage, where holes must be punched for the plant to develop. Some conventional growers in California use clear plastic, which warms the bed faster, stimulating early-season growth; these growers use fumigation to control most weeds. Black plastic is used in organic production, primarily for weed control. Because the black plastic prevents the sun's rays from penetrating, the beds remain cool, resulting in slower initial growth of the plants and reduced irrigation frequency compared with clear plastic mulch.

Drip irrigation is often used to provide water to the growing plants. Raised beds provide good drainage. Because they make the flowers and

Figure 24-5 Strawberries being grown in raised bed plasticulture. (Photo courtesy of USDA ARS)

fruit easier to see and reach, raised beds also help growers to forecast yields, while making harvesting easier and faster.

Plasticulture is an annual system of planting freshly dug plants in the fall. Plants and plastic are removed after spring harvest and the process begins again the next fall.

Matted Row System. The most common system is the matted row because it is easy to establish. The plants are set at two feet in rows three to four feet apart. Plants should not be allowed to form too many runner plants. An overcrowded matted row produces small berries. The runner plants should be 4–6 inches apart with the matted row 15–18 inches wide. The finished rows should be about 18 inches apart. This system is used most with June-bearing plants, but it is the least productive training systems for strawberries.

Hedgerow System. In the hedgerow system, plants are set 15–18 inches apart in rows spaced 24–30 inches apart. Only two runners are allowed to grow, one from each side of the plant. The runners are trained to grow in line with the row, not out to the sides. In the double hedgerow system, each plant is allowed to form four runners that are placed diagonally from the row. Looking down on a double hedgerow system row, the row would look like a row of Xs. The mother plant would be at the center of each X and the runners would form the four arms. In either hedgerow system, the runner plants end up being about one foot apart. This system produces high-quality fruit but considerable time is spent getting it established.

Spaced Bed System. To establish the spaced bed system, plants are set 24–36 inches apart in rows that are 42–48 inches apart. The runners are placed by hand to give a finished row 15–24 inches wide. Plants are spaced eight inches apart; after the bed is filled, all other runners are removed. A great deal of labor is involved in this system. Once established, the planting must be gone over to remove surplus runners. The spaced bed system produces larger berries and higher yield than the matted row.

Hill System. The hill system is used for everbearers. Plants are set 12–15 inches apart. Because all runners are removed, three rows can grow together to form one large row. The three rows are spaced one foot apart. Growers stagger the plants so they do not line up across the row. Space the triple rows two feet apart. All of the runners are removed before they are two weeks old. This system produces the maximum-sized berries.

Fertilization

Before planting, build up soil organic matter. Well-rotted manure is good for strawberries. Ten days after planting, growers fertilize with 12-12-12 at the rate of 2–3 pounds per 100 feet of row. The application is repeated in four to six weeks. Fertilizer should not be more than four inches away from the crown. Fertilizer can be broadcast over a row if the foliage is dry. Applying too much nitrogen causes the plants to make excessive leaf growth and causes soft berries that rot easily. Too much nitrogen will also delay ripening.

Bearing beds should not be fertilized in the spring. Growers wait until after the berries are harvested.

Removing Flower Stalks on New Plants

The flower stalks should be removed from new plantings. Stalks are removed as soon as they appear until the first of July. Allowing the plants to set fruit reduces the amount of runner formation. This reduces the number of plants, thus reducing the yield the following year. Varieties that form many runners may need the flower stalks removed only until the plants are established. Blossoms are taken off everbearing varieties for the first 60–80 days, then the plants can be allowed to set fruit. There is no benefit gained by allowing the fruit to form and then not to pick it.

Weed Control

In home plantings, the primary weed-control methods are shallow cultivation and hoeing. Problem perennial weeds are killed with herbicides before planting the strawberries. Mulching will help control weeds. Chemicals may control some weeds in strawberry plantings.

Mulching. Mulching strawberry plantings is strongly recommended. The mulch provides winter protection, eliminates dirty berries, delays blooming, reduces weed growth, conserves moisture, and decreases fruit rot. Mulch for winter protection is applied when the temperature is consistently about 20°F, normally in November. Plants may be damaged if the mulch is applied too early or too late. If applied too early, a number of warm days may injure the plants. If applied too late, some crown injury may occur from the cold temperatures. Crown injury occurs at temperatures below 20°F. The mulch should be put on before the ground freezes.

A two- to three-inch layer of straw or hay may be used for mulch material. A one-inch layer of sawdust may be used. The mulch should be left on as long as possible to delay bloom in the spring. Check the plants often. When the leaves begin to grow yellowish-green, growers remove the mulch. The plants will grow through a thin layer of mulch.

Watering

Strawberries need generous watering during and immediately following bloom. If the plants get too dry while developing fruit, the fruits will be smaller than they should be. Daytime watering should be about 1–1½ inches of water per week. This includes any rain that occurs during the week.

Frost Protection

Sprinkling can be used for frost protection. As water freezes on the plants it gives up heat. This heat can prevent plant injury even though they will be covered with ice. Sprinkling provides protection down to 22°F. Sprinklers are started as soon as the temperature at plant level reaches 32°F. The sprinklers will have to run as long as there is ice on the plants. Once all the ice has melted, the sprinklers may be turned off. Another method of frost protection is covering the plants with the recently removed winter protective mulch.

Harvesting

The first ripe berries appear about 30 days after the first blossoms open. Hot weather hastens ripening, shortens the harvest season, and makes frequent picking necessary. In moderate weather, harvesting every other day should be sufficient. In hot weather it may be necessary to harvest every day. Growers pick the berries in the morning when they are cool, and harvest only the ripe berries at a single picking. Cooling the berries removes field heat and increases shelf life. Berries to be frozen should be left on the plants until fully mature. When harvesting, diseased, rotted, or injured berries should be picked and thrown away to help control rot.

Renovating the Bed

A strawberry bed may last only two to three years. The best year will be the first year after planting, but by the third year the yield will be down by about two-thirds. A strawberry patch should be renovated only if the plants are vigorous and healthy. The first step, once harvest is done, is to mow off the foliage about ½ inch above the crown. The plant crowns are injured if the mower setting is too low. Then the rows are narrowed down 10–12 inches and thinned, leaving only the most vigorous looking plants.

For renovation to succeed, it must be done right after harvest. After renovation, manage the bed just as though it were a new planting. If the mowing and renovation is delayed too long, yields the next year will be reduced. The yearly application of fertilizer should be put on right after renovation.

BLUEBERRY

The lowbush blueberry is the low-growing type that grows wild. The highbush blueberry is the type usually planted in commercial plantings. The information in this chapter applies to the highbush blueberry. Blueberries do not need cross-pollination in order to produce a crop (Figure 24-6).

Figure 24-6 Highbush blueberry plant with ripe berries. (Photo courtesy of USDA ARS)

Some varieties are Bluecrop, Bluejay, Rubel, and Jersey. Grow varieties that mature at different times to prolong the harvest.

Site

Highbush blueberries need an average growing season of 160 days and are badly injured by temperatures of −20 to −25°F.

A loose soil is best. A mixture of sand and peat gives excellent results. Heavier-textured soils are suitable if acidic and high in organic matter. Peat soils may stimulate late growth that fails to harden before winter; such soils are often in frost pockets. Blueberries must have an acidic soil with a pH between 4.0 and 5.1. The pH can be lowered with applications of sulfur. If the pH is very high, it may not be practical to try to lower the pH. A soil with a constant moisture supply is best. The water table should be within 14–22 inches of the soil surface most of the time, but good surface drainage is needed.

Good air drainage reduces frost injury and helps control diseases. Problem weeds should be controlled before the blueberries are planted. Unless a home gardener has almost ideal conditions for blueberries, it may be wise to grow some other fruit.

Planting

The best planting stock is two or three years old. The three-year-old stock usually costs more. The best planting time is early spring. Set the plants four feet apart in rows that are ten feet apart. The plants should be set about two inches deeper than they were growing in the nursery. Growers can mix a shovelful of acidic peat into each hole at planting time if the soil is sandy and low in organic matter.

Cultivation

Blueberries are shallow-rooted, so cultivation should be no deeper than two to three inches. A cover crop may be sown after harvest.

Mulching

Most organic materials may be used, but they should be allowed to weather before being applied to the blueberries. The mulch should be about six to eight inches deep. Until the mulch has decomposed, the amount of nitrogen should be doubled. Leguminous mulches such as clover or soybean vines may be harmful.

Fertilizer

Growers avoid using nitrate forms of nitrogen and chloride forms of potassium. The nitrate and chloride may be toxic to blueberries. Growers use a blueberry fertilizer such as 16-8-8-4. The 4 refers to magnesium. New plants should be fertilized four weeks after planting. Established blueberries are fertilized in April before growth starts. The fertilizer should be spread evenly and kept off wet plants to prevent injury. A very sandy soil may need a second fertilization. A foliar analysis may be helpful in identifying specific nutritional needs.

Pruning

Fruit is produced on the previous season's wood. New plantings need no pruning until about the third year. Growers prune during the dormant season to remove the small, twiggy growth near the base of the plant. After the third year, growers remove dead or injured branches, fruiting branches that are close to the ground, spindly and bushy twigs on mature branches, and older stems of low vigor. Heavier pruning gives larger, but fewer, berries. Old black canes are removed at ground level.

Watering

Blueberries need one to two inches of water at ten-day intervals during dry weather.

Harvesting

Blueberries ripen over a period of several weeks. Three to five pickings are needed to harvest all of the berries. Pick only the ripe berries. A reddish tinge means the berry is not yet ripe; only ripe berries should be picked.

Pest Control

Diseases and pests of blueberries include mummy berry, other fungal diseases, scale insects, plum curculio, fruit worms, leaf hoppers, and the blueberry maggot. Occurrence of these depends on location. These can be controlled with insecticides and fungicides.

RASPBERRY

Red raspberries are considered a bramble. Other brambles include the loganberry, boysenberry, dewberry, and tayberry. Once established, a red raspberry planting should produce over a number of years. The most productive years are usually the third through the sixth (Figure 24-7).

Varieties

Some varieties include Latham, Heritage, Fall Gold, Taylor, September, Amber, Canby, Fall Red, Golden Queen, Hilton, Sentry, and Newburgh. Sodus and Clyde are varieties of purple raspberries. Black raspberry varieties include Logan (New Logan), Cumberland, Allen, Bristol, and Huron.

Site

A site with some slope to aid water and air drainage is preferred. Protection from wind will help. The junction of the cane and crown is quite weak, so strong winds may cause the cane to fall over, especially if

Figure 24-7 Raspberry plant with ripe berries.

the cane is carrying a heavy fruit load. If protection cannot be provided, a trellis will support the plants.

When possible, red raspberries should be 300 feet away from other cultivated or wild raspberries to help control virus disease. Raspberries should not follow tomatoes, eggplant, potatoes, peppers, and other raspberries. These crops are susceptible to verticillium wilt, a soil-borne disease that builds up if susceptible crops are grown. Black and purple raspberries are most susceptible.

The best soil is a well-drained, slightly acidic loam or clay loam, although raspberries are fairly tolerant. Growers avoid sites with subsoil that prevents good drainage or good root penetration.

Herbicides control problem perennial weeds before the raspberries are planted.

Planting

Red raspberries can be planted in early spring as soon as the soil can be worked. When the plants arrive from the nursery, they need to be kept cool, or, if they cannot be planted right away, they should be heeled in or stored. If stored, the storage area should have a temperature of about 35°F.

The plant spacing depends on which of the several possible training systems is used. Do not allow the plants to dry out before or during planting. Plant them about one inch deeper than they were growing in the nursery. The portion of the stem that was below ground is a different color. The hole should be big enough to allow the roots to spread out normally. The soil is firmed around the roots and the tops are cut back to about six inches. Cutting back may be done before or after planting. There can be a heavy loss of newly planted raspberry plants, but suckers produced by the surviving plants can be used to fill in gaps.

Training Systems

The hedgerow is the most common system for growing red raspberries. The plants are set two-and-one-half to three feet apart in rows from six to ten feet apart. Where space is limited, use closer spacings. Suckers will fill in the row, but do not allow rows to get wider than one to one-and-one-half feet. Wider rows are more difficult to spray and harvest. A two-wire trellis, with one wire running down each side of the row, can be used on windy sites. A single wire running down the middle of the row and to which plants are tied may also be used. Everbearing varieties especially need support.

To establish the hill system, plants are set five to six feet apart in rows that are also five to six feet apart. A stake is driven into the ground next to each plant. As the plant produces suckers, five to eight healthy, vigorous suckers, spaced at intervals around the stake, are kept. The canes produced by the suckers are tied to the stake with two or three ties during the spring pruning. The tied suckers form a roughly triangular outline. The canes are cut back to the height of the stake in the spring.

The linear system is established just as the hedgerow system. The only difference is that all the suckers produced by the plants are removed. The new fruiting canes come from the plant crown and are not the suckers. The suckers come up at a distance from the plants. The crown shoots all arise from the plant crowns.

Fertilizing

Growers fertilize 10–14 days after planting. Fertilizer should be three to four inches from the shoots and canes. Fertilization can be increased in the second year.

After the last cultivation, a cover crop may be sown. Any annual crop, such as oats or sudan grass, which dies during the winter may be used.

Watering

Raspberries use much water, especially when fruiting. They need about one inch of water per week and perhaps more during hot, windy weather. A lack of water is a serious problem during the time from just before fruiting through the fruiting period. Watering is most critical from the time the fruit begins to show color until picking has been completed. A good water supply in late summer enhances cane vigor and enhances productivity in the following year. Water the raspberries during the day.

Weed Control

Cultivation controls weeds but should not be deeper than three to four inches to avoid injuring raspberry roots. Growers cultivate until harvest, then once or twice after harvest. Herbicides labeled for use on raspberries may also be used.

Pruning

Red raspberries are pruned when dormant and after canes have fruited. The canes are biennial, so a cane emerges and grows during one year, then bears a crop of berries and dies the following year. The exception is everbearers. Remove canes that have fruited right after harvest. The early removal of these canes may help control pest problems and maximize the water and nutrients available to new canes. Everbearers produce a crop during the late summer and fall of the same year. The same canes have a crop the following spring, but it is not as large as the fall crop. The cane dies after the spring crop is harvested; canes can be removed at this point. This gives growers an opportunity to do different types of pruning. If the large, fall crop is enough to satisfy family needs, the canes can be cut to the ground during the winter or early spring. This effectively eliminates the spring crop. It also eliminates doing both an after-harvest and a dormant pruning. Only one pruning will be needed. If the second crop is wanted, prune off any winter-killed cane tips during the dormant season. Remove canes completely when they have finished bearing the spring crop.

Other types of red raspberries need a dormant pruning to remove weak or damaged canes. In the linear or hill systems, thin the canes to six to eight per hill. In the hedgerow system, the canes should be spaced eight inches apart. In the hill and linear systems, shorten the canes to about five and-one-half feet. In the hedgerow system, shorten the canes to four feet. If the canes are shorter than these heights, take off only the portion that has been injured in the winter. Dormant pruning is done before the buds swell in the spring. If pruned too early, winter kill may reduce the height further. Dormant pruning reduces the number of suckers to keep the rows from becoming jungles.

Pest Control

Insects are not usually as destructive to raspberries and other bramble fruits as are diseases. Control of disease introduction is most important, but some spraying may be necessary depending on location. Growers should consult a local county extension agent or university for details on spraying and controlling diseases in their location.

Harvesting

Raspberries for the fresh market are hand-harvested. Machine-harvested berries are generally taken to a processor for freezing or juicing. Mechanical harvest requires a business that can handle large volumes of berries in one day.

BLACKBERRY

The two types of blackberries are erect and trailing. The trailing blackberry is also called dewberry. These are usually tied to trellises and ripen earlier than the erect types.

Varieties

Trailing varieties may not be as hardy as the erect types in some climates. Lucretia is a trailing blackberry variety. Some erect blackberry varieties include: Alfred, Baily, Darrow, and Hedrick (Figure 24-8).

Figure 24-8 Blackberries on branch at various stages of ripeness. (Photo courtesy of USDA ARS)

Site

A good moisture supply is needed, especially when the fruit is ripening. If the site is low or poorly drained, winter hardiness of the plants is reduced.

Planting

The erect types are propagated by suckers or root cuttings, with root cuttings preferred. Trailing types may be propagated from root cuttings or tip layers. Thornless varieties are only propagated from tip layers.

Plants that cannot be planted immediately can be heeled in. If the plants are dry, soak them in water for several hours prior to planting. Before or just after planting, cut the tops back to six inches. The planting depth should be the same as it was in the nursery. Plant erect types five feet apart in rows that are eight feet apart; plant trailing varieties four to six feet apart in rows eight feet apart. Vigorous trailing varieties are spaced at 8- to 12-foot spacings in rows 10 feet apart.

Training

Erect blackberries are most easily trained to a single wire suspended 30 inches from the ground. As canes grow, they are tied to the wire when they cross it. The trailing types are trained to a two-wire trellis. The first wire is three feet from the ground, and the second wire is five feet from the ground. The canes are tied horizontally along the wire, or the canes are fanned out then tied to a wire where they cross. Sometimes the canes fruiting in the current year are tied on one wire and canes fruiting the following year are tied on the other wire.

Pruning

The cane tips of erect blackberries are pinched off in summer when the canes are 30–36 inches tall. The pinching stimulates the formation of laterals. In winter, cut the laterals back to 12 inches. Fruiting canes are removed once the crop has been harvested. When the fruiting canes are being removed, thin out the new canes. Leave three to four new canes on each plant or five to six canes per linear foot or row. Remove all suckers that appear between the rows.

Pruning trailing types is not as complicated as that of erect types. When the fruit has been harvested, remove the canes that produced the crop. At the same time, thin the new canes. Leave 8 to 12 canes; if the variety is a semitrailing, leave 4 to 8 canes.

Fertilizing

Blackberries are fertilized when they blossom. Growers use 5-10-5 fertilizer at 5–10 pounds per 50 feet of row.

Weed Control

Cultivation is the primary means of weed control. Growers cultivate only two to three inches deep near the row, and they stop cultivating about a month before freezing weather arrives.

Harvesting

Blackberries for the fresh market are hand-harvested. Machine-harvested berries are generally taken to a processor for freezing or juicing. Mechanical harvest requires a business that can handle large volumes of berries in one day.

CURRANT AND GOOSEBERRY

Currants and gooseberries cannot be planted in some areas unless a permit is obtained because black currants can be a host for white pine blister rust.

A planting of currants or gooseberries should last for 10–12 years.

Varieties

Some red currant varieties (Figure 24-9) include: Red Lake, Wilder, Cascade, Prince Albert, and Viking. White grape is a white currant variety.

Gooseberry Varieties

Two types of gooseberries (Figure 24-10) are available—American and European. The European varieties have larger and more flavorful fruits. The American varieties are healthier and hardier. Some American gooseberry varieties are Downing, Houghton, and Poorman. European gooseberry varieties include Fredonia, Chautauqua, and Industry.

Site

Currants and gooseberries prefer cool, moist growing conditions. The soil should be well drained and high in organic matter, avoiding light sandy soil. A dry soil may cause premature leaf drop on gooseberries, causing the fruit to sunscald from lack of shading. Gooseberries like a partially shaded growing area. Select a site with good air circulation and avoid frost pockets.

Figure 24-9 Ripe red currants. (Photo courtesy of USDA ARS)

Figure 24-10 Gooseberries.

Planting

Growers control any problem perennial weeds before planting. The soil should be worked the fall before planting; if available, work in well-rotted manure in the fall or early spring prior to planting. Fall planting right after the plants go dormant is best. Spring planting will have to be very early, as currants and gooseberries are quick to begin growth. Use 1- or 2-year-old plants and space them 4–5 feet apart in rows 8–11 feet apart. Remove any broken or injured roots or branches. Prune the top to within 6–10 inches of the ground. Set the plants so the lowest branch is just below the soil surface. Make sure the soil is firmed down around the roots.

Cultivation and Mulching

Cultivation should be shallow and continued until the harvest is completed. Mulches may be used as a substitute for cultivation; however, a six-inch layer of mulch can attract mice. Any natural material may be used as mulch.

Fertilizer

The best fertilizer for currants and gooseberries is manure. If applied annually, the plants will be productive. On infertile soil use a fertilizer with a 1-1-1 ratio such as 12-12-12. Fertilize in the fall after growth stops or in the spring before growth starts. If fresh sawdust or straw is used as a mulch, double the fertilizer rates the year the mulch is applied.

Pruning

Growers prune when the plants are dormant in late winter or early spring. At the end of the first season, remove all but six or eight of the most vigorous shoots. At the end of the second season leave four or five one-year shoots and three to four of the two-year shoots. At the end of the third season, leave three to four first-, second-, and third-year shoots. Each plant should have a total of 9–12 canes.

Growers prune older plants so they have 6 to 10 fruiting canes and 3 to 4 replacement canes. Leave only enough new canes to replace the older canes that are removed. Wood older than three years produces inferior fruit. Remove all branches that lie on the ground. The center of the bush should be fairly open.

Harvesting

Harvesting may last as long as a month.

PRUNING TOOLS

The basic tools for pruning are shears and saws. The exact type and size is determined by the kind of pruning. This discussion includes the most commonly used hand tools, but an endless list of air, electric, and hydraulic power equipment is available. Also, many specialized machines exist for specific types of pruning or hedging operations.

The primary tool for pruning fruit trees or grapevines is the pruning shear. Probably the most common size used for pruning fruit trees is a shear with a 24- to 36-inch handle. For grapevines, a shorter handle is more comfortable and more efficient. For some types of vines, a hand shear may be preferred.

All shears must be kept sharp and in good working order. A small pocket stone is convenient when sharpening is necessary in the field. When sharpening, use a small amount of light oil on the stone (which floats away the filings). Sharpen only the outside of the flat blade of the shears. Never remove any metal from inside the horn or the flat blade. Keep the bevel on the shear blade at a 45-degree angle.

The accumulation of sap and dirt can be removed with steel wool. Blades should be wiped with an oily cloth after buffing them with steel wool.

SUMMARY

Grapes, blueberries, strawberries, and the brambles are considered small fruits. Brambles include red, black, and purple raspberries, and the erect and trailing blackberries. Also included as brambles are loganberry, boysenberry, dewberry, and tayberry. Small fruits are grown in a variety of locations and climates, depending on the type grown. Many are able to grow in cooler climates. They require a wide variety of cultural practices, depending on type, use, and location. They are propagated by asexual methods. After harvesting, which is often by hand, the small fruits may be eaten fresh or they may be preserved or further processed.

Review

True and False

1. Grapes are commercially propagated by grape seeds from the dried grape.

2. Grapes are pruned just after new fruit has set.

3. Strawberries grow best at a soil pH of 5.0–5.5.

4. Strawberries for the fresh fruit market are generally picked with machines.

5. Highbush blueberries are propagated by budding.

6. The best planting depth for red raspberries is two inches.

7. Red raspberries are propagated by shoots.

Short Answer

8. How often should strawberries be picked?

9. Name a trellis type and training system used in grape production.

10. What is the best test for ripeness in table grapes?

11. When do everbearing strawberries form fruit?

12. Name two common systems for growing strawberries.

13. Why should strawberries not be planted in soil recently planted with tomatoes, potatoes, peppers, or eggplant?

14. What is the best pH range for the cultivation of blueberries?

15. Why are blueberries pruned?

16. List five key points to consider when selecting a site for brambles.

17. Identify the two types of blueberries.

18. List five key points to consider when selecting a site for brambles.

Critical Thinking/Discussion

19. Describe the characteristics of ripe grapes.

20. What is the best site for a vineyard?

21. Discuss eight steps for better strawberry yields.

22. Outline a fertilizer and pruning program for one of the bramble fruits.

23. What type of growing conditions do currants and gooseberries prefer?

24. Explain the importance of watering raspberries.

Knowledge Applied

1. Grow a grape plant from a hardwood cutting.

2. Develop a presentation or chart describing all of the products from grapes.

3. Grow several strawberry plants and observe the formation of runners.

4. Develop a presentation or chart describing the products and uses of strawberries.

5. Make a presentation on how to prune a bramble.

6. Develop a list of all the brambles grown in your area and describe how these brambles are used.

Resources

Brown, L. (2002). *Applied principles of horticultural science* (2nd ed.). Boston: Butterworth-Heinemann.

Gough, R. E., & Crandall, P. C. (1995). *Bramble production: The management and marketing of raspberries and blackberries*. Boca Raton, FL: CRC Press.

Gough, R. E., & Poling, E. B . (1997). *Small fruits in the home garden.* Boca Raton, FL: CRC Press.

Hackett, C., & Carolane, J. 1982). *Edible horticultural crops: A compendium of information on fruit, vegetable, spice and nut species.* New York: Academic Press.

Hancock, J. F. (2000). *Strawberries (crop production science in horticulture).* Cambridge, MA: CABI Publishing.

Reader's Digest. (2000). *Illustrated guide to gardening.* Pleasantville, NY: The Reader's Digest Association, Inc.

Rieger, M. (2006). *Introduction to fruit crops.* Binghamton, NY: Food Products Press.

Internet

Internet sites represent a vast resource of information. The URLs (uniform resource locators) for World Wide Web sites can change. Using one of the search engines on the Internet such as Google or Yahoo!, find more information by searching for these words or phrases: grapes, strawberries, blueberries, brambles, raspberries, blackberries, small fruits, loganberry, boysenberry, dewberry, tayberry, and vineyards.

Many Web sites maintained by cooperative extension service in each state provide growing guide for small fruits, and these can be found by doing a search for a specific small fruit. The National Sustainable Agriculture Information Service Web site (ATTRA; http://attra.ncat.org/) provides many guides to growing and selling small fruits.

Additionally, the PLANTS database (http://plants.usda.gov/index.html) provides standardized information about the vascular plants of the United States and its territories. It includes names, plant symbols, checklists, distributional data, species abstracts, characteristics, images, crop information, automated tools, onward Web links, and references. The PLANTS database is a great educational and general use resource.

To find a good book on a specific crop, search Amazon.com.

Chapter 25
Fruit and Nut Production

FRUIT OPERATIONS provide high-quality and bountiful varieties of fruits and nuts. Homegrown fruits can be enjoyed at the peak of ripeness. Supermarket fruits aim for this quality. Commercial fruit and nut enterprises supply the fruits and nuts for supermarkets nationwide.

After completing this chapter, you should be able to—

- Identify the benefits of fruit or nut production as a personal enterprise
- Name fruit and nut crops
- Describe how to plan and prepare a site for fruit or nut production
- Identify how to plant fruit and nut trees
- Describe appropriate cultural practices
- List some procedures for harvesting and storing a fruit or nut crop
- Discuss how to prune and why prune

FRUIT AND NUT PRODUCTION BUSINESS

Today, the production of fruit and nut crops is more a business than a way of life. Production requires the grower to be a good financial manager, as well as a scientist and farmer. The person who is a fruit grower or fruit scientist is called a pomologist. People who work in the industry must be able to propagate fruit and nut trees, as well as plant and transplant, prune, thin, train, and fertilize them.

The production, harvest, and marketing of fruits and nuts is a large industry. The United States accounts for a sizable portion of the combined world crops of apples, pears, peaches, plums, prunes, oranges, grapefruit, limes, lemons, and other citrus fruits. Table 25-1 lists the production

Table 25-1 Noncitrus Fruit and Nut Production

Crop	Average production (1,000 tons)[a]	Crop	Average production (1,000 tons)[a]
Fruit		**Tree Nuts**	
Apples	4,817	Almonds (CA)	966
Apricots	72	Hazelnuts (OR)	36
Cherries (sweet and tart)	210	Macadamias (HI)	25
Dates (CA)	18	Pecans	139
Figs (CA)	50	Pistachios (CA)	155
Kiwifruit (CA)	29	Walnuts (CA)	340
Nectarines	245		
Olives	99		
Peaches	1,102		
Pears	849		
Plums and prunes	196		

[a] Production of three recent years, 2005, 2006, 2007
Source: U.S. Department of Agriculture National Agricultural Statistics Service.

Figure 25-1 Citrus production/use in the United States. (Image courtesy of USDA NASS)

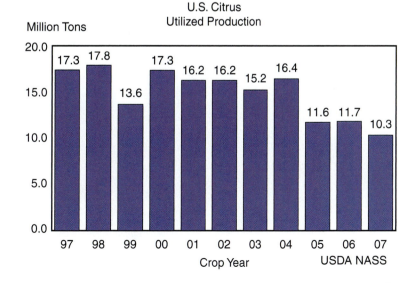

U.S. Citrus Utilized Production

of tree fruits and nuts in the United States. Figure 25-1 shows citrus production in the United States for the most recent 10 years.

TYPES OF FRUITS AND NUTS

The tree fruits and nuts include many types and varieties. Temperate fruit trees and their common and scientific names are listed in Table 25-2. Subtropical and tropical fruit trees are listed in Table 25-3.

Besides their family and scientific names, fruit trees can be identified as drupe or pome (Table 25-2). A **drupe** is a fruit with a large, hard seed called a stone. **Pome** fruits have a core and embedded seeds.

Fruit trees can also be grouped according to their growth habit—standard, **semidwarf**, and **dwarf**. A standard tree has its original **rootstock** and grows to normal size (Figure 25-2). For example, a standard apple tree can grow to be 30 feet high. Semidwarf and dwarf trees are standard varieties of trees grafted onto dwarfing rootstocks. These rootstocks cause the tree to produce less annual growth and the size of the tree produced determines whether it is called a dwarf or semidwarf (Figure 25-2). Semidwarf trees reach about 10–15 feet in height and dwarf trees only reach 4–10 feet.

Table 25-4 identifies the common nut trees in the United States.

Table 25-2 Temperate Fruit Trees

Group	Common Name	Scientific Name
Pome fruits	Apple	Malus domestic
	Pear	Pyrus communis
Drupe or stone fruits	Peach	Prunus persica
	Nectarine	Prunus persica
	Plum	Prunus spp.
	Cherry	Prunus spp.
	Apricot	Prunus armeniaca

Table 25-3 Subtropical and Tropical Fruit Trees

Group (Family)	Common Name	Scientific Name
Citrus	Orange	Citrus sinensis
	Grapefruit	Citrus paradisi
	Lemon	Citrus limon
	Tangerine	Citrus reticulata
	Limes	Citrus aurantifolia
	Tangelos	C. reticulata x C. paradis
Lauracea	Avocado	Persea americana
Oleacea	Olive	Olea europaea
Palmae	Date	Phoenix dactylifera
Moraceae	Fig	Ficus carica
Musaceae	Banana	Musa spp.
Caricaceae	Papaya	Carica papaya
Anacardiaceae	Mango	Mangifera indica

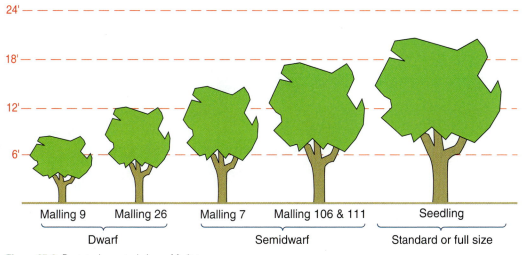

Figure 25-2 Rootstocks control sizes of fruit tree.

Table 25-4 Nut Trees

Family	Common Name	Scientific Name
Rosaceae	Almond	Prunus amygdalus
Juglandaceae	English Walnut	Juglans regia
	Pecan	Carya illinoensis
Betulaceae	Filbert	Corylus avellana
Proteaceae	Macadamia	Macadamia ternifolia

SELECTING FRUIT OR NUT TREES

Besides being grown commercially, fruit and nut trees may be grown in home gardens or as a part of a landscaping plan. Growers need to consider the climate, rootstock selection, frost susceptibility, fertility, pollination, and growth patterns and fruiting dates.

Climate

Some fruit and nut trees are more suited for one type of climate than another. For example, the citrus fruits grow only in tropical and subtropical climates such as Florida, Texas, and California (Table 25-2). Growers should seek advice on fruit and nut trees that will grow in their area.

Rootstock

Rootstock selection is important because many of the fruit and nut trees are propagated by grafting. Disease-free, hardy rootstock should be sought for fruit and nut trees (Figure 25-3). Nurseries will provide this information about their trees. Tables 25-5 through 25-7 provide some suggestions for fruit and nut varieties.

Figure 25-3 English walnut trees grafted onto the more disease-resistant black walnut rootstock.

Frost Susceptibility

Locations for fruit and nut trees should be protected from frost. If an area is susceptible to late frosts, growers should select late-blooming varieties. Planting trees on hillsides, near ponds, or near cities will help prevent frost damage. Commercial growers use irrigation sprinklers, heaters, and fans to help reduce frost damage.

Table 25-5 Fruit and Nut Tree Production Guidelines[1]

Crop	Space per Trees (feet)[2]	Pollenizer Tree Needed	Approximate Years to Bearing	Sprays Usually Required to Control
Apples	5–40	Sometimes	2–10	Codling moth,[3] scab
Apricots	15–25	No	6–7	Brown-rot bacterial canker
Butternut	30–40	Yes	3–5	None
Cherries, sour	14–20	No	3–5	Fruit fly[3]
Cherries, sweet	20–35	Yes	6–7	Fruit fly,[3] bacterial canker
Chestnut	20–40	Yes	5–7	None
Figs	12–20	No	5–6	None
Filberts	15–20	Yes	5–6	Filbert moth,[3] bacterial blight
Hickory	20–40	Yes	10–14	None
Papaw	15–20	Yes	10–14	None
Peaches, nectarines	12–15	No	4–5	Leaf curl, borers, coryneum blight, brown rot
Pears	10–20	Yes	5–7	Fire blight, scab, codling moth[3]
Persimmons	15–20	Yes	8–10	None
Plums/Prunes	10–20	Some varieties	3–5	Crown borers, brown rot
Walnuts, black	30–40	No	10–12	Husk rot[3]
Walnuts, English	40–50	No	10–12	Husk fly,[3] blight

[1]. Source: Oregon State University Extension
[2]. The vigor of the variety and the rootstock, and the amount of pruning, also determine space requirements.
[3]. Insect, if uncontrolled, causes wormy fruit or nuts

Fertility

Producing high-quality fruit and nuts depends on the fertility of the soil. Generally, a pH of 5.5–6.0 is required. Fruit crops need to be fertilized annually, but too much fertilizer can damage the roots and create other problems. Depending on the type of fruit or nut tree grown, growers should seek expert advice for their location.

Pollination

Most fruit crops require pollination. It is necessary for the fruit to set or form. Depending on the type of fruit or nut tree grown, the pollination requirements can vary. Sometimes trees need to have several varieties that bloom at the same time for adequate cross-pollination. Other trees require just themselves for pollen. Table 25-5 indicates some of the fruit and nut trees that need a pollinizer. Bees also help with pollination.

Growth Patterns and Fruiting Dates

As Tables 25-5 and 25-6 indicate, growing fruit or nut trees is a long-term commitment. Some trees can take from 2 to 12 years to bear sufficient fruit or nuts, but they will then continue to bear for years (Table 25-6).

Most fruit trees go through a vegetative and productive period of growth. During the vegetative period the tree grows vigorously and rapidly. In some trees this may be four to five feet in a single season.

Table 25-6 Starting and Bearing Time for Fruit Trees

Tree	Years to Start	Years Bearing
Pear	3–5	25–75
Plum	3–4	10–20
Quince	3–4	10–20

Table 25-7 Some Fruit and Nut Varieties and Approximate Time of Maturity[1]

Variety	Approximate Time of Maturity	Comments
Apples		
Lodi	July 15–30	Yellow, won't keep
Earlygold	Aug. 1–15	Yellow, crisp
Stark Summer Treat	Aug. 1–15	Red, good flavor
Summerred	Aug. 1–15	Red, good flavor
Gravenstein	Aug. 15–30	Pollenized by Lodi, not hardy, best sauce apple

(continued on the following page)

Table 25-7 Some Fruit and Nut Varieties and Approximate Time of Maturity[1] *(continued)*

Variety	Approximate Time of Maturity	Comments
Apples *(continued)*		
Jonamac	Sep. 1–10	Red, McIntosh-like
Elstar	Sep. 10–20	Tart, good flavor, cool climate
Gala	Sep. 15–25	Sweet, good flavor, heat tolerant
Jonagold	Sep. 15–30	Big, good flavor, cool climate, needs pollenizer
Spartan	Sep. 20–30	Red, productive
Delicious	Sep. 25–Oct. 5	Standard red, scabs badly
Golden Delicious	Oct. 1–10	Yellow, flavorful, very productive
Empire	Sep. 20–30	Small, red, flavorful
Braeburn	Oct. 5–15	Flavorful, stores well, productive
Fuji	Oct. 10–25	Sweet, flavorful, stores well
Granny Smith	Oct. 15–30	Tart, stores well
Newtown Pippin	Oct. 10–20	Green, tree vigorous, slow to produce
Apples, scab-resistant varieties		
Redfree	Aug. 5–15	Small, red, mild
Chehalis	Aug. 15–25	Yellow, big, long picking season
Prima	Sep. 1–10	Big, red, pits
Nova Easygro	Sep. 10–20	Good flavor
Liberty	Sep. 20–30	Best flavor, red
Jonafree	Sep. 25–Oct. 5	Medium, good flavor
Apricots		
Puget Gold	July	None
Rival	July	Mild flavor
Royal (Blenheim)	July	Self-fruitful
Moongold	July	Cold hardy, pollenized by Sungold
Sungold	July	Pollenized by Moongold, hardy
Chinese	July	Resists frost
Cherries, sour varieties		
Montmorency	July	Michigan strain best
North Star	July	Dwarf variety
Cherries, sweet varieties		
Royal Ann	Mid-summer	White, pollenized by Corum
Bing	Mid-summer	Black, pollenized by Van, Corum
Lambert	Late summer	Black, pollenized by Van, Corum
Van	Early summer	Black, pollenized by Bing, Lambert
Sam	Mid-summer	Black, pollenized by Lambert
Bada	Mid-summer	White, pollenized by Royal Ann, Bing, Lambert, semidwarf
Stella	Mid-summer	Self-fruitful, black
Compact Stella	Mid-summer	Smaller than Stella

(continued on the following page)

Table 25-7 Some Fruit and Nut Varieties and Approximate Time of Maturity[1] *(continued)*

Variety	Approximate Time of Maturity	Comments
Chestnuts		
Revival	Sep.	Pollenized by Carolina
Carolina	Sep.	Pollenized by Revival
Layeroka	Sep.	Reliable producer
Chinese seedling	Sep.	Pollenizer for Layeroka
Figs		
Brown Turkey	Aug.	Large, brown
Desert King	Aug.	Green, large, sweet
Lattarula	Aug.	Green, golden inside
Filberts (some regions too cold)		
Barcelona	Oct.	Standard variety, pollenized by Davianna
Davianna	Oct.	Light producer, pollenized by Barcelona
Nectarines (fuzzless peaches)		
Stark Red Gold	Aug.	Southern and northeastern Oregon only
Harko	Aug.	Better fruit set
Genetic dwarfs	Aug.	Grown in pots, take indoors for winter
Peaches		
Veteran	Aug. 20–25	Regular bearer
Red Haven	Aug. 5–10	Most popular, clingstone until fully ripe
July Elberta	Aug. 1–20	Old favorite
Early Elberta	Aug. 24–28	Old favorite
Rochester	Aug. 24–30	Old favorite
Reliance	Aug. 5–10	Resistant to cold
Frost	Aug.	Resists leaf curl
Genetic dwarfs	Summer	Very small trees, grown in pots, take indoors for winter
Pears, European varieties		
Bartlett	Aug. 15–30	Pollenized by Anjou, Fall Butter
Anjou	Sep. 5–20	Pollenized by Bartlett, needs 45–60 days of cold storage before ripening
Bosc	Sep. 10–30	Pollenized by Comice
Cascade	Sep. 10–30	Pollenized by Bosc, needs 45–60 days of cold storage before ripening
Seckel	Aug. 30–Sep. 10	Pollenized by Anjou, Bosc, Comice

(continued on the following page)

Table 25-7 Some Fruit and Nut Varieties and Approximate Time of Maturity[1] *(continued)*

Variety	Approximate Time of Maturity	Comments
Pears, red varieties		
Red Bartlett	Aug. 15–30	Pollenized by Anjou, Fall Butter
Reimer Red	Sep.	Pollenized by Bartlett
Red Anjou	Sep.	Pollenized by Bartlett
Starkrimson	Aug. 1–15	Pollenized by Bartlett
Pears, Oriental varieties		
Shinseiki	Aug.	Pollenized by Nijisseiki, Chojuro
Chojuro	Sep.	Pollenized by Nijisseiki, Shinseiki
Nijisseiki (20th Century)	Sep.	Pollenized by Chojuro, Shinseiki
Kikusui	Aug.	Pollenized by Chojuro, Nijisseiki
Persimmons		
Fuyu	Nov.	Seedless Japanese
Garrettson	Nov.	American, small
Early Golden	Nov.	American, small
Plums, cold-resistant varieties		
Mount Royal	Sep.	Self-fruitful
Superior	Sep.	Pollenized by Pipestone
Ember	Oct.	Pollenized by Superior
Plums, European varieties (prunes when dehydrated)		
Italian	Sep. 10–30	Tart, "purple plum"
Brooks	Sep. 20–30	Bears regularly, large
Parsons	Sep. 1–15	Pollenized by Stanley, sweet
President Plum	Sep. 20–30	Pollenized by Stanley
Moyer Perfecto	Oct. 1	Best dried, sweet
Stanley	Sep. 1–15	Bears but brown rots
Plums, Oriental varieties		
Early Golden	July	Apricot-like flavor
Red Heart	Sep.	Pollenized by Shiro
Shiro	Aug.	Pollenized by Red Heart
Burbank	Aug.	Pollenized by Elephant Heart
Walnuts, black varieties		
Thomas	Oct.	Seedlings inferior
Ohio	Oct.	None
Myers	Oct.	None
Walnuts, English varieties		
Franquette	Late Oct.	Standard variety, limited hardiness
Spurgeon	Late Oct.	Late bloomer, hardy
Chambers #9	Late Oct.	Heavy producer, moderately hardy

[1] Source: Oregon State University Extension Service

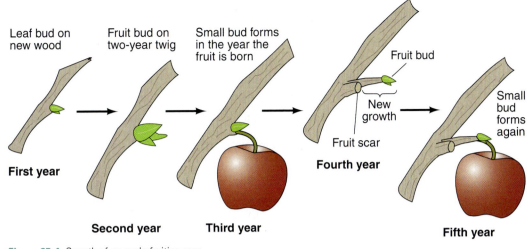

First year

Leaf bud on new wood

Second year

Fruit bud on two-year twig

Third year

Small bud forms in the year the fruit is born

Fourth year

Fruit bud

New growth

Fruit scar

Fifth year

Small bud forms again

Figure 25-4 Growth of an apple fruiting spur.

When the tree enters the productive period, growth slows and the fruit buds become larger and plumper (Figure 25-4).

Besides growth patterns, different varieties of the same fruit mature at different times of the year. Selection should be based on this and the variety that will satisfy market demands (Table 25-7).

SITE AND SOIL PREPARATION

The site should allow for maximum exposure to the sunlight, because most trees require full sunlight for maximum production. Land with a slight slope and good air circulation helps prevent frost damage. Soil at the site needs to be well drained and of medium texture.

Before planting, the soil should be sampled and tested for fertilizer requirements and pH. Also, the soil needs to be deeply plowed so the root systems can penetrate into the soil.

PLANTING

Trees from nurseries can be planted in the spring or the fall. Both seasons have their advantages. Each type of fruit or nut tree will have its own planting distance. Table 25-5 suggests the space per tree needed for some of the common fruit and nut trees.

Growers need to follow the guidelines for the actual planting. Each fruit or nut tree will have its own specifications. In general, the hole should be dug to be one-and-one-half to two feet wide and about one-and-one-half feet deep. The trees are planted so that the uppermost root is no more than two inches below the ground level. With dwarf trees, the graft union should be two to three inches above ground level. The roots need to be spread out in the hole, and dead parts need to be trimmed. Newly planted trees die because:

+ Roots suffocate by too-deep planting
+ Water standing in the hole
+ Late planting, leading to the top growing before the roots
+ Drought from lack of irrigation or weed competition
+ Fertilizer placed in the hole

Mulching newly planted trees with several inches of sawdust, bark, gravel, or plastic will help establishment and early growth. Fertilizers and herbicides should not be applied the first year. Pruning the top immediately after planting to restore the normal ratio of roots to top will help trees become established.

CULTURAL PRACTICES

For successful production, trees need fertilizer, irrigation, pruning, and disease and pest control.

Fertilization

After the first season, some trees will need nitrogen (N) to hasten growth. Fertilizer should be scattered under the branches, away from the trunk, after leaf fall and before bloom. Peach and filbert trees require more fertilizer than other fruits and nuts. Trees in grass sod will require much more nitrogen than areas in which the ground is clean-cultivated.

Irrigation

Depending on the type of tree and the location, trees will need irrigation, even in wet climates. How the trees are irrigated will depend on the area. Some trees may be irrigated by flooding and others by drip or sprinkler irrigation.

Pruning

Pruning and training of young trees establishes a strong framework of branches that will support fruit. Each tree type has a specific way that it should be pruned. For example, peach trees are pruned for an open center and V-shaped pattern. Apple trees are pruned into a Christmas-tree scaffold (Figures 25-5 and 25-6).

Disease and Pest Control

Control of diseases and pests is a major concern of fruit and nut growers. Because the crops are so diverse, so are the control measures. Table 25-5 lists some of the pests that need to be controlled in some fruits and nuts. Proper pesticides must be selected and used correctly for the fruit or nut tree being sprayed.

Timely and thorough spraying is required to control diseases and insects. Some growers will spray only if excessive damage appears imminent. Insect predators can help keep populations under control.

Figure 25-5 A peach tree pruned to the open center or V-shaped pattern.

The Religious Fig

The fig is considered sacred by the followers of Hinduism, Jainism, and Buddhism. The name "Sacred Fig" has been given to it. Siddhartha Gautama (Buddha) is said to have been sitting underneath a bo tree when he was enlightened or awakened. The bo tree is a well-known symbol for happiness, prosperity, longevity, and good luck.

The Sacred Fig (*Ficus religiosa*) is a species of banyan fig native to Sri Lanka, Nepal, India, southwest China, and Indochina east to Vietnam. Locally it has many names: bo or pou (from the Sinhalese bo), bodhi (in Thai language), pipal, arali, or ashvastha tree.

The bodhi tree and the Sri Maha Bodhi propagated from it are famous specimens of Sacred Fig with a known planting date of 288 BC. Today in India, Hindu sadhus still meditate beneath this tree, and in Southeast Asia, the tree's massive trunk is often the site of Buddhist and animist shrines.

The fig tree and its fruit, the fig, are mentioned several times in the New Testament, and in the Old Testament as well. The parable of the barren fig tree is recorded in the Gospel of Luke. A parallel is also found in Matthew and Mark in the New Testament. There is also a parable of the budding fig tree found in Matthew, Mark, and Luke. The fig tree and figs are featured in the Old Testament Book of Jeremiah.

In Genesis, in the Old Testament, after Adam and Eve eat of the forbidden fruit, both become aware of the fact that they are naked and cover themselves with garments made of fig leaves (Genesis 3:7).

Figure 25-6 An apple tree showing the cuts to be made for proper pruning .

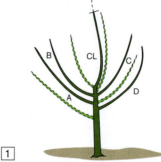

1

Sketch of a young apple tree after one year's growth. All limbs with broken lines should be removed. The central leader (CL) should be tipped if it is more than two feet long.

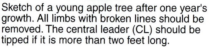

2

The same tree as in 1, after pruning.

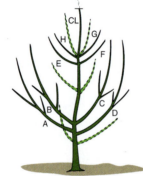

3

Sketch of how the tree in 2 may look after the second year. Branches with broken lines should be removed. The central leader (CL) should be tipped if it grew more than two feet.

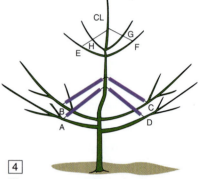

4

The same tree as in 3, after pruning and spreading of branches. All limbs in the first tier of branches (A, B, C, & D) have been spread with wooden spreaders with a sharp-pointed nail in each end, to illustrate the beginning of the Christmas-tree shape. Limbs E and F only, in the second tier of branches, have been spread with wire with sharpened ends.

HARVESTING AND STORAGE

Apples are mature when they easily separate from the tree when twisted upward; it is at this stage that they taste good. They should be picked before the core gets areas with a glassy appearance (water core).

Sweet cherries, apricots, figs, plums, prunes, and peaches taste ripe when ready for picking. Ripening will continue after harvest. For canning or drying they should be left on the tree until completely ripe. Sour cherries are ripe when they come off the tree easily, leaving behind the stems.

European pears should be picked when still green but when they easily separate from the tree. Most pear varieties other than the Bartlett require a month or more of cold storage before they will ripen properly. Oriental pears can be picked when fruits reach full size, color develops, and a taste test indicates the fruits are crisp, sweet, and juicy.

Persimmons ripen late in the fall when they become soft and a taste test reveals they have lost their astringency.

Nuts fall to the ground when they are mature. They can be gathered after falling to the ground or they may be shaken to the ground or into a harvester. Nuts are dried for storage; dried nuts can be frozen.

Fruit needs to be stored cool (near 32°F) but not frozen. A good storage room is insulated against heat and freezing. Humidity in the storage room should be moderate.

FIGS

Figs comprise a large genus, *Ficus*, of deciduous and evergreen tropical and subtropical trees, shrubs, and vines belonging to the mulberry family, Moraceae. Commercially, the most important fig is *Ficus carica*, the tree that produces the edible fig fruit. Among the most ancient cultivated fruit trees, the fig is native to the eastern Mediterranean and the southwest region of Asia, where its cultivation probably began.

The fig tree is characterized by its dark green, deeply lobed leaves. It bears no visible flowers; instead, its flowers are borne within a round, fleshy structure, the syconium, which matures into the edible fig. The common fig bears only female flowers but develops its fruits without pollination. Varieties of the Smyrna type also bear only female flowers, but in order to produce fruit, they must be pollinated by a process known as caprification.

Caprifigs

The caprifig is a wild form of fig tree whose male flowers produce inedible fruits that are host to the fig wasp, *Blastophaga psenes*. Fig wasps lay their eggs in the caprifig flowers; the eggs hatch within the developing caprifig, and the mature female wasps seek new flowers in which to lay their eggs. When caprifigs are hung among the branches of a cultivated fig, the pollen-dusted wasps squeeze through the narrow openings at the ends of the syconia and pollinate the flowers inside. The wasps die within the syconia and their bodies are absorbed into the developing fruit. Figs produced by caprification are usually larger than the common fig.

Fig trees are propagated through rooted cuttings taken from the wood of older trees. They grow best in somewhat dry areas that have no rain during the period of fruit maturation. Fresh market or canning figs are picked from the trees as they become fully colored and while they are still firm. Fresh figs are highly perishable, so timely transport to cold storage is important. Dried figs, most of the California crop, may be sun-dried or dried in a dehydrator.

Figs are classified either as Smyrna type, Common type, or San Pedro type figs. Smyrna figs produce only a summer crop. Common and San Pedro figs may also produce a spring crop, which requires caprification. Varieties of figs grown in California include Calimyrna, Mission, Adriatic, and Kadota.

DATES

The date palm, *Phoenix dactylifera*, of the palm family Arecaceae, has long been a staple carbohydrate food for Arab people. It originated in the hot, dry areas of the Old World. The date palm was taken to California by the Spanish in the seventeenth century. Date palms require a dry, hot climate. Dates are an important traditional crop in Iraq, Arabia, and

North Africa west to Morocco. Dates and palms are mentioned in many places in the Quran and the Bible.

Fruiting occurs after about five years. Loosely branched male and female blossoms grow from the crowns of separate trees. In commercial production, the pollen is collected and blown to the female and blossoms up through a tube. Pollen from different males differs in its effects on the color, size, and ripening qualities of the fruit. Most plantation palms (Figure 25-7) are female because the pollen output of one male plant is usually sufficient to pollinate up to 100 female plants. Female blossoms are pruned to increase fruit size and they are often covered with a bag to reduce damage by birds and insects. Adequately fertilized, a palm may yield more than 200 pounds of dates.

Dates are sweet, fleshy drupes, marketed in three categories: soft, semi-dry, and dry. Soft dates are harvested when still soft and unripe. Most of the world trade is in soft dates. Semidry dates, from firmer-fleshed cultivars, are also picked before maturity. Dry dates are sun-dried on the trees. Many cultivars of dates exist, but two important cultivars in the United States are Deglet Noor and Medjool, which are marketed as semidry dates.

Figure 25-7 Medjool date palm plantation near Yuma, Arizona.

SUMMARY

Fruit and nut trees are grown on a commercial scale or by home gardeners. Trees are adapted to a variety of climatic conditions but all trees are susceptible to frost damage. Trees can be standard size, semidwarf, or dwarf. Fruit and nut trees can be found for temperate climates and subtropical or tropic climates. Depending on the type of tree selected, a variety of cultural practices are necessary. Proper site selection, soil type, and fertility need to be considered when planting trees. Many trees require pruning. Trees pollinate by wind or insects; some growers bring in bees to increase pollination. Commercial dates are hand-pollinated, and figs rely on a process called caprification. Harvest and storage depend on the crop. With proper storage conditions, the fruit or nut can be kept fresh for a longer time.

Review

True or False

1. A semidwarf tree averages a mature height of 30–40 feet.

2. Frost susceptibility is not a factor to fruit crops.

3. An apple is an example of a citrus crop.

4. Insects and diseases are not a problem for fruit and nut trees.

5. A cherry is an example of a pome fruit.

6. A wasp is important in the pollination of figs.

Short Answer

7. What type of sunlight requirements do most fruits require?

8. How deep should the hole be in which to plant a fruit tree from a nursery?

9. Name five fruit trees and five nut trees.

10. List four drupe fruits.

11. Identify two temperate fruit trees and two subtropical fruit trees.

12. Name two cultivars of date palms important to production in the United States.

Critical Thinking/Discussion

13. Describe the storage of apples.

14. Why are fruit trees pruned?

15. How are fruits and nuts harvested?

16. Compare standard, semidwarf, and dwarf trees.

Knowledge Applied

1. Make a collection of all the fresh tree fruits sold in a grocery store. Identify where the fruits were produced and how they were stored fresh. Report on your findings.

2. Make a seed display by collecting all the seeds from fresh fruit sold in a local grocery store.

3. Select one of the fruit trees. Develop a presentation on how to prune this fruit tree.

4. Develop a display of all the tree-grown nuts sold in a local grocery store. As part of the display show the nut shell as well as the edible portion.

continues

Knowledge Applied, *continued*

5. Diagram the leaves from all of the fruit trees or all of the nut trees.

6. Plant a fruit or nut tree and provide the proper care for it.

7. Purchase some dates and figs to sample for taste and acceptance by class members.

Resources

Acquaah, G. (2004). *Horticulture: Principles and practices* (3rd ed.). Englewood Cliffs, NJ: Prentice Hall.

Brown, L. (2002). *Applied principles of horticultural science* (2nd ed.). Boston: Butterworth-Heinemann.

Childers, N. F., Morris, J. R., & Sibbett, J. R. (1995). *Modern fruit science* (10th ed.). New Brunswick, NJ: Horticultural Publications, Rutgers University.

Ensminger, A. H., Ensminger, M. E., Konlande, J. E., & Robson, J. R. K. (1983). *Foods & nutrition encyclopedia*. Clovis, CA: Pegus Press.

Hackett, C., & Carolane, J. (1982). *Edible horticultural crops: A compendium of information on fruit, vegetable, spice and nut species*. New York: Academic Press.

Litz, R. E. (2005). *Biotechnology of fruit and nut crops*. Cambridge, MA: CABI Publishing.

McMahon, M., Kofranek, A. M., & Rubatzky, V. E. (2006). *Hartmann's plant science: Growth, development, and utilization of cultivated plants* (4th ed.). Englewood Cliffs, NJ: Prentice Hall Career & Technology.

Poincelot, R. P. (2004). *Sustainable horticulture: Today and tomorrow*. Englewood Cliffs, NJ: Prentice Hall.

Reiley, H. E. (2006). *Introductory horticulture* (7th ed.). Albany, NY: Delmar Cengage Learning.

Schaffer, B., & Andersen, P. C. (1994). *Handbook of environmental physiology of fruit crops*. Boca Raton, FL: CRC Press.

Westwood, M. N. (1993). *Temperate-zone pomology: Physiology and culture* (3rd ed.). Portland, OR: Timber Press, Inc.

Internet

Internet sites represent a vast resource of information. The URLs (uniform resource locators) for World Wide Web sites can change. Using one of the search engines on the Internet such as Google or Yahoo!, find more information by searching for these words or phrases: fruit production, fruits and nuts, nut production, fruit pests, fruit diseases, apples, apricots, cherries, figs, date palms, dates, olives, pears, peaches, plums, prunes, almonds, hazelnuts, pecans, walnuts, and orchards.

The following national grower/horticultural organizations and publications address growers' concerns nationwide through dialogue, annual meetings, and publications:

+ American Pomological Society: http://americanpomological.org
+ American Society for Horticultural Science: http://www.ashs.org
+ U.S. Apple Association: http://www.usapple.org
+ International Dwarf Fruit Tree Association: http://www.idfta.org
+ National Peach Council: http://www.nationalpeach.org
+ American and Western Fruit Grower: http://www.americanfruitgrower.com
+ Walnut Marketing Board: http://www.walnuts.org

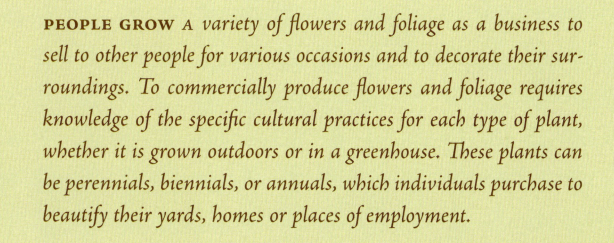

Chapter 26
Flowers and Foliage

PEOPLE GROW A variety of flowers and foliage as a business to sell to other people for various occasions and to decorate their surroundings. To commercially produce flowers and foliage requires knowledge of the specific cultural practices for each type of plant, whether it is grown outdoors or in a greenhouse. These plants can be perennials, biennials, or annuals, which individuals purchase to beautify their yards, homes or places of employment.

After completing this chapter, you should be able to—

- Name four types of flowers or foliage plants that people purchase at different times of the year

- Describe range of time from planting until flowers produce income

- Identify the pH range for perennial plants

- Describe watering precautions for some flowers and foliage

- Discuss planting of perennials and annuals

- Explain the winter care of perennials

- Name two methods of asexually propagating perennials

- Discuss considerations when selecting seeds

- Describe an indoor starting media

- Explain how to transplant flowers or foliage outdoors

- Identify four types of bulbs

- Discuss how bulbs can be forced

- Identify the light requirements of flowering houseplants

- Name four common indoor flowering plants

- Describe the greenhouse environment

- List five pests of bulbs

- Discuss the management of a commercial greenhouse

KEY TERMS

Bare-Root
Cold Frame
Container-Grown
Deadheading
Drifts
Dusting
Forcing
Media
Rootbound
Staking

FLOWER BUSINESS

People buy flowers and foliage for various reasons and at different times of the year, as Table 26-1 shows. Successful growers recognize the specific times certain flowers and foliage are purchased and how long these take to produce so they can have product ready when the customer is ready to buy. Table 26-2 shows the average time from planting to sales for different flower and foliage crops.

Flowers and foliage are adapted to a variety of growing conditions and climates, as Table 26-3 shows. These conditions need to be considered unless a greenhouse is used.

Table 26-1 Flowers and Foliage Use and Peak Seasons for Sales

Category	Crops	Peak Seasons
Personal and environmental adornment	Cut flowers and cut foliage or greens	Christmas, Valentine's Day, Easter, Mother's Day, and Memorial Day
Personal environment, intimate	Flowering potted plants	Christmas, Easter, and Mother's Day
	Tropical foliage plants	Holidays and January
	Hanging plants	Holidays, spring, and fall
	Landscape plants, outdoors	Spring, fall
	Patio container plants	Spring, fall
Personal environment, background	Bedding plants	Spring, fall
	Woody ornamentals	Spring, fall
	Herbaceous perennials	Spring, fall
Sustenance	Vegetable transplants	Spring, fall

Table 26-2 Time from Planting of Flowers to First Income

Crop	Time to Income
Cut flowers	3–4 months
Potted flowering plants	3–4 months
Potted foliage plants (6 inch)	4–6 months
Woody plants in containers	3–18 months
Cut foliage	1–2 years

Table 26-3 Temperature Requirements and Crop Examples

Crop Requirements	Crop Examples
Cool night temperatures	Cut flowers like carnations, snapdragons, stock heather; flowering potted plants like cyclamen, fuchsia, tuberous begonia
Medium night temperatures	Cut flowers like roses
Warm night temperatures	Foliage plants
Cold winters	Deciduous shrubs
Warm summers	Ornamental plants like junipers

VALUE OF FLORICULTURE

Not counting the value of personal flower gardens, floriculture in the United States has a total value of about $6 billion annually. Bedding plants, pulled flowers, foliage, cut flowers, and cut greens all contribute to the value of floriculture. As Table 26-4 shows, some states lead in floriculture. Figure 26-1 illustrates the trends in floriculture production over several years.

Table 26-4 Floriculture crops: Top five states by value of sales in 2006 for operations with $100,000 + sales.

Commodity	Rank	Value ($1,000)	1	2	3	4	5
Total value wholesale		15 States 3,834,912	CA 1,007,463 26.3%	FL 786,614 20.5%	MI 363,158 9.5%	TX 256,468 6.7%	NC 190,334 5.0%
Annual bedding/garden plants	1	15 States 1,281,113 33.4%	CA 231,199 18.0%	MI 192,861 15.1%	TX 156,058 12.2%	FL 118,776 9.3%	NC 108,712 8.5%
Potted flowering plants	2	15 States 619,925 16.2%	CA 205,990 33.2%	FL 79,791 12.9%	NY 50,288 8.1%	PA 39,162 6.3%	NC 38,356 6.2%
Foliage plants for indoor or patio use	3	15 States 542,533 14.2%	FL 366,414 67.5%	CA 97,893 18.0%	TX 19,301 3.6%	HI 15,254 2.8%	IL 7,600 1.4%
Cut flowers and cut cultivated greens	4	15 States 520,725 13.6%	CA 338,666 65.0%	FL 96,302 18.5%	WA 20,088 3.9%	HI 18,051 3.55%	OR 15,306 2.9%
Potted herbaceous perennial plants	5	15 States 507,346 13.2%	CA 70,186 13.8%	SC 67,484 13.3%	MI 45,970 9.1%	NJ 45,834 9.0%	OH 39,394 7.8%
Propagative floriculture materials	6	15 States 363,270 9.5%	FL 95,489 26.3%	MI 81,587 22.5%	CA 63,529 17.5%	WA 26,017 7.2%	PA 20,673 5.75%

Source: U.S. Department of Agriculture National Agricultural Statistics Service, July 2007.

FLOWERING HERBACEOUS PERENNIALS

Perennial plants live for many years after reaching maturity, producing flowers and seeds each year. Perennials are classified as herbaceous if the top dies back to the ground each fall with the first frost or freeze, and new stems grow from the roots each spring. They are classified as woody if the top persists, as in shrubs or trees. Most garden flowers are herbaceous perennials Any plant that lives through the winter is said to be hardy (Table 26-5).

There are advantages to perennials; the most obvious being that they do not have to be set out every year, as do annuals. Some perennials such

Figure 26-1 Trends in floriculture production. (Chart courtesy of USDA NASS)

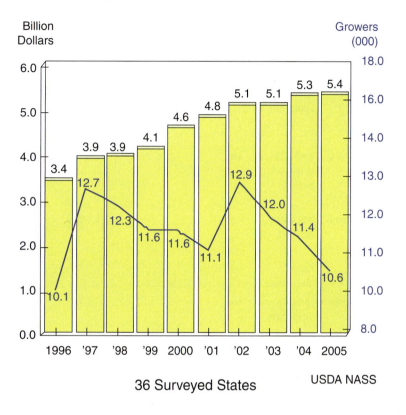

Floriculture Crops

Wholesale Value & Number of Growers, 1996–2005
Operations with $10,000+ Sales

36 Surveyed States   USDA NASS

as delphiniums have to be replaced every few years. Another advantage is that with careful planning a perennial flower bed will change colors, as one type of plant finishes and another variety begins to bloom. Also, because perennials have a limited blooming period of about two to three weeks, **deadheading,** or removal of old blooms, is not as frequently necessary to keep them blooming. However, they do require pruning and maintenance to keep them attractive. Their relatively short bloom period is a disadvantage, but by combining them with annuals, a continuous colorful show can be provided. Most require transplanting every three years.

Growing Perennials

Although perennials do not require replanting each year, as do annuals, they still require care. For best results, initial planning, proper soil preparation, and occasional maintenance are necessary. With proper attention to these details, a perennial garden can provide color and interest in the landscape throughout the growing season.

Selection

The site will influence what species of perennials can be grown. Most flowering perennials prefer six to eight hours of sun per day, making a

Table 26-5 Various Flowering Perennials

Name	When to Plant Seed	Exposure	Germination Time (Days)	Spacing	Height	Best Use	Color	Remarks
Achillea mille folium (yarrow)	early spring or late fall	sun	7–14	36"	24"	borders, cut flowers	yellow, white, red, pink	Seed is small. Water with a mist. Easy to grow.
Alyssum saxatile (golddust)	early spring	sun	21–28	24"	9"–12"	rock gaden, edging, cut flowers	yellow	Blooms early spring. Good in dry and sandy soils.
Anchusa italica (Alkanet)	spring to September	partial shade	21–28	9"–12"	48"–60"	borders, background, cut flowers	blue	Blooms June or July. Refrigerate seed 72 hours before sowing.
Anemone pulsatil la (windflower)	early spring or late fall for tuberous	sun	4	35"–42"	12"	borders, rock garden, potted plant, cut flowers	blue, rose, sarlet	Blooms May and June. Is not hardy north of Washington, DC
Anthemis tinctoria (golden daisy)	late spring outdoors	sun	21–28	24"	24"	borders, cut flowers	yellow	Blooms midsummer to frost. Prefers dry or sandy soil.
Arabis alpina (rock cress)	spring to September	light shade	5	12"	8"–12"	edging, rock garden	white	Blooms early spring.
Armeria maritima alpina (sea pink)	spring to September	sun	10	12"	18"–24"	rock garden, edging, borders, cut flowers	pink	Blooms May and June. Plant in dry sandy soil. Shade until plants are well established.
Aster alpinus (hardy aster)	early spring	sun	14–21	36"	12"–60"	rock garden, borders, cut flowers	white	Blooms June.
Astilbe arendsii (false spirea) 'Europa' Fanal' 'Deutschland' 'Superba'	early spring	sun	14–21	24"	12"–36"	borders	pink, red, whote	Blooms July and August. Gives masses of color.
Begonia evansiana (hardy begonia)	summer in shady, moist spot	shade	12	9"–12"	12"	flower bed	yellow, pink, white	Blooms late in summer. Can be propagated from bulblets in leaf axils.
Bergenia purpurascens (bergomot)	late winter	light shade	10	18"	2'–3"	medicinal	pink, red	Hummingbirds love it.

Source: USDA

southeast exposure ideal. Areas of shade will reduce the numbers of species that can be grown. Exposure to wind will vary, depending on the site, and thought should be given to wind protection, particularly if growing taller perennials such as delphinium or lilies.

Planning

The perennial garden is important to ensure continued bloom and desired combinations of color, texture, and height. Growers should

Figure 26-2 Rose, a perennial.

draw a plan of the garden to scale using graph paper. Most perennials have a limited period of bloom, and growers choose plants to give a succession of flowering. This requires selecting a range of perennials that have flowering periods that collectively cover the whole growing season. The blooming period of a particular species can usually be classified as spring, early summer, midsummer, or late summer/fall. A properly planned perennial garden will include plants from each of these flowering groups to create a season-long succession of bloom throughout the garden (Figure 26-2).

Perennials look best when planted in "drifts" of several plants, rather than as single plants or in rows. A rough guide for spacing would be 6–12 inches between dwarf plants, 12–18 inches between medium-sized plants, and 18–36 inches between tall perennials. Plant taller species toward the back of a flower border.

Soil Preparation

Soil preparation is probably the most important factor in determining the success of a perennial planting. Soil with good water drainage is necessary. It is particularly important that the soil not stay excessively moist during the winter dormant period. Incorporating organic matter such as compost or peat moss helps improve soil drainage. Spading or rototilling the soil to a minimum depth of eight to ten inches is also important. Soil preparations should be done in the fall or even one year ahead to remove all weeds that may germinate. Raised beds containing improved soil are better if there are poor soil conditions. Most perennials grow best in soil with a pH range of 6.5–7.0. A soil test can be made to determine the soil pH. Make any needed adjustments before planting.

Application of total vegetation killers can be added to beds before planting also. These chemicals control quack grass, thistle, and other perennial grasses and broadleaf weeds as well. Read and follow all label directions.

Planting

Perennials are generally planted in the spring (April to May). Plants bought from mail order nurseries are often shipped bare-root, not potted. Growers should plant these as soon as possible after receiving them. Container-grown (potted) perennials can be planted throughout the growing season, but spring is generally preferred. The proper time to plant will depend on how these perennials were produced by the grower.

Container-grown plants that have been exposed to outside temperatures throughout the winter can be planted as soon as the soil can be worked, about the same time as trees and shrubs. Perennials produced in a greenhouse during winter should not be planted until after danger of frost is past (below 32°F), much like annual bedding plants and vegetable transplants.

If planting in loose soil, plant crowns may end up higher than planned. If soil does settle after planting, add mulch or additional soil to cover exposed crowns. Growers need to get rid of weeds before planting. Once the plants are in, the weeds are more difficult to remove.

Perennials can be planted in spring or fall, any time the soil is workable and plants are available. Dormant plants are set out in early spring, four to six weeks before the last frost. Actively growing plants can be planted spring through fall. Perennials do best if planted before October 1 because roots can establish themselves before the ground freezes. Use stakes to mark where the plants will be set. A hole should be dug large enough to provide space for the roots. If the weather does not allow for immediate planting, growers need to keep plants in a cool, dark spot and make sure the peat moss around the roots does not dry out.

Watering

Although water requirements of perennials can vary greatly from species to species, most require supplemental watering until well established. One inch of water per week is a general rule. Once established, many species require watering only during prolonged dry periods. Selection of species adapted to drier climates will help reduce the need for supplemental watering.

Occasional overhead watering can harm perennials by causing disease and resulting in a shallow root system that is less able to withstand periodic dryness. Less frequent, more thorough watering, applied directly to the soil, is better.

Fertilization

Fertility of the soil can be improved before planting with the incorporation of a complete fertilizer, such as 4-12-4 or 5-10-5 at a rate of 2–3 pounds per 100 square feet. Avoid turf fertilizers high in nitrogen as these may promote excessive foliage production at the expense of a strong root system and good flower production. With proper soil preparation and improvement of soil fertility before planting, most perennials require little additional fertilization. Application of a "starter" fertilizer when perennials are first planted may aid in more rapid establishment of a good root system. For established plants, a little bone meal or super phosphate (0-19-0) worked into the soil around the plant in the spring can be beneficial.

Maintenance

Once established, most perennials require only routine maintenance. Taller species may require **staking,** particularly in windy areas. Staking is best done when plants are first sending up new growth, before they have become top-heavy.

Pinching back new growth may produce bushier plants, which are less likely to require staking.

A mulch applied to the soil will help suppress weeds while conserving moisture. Do not place the mulch too close to the crown as this may hold excess moisture and result in disease problems.

Insects and diseases are generally not severe threats to most perennials, particularly if locally adapted species are selected and proper care is given to the plants.

Winter Protection

Perennials damaged or killed during the winter are not usually injured directly by cold temperatures, but indirectly by frost heaving. Frost heaving occurs when the soil alternately freezes and thaws, resulting in damage to the dormant crown and root system. This action can be reduced by the application of winter mulch, which helps prevent rapidly fluctuating soil temperatures. Mulch about three inches thick is best. Evergreen boughs, clean straw, or other loose, coarse materials are good mulches. Materials such as tree leaves or grass clippings may compact too much around the plant, inhibiting water drainage and promoting disease development. Apply winter mulches after the ground freezes, usually in late November. Unless the perennials are evergreen, cut back the dead foliage to about six inches before applying the mulch, and discard the dead foliage and other debris. In the spring, mulch is gradually removed as new growth develops.

Dividing and Transplanting

Most perennials can be divided. Periodic division is needed to maintain vigor and maximum flower production. It is usually done every three to four years. Some perennials should never be divided.

The time of year when perennials are divided is a major factor in determining the success of this procedure. Species that bloom from midsummer to the fall are best divided in the early spring before much new growth has begun. Perennials that bloom in the spring or early summer should be divided in the fall or after the foliage dies. Exceptions are iris and daylilies, which are divided immediately after flowering.

To divide a perennial, the plant is removed from the soil by digging around and under the entire plant and lifting it carefully from the soil to avoid as much root damage as possible. All adhering soil is dislodged by hand or with a gentle stream of water from a hose. Growers should remove and discard diseased parts and cut back the top of the plant (stems, shoots, leaves) to about six inches. Divisions are usually taken from the outer perimeter of the plant, because this younger area tends to produce more vigorous growth. The plant can be divided by carefully breaking it apart by hand or by cutting with a heavy, sharp knife. Divide the plant in such a way that each new division has three to five "eyes"—buds that will produce new shoots.

Growers should replant the new divisions as soon as possible. They also should rework the soil if necessary to improve drainage and structure. Winter mulch is needed for divisions that are replanted in late summer or fall to help prevent frost heaving.

Ethylene Injury in Floriculture

Ethylene (C2H4) is considered to be a plant hormone, a growth regulator, and a potentially harmful pollutant of ornamental crops. It has sometimes been called the death hormone because it promotes the aging and ripening of many fruits and flowers. It is a simple organic substance that is active at very low concentrations. Ethylene and related substances are produced when almost any material is incompletely burned. It also evolves naturally from plant materials that are aging, ripening, or rotting. Many ripening fruits and vegetables generate ethylene gas as do certain microorganisms.

Ethylene toxicity and damage are of particular importance in the shipping and handling of floral produces. Ethylene gas is often added to banana-ripening rooms at food wholesalers. This can cause problems for floral products if they are handled and distributed from the same building.

Symptoms of excess ethylene include malformed leaves and flowers, thickened stems and leaves, abortion of leaves and flowers, abscission of leaves and flowers, excessive branching, and shortened stems. Hastened senescence (aging) of cut flowers occurs when they are exposed to ethylene. Flower corms or bulbs should never be stored with apples because serious injury occurs. *Fusarium* spp. disease organisms give off ethylene and can cause flower abortion of tulips and delayed bud activity in roses.

Many flowers, such as carnations, will begin to age in a matter of hours, and snapdragon and geranium florets will prematurely shatter after exposure to only 20 parts per billion. Some plants are far more sensitive to ethylene than others. Orchids, carnations, gypsophila, delphinium, and antirrhinums are extremely sensitive, whereas roses and chrysanthemums appear to be more tolerant.

Fumes from welding, auto exhaust, and local trash burning can cause extensive damage to greenhouse crops. Natural and propane gases are normally free of impurities and burn very cleanly. However, near-perfect or complete combustion must occur when gas burners are used in or near greenhouses. Burners must have proper exhaust venting with no leaks, and they must have adequate air (oxygen) intake for proper combustion.

Not all harmful forms of ethylene are from external sources. At certain times, endogenous ethylene products—that is, ethylene produced within the plants themselves—can contribute to premature senescence of flowers. This can happen whenever plants or cut flowers are mishandled or stored improperly.

Finally, ethylene is not all bad for floriculture. It is used to induce flowering in bromeliads and has been used successfully as a growth regulator and a chemical pinching agent in some floriculture crops.

Diseases and Insects

Perennials in general are healthy plants. Producers should select resistant varieties. They plant perennials in conditions of light, wind, spacing, and soil textures that are suited to them. After planting, the spent flowers, dead leaves, and other plant litter should be removed as these serve as a source of reinfestation. Growers need to know the major insect and disease pests (if any) of each specific plant type grown so that problems can be correctly diagnosed and treated as they arise.

Propagation

Plants can be propagated from tip or root cuttings (Figure 26-3).

FLOWERING ANNUALS

Annual flowers live only one growing season, during which time they grow, flower, and produce seeds, thereby completing their life cycle. Annuals must be set out or seeded every year because they do not persist. Some varieties will self-sow, or naturally reseed themselves. This may be undesirable in most flowers because the parents of this seed are unknown and hybrid characteristics will be lost. Seeds will scatter everywhere and plants will appear outside their designated place. Examples are alyssum, petunia, and impatiens. Some perennials—plants that live from year to year—are classed with annuals because they are not winter hardy and must be set out every year; for example, begonias and snapdragons. Annuals have many positive features. They are versatile, sturdy, and relatively inexpensive. Plant breeders have produced many new and improved varieties. Annuals are easy to grow, produce instant color, and, most importantly, they bloom for most of the growing season.

There are a few disadvantages to annuals. They must be set out as plants or sown from seed every year, which involves some effort and expense. Old flower heads should be removed on a weekly basis to ensure continuous bloom. If they are not removed, the plants will produce seeds, complete their life cycle, and die. Many annuals begin to look worn out by late summer and need to be cut back for regrowth or replaced.

Figure 26-3 Irises, a perennial, can be propagated by dividing the rhizome.

Annuals offer the gardener a chance to experiment with color, height, texture, and form. If a mistake is made, it is only for one growing season. Annuals are useful for:

+ Filling in spaces until permanent plants are installed
+ Extending perennial beds
+ Filling in holes where an earlier perennial is gone or the next one has yet to bloom
+ Covering areas where spring bulbs have bloomed and died back
+ Filling planters, window boxes, and hanging baskets
+ Planting along fences or walks
+ Creating seasonal color (Figure 26-4)

Site Selection and Preparation

Different annuals perform well in full sun, light shade, or heavy shade. Light, soil characteristics, and topography should be considered. The slope of the site will affect temperature and drainage. Also, the texture, fertility, and pH of the soil will influence the plant's performance.

Preparing the soil in the fall is the best time. Proper preparation of soil will increase success in growing annuals. Growers first have the soil tested and adjust the pH if needed. Check and adjust drainage. If drainage is poor, growers may plan to plant in raised beds. The next step is to dig the bed. Often growers will add four to six inches of organic matter to heavy clay to improve soil texture. They dig to a depth of 12–18 inches and leave until fall or early spring.

In spring, fertilizer is added and the area is spaded again and the surface raked smooth.

Seed Selection

To get a good start toward raising vigorous plants, growers always buy good viable seeds packaged for the current year. Seeds saved from previous years usually lose their vigor. They tend to germinate slowly

Figure 26-4 Bachelor's button is a flowering annual that grows early in the spring and easily re-seeds itself.

and erratically and produce poor seedlings. Seeds need to be kept dry and cool until planted. If seeds must be stored, they should be placed in an air-tight container, refrigerated, and stored with a material that will absorb excess moisture. Growers should buy hybrid varieties. Plants from hybrid seeds are more uniform in size and more vigorous than plants of open-pollinated varieties. They usually produce more flowers with better substance.

Starting Plants Indoors

The best media for starting seeds is loose, well-drained, fine-textured, low in nutrients, and free of disease-causing fungi, bacteria, and unwanted seeds. Many commercial products meet these requirements.

Clean containers are filled about two-thirds full with potting medium. The medium is leveled and moistened evenly throughout. It should be damp but not soggy. A furrow is made 1/4-inch deep. Large seeds are sown directly into the bottom of the furrow. Before sowing small seeds, growers fill the furrow with vermiculite and then sow small seeds on the surface of the vermiculite. Seeds may be sown in flats following seed package directions or directly into individual peat pots or pellets, two seeds to the pot.

After seeds are sown, all furrows are covered with a thin layer of vermiculite, then watered with a fine mist. A sheet of plastic may be placed over seeded containers, which are then set in an area away from sunlight where the temperature is between 60 and 75°F. Bottom heat is helpful.

As soon as seeds have germinated, the plastic sheeting is removed and seedlings are placed in the light. If natural light is poor, fluorescent tubes can be used. Seedlings should be placed close to the tubes. After plastic is removed from container, the new plants need watering and fertilizing as most planting material contains little or no plant food. Growers use a mild fertilizer solution after plants have been watered.

When seedlings develop two true leaves, plants are thinned in individual pots to one seedling per pot. Those in flats are transplanted to other flats and spaced one-and-one-half inches apart, or to individual pots (Figure 26-5).

Planting Times

As a general rule, growers delay sowing seeds of warm-weather annuals outdoors or setting out started plants until after the last frost date. Seeds of warm-weather annuals will not germinate well in soils below 60°F. If soil is too cold when seed is sown, seeds will remain dormant until soil warms and may rot instead of germinating. Some cold-loving annuals should be sown in late fall or very early spring.

Sowing Seed Outdoors

To seed annuals successfully, growers sow seeds in vermiculite-filled furrows. Annuals seeded in the garden frequently fail to germinate properly because the soil hardens on the surface, keeping the water out. The furrows in soil are about 1/2-inch deep. If soil is dry, producers water the furrow and then fill it with fine vermiculite and sprinkle with water.

Figure 26-5 Small seedlings being transplanted in to separate containers.

Make another shallow furrow in the vermiculite and sow the seeds in this furrow. Seeds should be sown at the rate recommended on the package. Mulch can be used until the plants are receiving enough sunlight.

Setting Out Transplants

By setting out started plants in the garden, producers can have a display of flowers several weeks earlier than if sown by seeds of the plants. This is especially useful for annuals that germinate slowly or need several months to bloom. Started plants can be purchased or produced. Buy only healthy plants that are free of pests and diseases.

Before setting out transplants, growers harden them off by setting the plants outside during the day. Annual plants may be set out after the last frost date. A hole is dug for each plant, large enough for the root system to fit comfortably. Plants are lifted out from the flat with a block of soil surrounding their roots.

Setting Plants

If plants are in fiber pots, growers remove the paper from outside the root mass and set the plant in a prepared planting hole. When setting out plants in peat pots, growers set the entire pot in the hole but remove the upper edges of the pot so that all of the peat pot is covered when soil is firmed around the transplant. If a lip of the peat pot is exposed above the soil level, it may produce a wick effect, pulling water away from the plant and into the air. After setting the plants, growers water them and provide protection against excessive sun, wind, or cold if needed, while the plants are getting settled in their new locations. Inverted pots, newspaper tunnels, or cloaks can be used to provide protection.

Thinning

When most outdoor-grown annuals develop the first pair of true leaves, they should be thinned to the recommended spacing. This spacing allows plants enough light, water, nutrients, and space for them to develop fully above and below the ground. If they have been seeded in vermiculite-filled furrows, excess seedlings can be transplanted to another spot without injury. Zinnias are an exception to this rule of thinning. In many varieties of zinnias, flowers will appear with a large nearly naked corolla and few colorful petals. This phenomenon is sometimes referred to as Mexican hats. To avoid such plants, growers sow two or three seeds at each planned location. Then they wait until the plants bloom for the first time and then remove the plants with this undesirable characteristic. Next, the remaining plants are thinned to the recommended 8- to 12-inch spacing.

Watering

Growers do not rely on summer rainfall to keep flower beds watered. They plan to irrigate them from the beginning. The entire bed needs to be moistened thoroughly, but not watered so heavily that the soil becomes soggy. After watering, soil is allowed to dry moderately before watering again. Drip systems are good for watering.

Sprinklers are not as effective as soaker hoses. Water from sprinklers wets the flowers and foliage, making them susceptible to diseases. Structure of the soil may be destroyed by impact of water drops falling on its surface. The soil may puddle or crust, preventing free entry of water and air.

Mulching

Mulches help keep the soil surface from crusting and aid in preventing growth of weeds. Organic mulches can add humus to the soil. Grass clippings make good mulch for annuals, if they do not mat. Sheet plastics also may be spread over the soil surface to retard evaporation of water and to prevent growth of weeds.

Cultivation

After plants are set out or thinned, cultivate only to break the crusts on the surface of the soil. When the plants begin to grow, stop cultivating and pull weeds by hand. As annual plants grow, feeder roots spread between the plants. Cultivation may to injure these roots. In addition, cultivation stirs the soil and uncovers weed seeds that germinate.

Deadheading (Removing Old Flowers)

To maintain vigorous growth of plants and assure neatness, growers remove spent flowers and seed pods. This step is particularly desirable if growing ageratum, calendula, cosmos, marigold, pansy, scabiosa, or zinnia.

Staking

Tall-growing annuals like larkspur, or tall varieties of marigold or cosmos need support to protect them from bad weather. Tall plants are supported by stakes of wood, bamboo, or reed that are large enough to hold the plants upright but not large enough to show. Stakes should be about six inches shorter than the mature plant so the blossom can be seen. Staking begins when plants are about one-third their mature size. Stakes are placed close to the plant but not so close as to damage the root system. The stems of the plants are secured to stakes in several places with paper-covered wire or other materials that will not cut into the stem. Plants with delicate stems can be supported by a framework of stakes and strings in criss-crossing patterns.

Fertilizing

When preparing beds for annuals, fertilizer should be added according to recommendations given by soil sample analysis, or by seeing plants that have grown on the site. Lime may also be needed if the soil test results so indicate. Dolomitic limestone should be used rather than hydrated lime. Ideally, lime should be added in the fall so it will have time to change the pH. Fertilizer should be added in the spring so it will not leach out before plants can benefit from it.

Additional fertilizers may be needed after annuals have germinated and started to grow. This is especially true if organic mulches are added because microorganisms decomposing the mulch take up available nitrogen. A fertilizer high in nitrogen should be used in these situations. Work the fertilizer into the soil around the plants, being careful not to touch the stems. Fertilizers should be applied to damp soil.

BULBS

People use the term bulbs to refer to corms, tubers, and rhizomes. All of these structures contain an embryonic plant and stored plant food, but they are all different in appearance and in their method of propagation.

Bulbs are also used in wooded areas with evergreen ground covers, in rock gardens, with evergreen shrubs to add color, and as cut flowers.

Site Selection and Soil

Bulbs grow well in well-drained loam. To improve soil texture, organic matter can be added to the soil in the form of compost, bark, and manure. Neutral soil (pH 7) is best for bulbs. Limestone can be added if the pH is below 6.0.

Planting Bulbs

Growers plant bulbs with a bulb planter, nursery spade, or hand trowel. Bulbs such as crocus, narcissus, and hyacinth are planted in the fall, while dahlia, amaryllis, gladiolus or similar bulbs are planted in the spring.

Each type of bulb has a recommended planting depth and spacing. As a general rule, bulbs should be planted the same distance apart as their planting depth (Figure 26-6).

Fertilizing

Bulbs should be fertilized by adding a small amount of fertilizer (5-10-5) to the bottom of the bed and covering it with soil before planting the

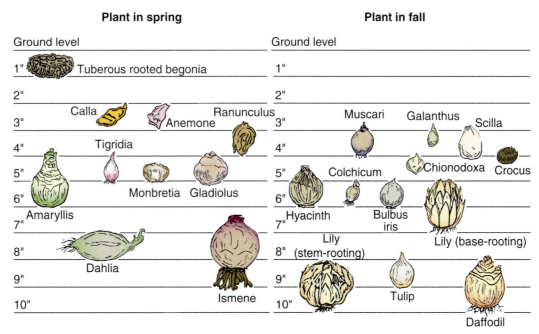

Figure 26-6 Planting depth of bulbs. (Chart courtesy of Brooklyn Botanic Garden)

bulb. Just before growth starts each spring, bulbs should receive a light application of a complete fertilizer; for example, 5 pounds of 5-10-5 for every 100 square feet of bed area.

Care after Flowering

Bulbs flowering in the spring should be dug up after the foliage turns yellow and dies. Some bulbs, such as tulips and hyacinths, need to be dug every year and replanted for high-quality flowers. Other bulbs, such as daffodil, crocus, lily, and colchicum need to be dug up only every three to five years, thinned, and replanted. Digging the bulbs after all of the foliage dies and no green remains in the foliage ensures that the bulb has completed its growth cycle and all the food is stored in the bulb.

Dusting the bulbs with a pesticide prevents insects and rodents from attacking bulbs during storage. Any bulbs showing signs of disease or damage are removed before storage. Bulbs are stored in peat moss or sawdust.

Bulb Pests and Diseases

Good cultural practices can control many of the pests. This includes such practices as weeding and keeping the bed free of trash that provide a home for insects and disease organisms. Chemical control is also effective and necessary. Growes should seek expert advice before applying chemical to bulbs. Table 26-6 lists the most common types of bulbs, pests, damage, and control measures.

Forcing Bulbs

To enjoy the color and fragrance of flowering bulbs in the winter, bulbs can be forced. Forcing bulbs artificially breaks dormancy so they will flower when brought into a warm room. High-quality bulbs can be

Table 26-6 Common Bulb Pests and Diseases and Their Control[1]

Host, Pest, Disease	Damage	Control
Amaryllis		
Spotted Cutworm	Feeds on flowers at night.	Scatter cutworm bait or spray with Sevin.
Bulb Mites	Rotting bulbs. See Hyacinth.	Discard soft bulbs.
Narcissus Bulb Fly	Decaying bulbs. See Narcissus.	Discard soft bulbs.
Leaf Scorch, Red Blotch	Reddish spots on flowers, leaves, bulb scales; stalks deformed.	Discard bulbs or remove diseased leaves. Avoid heavy watering.
Gladiolus		
Thrips	Leaves silvered, flowers streaked, deformed.	Spray with lindane in spring. Dust corms before storing.
Botrytis and other flower blights	Flowers, leaves, stalks spotted, then blighted.	Spray with zineb (Dithane Z 78 or Parzate).
Corm Rots, Scab	Lesions on corms, spots on leaves.	Dust with Arasan before planting.
Yellows (due to a soil fungus)	Plants infected through roots, turn yellow and wilt.	Choose resistant varieties.
Hyacinth		
Bulb Mites	Minute; less than 1/25 inch, white mites in rotting bulbs.	Discard infested bulbs.
Aphids, several species	Leaves are curled; virus diseases may be transmitted.	Spray with malathion, rotenone, or nicotine.
Bulb Nematode	Dark rings in bulbs.	Discard.
Soft Rot	Vile-smelling bacterial disease; often after mites.	Discard.
Iris (Bulbous)		
Tulip Bulb Aphid	See Tulip.	See Tulip.
Gladiolus, Iris Thrips	Leaves russeted or flecked, flowers speckled or distorted.	Spray or dust with malathion or lindane.
Leaf Spot	Light brown foliage spots with reddish borders.	Spray with zineb or bordeaux mixture; clean up old leaves.
Lily		
Aphids (Lily, Bean, Melon, Peach, other species)	Curl leaves, transmit mosaic and other virus diseases.	Spray with malathion, being sure to cover underside of leaves.
Botrytis Blight	Oval tan spots on leaves, which turn black, droop.	Spray with bordeaux mixture.
Host, Pest, Disease	Damage	Control
Mosaic and other virus diseases	Plants mottled, stunted.	Rogue infected plants. Start lilies from seed in isolated portion of garden.
Narcissus		
Narcissus Bulb Fly	Fly resembling bumblebee lays eggs on leaves near ground in early summer. Larva, fat, yellow maggot 1/2 to 3/4 inch long, tunnels in rotting bulb.	Sprinkle naphthalene flakes around plants to prevent egg-laying. Before planting, dust trench with 5% chlordane and dust over bulbs after setting.
Bulb Nematode	Dark rings in bulb.	Discard bulbs. Commercial growers treat with hot water, adding formalin to prevent rot.
Basal Rot	Chocolate-colored dry rot at base of bulbs.	Inspect bulbs before planting.
Smoulder (Botrytis Rot)	Plants stunted or missing; masses of black sclerotia on rotting leaves or bulbs. Yellow, red, or brown spots blight tips of leaves.	Remove diseased plants. Put new bulbs in new location.

Table 26-6 Common Bulb Pests and Diseases and Their Control[1] *(continued)*

Host, Pest, Disease	Damage	Control
Narcissus (*continued*)		
Scorch		Spray or dust with zineb, maneb, or copper.
Tulip		
Tulip Bulb Aphid	Powdery white or grayish aphids common on stored bulbs.	Dust with 1% lindane before storing.
Green Peach, Tulip Leaf, and other Aphids	Transmit viruses to growing plants.	Spray or dust with malathion or lindane.
Botrytis Blight, Fire	Plants stunted, buds blasted, white patches on leaves, dark spots on white petals, white spots on colored petals, gray mold, general blighting. Small, shiny black sclerotia formed on petals, foliage rotting into soil and on bulbs.	Discard all infected bulbs. Plant new tulips in new location. Spray with ferbam or zineb, starting early spring. Remove flowers as they fade, remove all tops as they turn yellow.
Cucumber Mosaic	Yellow streaking or flecking of foliage.	Do not grow near cucurbits or gladiolus.
Lily Mottle Viruses	Cause broken flower colors, mottled foliage, in tulips.	Do not plant near lilies. Control aphids.

[1] Courtesy of Brooklyn Botanic Garden

forced by potting them and storing them at 40–50°F for 10–12 weeks. Next, the potted bulbs are brought into a cool, partially lit room. Bulbs then grow and bloom within five weeks after being taken from the cool storage.

The process is challenging and provides plants for another market.

FLOWERING HOUSEPLANTS

Indoor plants have added color and variety to many places of work and play for hundreds of years. There are several factors to consider to grow indoor plants successfully. These considerations are light, temperature, water/humidity, and general care. Of these factors, light, and temperature are the most critical.

Light

Light intensity and duration influence the growth and development of indoor plants (see Chapter 12). Too little light causes some indoor plants to grow tall and leggy and lose their leaves because the long, thin, weak stems cannot support the plant because of limited photosynthesis.

Five different categories describe the light requirements of indoor plants:

1. Full sun: at least five hours of direct sunlight per day
2. Some direct sun: brightly lit but less than five hours of direct sunlight
3. Bright indirect light: considerable light but no direct sunlight
4. Partial shade: indirect light of various intensities and durations
5. Shade: poorly lit and away from windows

Temperature

Indoor plants survive best at constant temperatures. For optimum indoor growth temperature should remain in a range from 60° to 68°F. Temperature in an area also interacts with light, humidity, and air circulation. Plants have various temperature requirements (see Chapter 11) and should be placed in areas that matches their requirements.

Water/Humidity

Water is essential. But indoor plants do not require as much water as many individuals assume. Many indoor plants die from overwatering. Like the requirements for light and temperature, each plant varies in its requirement for water. Without knowing the specific requirements, indoor plants should be watered when they are a little on the dry side.

Indoor plants are watered from the top, or from the bottom up by placing water in a second container at around the bottom of the pot. The water used should not be too hot or too cold.

Besides water, plants need moisture in the air (relative humidity). Finally, good drainage is essential for potted indoor plants.

General Care

Under general care, houseplants need fertilizing on a regular basis for good health. A balanced fertilizer such as 5-5-5 should be applied at regular intervals. There are two types of fertilizers; slow-release and soluble. With slow-release type, the plant absorbs the fertilizer as needed, which eliminates the possibility of overfeeding or fertilizer burn. The soluble fertilizer must be dissolved in water before applied. Be sure to read manufacturer's directions before applying as these fertilizers may also be concentrates. Do not fertilize when the plant is dormant or very dry. Flowering plants need more fertilizer. Groom a plant by dusting, collecting dead leaves, and loosening dirt. Misting can be done with mild dishwashing soap and water to control insects.

Repotting. After time, the root systems of potted plants become restricted by the pot or other container. This creates a condition called **rootbound,** and the plant must be repotted to stay healthy. In general, repotted plants are moved to a pot no more than two inches wider than the original one. Flowering plants should be repotted after the flowers have faded. Repotting is also a time to check the health of the root system and remove any damaged or unhealthy roots.

Propagation. The method chosen for propagation depends on the type of plant—herbaceous or woody, flowering or foliar. Both sexual and asexual methods of propagation are used. Chapter 17 covers the methods of plant propagation. Common asexual methods used with indoor plants include leaf and stem cuttings, removal of plantlets, and air layering.

Common Flowering Indoor Plants

A few common indoor flowering plants are African violets, fuchsias, gardenias, geraniums, and impatiens. Some that grow outdoors as well as indoors are wax begonias, ageratums, verbenas, and petunias.

THE COMMERCIAL GREENHOUSE ENVIRONMENT

Three purposes of a greenhouse include:
1. Provide a controlled environment for plants grown on a large scale (Figure 26-7).
2. Grow plants in areas where outdoor growth during winter seasons is not possible.
3. Extend the growing season for plants that would normally go dormant.

Basically, greenhouses are structures used to start and to grow plants year-round. It should be located to receive the maximum amount of sunlight. Further, greenhouses control temperature, moisture, ventilation, and climate.

For optimal growth of plants, the temperature in greenhouses is controlled and monitored. Great variations in temperature cause fast vegetative growth or the lack of plant growth.

Plant growth in a greenhouse also depends upon moisture (humidity). Moisture aids in helping plants maintain their shape and nutrient transport. Watering plants controls growth and helps maintain the humidity. The amount of water needed by plants depends on the type of plants and the conditions outside the greenhouse (Figure 26-8).

Ventilation or the movement and exchange of air, is important for optimum growth. Ventilation helps ensure the proper temperature and humidity in the greenhouse.

Figure 26-7 Commercial greenhouses provide optimum growing conditions, in this case for hundreds of poinsettias.

Figure 26-8 Huge commercial greenhouses provide a controlled and monitored environment for thousands of plants in various stages.

When considering a greenhouse, the climate of the area must also be considered. The climate will directly influence the structure and the heating and the cooling systems. Potential greenhouse growers should seek the best advice possible before construction or purchase of a greenhouse.

Greenhouse Features

Greenhouses can be constructed of several different types of materials. They are aluminum, iron, steel, concrete blocks, or wood. Materials used to cover the greenhouse can be glass, soft plastic (polyethylene, vinyl or polyvinyl fluoride), fiberglass, shade fabric, acrylic rigid panels, or polycarbonate rigid panels. Some shapes for greenhouses are detached A-frame

Figure 26-9 Greenhouse on open ground.

truss, Quonset style, and ridge and furrow. There are also **cold frames,** hotbeds, and lath houses. Types of greenhouse structures used for floriculture in recent years include glass, fiberglass and other rigid plastic, plastic film, shade and temporary cover, and open ground (Figure 26-9).

Greenhouse Management

Greenhouse management involves a highly specialized and unique area of horticulture, requiring a strong understanding of plant growth and development, crop production, growing media, fertilization, mineral nutrition, water quantity and quality, growth regulators, and disease and pest management (refer to Chapters 7, 8, 9, 11, 12, 17, and 18). Greenhouse managers must also have a good understanding of environmental management, heating and cooling systems, lighting systems, and computer control systems. It is likely that no other area of horticulture uses as much technology as does greenhouse management.

Other issues of greenhouse management include sanitation, transplanting, scheduling, maintaining inventory, marketing, sales, transportation, and equipment care.

Sanitation. Sanitation and disease prevention are extremely important to keeping plants alive, keeping healthy plant stock, reducing the loss of income, and stopping disease from spreading to other plant material. Plant diseases are spread by the following methods: propagation of diseased plant stock; contaminated ground, benches, pots, flats, tools, or equipment; contaminated soil; insects or weeds acting as hosts; and improper watering.

Steps and methods to prevent diseases include preventing diseases from starting, keeping facilities clean, and eliminating pathogens from the soil. Soil pathogens can be eliminated by heating the soil to 180°F for 30 minutes or 120–130°F for 30–45 minutes or by using an approved chemical fumigant.

Cleaning all tools, benches, floors, used pots, and equipment with a 10 percent mixture of chlorine bleach and water helps prevent disease. Some areas of diseases can be controlled by applying a chemical fungicide as a soil drench.

Transplanting. Transplanting can be done at several different growing stages of nursery stock; for example, from seedling flats to Pony Paks or small containers; and from rooted cuttings to successively larger containers.

Although transplanting is an important part of growing nursery stock, it is a labor-intensive activity. Because of the associated high labor cost of transplanting, speed of work with high plant survival rates are essential. Careful transfer of the plant from one container to another and avoiding damage to the roots ensures survival.

Scheduling. When scheduling the sale of plants from a greenhouse, the first thing to do is to decide on a sale or shipping date and then count back the number of days required to grow the plants in the size of container desired. For plants started as seeds, most seed catalogs or seed packages list the number of days from planting to sale. When the number of days required for seed germination is added in, this should give the planting day. For some examples of the numbers of days required for scheduling, refer to Tables 11-1, 11-2, 23-2, 26-2, and 26-5.

Inventory. Keeping an accurate inventory of the plants is critical to meeting market demands. Also, knowing how many plants, what type of plants, and what stages the plants are in is helpful for scheduling production.

Marketing. Before growing any plants in a greenhouse, managers must determine the market potential for a crop. Managers should conduct a market analysis to determine what specific crops, product sizes, and quantities are in demand. Many markets are determined by location and season of the year (Tables 26-1 and 26-4).

Sales. Sales of greenhouse crops can occur through discount chains, retail nurseries, and large retail nursery chains. Discount chains include stores like Home Depot, K-Mart, Target, Wal-Mart, and others. These stores often have small inventory and limited selection of plants, and often a knowledgeable salesperson is not available in the nursery section. These stores usually are not a year-round or complete nursery; they sell plants seasonally, based on low prices.

Retail nurseries sell a medium-to-large selection of plants. They need repeat customers to do well, so knowledgeable salespeople are present to help the customer. Their sales are based on quality and service.

Large retail nursery chains carry a large selection of plants at economical prices, and they employ knowledgeable salespeople. These stores have the advantages of both the discount stores and the standard nursery.

When greenhouse plants are sold, the sales display might be nursery beds, mass display, or demonstration plantings. Nursery beds are the standard layout of a rectangular area on the ground where plants are placed in rows. Mass display is a large number of the same kinds of plants displayed near a main walkway. Demonstration plantings are displays in which the plants are placed in a sample landscape so the customer can see how the plants might be used. Regardless of the type of display, plants with color should be located to the front or near walkways, and nearby signs should contain information about the plants.

Transportation. Transportation of the plants to market is essential for the success of a greenhouse (Figure 26-10). Customers do not just show up at a greenhouse and load 10,000 flats of flowers or vegetables into their trucks. Greenhouse management requires the arrangement of some type of large-scale transportation method to the customer. Some growers own large trucks, sometimes called box trucks, that are adapted with adjustable, removable shelves so many different crops can be delivered. Other larger growers use greenhouse carts that roll up and fit into the large trucks, thus minimizing labor. Still others rely on large trucking companies.

Figure 26-10 Plants being readied for shipment from a commercial greenhouse.

Equipment. The operation of a commercial greenhouse also entails the use of a variety of tools and equipment. Managers make certain that all equipment is clean, operational, and maintained according to the manufacturer's specifications.

SUMMARY

Flowering plants and foliage plants are produced commercially and grown for personal enjoyment and decoration. These plants may be grown outdoors or indoors. People grow their own or purchase these plants from nurseries and greenhouses. Perennials have the advantage of regrowing year to year. When gardeners or growers understand the features and characteristics of each type of flowering plant or foliage, they combine plantings to give the best display of color. Even though perennials do not require planting each year, they still require care. Soil preparation, watering, and proper fertilization are necessary for flowering and foliage plants. Perennials require winter protection. For the best results from annuals, growers should select hybrid seeds. Some annuals do best when thinned after planting. Watering methods need to be checked because some plants become susceptible to disease if sprinklers are used. Bulbs are another way to produce flowering plants. Some bulbs are forced after cold storage.

Many flowering and foliage plants are grown outdoors, but many are produced in greenhouses. Here the lighting, temperature, and watering can be closely controlled and different types of plants produced on a year-round basis. These are sold for resale and for home use indoors and outdoors.

Review

True or False

1. Hybrid seeds are more vigorous.
2. Some annuals are propagated by division.
3. Zinnias grow best when planted from bulbs.
4. The best watering system for disease prevention in flowering plants is a sprinkler.
5. Flowering perennials require no fertilizer.

Short Answer

6. Name four common indoor flowering plants.
7. List four pests of bulbs.
8. What is the best pH range for flowering perennials?
9. How should bulbs be protected from insects and diseases?
10. What is transplanting?

Critical Thinking/Discussion

11. How are bulbs forced?
12. Why do people purchase flowering and foliage plants?
13. Describe a greenhouse environment.
14. Identify the features of a starting media.
15. Describe the time from planting to the time of scale for different types of plants.
16. Discuss the concerns of a commercial greenhouse manager.

Knowledge Applied

1. Make a collection of the seeds from flowering annual plants used in your area.
2. Grow a flower or foliage plant from seeds or from some form of asexual propagation.
3. Visit a florist. Develop a report on the types of flowers sold, the season in which they are sold, and what flowers are sold for special occasions.
4. Visit a greenhouse. Diagram the layout of the greenhouse, identifying the types of plants in each area. Find out the types of plants sold at different times of the year.
5. Try forcing some bulbs such as tulips, daffodils, hyacinths, or crocuses. Report on your success or failure.

Resources

Boodley, J., & Newman, S. E. (2008). *The commercial greenhouse* (3rd ed.). Albany, NY: Delmar Cengage Learning.

Dole, J. M., & Gibson, J. L. (2006). *Cutting propagation: A guide to propagating and producing floriculture crops.* Batavia, IL: Ball Publishing.

Griner, C. (2000). *Floriculture design & merchandising.* Albany, NY: Delmar Cengage Learning.

Hamrick, D. (Ed.). (2003). *Ball RedBook: Crop production* (Vol 2). (17th ed.). Batavia, IL: Ball Publishing.

Hamrick, D. (Ed.). (2003). *Ball RedBook: Greenhouses & equipment* (Vol 2). (17th ed.). Batavia, IL: Ball Publishing.

Reader's Digest. (2000). *Illustrated guide to gardening.* Pleasantville, NY: The Reader's Digest Association.

Reader's Digest. (2004). *Flower gardening: A practical guide to creating colorful gardens in every yard.* Pleasantville, NY: The Reader's Digest Association.

Internet

Internet sites represent a vast resource of information. The URLs (uniform resource locators) for World Wide Web sites can change. Using one of the search engines on the Internet such as Google or Yahoo!, find more information by searching for these words or phrases: greenhouse production, floriculture, greenhouse construction, flower bulbs, annual flowers, potted plants, perennial flowers, indoor plants, and houseplants.

Additionally, the PLANTS database (http://plants.usda.gov/index.html) provides standardized information about the vascular plants of the United States and its territories. It includes names, plant symbols, checklists, distributional data, species abstracts, characteristics, images, crop information, automated tools, onward Web links, and references. The PLANTS database is a great educational and general use resource.

Chapter 27
Sod Production

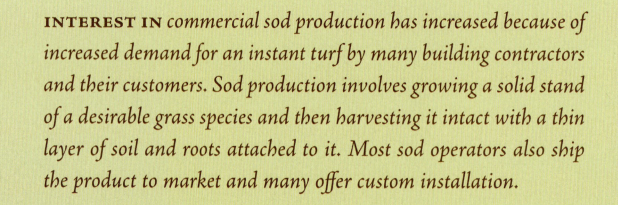

INTEREST IN *commercial sod production has increased because of increased demand for an instant turf by many building contractors and their customers. Sod production involves growing a solid stand of a desirable grass species and then harvesting it intact with a thin layer of soil and roots attached to it. Most sod operators also ship the product to market and many offer custom installation.*

After completing this chapter, you should be able to—

- Describe the soil preparation for growing sod
- Identify the best soil types for growing sod
- Discuss the water requirements of sod
- Name four turfgrasses
- List four factors to consider when selecting a grass species
- Describe the fertilizer needs of growing turfgrass
- List the micronutrients that may need to be considered for grass
- Identify three types of mowers used on turfgrass
- Describe the harvesting of sod
- List three factors to consider before installing sod
- Explain the purpose of fumigation
- Name three insect pests of turfgrass
- List three diseases of sod
- Describe nematode damage and control
- Identify six weed problems of sod production

THE SOD FARM

Ideally, a site to be chosen for a **sod** farm (Figure 27-1) should be based on several criteria: location (distance) in relation to targeted market, accessibility to major roads and highways, available water quantity and quality, soil type, land costs, and preparation requirements.

Figure 27-1 Sod farm showing sprinkler lines for irrigation.

In order to reduce shipping costs and because sod is a perishable product, a sod farm should be as near to an urban area as is practical. Sod that is stacked on pallets should be unstacked and laid within 72 hours after harvest, preferably within 24 hours. This is especially critical during summer months. Refrigerated trucks have been used to prevent sod deterioration when high-quality sod is transported over long distances. Sod on pallets waiting to be loaded or unstacked should be kept as cool as possible. Placing pallets in a shaded environment such as under trees or under shade cloth prolongs the sod's life.

PRODUCTION

Production practices are divided into several areas: establishment, primary cultural practices, pest management, and harvesting. Establishment involves land preparation, soil improvement, irrigation installation, and turf planting.

Land Preparation and Establishment

Prior to planting, the new turfgrass site should be prepared to correct any present problems and to avoid harvesting difficulties. Preparation includes land clearing, removal of trash, land leveling, tilling, installation of drainage and irrigation systems, roadway and building site selection, soil fumigation, and land rolling. The cutter blade on the sod harvester rides on a roller, allowing the unit to bridge the little hills, valleys, and holes in the field. If the surface irregularities left by poor soil preparation are too severe, the blade will not uniformly cut the sod. Then the yield will be reduced. Proper soil preparation also eliminates layers or **hardpans**, provides better air and water movement, and enhances deep rooting.

Growers test the soil in the area under consideration to determine lime and fertilizer nutrient requirements. Amendments should be applied and incorporated prior to turf establishment.

Where hardpan is a problem, the subsoil is broken up to get rid of any hardpans, and then plowed with either a moldboard or chisel plow to a depth of ten inches. This practice of breaking the subsurface hardpan should not be followed if subsurface irrigation is being used. Subsoiling is followed with soil incorporation of preplant fertilizer or liming material. The seedbed is firmed with a cultipacker roller. The surface must be as smooth and uniform as possible so that maintenance and harvesting problems are minimized. After cultipacking, the use of a laser plane for land leveling is suggested. The field should be **planed** (leveled or smoothed) in several directions to eliminate as many surface irregular spots as possible. After planing, dry soil is considered too fluffy if footprints are more than 1 inch deep. In this case, the field should be firmed by rolling it.

Preplant fumigation is strongly recommended where previous weed, disease, and nematode problems existed. Major weeds in sod production include common Bermuda grass, nutsedge, torpedograss, sprangletop, and crabgrass.

Sod is grown in several general soil groups. These include clay, sands, and **muck soils**. The agricultural suitability of these soils is determined by their ratio of sand, silt, clay, and organic matter fractions.

Clay soils do not drain well and stay wet for extended periods. Precious harvest days may be lost because of the wet ground. Also, because these soils hold so much water and have high bulk densities, clay soils are heavy to haul. Loam soils, in general, have good moisture-holding capacity, drain well, are easy to work, and are relatively light in weight for transport. These contain approximately 40 percent sand, 40 percent silt, and 20 percent clay. Next to muck soils, loam soils are most desirable as growing media. Ideally, these soils should have at least 5 percent organic matter and 15 percent or less clay. Sandy loams are desirable because of good drainage. Traffic and harvest operations may be performed sooner after water application.

Muck soils are found in old bogs, river deltas, and lake beds. They contain high organic matter and have good water-holding capacity. Nitrogen is also readily available through mineralization of organic matter. Muck soils are typically low in potassium, phosphorus, and various minor elements. Length of sod production on muck soil is usually shorter and production costs are less. Muck soils have less bulk density than sandy or clay soils, so they weigh less on a unit basis and are cheaper to transport. Muck soils are the most desirable for sod production.

Sod production is not recommended for deep, pure sandy soil (for example, sand dune-type sand), because of the difficulty of maintaining adequate soil moisture and nutrient levels. Furthermore, such soils typically have high levels of nematodes, which adversely affect soil quality and handling.

Proper soil water management is an important key to successful (and profitable) sod production. Poorly drained fields are unsuitable for competitive sod production. These fields often remain saturated, thus unworkable, for extended periods following substantial rainfall. Fields that are poorly drained need to be designed so that individual beds are crowned before planting. Lateral drain lines or ditches also need to be installed to intercept this surface drainage and to lower the water table to manageable levels.

Irrigation

Irrigation is required for quality sod production. Ample water of good quality should be a priority during the planning stage. Water sources include wells, sink holes, ponds, streams, and canals, as well as **effluent** sources from nearby municipalities and industrial sites. Effluent or gray water can be an excellent and inexpensive source of irrigation. These water sources may fluctuate widely in pH, salt, and nutrient levels. Many municipalities also require a contract stating that the grower must accept a certain number of gallons per given time whether or not irrigation is needed. These are issues that should be addressed early in the planning stage if effluent water is to be used. Irrigation systems normally involve center pivots, lateral pivots, walking or traveling guns, or subirrigation (raised water tables).

A **soil probe** is a very useful tool in irrigation management. The depth at which the soil is dry or wet can easily be measured with this, and irrigation scheduling adjusted accordingly. Tensiometers are soil moisture-sensing devices that measure the suction created by drying soil. If used correctly, the data gathered from these instruments' gauges can be used to determine irrigation scheduling. After the grass is planted, irrigation becomes the most important single factor for successful stolon establishment (Figure 27-2). It is critical not to plant more area than can be easily irrigated at one time.

Figure 27-2 Deep watering promotes deep, healthy root growth. Shallow watering promotes shallow rooting and leaves the grass susceptible to injury by drought.

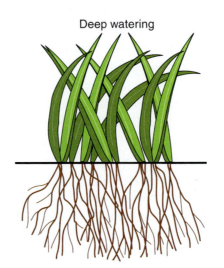

Deep watering

Shallow watering

Turf Selection and Planting

Determining which turfgrass is best for a particular situation is based on several factors. Table 27-1 lists the major warm- and cool-season grasses. In sandy soil, a deep-rooted grass is necessary. If properly maintained, Bahia grass and St. Augustine grass provide deep rooting and increased drought resistance. If the purchaser is willing to allot more time, energy, and economic resources to turf maintenance, a finer-textured species is suggested. Included in the suggestion is one of the Bermuda grass or zoysia grass cultivars. In addition, centipedegrass is available for those regions with heavier, acidic soils and for those persons with less resources and time available for upkeep. Other considerations for selecting a grass species include insect and disease resistance, nematode susceptibility, seed head/shoot growth rate, and frost and shade tolerance (Figure 27-3; Tables 27-1 and 27-2).

Fertilization

Proper fertilizing for sod production normally reflects the need for grass regrowth following establishment or cutting of the prior crop. Nitrogen is the most important nutrient regulating this regrowth. Generally, higher rates and frequencies of nitrogen application reduce the production time

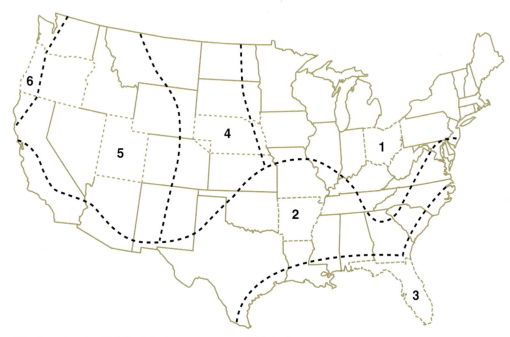

Climatic regions in which the following grasses are suitable for lawns:

1. Kentucky bluegrass, red fescue, and colonial bentgrass. Tall fescue, Bermuda, and zoysia grasses in the southern part.

2. Bermuda and zoysia grasses. Centipede, carpet, and saint augustine grasses in the southern part; tall fescue and Kentucky bluegrass in some northern areas.

3. Saint augustine, Bermuda, zoysia, carpet, and Bahia grasses.

4. Nonirrigated areas: crested wheat, buffalo, and blue grama grasses. Irrigated areas: Kentucky bluegrass and red fescue.

5. Nonirrigated areas: crested wheatgrass. Irrigated areas: Kentucky bluegrass and red fescue.

6. Colonial bentgrass, Kentucky bluegrass, and red fescue.

Figure 27-3 Regions of grass adaptations. (Image courtesy of USDA)

Table 27-1 Some Major Warm- and Cool-Season Turfgrasses in the United States

Cool-Season	Warm-Season
Colonial Bentgrass (*Agrostis tenuis*)	Bermuda grass (*Cynodon dactylon* L)
Creeping Bentgrass (*Agrostis palustris*)	Zoysia grass (*Zoysia japonica* Steud)
Kentucky Bluegrass (*Poa pratensis*)	Buffalo grass (*Buchloe dactyloides*)
Tall Fescue (*Festuca arundinacea*)	St. Augustine grass (*Stenotaphrum secundatum*)
Fine Fescue (various *Festuca* species)	
Red Fescue (*Festuca rubra* L)	
Perennial Ryegrass (*Lolium perenne*)	
Crested Wheatgrass (*Agropyron cristatum*)	

Table 27-2 Seeding Rates for the Major Turfgrass Species

Species	Pounds of Seed per 1,000 Square Feet	Species	Pounds of Seed per 1,000 Square Feet
Bahia grass	3–8	Fescue	
Bentgrass		Fine	3–5
Colonial	0.5–1.5	Tall	5–9
Creeping	0.5–1.5	Grama, blue	1.5–2.5
Bermuda grass (hulled)	1–2	Ryegrass	
Bluegrass		Annual	5–9
Kentucky	1–2	Perennial	5–9
Buffalo grass	3–7*	Wheatgrass, crested	3–6
Carpetgrass	1.5–5	Zoysia grass	1–3
Centipedegrass	0.25–2*		

* The higher rates are best, but lower rates are commonly used because the seed is expensive.

for a crop. Excessive nitrogen rates force excessive top growth at the expense of the roots, thus reducing the "liftability" of the sod. Economics also dictate, to an extent, the amount and frequency of nitrogen use.

A balance needs to be maintained between all major and minor elements because the unavailability of any nutrient may weaken or delay the production process. Sod managers should test all fields before planting and yearly thereafter to regulate pH and nutrient levels and the needs of the particular grass being grown.

Soils in some areas naturally provide adequate phosphorus and soil pH levels. Growers apply phosphorus and liming material (if necessary) prior to planting. Phosphorus is available as Super Phosphate (0–18–0) or Triple Super Phosphate (0–45–0). Growers commonly use one fertilizer containing both nitrogen and phosphorus. Examples of such sources include 16–20–0 or 11–48–0.

The optimum soil pH for St. Augustine grass, Bermuda grass, and zoysia grass is approximately 6.0–6.5. Centipedegrass and Bahia grass have an optimum soil pH of 5.0–5.5.

Following the first mowing, growers apply fertilizer at the rate of 40–45 pounds of actual nitrogen per acre. A fertilizer with a nitrogen-to-potassium ratio of 2:1 should be used to increase the turf's stress tolerance level and promote better rooting. Subsequent fertilizer applications should be made following the second mowing. Continue fertilizing every four to six weeks until the grass develops a complete ground cover.

Scheduling and Rates. Once the sod has covered, fertilizer scheduling is largely dictated by economics. Obviously, if sod orders are strong, the grass needs to be aggressively fertilized to minimize production time. If sales are slow, sod should be fertilized less to save on fertilizer and maintenance costs such as mowing and watering.

Bermuda grass and zoysia grass respond exceptionally well to ample fertilization. Quickest turnaround of these grasses occurs with monthly nitrogen application at the equivalent of 50 pounds of nitrogen per acre per application. This schedule should continue unless cold weather halts growth or economics dictate otherwise. A 2:1 or 1:1 ratio of nitrogen-to-potassium fertilizer should be used with each application to encourage strong rooting. Phosphorus should be applied as suggested by a yearly soil test (Table 27-3).

Table 27-3 Fertility Requirements and Some Recommended Fertilizers for Bermuda Grass and Zoysia Grass

Time to Apply	Nitrogen Applied per 1,000 Square Feet	Acceptable Fertilizer per 1,000 Square Feet of Area
April or May	1 lb.	8 lb. 12-4-8 **or** 10 lb. 10-6-4 **or** 5 lb. 20-10-10
June	1 lb.	8 lb. 12-4-8 **or** 10 lb. 10-6-4 **or** 5 lb. 20-10-10
July	1 lb.	8 lb. 12-4-8 **or** 10 lb. 10-6-4 **or** 5 lb. 20-10-10
August	1 lb.	8 lb. 12-4-8 **or** 10 lb. 10-6-4 **or** 5 lb. 20-10-10

St. Augustine grass is normally fertilized every six to eight weeks during the growing season. As with Bermuda grass and zoysia grass, St. Augustine grass should be fertilized with a 2:1 or 1:1 nitrogen-to-potassium ratio fertilizer and phosphorus added as suggested by a yearly soil test. If over-fertilized in summer with quickly available nitrogen sources, St. Augustine grass becomes more susceptible to chinch bug infestation and gray leaf spot disease. These problems can be minimized by using slow-release (or controlled-release) nitrogen sources and supplemental iron applications. These are discussed later.

Bahia grass and centipedegrass are fertilized less than the other sod-grown grasses. Bahia grass is fertilized yearly with 100 to 200 pounds of total nitrogen per acre. Again, economics and desired sod turnaround time dictate which rate range is used. Two applications per year are made if 100 pounds of nitrogen is used, with these being equally divided between early spring (April–May) and summer (July–August).

Centipedegrass has a very specific fertilization schedule. If overfertilized long term with nitrogen, centipedegrass will develop thatch and will have decreased winter survival and reduced rooting. The end result, referred to as "centipedegrass decline," is characterized as death or extremely weak spots roughly 2–20 feet in diameter that develop as the grass resumes growth in spring. Normally, centipedegrass decline does not develop until several years after establishment. So sod managers should fertilize centipedegrass similarly to St. Augustine grass for one year after establishment.

If the grass is not harvested within 18 months after establishment, then the fertility rate needs to be reduced to minimize the occurrence of centipedegrass decline. Established centipedegrass should be fertilized only two to three times yearly with 23–45 pounds of actual nitrogen per acre. In some areas, additional potassium should be considered in early fall to encourage proper rooting prior to winter. Supplemental iron or manganese application may be needed if unacceptable leaf chlorosis forms.

Micronutrients. Some soils are low in micronutrients. If recommended by soil testing, at least two applications of micronutrients are suggested per year, but more may be required. Several iron products are used. The least expensive and most commonly used source is ferrous sulfate. Ferrous sulfate contains 21 percent iron and is quick-acting, but color enhancement lasts only three to four weeks. Chelated iron products are more expensive but have been formulated to hold their greening effect for a longer period of time.

Iron should be sprayed on most turfgrasses to enhance color, especially near harvesting time. These are often injected into the irrigation system but may also be applied in a dry or spray solution form. Application of 20–40 pounds of elemental iron (e.g., 100–200 pounds of ferrous sulfate) may be timed approximately one to two weeks prior to harvesting to enhance color. To prevent burn, irrigations must be applied immediately after iron application during periods of high temperature.

Liquid fertilizers are often used by injecting them into the irrigation system. Ammonium nitrate is the primary nitrogen source used for this. The major problems with using fertilizer in irrigation systems involve difficulties in maintaining uniform distribution and concerns with possible fertilizer leaching.

Mowing

After irrigation as the first priority, mowing is perhaps the second most important turfgrass cultural practice for sod producers. Mowing helps control turfgrass growth and many undesirable weeds that are intolerant of close mowing. Sod fields require a mowing schedule similar to a well-maintained home lawn.

Three basic mower types include reel, rotary, and flail. A reel mower is most desirable because the highest possible mowing quality is achieved from a cleaner cut. Rollers on a reel-type mower also help smooth the sod field for easier, more uniform harvesting. Reel mowers should always be used the last four or five mowings before harvest. This produces the finest cut available and it maximizes sod quality. Rotary mowers are acceptable for St. Augustine grass, centipedegrass, and Bahia grass production if blades are properly sharpened and balanced. Flail mowers are widely used

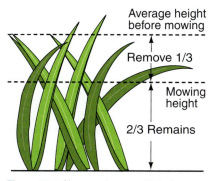

Figure 27-4 No more than one-third of the top growth should be removed per mowing.

in Bahia grass production until sod has a uniform dense stand, and then growers switch to a reel or rotary mower.

Regardless of the mower type, the blades should be well maintained and sharpened. Dull blades reduce turf quality by leaving grass tips shredded and bruised. Shredded tips dry easily, which leaves brown tissue that grows slowly, especially in hot weather. Mowers are big, heavy pieces of equipment. Ruts, which cause harvest losses, may develop if these machines are used when soils are too wet.

New sod fields are generally mowed once every one to two weeks until complete coverage is obtained, depending on grass growth and weed encroachment. Mowing frequency will vary for established sod, depending on the fertility level, season of year, species, and seed head production. Figure 27-4 and Table 27-4 show recommended mowing heights.

Table 27-4 Recommended Mowing Heights for Different Turfgrasses

Species	Mowing Height Range (Inches)	Species	Mowing Height Range (Inches)
Bahia grass	2–4	Colonial bentgrass	0.5–1.0
Bermuda grass		Fine fescue	1.5–2.5
Common	0.5–1.5	Kentucky bluegrass	1.5–2.5
Hybrids	0.25–1	Perennial ryegrass	1.5–2.5
Carpetgrass	1–2	Tall fescue	1.5–3.0
Centipedegrass	1–2	Crested wheatgrass	1.5–2.5
St. Augustine grass	1.5–3.0	Buffalo grass	0.7–2.0
Zoysia grass	0.5–2.0	Blue grama	2.0–2.5
Creeping bentgrass	0.2–0.5		

Growers establish a mowing frequency to ensure that no more than one-third of the leaf area is removed at any one mowing. Maintaining this schedule will allow for returning clippings to the field for nutrient recycling. An example of proper mowing frequency is a grass that is normally mowed at a height of one inch. In order not to remove more than 1/3 of the leaf area, it should be mowed before exceeding one-and-one-half inches. If that growth occurs in three days, then the field should be mowed every three days; if the growth requires two weeks, then that should be the mowing frequency. Established Bermuda grass and zoysia grass sod fields typically are mowed every three days, while centipedegrass, St. Augustine grass, and Bahia grass are mowed once every seven to ten days.

Grass clippings may or may not be picked up. If removed, sweepers and vacuums are used. The purpose of removing clippings is to prevent them from filtering down into the turf stand and turning brown. When the sod is delivered, the presence of these brown clippings may cause the sod to appear to have less density than it really has. Clipping disposal is a major problem. Restrictions on burning, dumping in landfills, and problems with odor are causes of disposal problems to many producers. Composting the clippings could be a viable option in some areas.

If clippings are removed, then the removal should begin during the first or second months before harvest. This timing will help prevent the browning effect that clippings may impose and prevent having disposal problems throughout the entire growing life cycle.

INSTALLATION

Proper soil preparation and turf maintenance procedures must be followed to ensure the survival and desirable aesthetics of sod. Generally sod is installed in the spring and summer (Figure 27-5). Year-round installation is possible in some southern areas if fall and winter temperatures remain conducive for turf growth. The following steps are suggested for laying sod:

1. Soil test to determine nutrient deficiencies.
2. Apply recommended nutrients, especially phosphorus and potassium, plus other soil amendments and incorporate these by tilling six to eight inches deep.
3. Allow soil to settle by irrigating or rolling. Rake or harrow the site to establish a smooth and level final grade. The finished grade should be about 1 inch below walks and drives.
4. Prior to sodding, irrigate the soil to cool the surface and provide initial moisture to roots. If this is not performed, the sod roots will be subjected to initial heat and water stress damage, resulting in lower sod survival.
5. When laying the sod, the first strip should be laid along a straight edge. For better knitting, stagger each piece of sod, similar to a bricklayer's running pattern, so that none of the joints are in a line. Each piece should be fitted against others as tightly as possible. Fill in gaps with clean soil to reduce weed encroachment.
6. Smooth the surface and encourage rooting by rolling.

Figure 27-5 Truck load of sod on pallets ready for delivery and installation.

7. Irrigate heavily to ensure good root zone moisture. This is especially necessary when laying a different sod-grown soil type over another. Provide good moisture for at least two weeks following planting. Gradually decrease the frequency between irrigations to an "as needed" basis.

8. Roll and/or top-dress with clean root zone soil to help smooth the sod surface.

9. Fertilize with nitrogen approximately 7–14 days following planting. Irrigate immediately after application.

10. Mow after the grass reaches the proper height.

After the grass is installed, problems that arise may or may not be the grower's responsibility. Sod with dead edges indicates that it became too dry before installation. Weeds within sod are probably brought in and are the responsibility of the grower. Weeds that grow between the seams of installed sod are due to faulty installation.

Improper irrigation after the installation of sod is always a problem. Enough water should be applied to thoroughly wet the root zone underneath the soil surface.

PEST MANAGEMENT

Preplant fumigation with materials such as methyl bromide, dazomet (Basamid), or metam sodium (Vapam) may be required when sod farms are established on land previously used for row crop farming. Fumigating will reduce perennial weed species such as Bermuda grass, nutsedge, torpedograss, and sprangletop. Soil sterilization will also reduce nematode populations, which are difficult to control once the grass is established. Depending on location, a sod field may be fumigated at least every five years to help control weeds, nematodes, and other pests.

Methyl bromide is expensive because of the plastic cover required to ensure activity and may be applied only by a certified applicator. This material provides better pest control and the treated area can be planted within 48 hours after the cover is removed.

Metam sodium and dazomet do not require a cover, but a certain amount of efficacy is sacrificed. If a cover is not used, metam sodium, once applied, requires incorporation into the soil. Incorporation is achieved by rolling, irrigation, and/or tilling the material to the depth of desired control (usually six to eight inches). Poor performance will result if this incorporation is not performed. A minimum waiting period of 14–21 days is required before planting in metam sodium- or dazonet-treated soil.

Weed Control

If preplant fumigation is not feasible, the use of a nonselective herbicide such as glyphosate is required on weed-infested fields. Weed-infested sod will reduce the salability of the product. Three applications of glyphosate spaced four to six weeks apart are necessary for postemergence control of

National Turfgrass Evaluation Program

Various turfgrass research programs contribute to the development of a nationwide database of unbiased information on cultivar performance. These programs participate in the National Turfgrass Evaluation Program (NTEP), a not-for-profit cooperative effort of the United States Department of Agriculture and the National Turfgrass Federation, Inc. Currently, more than 150 varieties of turf are being evaluated for spring greenup, density, drought tolerance, disease or weed activity, color, and overall quality. The trials include:

+ Tall fescues
+ Zoysia grasses
+ Bermuda grasses
+ Buffalo grasses

NTEP links the private and public sectors of the industry through the common goals of improving grasses, developing new cultivars, and establishing uniform evaluation standards. NTEP trials, which usually last five years, are replicated at many locations throughout the country and include most of the available varieties of each of the subject species.

Results from research facilities nationwide are collated, analyzed, and disseminated by NTEP annually. Seed companies and plant breeders use the NTEP information to determine grass adaptation and quality ratings.

Grasses in the trials are mowed weekly during the growing season, fertilized, and irrigated according to proper management for the area. No secondary management practices are used. Quality ratings are taken monthly and reported to NTEP yearly.

Other turfgrass research areas include:

+ Resource efficiency of water, nutrition, pest management, energy, and labor
+ Inputs of water and fertilizer in such areas as lawns, golf courses, parks, and grounds
+ Environmental enhancement of urban and suburban areas using turfgrass
+ Turfgrass persistence and performance with increased traffic and use in sports facilities

perennial weeds such as Bermuda grass or torpedograss. These should begin in spring after temperatures are consistently warm and weeds are actively growing.

Weeds can be introduced into a field in many ways. Irrigation water from open canals, ditches, or ponds often contains weeds. Soil introduced during soil preparation, such as a landplane pulling untreated soil into a field, leaves weeds. Birds, wind, soil erosion, and man also deposit weed seeds. Good housekeeping by keeping ditches and fence rows clean and by washing equipment before entering a weed-free field does benefit the sod producer.

Once the grass is established, weed management involves proper mowing, cultural practices to promote turf competition, and use of herbicides. Many upright-growing broadleaf weeds can be controlled effectively through the use of continuous mowing. These include ragweed, pigweed, cocklebur, and morning glory. Growers should mow these prior to seed head emergence to help prevent reinfestation from seeds.

Grassy weeds that are a problem in sod production include annual bluegrass, crabgrass, goosegrass, vasey grass, signalgrass, sprangletop, torpedograss, and Bermuda grass. Broadleaf weeds include purslane, betony, pusley, pennywort (dollarweed), oxalis, and spurge. Purple, yellow, annual, globe, cylindrical, and Texas nutsedges are also weed problems.

Herbicide recommendations are updated constantly. Growers should refer to publications providing the latest recommendations. Immature weeds (seedlings) are most susceptible to herbicides, and certain turf varieties can be damaged when air temperatures exceed 80–85°F at the time of herbicide application. The turf should not be under moisture or mowing (scalping) stress when treated with an herbicide.

Insect Control

Insect pests are generally grouped into three categories: shoot feeding, root feeding, and burrowing. Southern chinch bugs, spittlebugs, grass scales, and Bermuda grass mites suck plant juices.

Chinch bug damage is normally associated with St. Augustine grass. Damage is apparent as yellowish-to-brown patches in turf and appears sooner on turf under moisture and/or heat stress. Some cultivars provide some degree of resistance to chinch bugs.

Insect shoot feeders that eat grass leaves include sod webworms and armyworms. Armyworms feed during the day, and sod webworms feed at night. Injured grass has notches chewed in leaves, and grass has an uneven appearance.

Root-feeding and burrowing insects include mole crickets, white grubs, and billbugs. Mole crickets injure the turf through their extensive tunneling, which loosens soil, allowing desiccation to quickly occur. Mole crickets may be flushed out by applying water with two teaspoons of household soap per gallon per two square feet of fresh tunnels. If present, crickets will surface and die within several minutes. White grubs and billbugs are root feeders and are typically C-shaped. Grub damage is erratic, with patches of

Table 27-5 Common Turf Problems

Turf Insects	Turf Diseases	Other Problems
Ants	Anthracnose	Dogs
Armyworms	Brown patch	Gophers
Billbugs	Copper spot	Ground squirrels
Chinch bugs	Dollar spot	Mice
Cutworms	Fairy ring	Moles
Grubs	Fusarium blight	Human vandalism
Leaf hoppers	Leaf spots	Vehicles and equipment
Mites	Net blotch	
Mole crickets	Nematodes	
Periodical cicadas	Powdery mildew	
Scale	Pythium blight	
Sod webworm	Red thread	
Weevils	Rots	
Wireworms	Rusts	
	Slime molds	
	Smuts	

turf first showing decline and then yellowing. Under severe infestation, sod may actually be removed by hand. Monitoring these insect populations involves cutting three sides of a sod piece and laying this back. If there is an average of three or more grubs per square foot, an insecticide is needed.

Other insect pests that disrupt the sod surface or are a nuisance to man include ants, fleas, and ticks (Table 27-5).

Disease Control

Disease development requires three simultaneous conditions:

1. A virulent pathogen
2. A susceptible turfgrass
3. Favorable environmental conditions

Environmental conditions that favor incidence of most turf diseases include periods of high humidity, rain, heavy dews or fogs, and warm temperatures (but not always). Turf that is fast-growing and succulent from nitrogen overfertilization is typically more susceptible to disease and other pest invasion. Ideally, irrigate early in the day to minimize the time in which turfgrass remains moist.

If a disease problem is suspected, growers can prepare a sample for laboratory diagnosis by following these steps:

1. Sample the affected area before fungicide application.
2. Sample from marginal turf areas between diseased and healthy turf.
3. Cut a three- to four-inch plug from each area with symptoms.
4. Place plugs in paper bags or cardboard boxes and do not add water.
5. Submit the sample to the nearest county extension office.
6. Remember to complete a specimen data form with each sample.

The major diseases that occur on sod-grown grasses are dollar spot on Bermuda grass and Bahia grass, and gray leaf spot on St. Augustine grass. Dollar spot disease forms brown patches approximately the size of a silver dollar. On Bahia grass, dollar spot disease is generally more localized on individual leaves. Normally, dollar spot disease can be eliminated by a light nitrogen fertilization to encourage turf plants to outgrow the disease symptom.

Gray leaf spot disease of St. Augustine grass normally occurs during hot, humid weather. The use of excessive quick-release nitrogen or the use of atrazine or simazine during these conditions encourages this disease. If fertilized during the summer, use lower quick-release nitrogen rates or use a slow-release nitrogen source on St. Augustine grass. Foliar-applied iron also promotes desirable turf color without overstimulating disease occurrence.

Sometimes pythium and brown patch disease affect St. Augustine grass. Both diseases reduce rooting and turf appearance. Pythium normally occurs in poorly drained areas where water stands. Brown patch also occurs in wet areas and is most pronounced in spring and fall months when grass growth is slow.

Nematode Damage

Nematodes are small, microscopic worms that normally feed on or in plant roots. If populations become severe, plants wilt under moderate moisture stress, are slow to recover after rain or irrigation, and gradually decline or "melt out." Weeds that commonly become a problem in nematode-infested areas include spotted spurge and Florida pusley. Turf roots often become stubby, shortened, and turn black. Because of the extensive root damage, plants are not able to withstand stresses such as drought, insect, or disease invasion. Sampling of soils for a nematode assay is the only sure way to determine if they are in high enough populations to cause damage. Soil samples should be obtained similar to those in the disease control section. They should be submitted to a reputable nematode laboratory.

Control begins with those management practices that favor good turf growth. These include proper watering, fertilization, and mowing practices. Few nematicides are available. Proper turf management is becoming increasingly important to mask nematode presence.

HARVESTING

Turfgrass is harvested when sod has developed enough strength to remain intact with minimum soil adhering when cut. Time required to produce a marketable sod from initial establishment depends on turfgrass species, soil type, and growing conditions. Time typically required between harvests for most turf sod is listed as actual growing months in Table 27-6.

Several weeks prior to harvest, the turf should be conditioned in order to enhance its color. Suggested practices include mowing only with a reel mower, applying iron within two weeks of harvest, and applying no

Table 27-6 Time in Growing Months from Planting to Harvest for Some Grasses

| Cultivar | Growing Months | |
	Initial Establishment	After Harvest
Common centipedegrass	18	6-12
Centennial centipedegrass	18	9-15
Tifgreen Bermuda grass	6–12	3–6
Tifway Bermuda grass	6–12	4–8
Emerald zoysia grass	12–24	13–20
Matrella zoysia grass	12–24	15–20
Meyer zoysia grass	12–24	11–18
St. Augustine grass	10–18	10–18
Bahia grass	12–24	12–24

chemicals during the week prior to harvest. Using a sweeper or vacuum to remove mowing clippings the last three to four weeks leading up to harvest also improves the turf's appearance.

Sod must never be cut when it is under moisture stress. The cutter blade bounces out of the ground, the sod has little strength, and turf is under stress by the time the owner receives it (Figure 27-6).

Mechanical sod cutters harvest strips 12–16 inches wide and 2–3 feet long. Growers with less than 100 acres commonly use a small, hand-operated, walk-behind unit that has a 150–200 square yard cutting capacity per hour. Larger growers usually use tractor-mounted and/or self-propelled harvesters capable of cutting 600–800 square yards per hour. Sod is stacked on wooden pallets either in rolls or as flat slabs. The amount of sod harvested can be doubled if sod is rolled instead of stacked as flat slabs. Rolled, harvested sod must also be more mature. Approximately 400–500 square feet of sod is stacked per pallet, with a forklift required for placing pallets on transport trucks. A tractor-trailer

Figure 27-6 Cutting sod for placement in a new location as turfgrass. (Photo courtesy of University of Wisconsin Extension)

load typically consists of 10,000 square feet of sod. Forklifts that are rear-mounted on tractor-trailers provide a quick and easy method for unloading.

Recently, improvements allow larger rolls to be harvested. These big rolls of sod are typically cut as a continuous roll 42 inches wide and up to 100 feet long. This allows up to 24 100-foot rolls to be hauled on a semi-trailer totalling 8,400 square feet of grass. The roll lies like a carpet and generally is more stable and requires less water for establishment compared with traditional slab sod as fewer cut edges are exposed. Currently, the big rolls are being used for stabilization of roadsides and landfills. Less labor is involved in installing the big rolls on large-area jobs but they are more cumbersome on smaller sod-installation jobs such as lawns.

Thickness of soil removed during harvesting varies with turfgrass species. Removing the least amount of soil is the objective of an efficient sod harvest. Soil conservation must be a priority in order to ensure long-term productivity of the soil. Ideally, one-fourth to one-half-inch of root zone should be removed when sod is cut. Sod that is thin-cut is easier to handle, less expensive to transport, and knits in more quickly than thicker-cut sod, but it is more susceptible to drought injury.

Growers harvest up to 40,000 square feet per acre per cutting. Normal yields are generally between 28,000 and 38,000 square feet per acre. A two-inch ribbon of grass is typically left between harvested strips for re-establishment from stolons. Bermuda grass producers often clean-cut a field because Bermuda grass re-establishes from rhizomes, as well as from stolons. Centipedegrass and St. Augustine grass must re-cover the ground with stolons from ribbons left between harvested strips. Once harvesting has been performed, these strips should be lightly incorporated into the soil by rototilling and then rolled to smooth the soil surface. If this is not done, the remaining strips will provide a bumpy surface for mowing, fertilizing, and harvesting equipment. If practical, harvest the second crop at 90 degrees to the first one to minimize this uneven surface. For Bahia grass fields, ribbons may or may not be left. In either case, the fields are usually reseeded to hasten recovery.

Separating turfgrass cultivar areas in the field is necessary to prevent contamination from adjacent areas. Normally, this is achieved by careful planning, before establishment, with the use of service road or drainage ditches between cultivars. If these barriers are not used, a minimum of eight feet of tilled or bare soil must be maintained between grasses. A nonselective herbicide such as glyphosate may be used to maintain bare soil.

Marketing

Wholesale buyers for most sod producers consist of landscape maintenance/contractors, garden centers, building contractors, homeowners, and golf course/athletic field superintendents. Growers with small acreage and/or limited tractor-trailer shipping capabilities generally sell to homeowners and lawn care professionals.

Shipping costs generally limit the competitive range for most producers. Delivery charges are typically determined per load, per loaded mile, or per square yard. The weight of sod grown on mineral soils is about five pounds per square foot. Sod grown on muck soil is generally less expensive to produce and lighter in weight, so it can be transported over longer distances still at a competitive price.

Delivery means for growers will differ. For large producers, usually an 18-wheel tractor-trailer rig is preferred. Many job sites do not have unloading facilities, so rear-mounted portable forklifts are brought along with the sod. Smaller producers or smaller loads will best be served by appropriately sized trucks.

Sod pallets used normally are 48 inches square and are built from inexpensive lumber. Locating and maintaining adequate pallets can be a problem for the manager.

Costs and Returns

Costs and returns vary considerably with location, equipment, and labor available, and with management practices. Generally, prices for sod increase as the farm size decreases. Capital investments for sod farms include land, buildings, and equipment. Variable costs include labor, fuel, fertilizer, pesticides, repairs, and parts. Fixed costs include insurance, taxes, depreciation, land charge, management charges, and others.

Once sod is installed, fertilized, irrigated, and maintained, beautiful grass-covered areas are created (Figure 27-7).

Figure 27-7 Over half of the world's cool-season grass seed, about 500 million pounds, is produced in the U.S. Pacific Northwest. Parks and golf courses are important markets for turfgrass seed. (Photo courtesy of USDA ARS)

SUMMARY

Commercial sod production is expensive and labor-intensive farming. Growers must select the right grass to grow for their area. Proper preparation of the soil, watering, fertilizing, and management are essential to production. Once the grass is growing, constant monitoring for diseases and pests is also necessary. Keen competition, saturated markets, and a fluctuating economy make a thorough investigation of potential markets and costs of production necessary.

Review

True or False

1. Sod is shipped in plastic bags on trucks.

2. Sandy soil is the best for sod production.

3. Iron is the most important nutrient for growing grass.

4. Rotary mowers cannot be used on sod.

Short Answer

5. List three diseases of sod (grass).

6. Why is iron used for some sod production?

7. Name three insects and three diseases that affect sod production.

8. What four factors should be considered when selecting a grass species?

9. Name four turfgrasses.

10. List six weeds that can become a problem in sod production.

Critical Thinking/Discussion

11. Describe the site selection and soil preparation for growing sod.

12. How is sod harvested?

13. Discuss three factors to consider before installing sod.

14. Describe two of the best types of soil for growing sod.

15. Explain the purpose of fumigation.

16. Describe the fertilizer requirement for sod production.

Knowledge Applied

1. Visit a local county extension office and obtain copies of publications on turfgrass, sod, or lawn production and maintenance. Or search the Internet for extension publications on these topics. Report your findings to the class.

2. Obtain a grass identification key from the cooperative extension service. Collect ten specimens of different turf-grasses and identify them by using the key.

3. Add specimens of different turfgrasses to a pressed plant collection. See Appendix Tables A-20, A-21, and A-22.

4. Visit a golf course and discuss the duties of the turfgrass workers with the manager.

5. Establish turfgrass demonstration plots. Show the effect of different fertilizer applications and mowing heights on turfgrass health and vigor.

6. Visit with some local building contractors. Find out where they purchase sod or how they arrange for sod to be put around new homes or offices. Ask them what kind of a demand exists for sod around new buildings.

Resources

Emmons, R. (2007). *Turfgrass science and management* (4th ed.). Albany, NY: Delmar Cengage Learning.

Fry, J., & Huang, B. (2004). *Applied turfgrass science and physiology.* Indianapolis, IN: Wiley Publishing.

McCarty, L. B., Everest, J. W., Hall, D. W., Murphy, T. R., & Yelverton, F. (2008). *Color atlas of turfgrass weeds.* Indianapolis, IN: Wiley Publishing.

Turgeon, A. J. (2007). *Turfgrass management* (8th ed.). Englewood Cliffs, NJ: Prentice Hall.

Web Sites

The Web site for Seedland (www.seedland.com) provides many resources and many links to other resources for anyone wanting to know more about any type of grass. It is a commercial Web site but packed with information.

Some other specific Web sites include:

Maryland Department of Agriculture, Nutrient recommendations for sod production: www.mda.state.md.us/plants-pests/turf_seed/nutrient_recommendations_for_sod_production.php.

Turfgrass Producers International: www.turfgrasssod.org.

Ontario Ministry of Agriculture Food and Rural Affairs, Sod production: www.omafra.gov.on.ca/english/crops/facts/info_sodprod.html.

Michigan State University, Sod production research: www.turf.msu.edu/sod.html.

University of Florida, IFAS Extension, Electronic Data Information Source, Sod production: http://edis.ifas.ufl.edu/TOPIC_Sod.

Internet

Internet sites represent a vast resource of information. The URLs (uniform resource locators) for World Wide Web sites can change. Using one of the search engines on the Internet such as Google or Yahoo!, find more information by searching for these words or phrases: turfgrass, sod farms, preplant fumigation, turfgrass mowers, turfgrass sod production, grass species, soil, soil probe, warm season grass, cool season grass, Kentucky bluegrass, Bermuda grass, zoysia grass, St. Augustine grass, centipedegrass, Bahia grass, and turfgrass installation.

Additionally, the PLANTS database (http://plants.usda.gov/index.html) provides standardized information about the vascular plants of the United States and its territories. It includes names, plant symbols, checklists, distributional data, species abstracts, characteristics, images, crop information, automated tools, onward Web links, and references. The PLANTS database is a great educational and general use resource.

Chapter 28
Careers in Plant and Soil Sciences

THE PURPOSE OF *education and learning is to become employable and stay employable—to get and keep a job. People look for careers and careers look for people. Two broad categories of career opportunities in plant science and horticulture are (1) working for someone else or (2) working for yourself. Success in any career requires some general skills and knowledge as well as some very specific skills and knowledge unique to a chosen occupation.*

After completing this chapter, you should be able to—

- List the basic skills and knowledge needed for successful employment and job advancement in plant and soil science
- Describe the thinking skills needed for the workplace of today
- Identify the traits of an entrepreneur
- List six occupational areas of plant, soil, and crop science
- Describe the general duties of the occupations in four areas of horticulture
- Describe five ways to identify potential jobs
- List eight guidelines for choosing a job
- List 10 guidelines for filling out an application form
- Describe a letter of inquiry or application
- List the elements of a resume or data sheet
- Describe ten reasons an interview may fail
- Identify careers in plant science and horticulture
- Identify careers in soil science
- Discuss what research studies indicate about basic skills and thinking skills for the workplace
- Describe occupational safety from the employer's perspective and the employee's perspective

KEY TERMS

Career Clusters
Competencies
Creative Thinking
Data Sheet
Entrepreneur
Follow-Up Letter
Letter of Application
Letter of Inquiry
Resume
Visualization

GENERAL SKILLS AND KNOWLEDGE

Over the past few years, research study after study indicate that potential employees never receive some very basic skills and knowledge. Without these basic skills and knowledge, the specific skills and knowledge for employment in the plant science, horticulture, or soil science field is of little value. Also, the new workplace demands a better-prepared individual than in the past. Finally, those individuals working for themselves must develop a trait called entrepreneurship. This may also be a good trait for any employee.

Basic Skills

Success in the workplace requires that individuals possess skills in reading, writing, arithmetic, mathematics, listening, and speaking, at levels identified by employers nationwide.

Reading. Individuals ready for the workplace of today and the future demonstrate reading with the following **competencies**:

+ Locate, understand and interpret written information, including manuals, graphs, and schedules to perform job tasks
+ Learn from text by determining the main idea or essential message
+ Identify relevant details, facts, and specifications
+ Infer or locate the meaning of unknown or technical vocabulary
+ Judge the accuracy, appropriateness, style, and plausibility of reports, proposals, or theories of other writers

Reading skills in plant science, horticulture, and soil science are necessary to keep up with new information such as new research, new techniques, or new products (Figure 28-1).

Writing. Individuals ready for the workplace of today and the future demonstrate writing abilities with the following competencies:

+ Communicate thoughts, ideas, information, and messages
+ Record information completely and accurately
+ Compose and create documents such as letters, directions, manuals, reports, proposals, graphs, and flowcharts with the appropriate language, style, organization, and format
+ Check, edit, and revise for correct information, emphasis, form, grammar, spelling, and punctuation

As with any field of science, writing skills are necessary for such tasks as keeping records, describing disease conditions in plants, identifying and recording plant species (Figure 28-2), taking soil samples, describing soil types, or requesting a test or particular information.

Arithmetic and Mathematics. The workplace of today and the future requires individuals with competencies in arithmetic and mathematics. Arithmetic is computing with numbers by the operation of addition, subtraction, multiplication, and division. Mathematics is the study of the measurement, properties, and relationships of quantities and sets,

Figure 28-1 Utah State University research assistant pollinates Snake River wheatgrass on an Agricultural Research Service test plot. Becoming a plant breeder requires university training. (Photo courtesy of USDA ARS)

Figure 28-2 Identifying plant species for an inventory of native plants requires reading and record-keeping skills. (Photo courtesy of USDA NRCS)

using numbers and symbols. Important competencies of arithmetic and mathematics include:

+ Perform basic computations
+ Use numerical concepts such as whole numbers, fractions, and percentages in practical situations
+ Make reasonable estimates of arithmetic results without a calculator
+ Use tables, graphs, diagrams, and charts to obtain or convey information
+ Approach practical problems by choosing from a variety of mathematical techniques
+ Use quantitative data to construct logical explanations of real-world situations
+ Express mathematical ideas and concepts verbally and in writing
+ Understand the role of chance in the occurrence and prediction of events

Anyone not convinced of the value of arithmetic and mathematics to plant science, soil science, or horticulture should consider the skills required to figure mixing liquid spray, planting rates, using global positioning systems (GPS)/geographic information systems (GIS), field mapping soils (Figure 28-3), or calculating fertilizer needs.

Listening. Individuals working today and in the future need to demonstrate an ability to really listen. This means to receive, attend to, and interpret verbal messages and other cues such as body language. Real listening means the individual comprehends, learns, evaluates, appreciates, or encourages the speaker.

Speaking. Finally, individuals successful in the workplace of today and tomorrow demonstrate these speaking competencies:

+ Organize ideas and communicate oral messages appropriate to listeners and situations (Figure 28-4)
+ Participate in conversation, discussion, and group presentations
+ Use verbal language, body language, style, tone, and level of complexity appropriate for audience and occasion

Figure 28-3 Soil scientist field mapping using Global Positioning System technology. (Photo courtesy of USDA NRCS)

Figure 28-4 Anyone employed in plant science, soil science, or a horticulture career should be able to express ideas and communicate orally to listeners in various situations. (Photo courtesy of USDA ARS)

+ Speak clearly and communicate the message
+ Understand and respond to listener feedback
+ Ask questions when needed

Thinking Skills

Research studies indicate that employers in the new workplace want workers who can think. Employers search for individuals showing competencies in these areas: **creative thinking**, decision making, problem solving, mental **visualization**, knowing how to learn, and reasoning.

Creative Thinking. Creative thinkers generate new ideas by making nonlinear or unusual connections or by changing or reshaping goals to imagine new possibilities. These individuals use imagination freely, combining ideas and information in new ways (Figure 28-5).

Decision Making. Individuals who use thinking skills to make decisions are able to specify goals and limitations to a problem. Next, they generate alternatives and consider the risks before choosing the best alternative.

Figure 28-5 Rewarding careers allow individuals to seek creative solutions to problems or to conceive new products. Research on showy colors, bizarre shapes, or surprising textures of exotic lettuces may brighten salads of the future. (Photo courtesy of USDA ARS)

Problem Solving. Fundamentally, the first step to problem solving is recognizing that a problem exists. After this, individuals with problem-solving skills identify possible reasons for the problem and then devise and begin a plan of action to resolve it. As the problem is being solved, problem solvers monitor the progress and fine-tune the plan. Being able to recognize a disease condition and look for solutions is a good example of problem solving in plant or soil science or horticulture.

Mental Visualization. This thinking skill requires an individual to see things in the mind's eye by organizing and processing symbols, pictures, graphs, objects, or other information. For example, this type of individual sees a cornfield from a diagram or a greenhouse operation from a schematic.

Knowing How to Learn. Perhaps of all the thinking skills, this is most important with the rapid changes in available technology. The individual with this type of skill recognizes and can use learning techniques to apply and adjust existing and new knowledge and skills in familiar and changing situations. Knowing how to learn means awareness of personal learning styles—formal and informal learning strategies.

Reasoning. The individual who uses reasoning discovers the rule or principle connecting two or more objects and applies this to solving a problem. For example, chemistry teaches the theory of pH measurements, but the reasoning individual is able to apply this information to understand pH in soil acidity.

INTANGIBLE SKILLS

More and more employers seek employees with intangible skills or soft skills, like balance in a person's life, communicating effectively, problem solving, decision making, resolving conflict, working with others, planning, conducting effective meetings, professional growth, ethics, community service, and volunteerism. These skills are outside the technical skills needed in a job or career, yet they are often more important in determining the success of an individual. Employers have been asking schools and colleges to teach these soft skills. Some schools and colleges do teach these skills, but not in formalized classroom settings with prepared lessons and assessments. Many of these skills are often gained when the student takes an active role in a club or organization such as Future Farmers of America (FFA) or National Postsecondary Agricultural Student Organization (PAS). Leadership roles in an organization are particularly effective in developing these skills. Students who participate in contests such as the FFA Career Development Events (CDE) also seem to develop more of the soft skills that are important to success in the workplace.

Many companies recognize the need for these soft skills and have developed formalized lessons, training, and assessment. For example, the national FFA organization developed educational materials called LifeKnowledge that provide training in four areas of the soft skills: personal, organizational, career, and community. LifeKnowledge includes a set of 257 lesson plans structured to help FFA advisors and agricultural education teachers educate and prepare middle school and high school students to become proficient in 16 competency areas. In addition, another 100 lesson plans are available specifically for use at the collegiate level.

The 16 competency areas and precept statements include:

Premier Leadership

A—Action

A1. Work independently and in groups to get things done.
A2. Focus on results.
A3. Plan effectively.
A4. Identify and use resources.
A5. Communicate effectively with others.
A6. Take risks to get the job done.
A7. Invest in others by enabling and empowering them.
A8. Evaluate and reflect on actions taken and make appropriate modifications.

B—Relationships

B1. Practice human relations skills including compassion, empathy, unselfishness, trustworthiness, reliability, and listening.
B2. Interact and work with others.
B3. Develop others.
B4. Eliminate barriers in building relationships.
B5. Participate effectively as a team member.

C—Vision

C1. Contemplate the future.
C2. Conceptualize ideas.
C3. Demonstrate courage to take risks.
C4. Adapt to opportunities and obstacles.
C5. Persuade others to commit.

D—Character

D1. Live with integrity.
D2. Accurately assess personal values.
D3. Accept responsibility for personal actions.
D4. Respect others.
D5. Practice self-discipline.
D6. Value service to others.

E—Awareness

E1. Address issues important to the community.
E2. Perform leadership tasks associated with citizenship.
E3. Participate in activities that promote appreciation of diversity.

F—Continuous Improvement

F1. Implement a leadership and personal growth plan.
F2. Seek mentoring from others.

F3. Use innovative problem-solving strategies.
F4. Adapt to emerging technologies.
F5. Acquire new knowledge.

Personal Growth

G—Physical Growth

G1. Practice healthy eating habits.
G2. Respect one's body.
G3. Participate in a fitness program.
G4. Set goals for long-term health.

H—Social Growth

H1. Acknowledge that differences exist among people.
H2. Present self appropriately in various settings.
H3. Develop and maintain relationships.

I—Professional Growth

I1. Plan and implement professional goals and priorities.
I2. Make clear decisions in personal and professional life.
I3. Demonstrate professional ethics.
I4. Balance personal and professional responsibilities.
I5. Demonstrate exemplary employability skills.

J—Mental Growth

J1. Think critically.
J2. Think creatively.
J3. Practice sound decision making.
J4. Solve problems.
J5. Commit to life-long learning.
J6. Persuade others.
J7. Practice sound study skills.

K—Emotional Growth

K1. Cope with life's trials.
K2. Live a compassionate and selfless life.
K3. Develop self-assurance and confidence.
K4. Embrace the emotional development process.
K5. Establish emotional well-being.
K6. Seek appropriate counsel.
K7. Practice healthy expressions of love.

L—Spiritual Growth

L1. Nurture a spiritual belief system.
L2. Respect and be sensitive to others' beliefs.

Career Success

M—Communications

M1. Demonstrate technical and business writing skills.
M2. Demonstrate professional job-seeking skills.
M3. Make effective business presentations.

M4. Communicate appropriately with coworkers and supervisors.

M5. Operate effectively in the workplace.

N—Decision Making

N1. Demonstrate the decision-making process.

N2. Demonstrate problem-solving skills.

N3. Make ethical decisions.

N4. Choose a career based on passion, abilities, and aptitudes.

O—Flexibility and Adaptability

O1. Embrace emerging technology in the workplace.

O2. Manage change.

O3. React with openness to feedback and professional growth opportunities.

O4. Experiment and take risks.

P—Technical and Functional Skills in Agriculture and Natural Resources. Many high school agricultural programs and organizations such as 4-H, Block and Bridle, Alpha Zeta, PAS, FarmHouse, FFA, and Sigma Alpha endorse this training. Information about the training can be found on the LifeKnowledge Web site (www.CollegiateLifeKnowledge.org) or on the FFA Web site (www.ffa.org/ageducators/lifeknowledge). The goal of any of this type of training is to aid students in being successful in their careers and to become a society-ready graduate.

JOBS AND CAREERS

United States agriculture is the most productive and advanced in the world. This is because of the scientific advances in fertilizers, pesticides, and seed development. Product and crop management research is the key to this continued growth. Jobs and careers in plant science, soil science, and crop science continue to be promising and critical to the future of the world.

Careers in the Plant Science Field

Individuals working in plant science work for federal and state governments, universities, and the private sector. The job of a plant scientist includes collection of soil data, consultation, investigation, evaluation, interpretation, planning, or inspection relating to plant science. This career includes many different assignments and involves making recommendations about growing plants and using plants and plant products.

A plant scientist needs good observation skills to be able to analyze and determine the characteristics of different plants, plant diseases, and plant pests. Computer skills and geographic information systems are useful. These skills help the scientist to analyze the multiple facets of plant production.

Some of the varied occupations and careers in the broad classification of the plant sciences include:

- Plant biotechnology
- Biochemist
- Plant breeder
- Cytogeneticist
- Cytologist
- Ecologist
- Geneticist
- Marketing specialist
- Physiologist
- Production specialist
- Protection specialist
- Quality specialist
- Research scientist
- Utilization specialist
- Analyst
- Nutrition Specialist

Some additional related careers include:

- Agricultural climatologist
- Agricultural scientist
- Agronomist
- Biotechnologist
- Botanist
- Computer modeling specialist
- County extension service specialist
- District conservationist
- Entomologist
- Environmental scientist
- Erosion and sediment control specialist
- Fertilizer technologist/specialist
- Fertilizer use specialist
- Forage range manager
- Forest soil scientist/specialist
- Fruit or nut grower
- Plant geneticist
- GIS technician/specialist
- Hazardous waste management specialist
- Hydrologist
- Irrigation specialist
- Irrigation technician
- Sprinkler/drip irrigation technician
- Land management specialist
- Land use specialist
- Pedologist
- Pest management specialist
- Pesticide specialist/technician

✦ Pomologist
✦ Range management specialist
✦ Range soil scientist
✦ Range soil specialist
✦ Reclamation specialist
✦ Resource conservationist
✦ Seed production specialist
✦ Seed technologist
✦ Silviculturalist
✦ Statistician
✦ Surface mine reclamation specialist
✦ Tissue culture specialist
✦ Turfgrass manager
✦ Turfgrass specialist
✦ Waste management specialist
✦ Weed scientist
✦ GPS specialist/technician

Note: Many of the careers for plant science and soil science are similar or overlap.

Careers in the Horticulture Field

Many sources of employment exist for individuals trained in horticultural practices. Jobs can be found in such businesses as greenhouses, nurseries, garden centers, golf courses, parks, orchards, floral design shops, grounds maintenance operations, vegetable growers, and fruit growers. Table 28-1 is a job analysis chart that explains positions in these areas of employment.

Table 28-1 Job Analysis Chart[1]

Job Title[2]	Year-round Work	Hours	Outdoors/ Indoors	Variety	Fringe Benefits	Work with Others
Greenhouse worker	Yes	Regular but some overtime required	Indoors	Yes	Yes	Yes
Nursery worker	Most times	Regular but peak seasons	Mostly indoors	Yes	Some	Yes
Garden center employee	Yes	Regular but peak seasons	Both indoors and outdoors	Yes	Some	Yes
Golf course employee	No	Regular during the golf season	Outdoors	Yes	Not many	Yes
Assistant grounds keeper	Yes	Regular with some overtime	Mostly outdoors	Yes	Maybe some	Maybe some
Park employee	Yes	Regular	Outdoors	Yes	Not many	Yes
Vegetable grower	Depends on grower	Seasonal	Outdoors	Yes	Some	Yes
Orchard employee	Depends on grower	Seasonal	Outdoors	Yes	Some	Yes
Employee of small fruit grower	Depends on grower	Seasonal	Outdoors	Yes	Some	Yes
Employee of floral design shop	Yes	Regular with some overtime	Indoors	Yes	Yes	Yes

[1] Adapted from the Job Profile Chart, Ohio State University
[2] All jobs offer opportunity for promotion, work in one place, and require a high school diploma with a course in agriculture or horticulture.

The horticulture industry consists of four major divisions:

1. Pomology—The science and practice of growing, harvesting, handling, storing, processing, and marketing tree fruits.
2. Olericulture—The science and practice of growing, harvesting, storing, processing, and marketing vegetables.
3. Floriculture—The science and practice of growing, harvesting, storing, designing, and marketing flowering plants.
4. Landscape and Nursery Industry—The science and practice of propagating, growing, installing, maintaining, and using grasses, annual plants, shrubs, and trees in the landscape.

Careers in the Soil Science Field

Soil scientists work for federal and state governments, universities, and the private sector. The job of a soil scientist includes collection of soil data, consultation, investigation, evaluation, interpretation, planning or inspection relating to soil science. This career includes many different assignments and involves making recommendations about many resource areas.

A soil scientist needs good observation skills to be able to analyze and determine the characteristics of different types of soils. Soil types are complex and the geographical areas a soil scientist may survey are varied. Aerial photos or various satellite images are often used to research the areas. Computer and GPS/GIS skills are useful.

Individuals with jobs in soil science work in both the office and field. The work may require walking over rough and uneven land and using shovels and spades to gather samples or examine a soil pit exposure. Often the work is done with non–soil science professionals. Some of the work of those involved in soil science includes:

+ Conducting general and detailed soil surveys
+ Determining the wetness characteristics of the soil
+ Recommending soil management programs
+ Helping to design hydrologic plans in suburban areas
+ Monitoring the effects of farm, ranch, or forest activities on soil productivity
+ Giving technical advice used to help plan land management programs
+ Predicting the effect of land management options on natural resources
+ Preparing reports describing land and soil characteristics
+ Advising land managers of capabilities and limitations of soils (e.g., timber sales, watershed rehabilitation projects, transportation planning, soil productivity, military maneuvers, recreation development)
+ Training others
+ Preparing technical papers and attending professional soil science meetings
+ Conducting research in public and private research institutions
+ Managing soils for crop production, forest products and erosion control management
+ Evaluating nutrient and water availability to crops

+ Managing soils for landscape design, mine reclamation, and site restoration
+ Investigating forest soils, wetlands, environmental endangerment, ecological status, and archeological sites
+ Assessing application of wastes, including nonhazardous process wastes (residue and sludge management)
+ Conducting studies on soil stability, moisture retention or drainage, sustainability, and environmental impact
+ Assessing environmental hazards, including hazardous waste sites that involve soil-investigation techniques, evaluation of chemical fate and transport phenomena, and remediation alternatives
+ Regulating the use of land and soil resources by private and public interests (government agencies)

Well-trained plant and soil scientists are in high demand for a wide array of professional positions with public agencies or private firms. Here are some specific examples of positions available:

+ Wetland specialist
+ Watershed technician
+ Hydrologist with board of health
+ Environmental technician
+ State soil and water quality specialist
+ Soil conservationist (Figure 28-6)
+ County extension
+ Landscaping business
+ Farming
+ On-site evaluation
+ Crop consultant
+ Mapping and interpretation of soil
+ Research technician
+ Conservation planner
+ District marketing manager for an agricultural firm
+ County conservationist
+ Crop production specialist
+ Research scientist

Note: Many of the careers for plant science and soil science are similar or overlap.

Figure 28-6 Soil conservationists provide solutions to problems like water erosion. (Photo courtesy of USDA ARS)

EDUCATION AND EXPERIENCE

Requirements to begin working in plant science, horticulture, or soil science vary, depending upon the level of work. One requirement common to all is practical work experience in working with plants. To gain this practical experience, the new employee often begins at an entry-level job and then is advanced through the organization. The advancement depends on the skills and knowledge the employee brings to the job and the skills and knowledge gained on the job as well as productivity on the job.

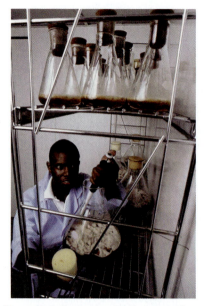

Figure 28-7 Student involved in a research project at the Agricultural Research Service Southern Regional Research Center in New Orleans, developing practical skills for future employment. (Photo courtesy of USDA ARS)

Entry-level educational requirements vary. Certainly all of the basic skills, thinking skills, and general workplace competencies discussed in the first part of this chapter are important. These skills should be obtained in high school and reinforced during additional training and schooling. More specialized education in the plant science, horticulture, or soil science field is offered in some high schools, community colleges, and universities.

Many high school programs provide the education necessary for lower-level entry positions. High school programs often provide students with supervised work experience in some aspect of plant science, horticulture, or soil science. This is invaluable in getting a job and in helping individuals determine if they wish to pursue additional education.

Some community colleges and other postsecondary schools provide specialized programs in plant science, horticulture, and soil science with practical experience as a part of the schooling. Programs at community colleges focus on entry-level technician jobs.

Universities and colleges offering bachelor's degrees, master's degrees, and doctorate programs provide some highly specialized education in the plant science, horticulture, and soil science areas.

Both colleges and universities provide internship opportunities in which students actually work in a career area during the semester, quarter, or over the summer. Also, many colleges and universities offer opportunities for students to become involved in research projects (Figure 28-7), which present great learning experiences and practical applications of classroom learning.

SUPERVISED AGRICULTURAL EXPERIENCE

A Supervised Agricultural Experience (SAE) program is designed to provide students the opportunity to gain experience in agricultural areas based on their interests. An SAE represents the actual, planned application of concepts and principles learned in agricultural education. Students experience and apply what is learned in the classroom to real-life situations. Students are supervised by agriculture teachers in cooperation with parents/guardians, employers, and other adults who assist them in the development and achievement of their educational goals. The purpose is to help students develop skills and abilities leading toward a career.

Planning and conducting an SAE for horticulture, plant science, or soil science could include areas of interest such as greenhouses, vineyards, golf courses, parks, floral design, biotechnology, soil mapping, soil testing, weed control, irrigation, genetics, fertilizers, orchards, or landscape. Students should work with their instructors to:

+ Identify an appropriate SAE opportunity in the community
+ Ensure that the SAE represents meaningful learning activities benefiting the student, the agriculture education program, and the community
+ Obtain classroom and individual instruction on the SAE
+ Adopt a suitable record-keeping system

+ Plan the SAE and acquire needed resources
+ Coordinate release time and visits to the SAE
+ Sign a training agreement along with the employer, teacher, and parent/guardian
+ Report on and evaluate the SAE and records resulting from it

Additional help and ideas for planning and conducting an SAE can be found through the national FFA Web site (www.ffa.org).

ENTREPRENEURSHIP

The most common view of an entrepreneur is one who takes risk and starts a new business. Although this may be true for some in the plant science field, some traits of entrepreneurship are desirable at many levels of employment in plant science. Within any organization, an entrepreneur may—

+ Find a better or higher use for resources
+ Apply technology in a new way
+ Develop a new market for an existing product
+ Use technology to develop a new approach to serving an existing market
+ Develop a new idea that creates a new business or diversifies an existing business

Anyone can be an entrepreneur. It is more of an attitude, but an attitude that incorporates many desired traits. The attitude of an entrepreneur includes—

+ Risk-taking with clear expectations of the odds
+ Focus on opportunities and not problems
+ Primary focus on the customer
+ Seek constant improvement
+ Impressed with productivity and not appearances
+ Recognize importance of example
+ Keep things simple
+ Open-door and personal contact leadership
+ Encourage flexibility
+ Purposeful and communicates a vision

Entrepreneurs are ready for the unexpected, differences, new needs, change, demographic shifts, changes in perception, and new knowledge. Entrepreneurs are good employees and good employers. Entrepreneurs keep plant science industries growing.

IDENTIFYING A JOB

Finding that first job or finding a different job can be difficult. Whole books, videos, and seminars are available on finding jobs. What follows are some suggestions. The Knowledge Applied section at the end of this chapter contains more information on how to find a job.

Sources for locating jobs include—

+ Classified advertisements of newspapers
+ Magazines or trade journals and publications
+ Personal contacts
+ Placement offices
+ Employment or personnel offices of companies
+ Public notices
+ Internet or other computerized online services

Newspapers, magazines, trade journals, and publications can be good resources for locating a job. By reading the advertisements in these publications the potential employee can determine the demand for his or her job skills. Also, the potential employee can compare his or her skills and training with those listed in the advertisement.

A different twist on the classified advertisement is the "situation wanted" section of newspapers, magazines, or trade journals. Many people secure excellent jobs by advertising their skills this way. An employer may read about an individual's skills and realize the job-seeker is the answer to the employer's needs.

Personal contacts are still a top source of jobs. This goes back to "it's not what you know but who you know." This is not really the "good ole boy" system but employers do not like to make mistakes. Some think that if a trusted acquaintance makes a recommendation, this lessens the chance of making a mistake in hiring. Also, personal contacts may know of jobs becoming available before being publicly announced. This gives the potential employee more time to prepare and research the job. Personal contacts include friends, relatives, teachers, guidance counselors, and employees of a company.

Placement offices provide vocational counseling, give aptitude and ability/interest tests, locate jobs, and arrange job interviews. Three types of placement offices are available: public, private, and school. These agencies work to match employers with prospective employees. An agency also often knows how to help a potential employee prepare and present himself or herself for a prospective interview.

Public placement offices are supported by federal and state funds. Their services are free. Private placement offices charge for services they provide. This charge is usually a percentage of the beginning salary, or the fee might also be paid by the employer seeking an employee. Individuals using private placement services sign a contract before services are provided. High schools, trade schools, and colleges may maintain a placement service for their students. They also provide help for individuals to identify their aptitude or interest for a job and help in preparation for job interviews.

Many companies support their own employment or personnel offices. Individuals seeking employment can fill out application forms and/or leave resumes for when a job becomes available.

Finally, some companies seeking new employees may issue some kind of public notice. This includes posters or fliers on bulletin boards around a community. Posters or fliers are sent to related businesses, which end up on their bulletin boards. Schools and colleges often receive public announcements of jobs.

Posting of jobs on the Internet is another type of bulletin board announcement. Some online information services maintain computerized databases of jobs. Interested individuals use an Internet connection to contact the computerized database to search for jobs that match their qualifications and desires. This type of job listing gives access to many potential jobs, but not necessarily local ones.

Career Clusters

Recently, a series of career clusters was developed to help students and instructors identify jobs and careers in 16 broad career areas. Career clusters provide a way for schools to organize instruction and student experiences around categories that encompass virtually all occupations from entry through professional levels. The clusters provide information about the knowledge and skills required. One of the 16 clusters is "Agriculture, Food & Natural Resources" and within that cluster are the occupations for plant science, soil science, and horticulture. More information on the career clusters can be found at CareerClusters.org (www.careerclusters.org) or AgrowKnowledge (www.agrowknow.org).

FINDING CAREER OPPORTUNITIES

Just the shear number of career opportunities in plant science, horticulture, and soil science creates an almost overwhelming task of selecting a career. Here are some places to begin investigating the choices:

- AgCareers: www.agcareers.com (Figure 28-8)
- American Society of Agronomy, Career Placement: www.agronomy.org
- Botanical Society of America: www.botany.org/bsa/careers
- Careers in Horticulture: www.careersinhorticulture.com
- Earthworks: www.earthworks-jobs.com/plant.html
- National Society of Consulting Soil Scientists: www.nscss.org/jobs.html
- Office of Personnel Management: www.usajobs.opm.gov
- Soil Science Society of America: www.soils.org
- The Ohio State University, enPlant: http://enplant.osu.edu
- U.S. Consortium of Soil Science Associations: soilsassociation.org
- U.S. Department of Agriculture, Agricultural Research Service, Careers in Plant Science: www.ars.usda.gov/careers
- Vocational Information Center: www.khake.com/index.html

Figure 28-8 AgCareers, a relatively new concept in connecting agriculture, food, and natural resources employers with employees. (Image courtesy of AgCareers.com)

The United States Department of Labor, Bureau of Labor Statistics, publishes the *Occupational Outlook Handbook* (www.bls.gov/oco). This handbook provides a detailed account of common jobs in the industry.

Additionally, career technical educators (vocational) at high schools and technical colleges can provide guidance to career selection. Major advisors at universities also can help with career choices.

GETTING A JOB

Once some job possibilities are identified, the work begins. Getting the job is difficult and requires some preparation. Again, there are entire books, videos, and seminars that teach how to get a job. A few tips follow.

Once a job is identified, do some research on the company and the job before applying. Know these things about the job and the company—

+ Name of the company
+ Name of personnel manager
+ Company address and phone number
+ Position available

- Requirements for the position
- Geographic scope of the company—local, county, state, regional, national
- Company's product(s)
- Recent company developments
- Responsibilities of the position
- Demand for the company's product(s)

Application Forms

If the company requires an application form, remember you are trying to sell yourself by the information given. Review the entire application form before you begin to complete it. Pay particular attention to any special instructions to print or write in your own handwriting. When answering ads that require potential employees to apply in person, be prepared to complete an application form on the premises. Take an ink pen. Prepare a list of information you will need to complete the application form. The information may include your social security number; the addresses of schools you have attended; names, phone numbers, and addresses of previous employers and supervisors; names, phone numbers, and addresses of references. The following guidelines will provide you with some direction when completing application forms.

- Follow all instructions carefully and exactly.
- If handwritten rather than typed, write neatly and legibly. Handwritten answers should be printed unless otherwise directed.
- Application forms should be written in ink unless otherwise requested. If you make a mistake, mark through it with one neat line.
- Be honest and realistic.
- Give all the facts for each question.
- Keep answers brief.
- Fill in all the blanks. If the question does not pertain to you, write "not applicable" or "N/A." If there is no answer, write "none" or draw a short line through the blank.
- Many application forms ask what salary you expect. If you are not sure what is appropriate, write "negotiable," "open," or "scale" in the blank. Before applying, try to find out what is the going rate for similar work at other locations. Give a salary range rather than exact figure.

Letters of Inquiry and Application

The purpose of a letter of inquiry is to obtain information about possible job vacancies. The purpose of a letter of application is to apply for a specific position that has been publicly advertised. Both letters indicate your interest in working for a particular company, acquaint employers with your qualifications, and encourage the employer to invite you for a job interview.

A Job Is More Than Money

Before taking a job, be certain that it is what you want. Although the salary or the wage is important, job satisfaction is something quite different and very important. Jobs quickly become routine and mundane. For example, for some people, it can easily become a chore just to go to a job if it has little fulfillment and challenge. Before taking a job or even while looking for a job, answer these questions:

1. Does the job description fit your interests?
2. Is this the level of occupation in which you wish to engage?
3. Does this type of work appeal to your interests?
4. Are the working conditions suitable to you?
5. Will you be satisfied with the salaries and benefits offered?
6. Can you advance in this occupation as rapidly as you would like? What are the advancement opportunities?
7. Does the future outlook satisfy you?
8. Is the occupation in demand now and in the foreseeable future?
9. Do you have or can you get the education needed for the occupation?
10. What type of training is available after taking the job?
11. Can you get the finances needed to get into the occupation?
12. Can you meet the health and physical requirements?
13. Will you be able to meet the entry requirements?
14. Do you know of any other reasons you might not be able to enter this occupation?
15. Is the occupation available locally or are you willing to move to a part of the country where it is available?

Also, before taking a job or looking for a job, do a little personality inventory of yourself. Consider the following:

1. Do I like to be alone or with people?
2. Am I mechanical or artistic?
3. Would I rather work independently or work under supervision?
4. Would I like to think or be active?
5. Could I take authority and responsibility for others?
6. Must I have freedom to express creativity?
7. What things do I like to do? Make a list.
8. At what time of day can I work best?
9. Can I work under pressure or stress?
10. Make a list of your strong points. Consider skills, hobbies, and leisure-time interests you can offer an employer.

Do your research and your job will be more rewarding and you will feel better about yourself.

Letters of inquiry and application represent you. They should be accurate, informative, neat, and eye-catching. Your written communications should present a strong, positive, professional image both as a job seeker and future employee. The following list should be used as a guide when writing letters of inquiry and application.

+ Be short and specific, one or two pages (details left to resume). Use 81/2 × 11-inch white typing paper, not personal or fancy paper.
+ Have the letter be neatly typed and error-free.
+ Use an eye-catching format, neat and free from smudges.
+ Write to a specific person. Use "To Whom It May Concern" if answering a blind ad.
+ Use logical organized paragraphs that are to the point.
+ Use carefully constructed sentences that are free from spelling or grammatical errors.
+ Use a positive tone.
+ Express ideas in a clear, concise, direct manner.
+ Avoid slang words and expressions.
+ Avoid excessive use of the word "I."
+ Avoid mentioning salary and fringe benefits.
+ Write a first draft and then make revisions.
+ Proofread the final letter yourself, and also have someone else proofread it.
+ Address and sign correctly. Type envelope addresses.

This information should be included in a letter of inquiry:

+ Specify the reasons why you are interested in working for the company and ask if there are any positions available now or in the near future.
+ Express your interest in being considered as a candidate for a position when one becomes available.
+ Because you are not applying for a particular position, you cannot relate your qualifications directly to job requirements. You can explain how your personal qualifications and work experience would help meet the needs of the company.
+ Mention and include your resume.
+ State your willingness to meet with a company representative to discuss your background and qualifications. (Include your address and the phone number at which you can be reached.)
+ Address letters of inquiry to the personnel manager unless you know his or her name.

A letter of application should include:

+ Your source of the job lead.
+ The particular job you are applying for and the reason for your interest in the position and the company.
+ How your personal qualifications meet the needs of the employer.

+ Your work experience relates to the job requirements.
+ Your resume.
+ A request for an interview and 2 statement of your willingness to schedule an interview. (Include your address and 2 phone number at which you can be reached.)

Resume or Data Sheet

Some jobs require a resume or data sheet (Figure 28-9). The following information should be considered when writing a resume or data sheet:

+ Name, address, and phone number
+ Brief, specific statement of career objective
+ Educational background—names of schools, dates, major field of study, degrees or diplomas—listed in reverse chronological order
+ Leadership activities, honors, and accomplishments
+ Work experience listed in reverse chronological order
+ Special technical skills and interests related to job
+ References

RESUME
Sheila Smith

Current Address
PO Box 1238
Webster, IA 5000
Tel: 000/888-8888; e-mail address: sasmith3@crst.com

Education
Local High School, IA: Graduated 2005.
Kirkwood Community College, Cedar Rapids, IA, 2005-2007: AAS, Plant Science.

Career Objectives
Obtain satisfying job in the horticulture industry that provides advancement opportunities during my career.

Activities and Honors
- Active member of 4-H Club for three years. Learned plant collection and identification.
- Member FFA for four years and was elected president during senior year.
- Member Postsecondary Agricultural Student (PAS) Organization 2005-2007.
- Advisor to local 4-H Club 2005 to present.

Employment and Work Experiences
January 1997 to Present: Sunrise Nursery, Arco, IA; general help; prepare bedding plants, clean greenhouse, work some in sales.
July 1995 to December 1996: ABC Grocery, Nevada, IA; restocked shelves; boxed groceries; worked into checker position.

References
Available on request.

Figure 28-9 A good resume presents information quickly and clearly.

◆ Limit to one page if possible
◆ Neatly typed and error-free
◆ Logically organized
◆ Honestly listed qualifications and experiences

Employers look for a quick overview of who you are and how you can fit into their business. On the first reading, the employer will spend 10–15 seconds reading a resume. Be sure to present relevant information clearly and concisely in an eye-catching format.

The Interview

The next step in the job-hunting process is the interview. Although there are many dos and don'ts of an interview available, perhaps the best advice comes from the interviewer's side of the desk. This list presents items that are common reasons interviewers give for not being able to place applicants in a job:

1. Poor attitude
2. Unstable work record
3. Bad references
4. Lack of self-selling ability
5. Lack of skill and experience
6. Not really anxious to work
7. "Bad mouthing" former employers
8. Too demanding (wanting too much money or to work only under certain conditions)
9. Unable to be available for interviews or canceling out of interviews
10. Poor appearance
11. Lack of manners and personal courtesy
12. Chewing gum, smoking, fidgeting
13. No attempt to establish rapport; not looking the interviewer in the eye
14. Being interested only in the salary and benefits of the job
15. Lack of confidence; being evasive
16. Poor grammar, use of slang
17. Not having any direction or goals

Follow-Up Letters

A follow-up letter is sent immediately after an interview. The follow-up letter demonstrates your knowledge of business etiquette and protocol. Always send a follow-up letter regardless of whether or not you had a good interviewing experience and regardless of whether or not you are still interested in the position. When an employer does not receive a follow-up letter from a job candidate, the employer often assumes that the candidate is not aware of the professional protocol needed to be demonstrated on the job.

The major purpose of a follow-up letter is to thank those individuals who participated in your interview. In addition, a follow-up letter reinforces your name, application, and qualifications to the employer and indicates whether you are still interested in the job position.

OCCUPATIONAL SAFETY

Employees with jobs or careers in plant science, horticulture, or soil science should expect safe and healthful workplaces. Even so, individuals may encounter such hazards as:

+ Toxic chemicals
+ Slippery floors
+ Unsafe equipment
+ Sharp objects
+ Heavy lifting
+ Stress
+ Harassment
+ Poorly designed work areas

To prevent or minimize exposure to occupational hazards, employers are expected to provide employees with safety and health training, including providing information on chemicals that could be harmful to an individual's health. If an employee is injured or becomes ill because of a job, many employers sometimes provide payment for medical care and lost wages.

Not only are plant industry employers responsible for creating and maintaining a safe workplace, employees must do their part, including:

+ Follow all safety rules and instructions.
+ Use safety equipment and protective clothing when needed.
+ Look out for coworkers.
+ Keep work areas clean and neat.
+ Know what to do in an emergency.
+ Report any health or safety hazard to the supervisor.

For additional information about personal and occupational safety practices in the workplace, contact the Occupational Safety and Health Administration (OSHA) on the Web at www.osha.gov, the National Institute for Occupational Safety and Health (NIOSH) at www.cdc.gov/niosh/homepage.html, or the Department of Labor at www.dol.gov.

SUMMARY

The goal of education and training is primarily to become employable and stay employable—to get and keep a job or run a successful business. The world of work still requires people who can read, write, do math, and communicate. Rapidly advancing technology makes these needs even more apparent. Also, the modern workplace looks for people who possess thinking skills. With a solid set of basic skills, future employees also need to relate to other people, be able to use information, understand the concept of systems, and use technology. Ideas like responsibility, self-esteem, sociability, self-management, and integrity are not out of date either.

Jobs or careers in the plant science, horticulture, or soil science areas range from those that are very specific to those that support these industries. In general, potential job areas include supplies and services, training, production, marketing, inspection, and research and development. Education and training for jobs in plant and soil science and horticulture vary from on-the-job training to high school to college degrees.

After training and education, finding and getting the right job or career may still be a challenge. Several good resources exist for locating a job. However, the best resource is personal contact. Letters of inquiry, letters of application, a resume, and being prepared for the job interview help secure a job.

Review

True or False

1. Employers spend, on average, ten minutes looking at a resume.

2. Dressing appropriately is important for an interview.

3. Personal contacts are still a top source of jobs.

4. Jobs in the plant science areas are only for highly skilled people.

5. Research is critically important in the plant science fields.

Short Answer

6. List the five basic skills necessary for success in the workplace.

7. Name five ways to identify a job.

8. List ten jobs or careers in any area of plant science and ten in the area of soil science.

9. Identify four types of jobs in the horticulture field.

10. List the parts of a resume.

Critical Thinking/Discussion

11. Discuss the difference between a letter of inquiry, a letter of application, and a follow-up letter.

12. Define an entrepreneur.

13. Why are thinking skills important in the workplace?

14. Why is on-the-job training important?

15. What questions should you ask yourself when looking for a job in the plant or soil science area?

16. Who is responsible for workplace safety?

Knowledge Applied

1. Gather sample resumes from local sources to share with other students. Have students develop their own resumes or data sheets.

2. Collect position announcements and classified ads for jobs in the plant science fields. Have students write a letter of job inquiry and a letter of job application for their selected job using this information.

3. Develop a list of questions frequently asked during an interview. Use the questions in role playing job interviews and videotape the interviews.

continues

Knowledge Applied, *continued*

4. Organize a field trip to a public or private placement office. Following the field trip, discuss the office's policies and how they affect job searchers and employers. Alternatively, invite a representative from a state employment agency to explain how employment agencies can help students gain employment.

5. Hold a plant and soil science career field day. Invite individuals currently employed in the industry to present a panel discussion on career opportunities. For example, invite representatives from research, education, and government.

6. Select one career in a plant or soil science area of interest and prepare a research paper on the career using a computer and word processing software. The paper should identify the knowledge and skills required and the employment opportunities.

7. Collect pictures or photographs of people engaged in various careers with plants or soils and prepare a bulletin board collage.

8. Invite a resource person such as a business owner or personnel manager to discuss what he or she looks for in resumes, application letters, forms, and during interviews.

9. Invite a panel of local agribusiness people to discuss the importance of employee work habits, basic skills, and attitudes and how they affect the entire business.

10. Develop a class presentation after collecting information found on one of the Web sites listed in "Finding Career Opportunities."

Resources

Aslett, D. (1993). *Everything I needed to know about business I learned in the barnyard.* Pocatello, ID: Marsh Creek Press.

Bolles, R. N. (2007). *What color is your parachute? 2008: A practical manual for job-hunters and career-changers.* Berkeley, CA: Ten Speed Press.

BCEL Newsletter for the Business & Literacy Communities Bulletin. (1987). *Job-related basic skills: A guide for planners of employee programs.* New York: Business Council for Effective Literacy.

Carnegie, D. (1936). *How to win friends & influence people.* New York: Simon and Schuster. (Downloadable PDF at www.targetitmarketing.com/ebooks/pdfs/how-to-win-friends.pdf)

Covey, S. R. (2004). *The 7 habits of highly effective people.* Salt Lake City, UT: FranklinCovey.

Parker, G. M. (2008). *Team players and teamwork, completely updated and revised: New strategies for developing successful collaboration* (2nd ed.). San Francisco: Jossey-Bass Publishers.

Ricketts, C. (2001). *Leadership: Personal development and career success* (2nd ed.). Albany, NY: Delmar Cengage Learning.

U.S. Department of Education. (1991). *America 2000: An education strategy. Sourcebook.* Washington, DC: Author.

Ziglar, Z. (2000). *See you at the top* (2nd ed.). Gretna, LA: Pelican Publishing Company.

Internet

Internet sites represent a vast resource of information. The URLs (uniform resource locators) for World Wide Web sites can change. Using one of the search engines on the Internet such as Google or Yahoo!, find more information by searching for these words or phrases: resume, letter of inquiry, letter of application, interview, workplace skills, entrepreneur, competencies, creative thinking, plant science, soil science, horticulture, or any specific word listed under jobs and careers such as plant geneticist, biochemist, water specialist, cooperative extension, and so forth.

Glossary

Like a foreign language, terms unique to plant science can be baffling to the newcomer. Any individual traveling to a foreign country and wanting to do business there would be expected to know the language of the country. The same is true for the individual wanting to learn about plants and crops. Indeed, the term glossary means obscure or foreign words of a field. Successful individuals use the glossary and learn the language. Words not found in the glossary may be listed in the index and defined within a chapter of the book.

A

Abiotic—Nonliving.

Abscisic acid—A plant hormone, it is involved in leaves, flowers, and fruits separating from the stems, also dormancy.

Abscission—Shedding leaves or fruit.

Absorb—To take up or receive by chemical or molecular action.

Absorption—(1) Transformation of any form of radiant energy as it passes through a material substance. (2) Process by which a liquid or gas is taken into the spaces of a porous substance.

Acid—A substance with a pH below 7.

Acid-forming fertilizers—Fertilizer compounds, generally N, P, and S materials that produce acidity when dissolved in a soil solution or during chemical breakdown in the soil.

Acidity, soil—Relative concentration of hydrogen (H+) ions in a soil system.

Active ingredient—The chemical compound in a product that is responsible for fulfilling the purpose for which the product was manufactured.

Adaptation—Changes a plant or animal makes to survive in a specific environment.

Adenine—One of the four bases in DNA. It pairs only with thymine.

Adsorb—To gather a gas, liquid, or dissolved substance on a surface.

Adventitious bud—A vegetative bud that develops along the root of a bramble plant and that will grow into a primocane.

Aerial roots—Roots that form freely on land or water plants; they may attach the plant to its host and can absorb water from the air. They may be fleshy or semifleshy, functioning as reservoirs for water storage.

Aerobic—With atmospheric oxygen.

Aerobic conditions—Environmental presence of oxygen. Occurring or growing only in the presence of molecular oxygen (in reference to microorganisms and chemical reactions).

Aeroponics—The process of growing plants in an air or mist environment.

Aggregate fruit—A type of fruit that develops from a single flower but consists of many smaller drupelets that fuse and ripen together.

Agricultural Adjustment Act (AAA)—This act, created in 1946, sought to change the imbalance between production and postproduction research. This act also set up the mechanism for contracting with private research facilities, permitting the government to draw on the expertise of private sector scientists as well as its own. This institutional framework supported research into the problems of particular areas, helping to supply quick answers to questions of national importance.

Air shear—Cutting with air; used to define a spray nozzle that uses air to break up the spray into the desired pattern and droplet size.

Aleurone—The outer protein-rich layer of cells of the endosperm of grass seed.

Algae—Simple rootless plants that grow in bodies of water in relative proportion to the amount of nutrients available. Algal blooms, or sudden growth spurts, can affect water quality adversely.

Alkaline—A soil that contains more hydroxyl ions (OH⁻) than hydrogen ions (H⁺) is said to be alkaline.

Alkalinity, soil—Relative concentration of hydroxyl (OH⁻) ions in a soil system.

Allele—(1) The alternative form of a gene having the same place in a homologous chromosome. (2) Genes on the same location of a pair of homologous chromosomes. (3) One of a pair of genes.

Allelopathy—The release of a substance by one plant that is toxic to another plant.

Alluvial fan—A fan-shaped alluvial deposit formed where flowing water slows down and "fans" out at the base of a slope.

Alluvial soil—A soil developed from mud deposited by running water.

Amino acid—An organic acid containing an amino group (–NH₂) and a carboxyl or acid group (–COOH). Some amino acids contain sulfur. Proteins are constructed from 20 different building blocks of amino acids.

Ammonia volatilization—The loss of urea nitrogen as ammonia from the soil to the atmosphere.

Ammonification—Biochemical process whereby ammoniacal nitrogen (NH₄-N) is released by nitrogen-containing compounds, such as organic matter, crop residues, and fertilizer materials.

Anaerobic—Environmental absence or deficiency of molecular oxygen. Conditions frequently encountered in flooded soils or following irrigation.

Anchoring—A tillage operation used to partially bury foreign materials, such as plant residue or paper mulches, to prevent movement of the materials.

Angiosperm—Plants with seeds enclosed in a fruit.

Anhydrous ammonia—Liquid ammonia under high pressure that does not contain water.

Animal pests—Any animal out of place—such as a deer in a hay field, stray dog in with sheep, livestock that break through a fence, and others—is a pest.

Anion—Ion or radical carrying a negative charge (X⁻).

Annual—A plant that completes its life cycle in less than one year and must be planted again.

Anther—Lobed, oblong, baglike appendage at the top of the filament, which produces the pollen grains that develop the male germ cells. Anthers are generally yellow and when young have from one to four cavities (cells) in which pollen grains arise. When mature, the anther usually contains two cavities from which the pollen grains are released by the formation of apical pores or longitudinal slits in the cavity wall.

Antigen—A substance (usually a protein or carbohydrate) that stimulates the production of an antibody when introduced into the body of animals.

Apex—The tip of a stem or root.

Apical meristem—A group of cells at the growing point of roots and shoots from which all vegetative or reproductive growth originates.

Apprenticeship—A job that involves working under the supervision of a professional for a variable amount of time. This type of job may or may not include a salary.

Aquaponics—The symbiotic cultivation of plants and fish in a recirculating environment. Plants use the metabolic waste of the fish and fish use the water cleaned by the uptake of plant nutrients. Each benefits from the other.

Aquifer—A sand, gravel, or rock formation capable of storing or conveying water below the surface of the land.

Aroma volatiles—Fragrant chemicals in fruit that quickly vaporize when exposed to air.

Asexual—Plant reproduction used to ensure that new plants are identical to the parents. These types of reproduction are cuttings, graftings, budding, layering, dividing, rhizomes, stolons, and tillers or suckers.

ATP—Adenosine triphosphate, an important energy-storage form in cellular metabolism.

Auricles—Appendages that surround the stem at the junction of the blade and sheaths on some grasses.

Autotoxicity—Old stands of alfalfa release a toxin that reduces germination and growth of new alfalfa seedlings.

Autotroph—An organism that obtains energy by oxidizing inorganic substances or by using the sun's energy in biochemical processes.

Auxins—Compounds that have the capacity to induce lengthening of cells.

Available nutrient—Quantity of a nutrient or compound in soil that can be readily absorbed and assimilated by plants.

Available water—Portion of water in a soil that can be readily absorbed by plant roots. Generally considered to be soil water potential greater than –15 bars (permanent wilting point) and less than –0.3 bars (field capacity).

Axillary bud—(1) A bud that develops in the axil of a leaf, the point where the leaf attaches to the shoot. (2) When flowers or clusters of flowers arise at the junction of the stem or axis and the leaf.

B

Bacterium—A single-celled, microscopic organism with cell walls and no chlorophyll; it reproduces by fission.

Balance sheet—Statement of the assets owned and liabilities owed in dollars. It shows equity or net worth at a specific point in time.

Band application—Placement of fertilizer in a concentrated strip within the rooting zone, including side band, deep band, seed band, and surface band.

Bare fallow—Complete incorporation of residues for maximum decomposition. Essentially no crop residue is left on the soil surface.

Bare-root—Not potted.

Bars, atmospheres—Measure of pressure, generally used with soil water potential.

Baseflow—The stream discharge composed of ground water drainage and delayed surface drainage.

Bases—Nitrogen-containing chemicals. The four bases of DNA are adenine, cytosine, guanine, and thymine.

Bed planting—A method of planting in which several rows are planted on elevated, level beds that are separated by furrows or ditches.

Bedrock—Unbroken solid rock, overlain in most places by soil or rock fragments.

Berry—One or more carpels developed within a thin covering, very fleshy within, with the seeds embedded in the common flesh of a single ovary.

Best Management Practices (BMP)—A proven or time tested standard way of doing things that multiple organizations can use.

Biennial—A plant that requires two growing seasons to complete its life cycle; vegetative growth occurs the first year, reproductive growth the second year, and then the plant dies.

Biennial weed—See biennial, apply to weeds.

Biocontrol—Another word for biological control.

Biological weed control—Using natural enemies for the control of certain weeds. The objective of biological control is reduction of the weed population, not eradication.

Biome—A large area occupying a given climatic zone and with a characteristic vegetation and associated groups of animal species.

Bioinformatics—The use of mathematical tools to extract useful information from data produced by high-throughput biological techniques such as genome sequencing.

Biopharmaceuticals—Drugs produced by living tissues.

Biotechnology—The application of biological and engineering techniques to microorganisms, plants, and animals; sometimes used in the narrower sense of genetic engineering. It is the application of scientific and engineering principles to the processing of materials by biological agents to provide goods and services. The three major techniques are genetic engineering, monoclonal antibody technology, and bioprocessing.

Biotic—Living.

Bipolar—Having two charged poles like a magnet.

Blade—(1) A nonrotating, soil-working tool, consisting of an edge and a surface, designed primarily to cut through soil. It may be operated at a steeply inclined angle to roll or bulldoze soil. (2) The broad, thin part of the leaf.

Blight—A disease that affects certain crops such as corn, cotton, or potatoes.

Boll—The seed capsule of cotton or flax.

Boot—One of the growth stages of small grains; during the boot stage, the head can be felt inside the upper leaf sheath and the flag (last) leaf has developed.

Boron (B)—An element that plays an important role in kernel quality and flavor of some grains.

Brambles—Raspberry, loganberry, boysenberry, dewberry, and tayberry are some of the bramble fruits.

Broadcast—Uniform application of fertilizer over soil surface, typically incorporated during a tillage operation.

Broadcast planting—Seeds scattered over a wide area.

Budding—A type of grafting that consists of inserting a single bud into a stock. It is generally done in the latter part of the growing season.

Budget—Allocating a crop's fertilizer or water requirement based on soil test, time of sampling, yield potential, cropping history, stage of growth, and weather.

Budstick—A shoot of the current season's growth used for budding. Leaves are removed, leaving ½ inch of leaf stem for a handle.

Bulb—A mass of overlapping membranous or fleshy leaves on a short stem base, enclosing one or more buds able to develop into new plants and constituting the resting stage of many plants such as tulip, onion, lily, and others.

Bulbous perennials—Perennials that propagate by bulbs and bulblets as well as seed.

Bulk density—Mass of oven-dry soil per unit volume, usually expressed as grams per cubic centimeter.

Bulk fertilizer—Dry or liquid fertilizer stored and handled in an unbagged form.

Bunch grass—Grass that grows in clumps rather than spreading out.

C

Calcareous soil—Soil containing sufficient free calcium carbonate (greater than 1% $CaCO_3$) to effervesce (fizz) visibly when treated with hydrochloric (HCl), sulfuric (H_2SO_4), or phosphoric acid (H_3PO_4).

Calvin cycle—Occurs in the stroma of chloroplasts; carbon dioxide is captured by the chemical ribulose biphosphate (RuBP). RuBP is a five-carbon (5-C) chemical. Six molecules of carbon dioxide enter the Calvin cycle, eventually producing one molecule of glucose. The first stable product of the Calvin cycle is phosphoglycerate (PGA), a three-carbon (3-C) chemical.

Calyx—The ring of sepals making up the outermost, leaflike part of the flower.

Calyx lobe—The free, unfused, projecting parts of the base of a flower below the petals.

Cambium—Thin-walled cells between the xylem and phloem that divide and give rise to new xylem and phloem cells.

Canes—Part of the vine left that will produce fruit the next growing season.

Capillary water—Water held in the capillaries of the soil. It moves freely in the soil and can move up and down, or horizontally, and it usually meets the water needs of plants.

Capital—The amount of money that can be obtained through borrowing or selling assets; used to promote the production of other goods.

Caprification—A method of assuring pollination of the Smyrna and other edible figs in which flower clusters of the caprifig are hung from trees of the edible fig, allowing wasps to carry pollen from the flowers of the caprifig to those of the edible varieties.

Carbohydrates—Any of a group of neutral compounds composed of carbon, hydrogen, and oxygen (CHO), including the sugars and starches. Plants produce carbohydrates. They are used immediately for growth or stored for future use.

Carbon sinks—Plants are considered carbon sinks because they remove carbon dioxide from the atmosphere and oceans by fixing it into organic chemicals.

C:N (carbon:nitrogen) ratio—The relative amounts by weight of carbon and nitrogen in plants or other organic materials. Material with a carbon:nitrogen ratio of 12:1 would contain 12 parts of carbon and 1 part nitrogen by weight.

Carpel—Refers to either a simple pistil or one of the segments of a compound pistil.

Cation—Ion or radical carrying a positive charge (X^+).

Cation exchange capacity (CEC)—Sum total of exchangeable cations (X^+) that a soil can hold on surfaces of clay minerals and organic matter (expressed in meq/100 g soil). Normally ranges from about 3 to 5 meq/100 g soil for sands and sandy loams with low organic matter to around 25 to 30 meq/100 g soil for productive clays and clay loams.

Catkins—Structures, such as cones, in which reproductive organs are borne.

Cell—The basic structural and physiological unit of crop plants, within which chemical reactions of life occur providing metabolites for plant life and for human use.

Cell wall—Structure surrounding a cell; gives the cell form and controls substances leaving and entering the cell.

Cellulose—A polymer (chain) of glucose molecules that forms the major structural component of cell walls. It is the most abundant biochemical in the world.

Cereal crop—A member of the grass family grown for its edible seed—for example, wheat, oats, barley, rice, rye, corn, and grain sorghum.

Certified plant stock—Plants from a nursery that have been visually inspected by a state representative and found to be free of virus and other disease symptoms.

Chelate—Type of chemical compound in which metal cations (iron, copper, manganese, and zinc) firmly combine with a molecule by means of multiple chemical bonds.

Chelating agents—Chemicals that combine with metal ions and remove them from their sphere of action, also called sequestrants.

Chemical availability—Relative amount of nutrient available for plant uptake based on chemical and physical soil properties.

Chemical fallow—The use of herbicides (contact and residual) for vegetation control during a period when the land is fallowed. The crop residue is left intact. Same as chemical tillage or no-till fallow.

Chemical name—Indicates the chemical composition of the compound and also the structure of the molecule in organic compounds—for example, 2,4-dichlorophenoxyacetic acid.

Chiseling—The breaking or shattering of compact soil or subsoil layers with chisels. A tillage operation in which a narrow tool is used to break up or loosen the soil, without inversion. It may be performed at other than the normal plowing depth. Chiseling at depths greater than 16 inches is usually termed subsoiling.

Chisel planting—Seedbed preparation by chiseling, without turning of the soil. This leaves a protective cover of crop residue on the surface for erosion control. The operation may or may not include seedbed preparation and planting.

Chlorophyll—Green pigmentation.

Chloroplast—A self-replicating organelle or organized cellular body that contains chlorophyll and is the site of photosynthesis and starch formation. Chlorophyll is responsible for the green color of plants.

Chlorosis—Yellowing of plant vegetation (primarily leaves) resulting when chlorophyll fails to develop, caused by essential plant nutrient deficiency, plant disease, or other environmental, soil chemical or physical problem.

Chromatids—As a result of the syntheses during the interphase stage of mitosis, each chromosome consists of two sister chromosomes, chromatids, that are identical in their structural and genetic organization. They become visible when mitosis begins.

Chromosome—A small, normally linear body containing DNA and found in the cell nucleus in higher organisms. Bacterial chromosomes are circular, highly folded, and free in the cell cytoplasm. Chromosomes maintain the genetic template and replicate through fission when a parent cell divides into two identical daughter cells.

Citrate-insoluble phosphorus—Amount of phosphorus (P) present in residue following filtration of P fertilizer material treated with 1 N ammonium citrate.

Citrate-soluble phosphorus—Amount of phosphorus (P) remaining in solution following filtration of P fertilizer material treated with 1 N ammonium citrate. Used as a measure of P availability in fertilizers.

Classification—An organizational system for describing related plants or animals.

Clay—(1) Soil separate consisting of particles less than .02 mm. (2) A textural class when soils contain more than 40 percent clay.

Clean cultivation—Physically disturbing the soil between plant rows at regular intervals to eliminate weeds.

Climate—The long-term average weather conditions.

Climatic zone—A large area with similar climatic features.

Cloning—In molecular biology, the process of producing many copies of a single ancestral gene or sequence of DNA by incorporating that DNA into a replicating microorganism. In plants and animals, the replication of cell lines, tissues, or organisms from somatic (nonreproductive) tissues.

Coarse-textured soil—A soil containing large amounts of sand. Feels similar to table salt.

Coarse aggregates—For horticultural soil-potting mixes, refers to large, inorganic particles used to create large pores in the mix. These materials include coarse sand, perlite, vermiculite, shredded plastics, and others.

Codons—Another word for bases in the DNA structure.

Coir—Fiber obtained from the husk of a coconut.

Cold frame—A frame house built for growing plants that has no inside heating provided.

Colloid—Particle consisting of organic or inorganic matter with small size and high surface area—for example, clay or soil humus.

Colluvium—A deposit of rock and soil resulting from materials sliding down a slope under the force of gravity.

Combine—A type of machine that harvests crops.

Common name—The accepted and approved common name for a given active ingredient, such as 2,4-D.

Compaction—The packing effect of a mechanical force on the soil. This packing effect decreases the volume occupied by pores and increases the density and strength of the soil mass.

Companion crop—A crop grown specifically with another crop because it recovers the land or protects another crop until it can get established; also called a nurse crop.

Competencies—Abilities or capabilities of employees.

Complete flower—Flower that contains both male and female structures.

Compost—Organic residues or a mixture of organic residues and soil that have been allowed to biologically decompose to increase plant nutrient availability from organic materials.

Composting—Piling organic materials under conditions that cause rapid decay. Reduces the carbon-nitrogen ratio and destroys many weed seeds and disease organisms.

Compound fertilizer—Fertilizer containing at least two macronutrients (N, P, and K).

Compound leaf—A leaf that consists of several leaflets.

Conductivity meter—An instrument used to measure the soluble salts concentration in soil or potting mixes. It uses the principle that the more ions that are dissolved in water, the better the water will conduct electricity.

Conservation tillage—Any tillage system that reduces loss of soil or water, compared to clean or unridged tillage. Conventional tillage system includes moldboard plowing, followed by disking and dragging.

Container-grown—Plants grown in containers.

Contour-cropping—A method of farming that reduces erosion and is most effective on deep, permeable soils and on gentler slopes (2–6 percent) that are less than 300 feet long.

Cool-season—Refers to plants that can survive mild spring frosts and may be planted early in the spring or in the fall.

Corm—A means of vegetatively propagating some plants. It is a short, fleshy, underground stem with few nodes and short internodes.

Corolla—The inner set of leaflike parts of a flower lying just within the calyx and composed of petals.

Corporation—A body of people recognized by law as an individual person having a name, rights, privileges, and liabilities distinct from the individual members.

Cotyledon—Embryonic leaves that serve as foodstoring organs or develop into photosynthetic structures as the seed germinates.

Cover crop—A crop planted to prevent erosion on a soil. Cover crops can be planted on soils not currently being farmed, between crop rows, or after main crop harvest.

Creative thinking—Ability to generate new ideas by making nonlinear or unusual connections or by changing or reshaping goals to imagine new possibilities; using imagination freely, combining ideas and information in new ways.

Creeping perennials—Plants that propagate by means of rhizomes (underground stems), stolons (aboveground stems), or spreading roots or tubers, as well as by seed.

Critical soil test level—Quantity of plant-available nutrient in a soil below which crops will suffer from a deficiency of that nutrient.

Crop rotation—Practice of growing a series of dissimilar types of crops in the same space in sequential seasons for various benefits such as to avoid the buildup of pathogens and pests that often occurs when one species is continuously cropped. Crop rotation also seeks to balance the fertility demands of various crops to avoid excessive depletion of soil nutrients.

Cross-pollination—Process by which pollen is transferred from an anther of one flower to the stigma of a different plant or cultivar.

Cross-protection—Infection of plant tissues by one virus suppresses the disease caused by another closely related strain of the virus. The protecting strain must have negligible impact on the host.

Crown bud—A vegetative bud that originates from just below the base of a bramble plant and that will develop into a primocane.

Crowns—Short and inconspicuous stems.

Cultipacker—A type of seeding equipment that presses the seed into the soil with corrugated wheels to give more soil-seed contact.

Cultivar—A cultivated variety; an entity of commercial importance that is given a name to distinguish it from other genetically different plants.

Cultural control—The use of production practices typically used to improve plant growth and yield to also control pests.

Cutin—Any of a group of lipid substances deposited both on the inside and outside of epidermal cell walls.

Cuttings—A portion of a plant that is removed and made to form roots.

Cytokinesis—Cell division without nuclear division as in mitosis or meiosis.

Cytokinins—A class of plant hormone that stimulates cell division in plants.

Cytoplasm—Cell contents other than the nucleus.

Cytosine—One of the four bases of DNA; it pairs only with guanine.

D

Dark reaction—A light-independent process that occurs when the products of the light reaction are used to form carbon-to-carbon (C-C) covalent bonds of carbohydrates.

Data sheet—Similar to a resume; contains pertinent information about a potential employee.

Day-neutral—A plant that may flower under any day length.

Deadheading—Removal of old blooms.

Deciduous—Characteristics of perennials whose leaves are shed, generally at the end of the growing season.

Deep band—Fertilizer placed well below the seed for growing and late season use—for example, 4 to 5 inches deep.

Degradation—(1) Breakdown of the tissues. (2) Rotting of harvested crops.

Dehiscent—Fruit that opens naturally when mature.

Delta—A usually fan-shaped alluvial deposit created where a stream or river enters a body of quiet water like an ocean or a lake.

Denitrification—Biochemical reduction of nitrate to gaseous nitrogen, either as molecular nitrogen or as an oxide of nitrogen, typically occurring under waterlogged (anaerobic) conditions; results in a loss of available N from the soil.

Deoxyribonucleic acid (DNA)—A compound of deoxyribose (a five-carbon sugar), phosphoric acid, and nitrogen bases.

Dermal system—Outermost layer of cells; the epidermis of floral parts, leaves, fruits, and seeds, and of stems and roots until they begin secondary growth.

Determinate—After a certain period of vegetative growth, the plant flowers, but clusters form at the shoot terminals so that most growth to the shoot elongation stops.

Dichotomous—Dividing into two parts or divisions.

Dicot—A flowering plant with two-seed leaves or cotyledons, with xylem and phloem cells separated into zones and nonparallel venation in leaves.

Differentiation—Physiological and morphological (structural) changes that occur in cells, tissues, or organs during development and maturation.

Diffusion—Spreading out of molecules in a given space.

Dioecious—Plants bearing staminate flowers on one plant and pistillate flowers on a different plant. They are called male and female plants.

Diploid—Having two complete haploid sets of chromosomes. This is considered typical or normal for a species.

Disease—Any condition of a plant that impairs normal physiological functions.

Disease-free seed—Seed that is high quality and has been tested to be nearly free from disease.

Dispersion, soil—Breakup of soil aggregates or groups of soil particles into individual component soil separates. This condition inhibits root and water penetration.

Division—(1) Within the taxonomic system of classifications, how plants are ordered concerning their characteristics; for example, the mycota or fungi family. (2) A method of asexual plant propagation.

Dolomite—A liming material.

Dominant—A character that is manifested in the hybrid of a plant to the apparent exclusion of the contrasting character from the recessive parent.

Dormant—Condition of live trees (or some plants) at rest in winter.

Double cross—Combination of two single-cross hybrids in hybridization between four inbred lines.

Double fertilization—Union of the two male gametes with the female gamete and the polar nuclei.

Drainage—Removal of water from soil.

Drifts—Color groupings of plants.

Drill—Machinery for evenly placing seeds (grain) in the soil at a uniform depth and giving good soil-seed contact.

Drip irrigation—A method of irrigation that conserves water by slowly releasing small amounts of water through emitters near the plant.

DRIS (Diagnostic and Recommendation Integrated System)—Interpretation of leaf or plant nutrient analysis of high-yield crops (on a ratio basis) for the purpose of improving fertilizer recommendations.

Drupe—A fruit with a large hard seed called a stone.

Drupelets—The small sections of a raspberry or blackberry fruit, each containing a seed.

Dry fruits—Pericarp is often hard and brittle at maturity; the food is largely confined to the seeds.

Dryland farming—Production of crops that require some tillage in subhumid or semiarid regions without irrigation. The system usually involves fallow periods between crops during which water from precipitation is absorbed and retained.

Dusting—Sprinkling flowers or plants with pesticides to protect them from insects and rodents.

Dwarf—A tree that grows to about 4 to 10 feet in height.

E

Eco-fallow—The use of chemicals, alone or in combination with mechanical tillage, to conserve moisture and control weeds during the fallow period.

Ecosystem—A natural community in which different kinds of plants and animals live closely together, getting the things they need for living from within the community and its natural resources.

Effluent—Water from industrial facilities, municipalities, and other such sources.

Electrical conductance (EC)—Ability of the soil solution to conduct an electrical current. The more salt in solution, the more dissolved ions, the higher the electrical conductance. Measured as electrical conductance per unit distance, mmhos/cm.

Electron transport—A series of biochemical reactions that move electrons through different energy levels. It is a part of the process of respiration.

Element—A simple form of matter that cannot be decomposed by ordinary chemical means. Nitrogen (N), phosphorus (P), potassium (K), carbon (C), and the like, are examples of elements.

Eluviation—Removal of a material, such as clay or nutrients, from a layer of soil by percolating water.

Embryo—A tiny plant as it exists in a seed.

Embryo rescue—The use of tissue culture techniques to propagate an embryo that otherwise would not develop into an individual. It allows crosses to be made between sexually incompatible species.

Emergency tillage—Roughening the soil surface by such methods as listing, ridging, duckfooting, or chiseling. (This practice is an emergency conservation measure and does not provide long-term benefits.)

Emitters—In drip or trickle irrigation systems, they are the water delivery mechanisms or outlets.

Endocarp—The inner layer of the fruit pericarp.

Endodermis—Single layer of cells between the cortex and the pericycle in the root.

Endogenous—A naturally produced substance; produced within the organism and acting within it.

Endophyte—Fungus.

Endoplasmic reticulum—Structure extending throughout the cytoplasm of a cell. It functions in the transport of cell products and as a surface for protein synthesis by the ribosomes.

Endosperm—This is the food source for the embryo.

Endosperm nucleus—The triploid (3n) structure that forms when the haploid sperm nucleus that was carried down the pollen tube to the ovule joins with the two polar nuclei. This divides many times to form a mass of tissue called the endosperm.

Energy—Capacity for doing work and for overcoming inertia.

Enterprise—A specific process or activity that requires a certain amount of risk to make a profit.

Entomology—The study of insects.

Entrepreneur—One who starts and conducts a business assuming full control and risk.

Enzyme—A protein that catalyzes a specific chemical change without being used up in the reaction.

Eolian Deposit—Wind-deposited soil material, mostly silt and fine sand.

Epicotyl—The part of the axis of an embryo above the region of attachment of the cotyledons.

Epidermis—The outer layer of cells on all parts of a young plant and on some parts of older plants—for example, the leaves and fruits.

Epigeal emergence—Growing point is above soil surface immediately upon emergence.

Erosion—The removal of soil material by wind or water moving over the land; erosion is a natural process. Most hills and valleys are the result of very slow erosion by water.

Erosion pavement—A layer of coarse fragments, such as sand or gravel, remaining on the surface of the ground after the removal of fine particles by erosion.

Essential nutrient—One of 16 elements required for plant growth. Essential nutrients are necessary for a plant to complete its life cycle; are directly involved in plant nutrition; are not considered essential simply by correction of a disease or soil chemical condition; and are considered essential when a nutrient deficiency is element-specific. For example, the deficiency can be prevented or corrected by supplying that specific element. Carbon (C), hydrogen (H), and oxygen (O) are absorbed primarily from air and water. The other 13 essential nutrients are absorbed from the soil.

Ethylene—A plant hormone, involved in fruit ripening, stem elongation, and stoma closing.

Etiolated—A white stem with little leaf growth.

Eukaryote—Genetic information or DNA contained in the nucleus like most organisms.

Eutrophic lake—A lake that has a high level of plant nutrients, a high level of biological productivity, and low oxygen content.

Eutrophication—Natural or artificial process of nutrient enrichment whereby a water body becomes filled with aquatic plants and low in oxygen content.

Evapotranspiration—Total water lost through evaporation from the soil surface and from transpiration through the plant.

Exchangeable Sodium Percentage (ESP)—Percentage of cation exchange capacity occupied by sodium. Used to evaluate the potential sodium hazard.

Expanded clay—Clays that expand greatly upon adsorption of interlayer water and shrink upon drying.

Explants—Small pieces of plant tissue.

F

Fall-fruiting—A raspberry that forms fruit on the tops of first-year vegetative canes near the end of the growing season; the same as primocane-fruiting.

Fallow—Ground that is plowed but not seeded.

Family—A group of closely related genera.

Fecal coliform bacteria—A group of bacteria found in the intestinal tract of humans and animals, and also found in soil. Although harmless in themselves, coliform bacteria are commonly used as indicators of the presence of pathogenic organisms.

Fermentation—Decomposition of organic substances, especially carbohydrates, under anaerobic conditions. These conditions are often created by the enzymes produced by microorganisms such as yeast, molds, and bacteria.

Fertigation—Application of fertilizers through irrigation water.

Fertile Crescent—An area at the eastern end of the Mediterranean Sea and Central Asia where a large variety of crops are grown.

Fertility, soil—Soil's ability to supply essential nutrients in adequate amounts and proportions for plant growth throughout the growing season.

Fertilization—(1) Practice of adding nutrients to soil or plants for use by plants. (2) The union of the egg and sperm.

Fertilizer—Any natural or manufactured material added to the soil to supply one or more essential plant nutrients.

Fertilizer, fluid—Any material, either dissolved or suspended in water, applied as a liquid to soil to supply one or more essential plant nutrients.

Fertilizer, organic—Natural material containing carbon, hydrogen, oxygen, and one or more essential plant nutrients.

Fertilizer, placement—Where fertilizer materials are added to soil.

Fertilizer blend—Mixture of different fertilizer materials to obtain a desired fertilizer grade or application rate of actual nutrients.

Fertilizer bulk density (granular)—Mass of dry fertilizer per unit volume; typical units are lbs/ft^3.

Fertilizer grade—Guaranteed minimum analysis (in percent) of the major plant nutrient elements contained in a fertilizer (refers to percent of N, P_2O_5, K_2O, and S).

Fibrous roots—A type of root system characterized by many branches of fine roots.

Field capacity—Amount of water a soil can hold against gravity; expressed as a percentage of the dry weight of a soil.

Filament—The thin stalk that attaches the anther to the rest of the flower.

Fine-textured soil—Soils made up of mostly clay particles; feels sticky when wet.

Fixation—Any process in the soil that changes available nutrients into unavailable or "tied-up" forms. See nitrogen fixation.

Fleshy fruits—Usually juicy and brightly colored, contrasting with their background to make them more noticeable to animals, who are responsible for their dispersal. All fleshy fruits are indehiscent, and considerable fleshy tissue is developed as the ovary changes into the fruit.

Fleshy roots—Storage reservoirs for plants.

Flood irrigation—On level land, water enters through a head ditch or biplane and is released into the individual checks (areas bounded by levees running downslope) by siphons, gates, or valves.

Floodplains—Land near a stream that is commonly flooded when the stream is high. Soil is built from sediments deposited during flooding.

Floral bracts—Modified leaves that can simulate petals and add the conspicuous part to otherwise inconspicuous flowers.

Floricane—Second-year cane that overwintered and will fruit and die in the current year.

Floricane-fruiting—A raspberry or blackberry that forms fruit only on second-year canes in late spring or summer; the same as summer-fruiting.

Fluorescent—A cool white light that is high in the blue range of light quality and is used to encourage leafy growth. It is very good light for starting seedlings.

Foliar—(1) Having to do with the leaves. (2) Application of small amounts of fertilizer to foliage of growing plants.

Foliar diagnosis—Estimation of mineral nutrient deficiencies or nutrient or element excesses based on visual examination of plant leaves and selected plant parts.

Follow-up letter—Letter written immediately after a job interview.

Food and Agricultural Organization (FAO)—An agency of the United Nations that conducts research, provides technical assistance, conducts education programs, maintains statistics on world food, and publishes reports with the World Health Organization.

Forage—Feedstuffs from the leaves and stocks of plants. These could be grasses, legumes, or other cultivated crops.

Forbs—Any nongrasslike plant that is relatively free of woody tissue that an animal consumes.

Forcing—Making bulbs artificially break dormancy so that they will flower when brought into a warm room.

Forecast—To calculate before hand.

Foundation seed—Seed stock handled to maintain specific genetic identity and purity as closely as possible under supervised or approved production practices.

Free water—Water moving through the soil by the force of gravity.

Frost Wedging—Breakage of rocks caused by pressure created by water freezing in cracks in the rock.

Fumigant—An organic compound that is a gas at fairly low temperatures; it is used for insect and disease control.

Fungicide—A chemical used for controlling fungi.

Fungus—An organism (plant) with no chlorophyll that reproduces by means of structures called spores and usually has filamentous growth.

Furrow irrigation—Water runs down the furrows between plant rows. Water moves to all parts of the soil by capillary action or gravity.

G

Galls—Large swellings on plants.

Gametes—Eggs or pollen (sperm).

Gene—An element of the genetic material that controls the expression of an inherited characteristic by specifying the structure of a particular RNA molecule or protein or by controlling the function of other genetic material. Genes usually consist of specific sequences of purine and pyrimidine bases, usually in DNA. Genes are also called genetic factors.

Genetic code—Set of 64 different possible condons (codes) or nucleic acid triplets and their corresponding amino acids; determines which amino acids will be added during protein synthesis.

Genetic engineering—Alteration of the genetic components of organisms by human intervention.

Genome—One complete set of genes, containing all the genetic information to produce an individual.

Genomics—Study of an organism's entire genome. The field includes intensive efforts to determine the entire DNA sequence of organisms and fine-scale genetic mapping efforts.

Genotype—The genetic makeup of an individual determined by the group of genes it possesses.

Genus—A unit of scientific classification that includes one or several closely related species. The scientific name for each organism includes designations for genus and species.

Geotropism—Response of plants to gravity.

Germinate—To begin to grow or sprout.

Germination—Occurs when the seed coat is softened by water and favorable temperatures exist and the seed begins to grow.

Germ plasm—A collection of genetically diverse plants, including wild material, that can be used to improve cultivated plants through breeding.

Gibberellins—Compounds that stimulate cell division and lengthening of cells.

Ginning—The process of removing the seeds from cotton.

GIS (geographic information system)—Integrates hardware, software, and data for capturing, managing, analyzing, and displaying all forms of geographically referenced information. It allows individuals to view, understand, question, interpret, and visualize data in many ways that reveal relationships, patterns, and trends in the form of maps, globes, reports, and charts.

Glacial drift—General term for debris deposited by glaciers; common in northern tier of states.

Glacial outwash—Glacial drift deposited in water flowing away from a melting glacier. Outwash is sorted by the running water.

Glacial till—Accumulations of unsorted, unstratified mixtures of clay, silt, sand, gravel, and boulders seen as glaciers melt.

Glacial weathering—Mechanical processes caused by glaciers. Rocks exposed to the weather undergo changes in character and break down.

Glycolysis—The first metabolic pathway for respiration; the breakdown of glucose to pyruvic acid, which then enters the tricarboxylic acid cycle.

Goals—The end objectives or terminal points of a business or person over a specific time.

Golgi apparatus—Cell organelle, important for glycosylation and secretion in cells.

GPS (Global Positioning System)—A system of satellites, computers, and receivers that is able to determine the latitude and longitude of a receiver on Earth by calculating the time difference for signals from different satellites to reach the receiver.

Grafting—When a shoot or scion is removed from the desired plant and placed on another plant (the stock).

Grana—Thylakoids are stacked like pancakes in stacks collectively called grana.

Gravity water—Water in excess of capillary water.

Grazing capacity—The maximum stocking rate possible without inducing damage to vegetation, water, or related resources.

Green manure crop—A crop grown with the intention of plowing under while still green to improve soil structure, fertility, tilth, organic matter, and nitrogen level (if legume).

Groundwater—Water stored in the pores and cracks in rocks below the earth's surface; the source of water for wells.

Growing degree day (GDD)—Used to estimate the growth and development of plants and insects during the growing season. The basic concept is that development will occur only if the temperature exceeds some minimum developmental threshold, or base temperature (Tbase).

Growth—An irreversible increase in volume and/or weight. Plant growth occurs by an increase in cell numbers and cell size. Cell division and enlargement involve the synthesis of new cellular materials.

Guanine—One of the four bases of DNA; it pairs only with cytosine.

Gully—A channel resulting from erosion and caused by the concentrated flow of water during or immediately following heavy rains. A gully generally is an obstacle to farm machinery and is too deep to be obliterated by ordinary tillage. (A rill is of less depth and can be smoothed by ordinary tillage.)

Gymnosperm—Plants producing seeds but not true fruits because they have no carpel. Such seeds are said to be naked and are borne on the inside of the scales or cones.

Gypsum—A calcium material for plants.

H

Half-life—Time required for one half of a specified substance to degrade or become inert.

Haploid—Having one half of the complete set of chromosomes typical for a species.

Hardening—Many changes that occur in a plant as it develops resistance to adverse conditions, especially cold.

Hardening off—Treatment of tender plants to enable them to survive a more adverse environment.

Hardiness—The ability of a plant to withstand cold temperatures.

Hardiness zone—Hardiness zones determine the type of plants that will grow based on the average annual minimum temperatures.

Hardpan—An impervious layer in the soil that restricts root penetration as well as movement of air and water.

Harrowing—A secondary tillage operation that pulverizes, smooths, and packs the soil to prepare the seedbed or to control weeds.

Hatch Act—This act, initiated in 1887, increased research by setting up experimental stations at each of the land-grant colleges.

Heading—One of the growth stages of small grains; when the wheat or oat emerges; when the florets are fertilized and the kernels develop.

Heads—Edible or harvestable part of the grain.

Heavy metals—Chemical elements (transition metals) that can be toxic in excessive quantities: iron (Fe), cobalt (Co), arsenic (As), zinc (Zn), cadmium (Cd), lead (Pb), copper (Cu), and mercury (Hg).

Heeling in—Temporary planting in a trench.

Helix—A spiral; DNA is a long double-stranded molecule wound in a spiral.

Herbicide—A phytotoxic chemical used for killing or preventing plant growth.

Heterozygous—Different genes for the same trait.

Hill system—A method of planting strawberries; the hill system is used for everbearers. Plants are set 12 to 15 inches apart. Because all runners are removed, three rows can grow together to form one large row. The three rows are spaced one foot apart.

Homozygous—Possessing identical genes.

Horizon—Soil layer.

Hormone—Naturally occurring compound produced by the plant that affects plant growth.

Humidity—Moisture in the atmosphere.

Humus—Stable fraction of soil organic matter that remains in soil after the major portion of plant and animal residues have decomposed.

Hybrid—A first-generation cross between two genetically diverse parents.

Hydrocooling—Vegetables are immersed in cold water long enough to lower the temperature to the desired level.

Hydrolysis—Reaction by which fertilizer salts decompose into weak acids, bases, or both through the addition of water.

Hydrophytes—Plants classified on the basis of their water needs. Hydrophytes thrive in very wet or even flooded conditions.

Hydroponics—Cultivation of plants in water.

Hydroxyl ions—OH^-.

Hygroscopic water—Water that bonds to the soil particles.

Hypocotyl—Part of the plant that forms the lower stem and roots.

Hypogeal emergence—Growing point remains below the soil surface for a time after emergence.

I

Igneous rock—Rock formed from the cooling of molten rock from deep in the earth.

Illuviation—Deposition in a soil layer of materials transported from a higher soil layer by percolating water.

Immobilization—Converting an element from inorganic to organic forms in microbe or plant tissues, immediately rendering the nutrient temporarily unavailable. As nutrients recycle, elements may be released for plant uptake.

Incandescent—Light producing a great deal of heat and not a very efficient user of electricity. It emits very heavily in the red end of the spectrum.

Incomplete flower—Flowers that lack one or more of the four regular parts of a complete flower.

Incorporating (mixing)—Tillage operations that mix or disperse foreign materials, such as pesticides, fertilizers, or plant residues, into the soil.

Incubation—Period of development during which a pathogen undergoes changes to a form that can penetrate or infect the new host plant.

Indehiscent—Fruit that remains closed at maturity.

Indeterminate—Plants that bear the flower clusters laterally along the stem and in the axils of the leaves so that the shoot terminals remain vegetative. The shoot continues to grow until it is stopped by senescence or some environmental influence.

Index—Soil test value that tells how much of a nutrient present in soil is available for plant use at a given time (P and K).

Inert Material—Nonactive portion of a chemical formulation. Total formation minus the active ingredient equals the inert matter.

Infection—Pathogen invades the plant tissue.

Infectious disease—Pathogenic organisms present in the environment or carried by animals or other plants capable of causing and spreading a disease.

Infiltration—Downward entry of water into the soil. This is distinct from percolation, which is movement of water through soil layers of material.

Inhibitors—Compounds that retard a physiological or biochemical process in plants.

Inoculation—The introduction of the pathogen to the host plant tissue. Wind or rain or running water can move pathogens, as can birds, insects, people, or equipment, and introduce pathogens to a host plant. Some pathogens move themselves short distances, but most rely on other means.

Inorganic—Derived from matter not of plant and animal origin (rocks and minerals).

Input costs—Money required to begin production.

Installation—Refers to putting sod in a permanent location after having been grown elsewhere.

Insecticide—A chemical used in controlling insects.

Insoluble—A chemical compound that does not readily dissolve in water.

Integrated pest management (IPM)—The control of one or more pests by a broad spectrum of techniques ranging from biological means to pesticides.

Interest—Payment for the use of money or credit.

Internode—The region of the stem between any two nodes.

Ions—Charged chemical elements or disassociated (X^+, X^-) parts of chemical compounds.

Irrigation—Applying water to crops in such a way as to keep them wet but not too wet. Different irrigation methods depend on the land, sources of water, work involved, and so on.

J

Jointing—One of the growth stages of small grains; at jointing stage, the nodes begin to separate and can be felt in the lower part of the stem.

Juvenile—A plant that has not yet reached reproductive maturity.

K

Karst—Topography characterized by depressions without external drainage; sinkholes; underground caverns; solution channels.

Kingdom—Major divisions for living things: the plant kingdom and the animal kingdom.

Knees—Woody roots that protrude above the surface when the plant is cut off from air by water or the ground.

L

Lacustrine—Mineral sediments deposited in fresh water.

Lamellae—Membranous structures in the chloroplasts.

Laterals—When the cane tips of erect blackberries are pinched out in summer, it stimulates the formation of new growth.

Layering—A vegetative method of propagating new plants by producing adventitious roots before the new plant is cut from the parent. A portion of an attached shoot is partially buried underground where roots develop.

Leachate—Liquids that have percolated through a soil and that carry substances in solution or suspension.

Leaching—When water passes through the soil profile and removes chemical compounds or nutrients in solution from soil.

Leader buds—Buds at the base of a cane, often underground, that have the capacity to grow into a primocane.

Legume—Plants with the characteristic of forming nitrogen-fixing nodules on their roots, in this way making use of atmospheric nitrogen.

Lesions—Any visible alteration in the normal structure of organs, tissues, or cells.

Letter of application—Sent with a resume or data sheet when applying for a job.

Letter of inquiry—Sent to a potential employer requesting possibility of employment.

Leucoplasts—Organelles in cells, used for the storage of oil, starch, and proteins.

Levee—Earthen dike used to enclose water.

Liabilities—Just or legal responsibilities.

Light reaction—Occurs in the grana when light strikes chlorophyll a in such a way as to excite electrons to a higher energy state. In a series of reactions, the energy is converted (along an electron transport-like process) into ATP and NADPH. Water is split in the process, releasing oxygen as a by-product of the reaction.

Lignin—A complex organic polymer that provides rigidity, toughness, and strength to cell walls.

Ligule—A collarlike extension at the top of the sheath.

Lime requirement—Quantity of agricultural lime, or equivalent of other specified liming material, required to raise soil pH to a specific value under field conditions.

Liming—Adding chemical material to the soil to alter the pH.

Liquid fertilizer—Fertilizer in which all the nutrients are dissolved in solution.

Lister-planter—A combination tillage implement that is comprised of a lister and fertilizer and planter attachments. It combines listing with planting, with the seed normally being placed in the furrow.

Listing—A tillage and land-forming operation using a tool that splits the soil and turns two furrows laterally in opposite directions, providing a ridge-and-furrow soil configuration.

Lo-till—Any of several farming systems that use herbicides, alone or in combination with tillage, to maintain weed control in crops and during noncrop intervals. Same as eco-farming.

Lodging—Bending or falling over of cereal grain crops near the soil surface to the ground for various reasons.

Loess—Wind-deposited silt.

Long-day—Plants that require a day longer than its critical day length in order to flower; also called short-night plants.

Luxury consumption—When plants absorb nutrients in excess of immediate metabolic needs.

M

Macronutrient—Chemical element necessary in relatively larger amounts (usually greater than 500 parts per million in the plant) for plant growth. These elements are C, H, O, N, P, K, S, Ca, and Mg.

Maize—Corn.

Marine sediment—Parent material that settled to the bottom of old oceans and seas.

Market gardening—Refers to growing a wide variety of vegetables for local or roadside markets.

Mass flow—When plant nutrient ions move through flowing water; a result of transpirational water uptake by plants.

Master horizon—O, A, E, B, C, and R horizons of the soil.

Media—(1) Growing materials in which plants can be started that are loose, well drained, fine textured, low in nutrients, and free of disease. (2) Mixtures which are used to grow potted plants.

Medium culture—Hydroponic method in which the nutrient solution constantly flows past the roots.

Medium-textured soil—Soil that is high in silt. When moist, they feel smooth and pliable to the touch and not sticky or gritty.

Megaspore—The surviving cell from the process of meiosis.

Meiosis—The cell division that produces the gametes or sex cells. The steps in the cycle are prometaphase, metaphase, anaphase, and interphase.

Meristem—A region of a plant where cells are not fully differentiated and are capable of repeated mitotic divisions.

Mesic—Refers to the temperate zone.

Mesocarp—The center portion of the fruit pericarp.

Mesophilic—Organisms whose optimum temperature for growth is an intermediate range, between 59 and 95°F. Dominant microorganisms in early and late stages of composting.

Mesophytes—Plants that grow in moderately wet areas.

Messenger RNA (mRNA)—Transmits the protein-coding instructions from the genes.

Metabolism—Physical and chemical processes in an organism by which living matter is produced, maintained, and destroyed, and by means of which energy is made available.

Metamorphic rock—Rock that has been changed by heat or pressure in the earth.

Micronutrient—Chemical element necessary in relatively small amounts (usually less than 100 parts per million in the plant) for plant growth. These elements are B, Cl, Cu, Fe, Mn, Mo, and Zn.

Micropropagation—Tissue culture.

Micropyle—A small opening in the ovary through which a cell from the pollen grain passes through the pollen tube.

Microsphore—Haploid cells produced when diploid (2n) cells in the pollen sacs on the anther go through meiosis.

Microtubules—Organelles made from tubulin which compose centrioles and cilla.

Mineral soil—Soil whose properties are dominated by soil minerals, usually containing less than 20 percent organic matter.

Mineralization—When an element converts from organic to inorganic forms as a result of microbial decomposition (reversal of immobilization).

Minimum tillage—Minimum amount of soil manipulation necessary for crop production or to meet tillage requirements under the existing soil conditions.

Miracle products—Nontraditional fertilizer or soil amendment materials that are applied to soil in hopes of improving crop productivity through magical means.

Mitochondria—Cell organelles composed of an outer membrane and a winding inner membrane. A series of chemical reactions that occur on the inner membrane convert the energy of oxidation into the chemical energy of ATP.

Mitosis—Cell division; the steps in mitosis are early prophase, late prophase, metaphase, anaphase, telophase, and interphase.

Monocot—A flowering plant with one seed leaf or cotyledon, xylem and phloem contained within bundles, and parallel venation in leaves.

Monoecious—Bearing both staminate and pistillate flowers on the same plant.

Morphogenesis—Process of development.

Morphology—Science of the form and structure of animals and plants.

Muck soil—Soil found in old bogs, river deltas, and lake beds. It contains high organic matter and has good water-holding capacity. Nitrogen is also readily available through mineralization of organic matter.

Mulch—(1) Materials such as straw, sawdust, leaves, plastic film, and the like, spread upon the surface of the soil to protect the soil and plant roots from the effects of raindrops, soil crusting, freezing, evaporation, and so on. (2) To apply protective materials to the soil surface.

Mulch farming—A system in which the organic residues are not plowed into or otherwise mixed with the soil but are left on the surface as a mulch.

Mulch tillage—Tillage or preparation of the soil in such a way that plant residues or other mulching materials are purposely left on or near the surface.

Mutagens—Chemicals used to create mutations.

Mutation—A spontaneous change in the genetic makeup of a cell.

Mycoplasma—A microscopic bacterium-like organism that lacks a cell wall and appears filamentous.

Mycorrhizae—Specific fungi that grow in intimate association with plant roots.

N

Nanometer—One billionth (1/1,000,000,000) of a meter.

Natural erosion—Wearing away of the earth's surface by water, ice, or other natural agents. Synonymous with geological erosion.

Nematodes—Microscopic roundworms, usually living in soil, many of which feed on plant roots. They cause galls on roots, cause root lesions, injure root tips, and sometimes cause excessive root branching. Nematodes reproduce by eggs.

Neutral—Neither acid nor alkaline (pH=7).

Nitrification—Biological oxidation of ammonium (NH_4^+) to nitrate (NO_3^-) by Nitrosomonas and Nitrobacter bacteria.

Nitrogen cycle—Sequences of transformations in N forms among gaseous, inorganic, and organic compounds. These transformations occur in cycles and involve numerous compounds, organisms, or reactions.

Nitrogen fixation—When molecular nitrogen (gaseous N_2) biologically or chemically converts to organic combinations or forms available for biological processes. Biological fixation occurs with legumes, whereas chemical fixation involves the manufacture of ammonia. Legume roots convert nitrogen gas to nitrates by bacteria.

Nitrogen solution—A liquid form of nitrogen fertilizer.

Node—The region of the stem where one or more leaves are attached. Buds are commonly borne at the node.

Nodulation—Legume plants form nodules in the process of nitrogen fixation.

Nomenclature—A set or system of names.

Noninfectious diseases—Diseases caused by abiotic agents.

Nonpoint source pollution—Pollution arising from an ill-defined and diffuse source, such as runoff from cultivated fields, grazing land, or urban areas.

No-till—Planting a crop directly into an unprepared seedbed. The tillage involved in planting is nothing more than opening the soil for the purpose of placing seed at the intended depth. This usually involves opening a small slit or punching a hole into the soil. Usually no cultivation occurs during crop production. Weed control is achieved entirely by surface applied and contact herbicides. Also referred to as zero tillage or slot planting.

Noxious weeds—Weeds that are arbitrarily defined by law as being especially undesirable, troublesome, and/or difficult to control.

Nubbins—Ears too small to be harvested.

Nucleotides—Each strand of DNA is a linear arrangement of repeating similar units called nucleotides, which are each composed of one sugar, one phosphate, and a nitrogenous base.

Nucleus—A membrane-bounded cellular body that contains the principal hereditary material.

Nutrient availability—Amount of soil or fertilizer nutrient supply that can be immediately used by plants.

Nutrient diffusion—Movement of nutrients from a high concentration (bulk soil) to a low concentration (root surface) in soil. Nutrient availability throughout the growing season depends on this process.

Nutrient film technique—A hydroponic method of growing plants where a very shallow stream of water containing all the dissolved nutrients required for plant growth is recirculated past the bare roots of plants in a watertight gully, also known as channels.

Nutrient interaction—Refers to how one nutrient may help or hinder the uptake of another.

Nutrients—Fertilizer, particularly phosphorus and nitrogen—the two most common components that run off in sediment.

O

Oil crop—A crop grown for the oil content of its seeds—for example, soybeans, peanuts, sunflowers, safflowers, canola, flax, jojoba, and corn.

Oilseed—Crops that capture sunlight energy and store oil in their seed, which is extracted for human use. Some examples are canola, safflower, sunflower, and soybeans.

Olericulture—Vegetable production.

Opposite—Leaves occur two at a node on opposite sides of the stem.

Optimal temperature—Temperature needed for plants to germinate.

Organ—Any part of a plant that performs a specific function.

Organelles—The inside parts of a cell such as the Golgi apparatus, nucleus, ribosomes, microtubules, and storage particles.

Organic—Chemical compounds of carbon combined with other chemical elements and generally manufactured in the life processes of plants and animals. Most organic compounds are a source of food for bacteria and are usually combustible; derived from living organisms (plants and animals).

Organic farming—Pest and nutrient management are achieved with nonchemical methods.

Organic matter—Partially decomposed plant and animal residues in soil and soil humus.

Organic soil—Soil that contains a high percentage of organic matter or materials (greater than 15–20 percent) throughout the soil profile.

Organic standards—A framework of guidelines and regulations that govern the production of organic crops.

Oriented tillage—Tillage operations conducted in specific paths or directions with respect to the sun, prevailing winds, previous tillage, or field base lines.

Ovary—Enlarged, bulbous, basal part of the pistil that bears the ovules—the egg-containing units that, after fertilization, become the seeds attached either to its central axis or to its inner wall.

Oviposition—The laying of an egg by an insect.

Ovule—Contains the female gametes.

Oxaloacetic acid (OAA)—A four-carbon chemical formed during the C-4 pathway in the carbon fixation process.

Oxidation—Combining with oxygen to release energy.

Oxidation-reduction—Oxidation can be represented as involving a loss of electrons by one molecule and reduction as involving an absorption of electrons by another. Both oxidation and reduction occur simultaneously and in equivalent amounts during any reaction involving either process.

Oxidative phosphorylation—A series of chemical reactions that occur on the inner membrane converting the energy of oxidation into the chemical energy of ATP. In this process the predominant energy transfer molecule is ATP.

P

Palatability—Taste potential for an animal.

Palisade cells—Cells within the leaf may be formed into two layers: the upper, tightly packed with elongated palisade cells; and the lower, loosely packed with spongy tissue.

Palmate—In leaves, the principal veins extend from the petiole near the base of the blade similar to the bones in the hand.

Parallel—In leaves, the veins extend parallel from the base to the apex or tip of the leaf.

Parasitic seed plant—A higher plant with chlorophyll that lives parasitically on other plants. It causes general weakening of the host plant.

Parenchyma—Cells with thin cell walls and with large vacuoles. In leaves, parenchyma cells contain chloroplasts for photosynthesis.

Parent material—Unconsolidated and somewhat chemically weathered mineral or organic matter from which soils are derived by natural soil development processes.

Parthenocarpy—Fruit development without fertilization.

Parthenogenesis—Reproduction by development of a female germ cell without fertilization; can occur in aphids.

Parts per million (ppm)—A ratio similar to percent, the number of parts in one million parts; percent is the number of parts in one hundred parts.

Pathogens—Disease-causing organisms.

Pathology—Study of diseases and the organisms that cause them.

Pectin—A large carbohydrate contained in the cell walls of fruits and vegetables.

Pedicel—Stalk of an individual flower in a cluster.

Pedology—Study of the formation and classification of soil.

Pedon—The smallest soil body. A section of soil that extends to the root depth and has about 10 to 90 square feet of surface area.

Peduncle—(1) Stalk that bears the single flower at the top. (2) The main stem or axis of a flower cluster.

Peg—Involved in formation of pod and seed development of the peanut.

Penetration—Process of getting inside the plant; it may be an active or passive process.

Percent purity—Percentage inert matter + percentage other crop seed + percentage weed seed.

Perched water table—(1) A zone of saturated soil that, because of obstructions or other reasons, is maintained above the normal water table. (2) The layer of saturated soil in the bottom of a pot of soil, because excess water cannot drain to the normal water table in the soil.

Percolation—Movement of water through soil layers of material.

Perennial—A plant or plant part that lives for more than two years.

Perianth—The outer floral parts, composed of the calyx and the corolla.

Permanent roots—Crown area, from which the nodal or permanent roots eventually develop, will grow from 1 to 1½ inches below the soil surface with little variation due to planting depth.

Permanent wilting point—Point at which no more water is available to the plant.

Peroxisomes—Organelles in the plant cell that use oxygen to carry out catabolic reactions.

Persistence time—Time required for a pesticide to become inert. Arbitrarily assumed to equal four half-lives when measured persistence time is not available.

Pesticide—A chemical substance used to kill or control pests such as weeds, insects, fungi, mites, algae, rodents, and other undesirable agents.

Petiole—Stalk of the leaf.

pH, soil—Negative logarithm of the hydrogen ion concentration of a soil [pH = –log (H$^+$)]. Degree of acidity or alkalinity as determined by an electrode or indicator at a specified soil moisture content and expressed in terms of the pH scale (1–14); a low pH indicates acid soil, a pH of 7 is neutral, and a high pH indicates an alkaline soil.

Phenotype—Visual characteristics that are seen.

Phloem—One of the two components of the vascular system whose primary function is the transport of manufactured products.

Phosphate—Common term used to refer to the amount of P$_2$O$_5$ present in fertilizers. Multiply P$_2$O$_5$ by .44 to obtain actual P.

Photoelectric effect—Light energy creating electrical energy.

Photomorphogenesis—A developmental response to a nondirectional light stimulus.

Photoperiod—Length of daylight.

Photoperiodism—Response of the plant to the length of daylight.

Photophosphorylation—The process of converting energy from a light-excited electron into the pyrophosphate bond of an ADP molecule.

Photosynthates—Products of photosynthesis, such as carbohydrates such as sugars and starches (CHOs) and other complex compounds referred to collectively.

Photosynthesis—Process in a plant of making sugars for growth and respiration from the raw products of water, carbon dioxide, and sunlight, releasing oxygen.

Photosystems—Arrangements of chlorophyll and other pigments packed into thylakoids.

Phototropism—Tendency of plants to "lean" in the direction of the greatest light intensity.

Phylla—Refers to the four main divisions of the plant kingdom: Thallophyta, Bryophyta, Pterodophyta, and Spermophyta.

Physical weathering—Breakdown of rock particles by physical forces such as frost action or wind abrasion.

Phytochrome—Pigment receiving radiant (light) stimuli.

Phytotoxicity—Ability of a chemical substance to cause harm to a plant, often used to characterize the effect of a herbicide on crops.

Pinnate—In the leaf, the secondary veins extend from the midrib, like the divisions of a feather.

Pistil—Female portion of the flower responsible for the formation of seeds.

Planed—Leveled or smoothed.

Plant growth regulator—Organic compounds other than nutrients that affect plant growth.

Plant hormone—Naturally occurring plant growth regulators. The term plant growth regulator includes synthetic compounds as well as natural hormones.

Plantlets—Small plants at the tips of new plants. They are produced from stolons or runners. They may be rooted while still attached to the parent, or detached and placed in a rooting medium.

Plant populations—Number of plants planted in a particular area.

Plasmids—Small circles of DNA that are much smaller than bacterial chromosomes.

Plasmolemma—Plasma membrane or cytoplasmic membrane.

Plow down—Fertilizer applied to soil surface prior to inverting the plow layer.

Plowing—A primary tillage operation performed to uniformly shatter soil, usually with complete soil inversion.

Plow layer—Upper part of a soil profile disturbed by humankind by plowing or other disturbances; P horizon suffix.

Plow-plant—Plowing and planting a crop in one operation, with no additional seedbed preparation.

Plumule—Young shoot.

Point source pollution—Pollution arising from a well-defined origin, such as a discharge from an industrial plant.

Pollen—Contains the male gametes.

Pollen grain—The anther or male part of the flower produces pollen grains that are the male sex cells.

Pollination—Act of placing pollen from the male reproductive organ onto the female reproductive organ of a flower; often is carried out by bees or wind.

Polygamous—Plants that bear staminate, pistillate, and hermaphroditic (bisexual—both sexes present and functional in the same flower) flowers on the same plant.

Polypedon—A group of similar neighboring pedons that makes up a soil series.

Pome—Fruits that have a core and embedded seeds.

Pomologist—Fruit scientist.

Positional availability—Nutrient absorbed by plant roots depending on the roots' physical accessibility to the nutrient.

Postemergence herbicide—Herbicide that is applied after the emergence of a specified crop or weed.

Potash—(1) Common term used to refer to the amount of K$_2$O present in fertilizers. Multiply K$_2$O by .83 to obtain actual K. (2) General term used to refer to potassium fertilizers.

Potting mixes—Plant-growing media made up of some type of course aggregate, organic amendment, and possibly some soil.

Precipitation—Water in the forms of rain, snow, sleet, and hail.

Precooling—Process of rapidly removing heat from vegetables before shipping.

Preemergence herbicide—Herbicide that is applied prior to the emergence of the specified crop or weed.

Preplant fumigation—Fumigation of weeds before planting.

Preplant incorporated herbicide—Herbicide that is applied and incorporated into the soil prior to the planting of a crop.

Primary growth—This growth takes place in young, herbaceous organs, resulting in an increase in length of shoots and roots.

Primary root—Arises from the embryo. Also called the taproot.

Primocane—First-year cane of a raspberry or blackberry.

Primocane-fruiting—A type of raspberry that forms fruit on the tops of first-year canes near the end of the growing season; the same as fall-fruiting.

Productivity, soil—Capacity of a soil, in its normal environment, to produce a specific plant or sequence of plants under a specified system of management. Productivity emphasizes the capacity of soil to produce crops and is expressed in terms of yield.

Profile, soil—Vertical section of soil through all horizons extending into parent material.

Profit—Money that remains after all fixed and variable costs are deducted from income.

Prokaryote—Cell with DNA not enclosed in the nucleus.

Proteomics—Study of genetics which refers to all the proteins expressed by a genome; involves the identification of proteins in an organism and the determination of their role in physiological functions.

Pruning—Removing all the old wood and leaving growth that will produce next year's crop.

Pulses—Legumes grown for their seed.

Pyruvic acid—One of the biochemicals produced during glycolysis. It enters the tricarboxylic acid (TCA) pathway.

R

Radicle—Root.

Ratoon—Second harvest.

Reactant—A product in many metabolic processes.

Reaction, soil—Degree of acidity or alkalinity of a soil, usually expressed in terms of pH. Descriptive terms commonly used: extremely acid, 4.5; very strongly acid, 4.5–5.0; strongly acid, 5.1–5.5; moderately acid, 5.6–6.0, slightly acid, 6.1–6.5; neutral, 6.6–7.3; mildly alkaline, 7.4–7.8; moderately alkaline, 7.9–8.4; strongly alkaline, 8.5–9.0; very strongly alkaline, greater than 9.1.

Receptacle—Where the apex of the pedicel upon which the organs of a flower are developed.

Recessive—A gene that is not expressed in the presence of a dominant allele.

Recombinant DNA (rDNA)—Modified DNA produced by enzymatic breaking of the molecular chain and rejoining it after removing, modifying, or adding genes. The term recombinant DNA refers also to molecules resulting from the multiplication of such DNA, or to

the technique by which recombinant DNA is produced.

Recombination—(1) A new combination of alleles resulting from rearrangement following crossing over. (2) The alteration of a genotype by gene splicing techniques.

Reduced tillage—A system in which the primary tillage operation is performed in conjunction with special planting procedures in order to reduce or eliminate secondary tillage operations; less than conventional tillage.

Reduction—Chemical reaction removing oxygen.

Reflection—Part of light that is not absorbed but reflected, such as a red object reflects red light, a green object reflects green. The reflected portion of the spectrum is the color of an object.

Regeneration—Process of inducing single cells or groups of cells to produce an embryo, organ, or entire plant.

Relative humidity—Amount of water vapor present in the air expressed as a percentage of the maximum water vapor the air can hold at the same temperature and pressure.

Research and Marketing Act—This act created four regional research laboratories as part of what is now the Agricultural Research Service. Each laboratory specializes in the crops grown in its region.

Residual fertility—Available nutrient content of a soil carried over to the following crop.

Residual soil—Soil formed in place from bedrock, rather than from transported parent materials

Residues—Crop materials, including roots and tops, that remain on the soil following harvest.

Resistant—Treated so as not to get certain diseases.

Respiration—Process of converting sugars into carbon dioxide, water, and energy. Often, the energy is in the form of heat.

Resume—Written information for a prospective employer that may include any of the following: career objectives, work experience, education background, accomplishments, awards, or skills. Also called a data sheet.

Rhizobia—Bacteria capable of living symbiotically in roots of legumes; receive energy from the host and often use molecular (gaseous) N to create N forms available for host use. Collective common name for the bacterial genus Rhizobium.

Rhizomes—Underground stems.

Rhizosphere—Zone of soil directly affected by plant roots and where the microbial population is highly altered.

Ribonucleic acid (RNA)—A nucleic acid consisting of the sugar ribose, together with phosphate, and adenine, guanine, cytosine, or uracil.

Ribosomes—Where the RNA goes for translation into proteins.

Ribulose phosphate carboxylase—An enzyme in the first step of the CO_2 fixation process. It is the limiting reagent of photosynthetic CO_2 fixation.

Ribulose biphosphate—A chemical captured by carbon dioxide in the dark reaction.

Ridge planting—Planting seed on ridges that are formed with a lister or similar tillage tool.

Rill erosion—An erosion process in which numerous small channels, usually several inches in depth, are formed. It occurs mainly on recently cultivated soils.

Ripening—When a plant matures and is ready for harvest.

River terrace—A former river floodplain now at a higher elevation.

Rockwool—Inorganic fibrous substance that is produced by steam blasting and cooling molten glass or a similar substance and is used as an insulator and a filtering material.

Rodents—Mammals that interfere with people or cause harm to crops.

Rogue—(1) An off-type plant. (2) To remove off-type plants by pulling or cutting.

Roll—A system of cleaning and preparing ground to receive sod.

Root hairs—Specialized cell extensions that penetrate into the openings between soil particles.

Roots—Serve three functions in the plant: (1) anchor the plant; (2) absorb water and minerals from the soil and transport them to the stem; and (3) store food produced by the aboveground portion of the plant.

Rootstock—That part of a tree that becomes the root system of a grafted or budded tree.

Root wedging—Rocks forced apart by root pressure.

Root zone; rooting zone—Depth of soil penetrated by crop roots.

Rotary tillage—Tillage that employs power-driven, rotary actions to shatter and mix soil.

Rotation—Repetitive cultivation of an ordered succession of crops; one cycle may take several years to complete.

Roughage—Feedstuffs with a high fiber content.

Runners—New plants are formed at nodes by runners, which are stems from old plants. The stems grow along the ground.

Runoff—That portion of precipitation or irrigation water that flows off a field and enters surface streams or water bodies; water that flows off the surface of the land without sinking into the soil.

S

Saline—Containing soluble salt. Saline soil is nonsodic soil containing sufficient soluble salt to impair productivity (EC greater than 4 mmhos/cm).

Saline-sodic soil—Soil containing a combination of soluble salts and exchangeable sodium sufficient to interfere with growth of most plants (EC greater than 4 mmhos/cm and ESP greater than 15 percent).

Salt index—Index of fertilizer materials used to compare solubilities of chemical compounds and their potential damage to germinating seeds or young seedlings with material placed near seed or on foliage.

Scarification—Chemical or physical treatment of seeds to break or weaken the seed coat enough for germination to occur.

Scion—A piece of last year's growth with three or four buds; the part inserted on the understock.

Secondary growth—Follows primary growth in some plants and results in an increased girth as layers of woody tissue are laid down. Monocots and herbaceous dicots typically exhibit only primary growth.

Secondary roots—Branches of the primary (tap-) root are often fibrous and referred to as secondary roots; they become the permanent roots of many monocotyledons.

Sediment—Solid material that is in suspension, is being transported, or has been moved from its original location by air, water, gravity, or ice.

Sedimentary rock—Rock made of sediments hardened over time by chemicals or pressure.

Sediment yield—Quantity of sediment arriving at a specific location.

Seed—Unit of dispersal for the new plant. It provides some protection from injury and drying and some nourishment for the young plant until it can make its own food.

Seedbed—Soil prepared to receive seeds.

Seed band—Fertilizer placed with the seed. Also called pop-up when applied in low rates.

Seed coat—Ovule walls develop from the seed coat.

Seeding rate—Rate at which seeds are released from a planter; for example, 50 per 100 feet or 100 pounds per acre.

Seepage—Percolation of water through the soil from unlined canals, ditches, laterals, watercourses, or water storage facilities.

Seeps—Hillside springs or zones where water flowing laterally within the soil breaks out into the open.

Selective herbicide—A herbicide that is more toxic to some plant species than to others.

Self-pollination—Process by which pollen is transferred from an anther to a stigma of the same flower or another flower of the same plant or cultivar.

Semidwarf—A tree that grows to about 10 to 15 feet.

Senescence—The stage of development when deterioration occurs, leading to the death of an organism or organ.

Separation—A form of propagation by which plants that produce bulbs or corms multiply.

Sewage sludge—Solid end product of the sewage treatment process that can be applied to soil as a fertilizer material.

Sexual—Reproduction in plants requiring that flowers form, that pollination and fertilization occur, that seeds develop, and that the seeds grow into new plants.

Shank—A structural member that may be used for attaching a tillage tool to a beam or a standard.

Shear—Force acting at right angles to the direction of movement of a tillage implement.

Sheath—Base of a leaf that wraps around the stem, as in grasses.

Sheet erosion—Removal of a fairly uniform layer of soil material from the land surface by the action of rainfall and surface runoff.

Shoot bud—A bud on the aboveground portion of a plant.

Short-day—Plants requiring a day shorter than its critical day length or a night longer than its critical dark period in order to flower; also called long-night plants.

Side band—Fertilizer placed to the side and/ or below the seed—for example, 2 inches by 2 inches.

Sidedress—Apply fertilizer to the side of a row for best growth results.

Simple—(1) Stems without branches (sidegrowths). (2) A leaf consisting of only one blade.

Sinkhole—A depression in the landscape where limestone has been dissolved.

Slick spots—Small areas within a field that are slick when wet because of high salt or sodium content that disperses the soil.

Slot planting—See no-till.

Sod—Grass that has soil and roots attached.

Sod-forming—Grasses that produce a densely matted sod from the spread of stolons and rhizomes. It covers more uniformly than bunchtype grasses.

Sodic—Containing excessive amounts of sodium. Sodic soil contains sufficient exchangeable sodium to interfere with plant growth (ESP greater than 15 percent).

Sodium absorption ratio (SAR)—Ratio between soluble sodium (Na+) and soluble divalent cations (Ca++ + Mg++). Used to predict exchangeable sodium percentage (ESP) of a soil equilibrated with a given solution and the potential sodium hazard of irrigation water.

Sod planting—A method of planting in killed sod with little or no tillage.

Soilborne—Carried in the soil, such as diseases that come from the soil.

Soil classification—Soils based on three-dimensional entities that can be grouped together according to their similar physical, chemical, and mineralogical properties.

Soil conditioners—Materials added to soil that may affect properties to the point that some physical or chemical conditioning occurs.

Soil conservation—(1) Protection of the soil from erosion or chemical deterioration. (2) Prevention of excessive loss of fertility by either natural or artificial means. (3) A combination of land use and management methods that safeguard the soil against depletion or deterioration by natural or human-induced factors. (4) A division of soil science concerned with soil conservation by preventive action.

Soil extract—Solution separated from a soil at a particular moisture content by filtration, centrifugation, or displacement.

Soil genesis—(1) The mode of origin of the soil with special reference to the processes or soil-forming factors responsible for development of the solum, or true soil, from unconsolidated parent material. (2) The branch of soil science that deals with soil genesis.

Soil horizon—Layer of soil or soil material approximately parallel to land surface and differing from adjacent layers in physical, chemical, and biological properties or characteristics.

Soilless—Growing plants without soil.

Soil management—(1) Total of all soil management practices, including tillage, cropping practices, fertilization, use of lime, and other treatments for the production of plants. (2) A division of soil science concerned with efficient soil management and conservation.

Soil probe—Tool used to withdraw small cylindrical samples of soil for analysis or for determining moist soil depth.

Soil profile—Refers to the arrangement and properties of the various soil layers.

Soil separates—Mineral soil particles defined by specified size limits: sand (2.0–0.05 mm), silt (0.05 mm–0.002 mm), and clay (less than 0.002 mm).

Soil solution—Water held by soils and the nutrients it contains.

Soil structure—Combination or arrangement of primary soil particles into secondary particles, or units such as crumbs, peds, granules, or aggregates.

Soil test—Analysis of nutrient-supplying properties of a soil sample to determine the capacity of that soil to support crop growth (see nutrient availability and index).

Soil texture—Relative proportions of various soil separates (sand, silt, and clay) in a soil.

Solarization—Vegetation is removed, soil is broken up and then wetted to activate the nematode population. Next the soil is covered with a sturdy clear plastic film during the warmest six weeks of summer. High temperatures (above 130°F) must be maintained during this time for best results.

Soluble—Able to be dissolved.

Solum solution—The upper, weathered part of the soil profile; the A, E, and B horizons.

Solution—Water with ions dissolved in it. Applied to soil, the soil solution is the water in the soil with nutrients and other materials dissolved in it.

Solvency—Having sufficient means to pay all debts.

Solvent—A liquid capable of dissolving. Water is the universal solvent.

Somoclonal variation—Production of new genotypes and phenotypes following propagation of cell cultures into whole plants.

Species—(1) The largest group of similar individuals that actually or potentially can successfully interbreed with one another but not with other such groups. (2) A systematic unit including geographic races and varieties, and included in a genus.

Splash erosion—The spattering of small soil particles, caused by the impact of raindrops on very wet soils. The loosened and spattered particles may or may not be removed subsequently by surface runoff.

Split boot—Dual-band placement on either side of seed.

Sprinkler irrigation—An irrigation system that has a pumping unit, control head, mainline and perhaps submain pipe(s), and laterals. Three basic types of sprinklers are used: rotating sprinklers, stationary spray-type nozzles, and perforated pipe.

Staking—Keeping plants in the correct growing position by using wires, wooden posts, or similar supports.

Stamen—Male part of a flower; it produces pollen.

Standard (beam)—An upright support that connects the shank to a tillage implement frame.

Starch—A polysaccharide, one of the products of photosynthesis.

Stele—Where water and minerals pass through the cortex, or root wall, into the center of the root. Also called the vascular cylinder.

Stem—Forms the major aboveground structural part of the plant; also is the attachment point for leaves, flowers, and fruit. It also contains the water and food distribution system.

Sterile—(1) Free from living microorganisms. (2) Unable to reproduce.

Stigma—Tip of the style or pistil, especially adapted to receive the pollen grains, which is expanded into a bulb or disk or divided into two or more slender parts.

Stipules—Pair of small leaflike appendages at the base of the petiole.

Stolons—Aboveground stems.

Stomata—Pores on the bottom of a leaf through which carbon dioxide enters the plant and water vapor exits.

Stone fruit—A simple fruit produced from a single carpel, usually one-seeded, with an outer fleshy layer of tissue called the pericarp and an inner, heavy stony layer called the endocarp.

Strip cropping—Practice of growing crops that require different types of tillage, such as row and sod, in alternate strips, along contours or across the prevailing direction of wind.

Strip incorporation—Application of fertilizer in a relatively concentrated strip on the soil surface and incorporated by tillage or cultivation.

Strip tillage system—A system in which only alternate bands of soil are tilled.

Stubble mulch—Stubble or other crop residues that are left to cover the land surface during the seedbed preparation and growing of a succeeding crop.

Style—Elongated stalk or neck connecting the ovary with the stigma.

Suberin—A substance that cell walls are made of.

Subsoil—The layer of soil just under the topsoil.

Subsoiling—Deep chiseling; breaking compact subsoils, without inverting them, with a special knifelike instrument (chisel) that is pulled through the soil at depths of 12 to 24 inches and at spacings of 2 to 5 feet.

Subsurface tillage—Tillage with a special sweeplike plow or blade that is drawn beneath the surface at depths of several inches to cut off plant roots and loosen the soil without inverting it.

Sucker shoot—A young, vegetative cane originating from an adventitious bud on the root.

Sugar—An energy-yielding chemical produced by photosynthesis.

Sulfur cycle—Sequence of transformations whereby sulfur is oxidized or reduced through organic or inorganic processes.

Summer annuals—Plants that germinate in the spring and die in the fall after the first frost.

Seeds mature and are left scattered on the ground. Some seed may germinate in the early fall, but most lie dormant until spring.

Summer-fruiting—A raspberry or blackberry that forms fruit only on second-year canes in late spring or summer; the same as floricane-fruiting.

Superphosphate—A triple phosphorus source.

Surface band—A fertilizer band applied to the soil surface.

Surface layer—Soil ordinarily moved in tillage, or its equivalent in uncultivated soil, ranging in depths from 4 to 10 inches. Also called the plow layer.

Syconium—Fruit, such as the pear or strawberry, that develops from a ripened ovary or ovaries but includes a significant portion derived from nonovarian tissue; also called false fruit, pseudocarp.

Symbiosis—When two different organisms associate or live together with a resultant mutual benefit (such as rhizobia with legumes).

Synthetic hormone—Compounds produced outside of the plant that act in a manner similar to naturally produced compounds inside the plant.

Systemic—To be absorbed by plant roots, leaves, or stems and translocated to the whole plant.

T

Talus—Deposits of dry rock and soil that have slid to the base of a slope under the force of gravity.

Taproots—Prominent primary roots from which all other lateral rootlets or secondary roots grow. They may divide, become fleshy, and often penetrate deeply into the soil.

Taxonomy—Organizational system for descriptive classification of plants.

Temperate—Areas of the world where temperatures are not too hot or too cold.

Tenderometer—A device to determine the time of harvest for a plant or crop.

Terminal—When flowers or clusters of flowers are carried on the ends of the axis or branches.

Terraces—Low dams or dikes built across slopes to catch runoff water and eroded soil before they leave the field.

Tetraploids—Plants with four sets of chromosomes.

Texture—Amount of sand, silt, and clay in the soil.

Thermoperiod—Refers to a daily temperature change.

Thermophilic—The description of an organism that thrives at high temperatures.

Thermophilic organism—Heat-loving organism whose optimum temperature range for growth is above 95°F. Dominates composts during middle stage of decay.

Thylakoid—Structural unit of photosynthesis.

Thymine—One of the four bases of DNA; it pairs only with adenine.

Tillage—Mechanical manipulation of soil for any purpose. In agriculture the term is usually restricted to the changing of soil conditions for crop production.

Tillage, deep—A primary tillage operation that manipulates soil to a greater depth than normal plowing. A heavy-duty moldboard or disk plow that inverts the soil, or a heavy-duty chisel plow that shatters the soil may be used.

Tillage implement (machine)—Single or multiple soil-working tools, together with a power unit. Control and protection systems are an integral part of the machine.

Tillage, primary—Initial major soil-working operation. The tillage is designed to reduce soil strength, break up and cover plant materials, and rearrange aggregates.

Tillage, secondary—Any tillage, following primary tillage, designed to improve seedbed conditions.

Tillage tool—An individual element of soil-working equipment.

Tillage tools, dynamic—Tillage tools that are powered so that some parts move in a different direction than the line of travel.

Tillage tools, multipowered—Tillage tools that utilize more than one form of power, such as draft and rotating, or draft and electrical.

Tillage tools, simple—Tools used in a fixed position in relation to the soil, with no movement of individual parts.

Tillering—Development of secondary shoots.

Tillers—First side shoots in small grains.

Till plant—Seedbed preparation by pushing soil and residue aside, leaving a protective cover of crop residue on the soil surface and mixed in the surface layer. Seedbed preparation and planting are completed in the same operation. It is considered minimum tillage in some areas.

Tilth—Physical condition of soil related to its ease of tillage, fitness as a seedbed, and degree of impedance to seedling emergence and root penetration.

Tip-layering—A method of propagation in which the ends of a plant are buried in the soil and new plants arise from them.

Tipping—Process of removing the tops of primocanes for the purpose of stimulating lateral branching; sometimes called pinching.

Tissue—Large groups of organized cells of similar structure to perform specific functions in the plant. The two generalized types of tissues are meristematic and permanent.

Tissue culture—Process or technique of making plant or animal tissue grow in a culture medium outside the organism.

Tissue testing—Testing the status of plants by collecting a sample of the leaves or stems.

Top-dress—Uniformly apply fertilizer over the field, generally with P on established forage and N on small grains during the growing season.

Topography—Slope of the land and the position on the landscape, such as the top of a hill, a hillside, or the foot of a slope.

Topping—Removal of the top portion of cane after it has overwintered.

Top soil—Layer of soil moved in normal cultivation.

Topworking—Operation of cutting back the branches and top of an established tree and budding or grafting part of another tree on it.

Torus—Center of a fruit to which the druplets attach, also called the receptacle.

Trade name—Proprietary or brand name for a chemical; for example: "Formula 40" is one trade name for 2,4-D.

Training—Directing plants to grow certain directions by using different methods.

Transfer RNA (tRNA)—Form of RNA that brings specific amino acids to the ribosome for protein synthesis.

Transgenetic—Plants and animals that result from genetic engineering experiments in which genetic material is moved from one organism to another, so that the latter will exhibit a desired characteristic.

Translocation—Movement of water and dissolved compounds through the plant.

Transpiration—Process of water exiting the plant through the stomata.

Transplantability—Success of plants being transplanted after starting indoors.

Transported soil—Soils formed in parent materials brought to the final location of soil formation by transporting agents such as gravity, flowing water, wind, and glacial ice sheets.

Treated seed—Seed that has been coated with a chemical before being planted to control diseases or insects.

Trellis—A support system of wires or wood for growing certain plants.

Tricarboxylic acid (TCA)—Produces electrons that enter the electron transport system and produce ATP's.

Triploid—Three sets of chromosomes.

Truck cropping—Refers to commercial, largescale production of selected vegetable crops for wholesale markets and shipping.

Tuber—Edible portion of the plant, and botanically, stems not roots. They are stems because they contain all the morphological features of stems.

Turgid—Condition in which a cell or plant is fully expanded by hydrostatic pressure exerted on the cell wall by the protoplast.

Turgor—Stiffness in the cells.

U

Umbel—Type of inflorescence.

Understock or stock—Part on which the scion is inserted; the part below the graft.

V

Vacuoles—Occupies the major volume of the cell and contains water solution and dissolved substances—sugars, organic acids, and pigments. The vacuole is the storage reservoir for water, sugars, salts, and other biochemicals.

Vapor concentration—The gradient equals the water vapor concentration inside the product minus that of the surrounding air.

Variety—A plant group different in the wild from the general species. It is often used for varieties named from the general species.

Vascular system—Conducting tissues of a plant composed of xylem and phloem.

Vegetative—Period when the plant grows vigorously and rapidly.

Venation—Arrangement of the veins of a leaf.

Vermiculite—A substance used in growing media that holds water, is very lightweight, and expands when heated.

Vernalization—Promotion of flowering by cold treatment given to plants or imbibed seeds.

Vertical mulching—A subsoiling operation in which mulching material is injected into the slit immediately behind the chisel to leave a vertical wind barrier.

Viroid—A viruslike particle that lacks the outer protein coat of a virus particle.

Viruses—Submicroscopic, subcellular particles that require a host cell in which to multiply.

Visible spectrum—Colors on the spectrum that can normally be seen by the eye. Ultraviolet and infrared colors are on the ends of the spectrum that cannot be seen.

Visualization—Ability to "see" abstract or concrete ideas or concepts in one's mind.

Volatilization—Diffusion into the atmosphere.

Volunteer plants—Plants that may grow following harvest or the next season without being planted.

W

Warm-season—Refers to plants that are usually killed by frosts and require much warmer temperatures to grow properly. They are planted later in the spring.

Water budgeting—This is a method used to balance the available soil moisture. Rainfall and irrigation amounts represent credit entries. Evapotranspiration is a debit entry. Rainfall and irrigation should be measured.

Watershed—Surrounding land area that drains into a lake, river, or river system.

Water soluble—Readily dissolves in water.

Wavelength—Distance between consecutive crests of waves.

Weathering—The natural process by which rock is broken into smaller pieces such as wind, heat, cold, and the like.

Weed—A plant growing where it is not wanted.

Wetting agents—Substances that help other substances absorb water (such as soil).

White light—Portion of radiant energy; white light is made up of the colors red, orange, yellow, green, blue, and violet.

Whorled—Three or more leaves at each node.

Wilt—When plants lose water more rapidly than they take it up, they wilt. Life processes slow, and growth may even stop.

Windbreak—Trees, shrubs, or other vegetation, usually nearly perpendicular to the principal wind direction, planted to protect soil, crops, homesteads, roads, and the like, against wind, wind erosion, and drifting soil and snow.

Winter annuals—Plants that germinate in the fall or winter, produce their seed in the spring or early summer, and then die. The seed lies dormant in the soil during most of the summer months.

Winter hardy—Seeds that can be seeded even where temperatures frequently fall to 40°F, below zero, or where mean winter temperatures are about 0°F.

X

Xeric—Referring to the tropical zone.

Xerophyte—A plant that normally requires relatively small amounts of water for growth; a hydrophyte is just the opposite.

Xylem—One of two components of the vascular system whose primary function is to transport water and soil nutrients.

Y

Yield—Amount of crop produced in response to cultural practices.

Yield potential—Level of crop productivity that can be obtained under specific physical, chemical, and environmental conditions of the soil.

Z

Zero tillage—Same as no tillage.

Zygote—Cell formed by the union of the male and female gametes, the new organism developing from this cell.

Appendix A

Due to its location in a book and because of the implications of its name, an appendix is often ignored by the reader. But an appendix contains valuable information that can enhance a reader's understanding and learning. Moreover, the information in an appendix is quick and easy to find.

The information in this appendix includes a variety of useful conversions, conversion factors, measurement standards, common measures, mixing guides, seeding guides, and row spacings. This appendix also includes a guide to collecting, pressing, and mounting plant specimens. Finally, this appendix contains some Internet sites (URLs) that lead to many sources of data and information.

By making full use of this appendix, the reader can understand more, plan better, do more, and learn more.

Table A-1 **Conversion Tables for Common Weights and Measures**

Metric Conversions	Equaled Amount
1 pound	454 grams
2.2 pounds	1 kilogram
1 quart	1 liter
1 gram	15.43 grains
1 metric ton	2.205 pounds
1 inch	2.54 centimeters
1 centimeter	10 millimeters or .39 inch
1 meter	39.37 inches
1 acre	.406 hectare

Table A-2 Weight and Volume Conversions

Measurements	Equaled Amount
8 tablespoons	¼ pound
3 teaspoons	1 tablespoon
1 pint	1 pound
2 pints	1 quart
4 quarts	1 gallon or 8 pounds
2,000 pounds	1 ton
16 ounces	1 pound
27 cubic feet	1 cubic yard
1 peck	8 quarts
1 bushel	4 pecks
Other Conversions	
1%	.01
1%	10,000 ppm
1 Megacalorie (M-cal)	1,000 calories
1 calorie (big calorie)	1,000 calories (small calories)
1 M-cal	1 therm

Table A-3 **Standard Weights of Farm Products per Bushel**

Product	Pounds	Product	Pounds
Alfalfa	60	Lespedeza	40–50
Apples (average)	42	Millet (grain)	50
Barley (common)	48	Oats	32
Beans	60	Onions	52
Bluegrass (Kentucky)	14–28	Peas	60
Bromegrass, orchard grass	14	Potatoes	60
Buckwheat	50	Rape	60
Clover	60	Ryegrass	24
Corn (dry ear)	70	Rye	56
Corn & cob meal	45	Soybeans	60
Corn (shelled)	56	Spelt	30–40
Corn kernel meal	50	Sorghum	56
Corn (sweet)	50	Sudangrass	40
Cotton	32	Sunflower (oil type)	24–32
Cowpeas	60	Timothy	45
Flax	56	Trefoil, bird's-foot	60
Grass (blue, brome, fescue, orchard)	14	Wheat	60
Grass (timothy)	45	Vetch	60

Table A-4 Calculating Approximate Weight of Grain by Volume

Standard Bushel Weight (Pounds)	Pounds/Cubic Feet
60	48.18
56	44.97
50	40.15
48	38.54
45	36.14
28	22.48
14	11.24
70 (ear corn)	28.00[1]

[1] Varies greatly with ear size and moisture content.

Notes on Using the Information in Table A-4

Step 1: Measure cubic feet

Width in feet multiplied by length in feet multiplied by depth of grain in feet equals cubic feet in square or rectangular enclosures.

Example. 10 ft. width × 14 ft. length × 9 ft. grain depth = 1,260 cu. ft.

In circular bins, the formula is: *pi* (3.14) multiplied by the radius squared, multiplied by the depth of grain = cubic feet.

Step 2: Obtain total grain weight

Multiply cubic feet of volume by the appropriate figure from Table A-4 under Pounds/Cubic Feet.

If actual bushel weight (test weight) is available, multiply actual bushel weight by 0.803. This calculation will give a more accurate figure for lbs./cu. ft. than you can get from the table.

Table A-5 **Bushel Weights and Volumes**

Item	Pounds/Cubic Feet	Cubic Feet/Ton
Oats at 32 lb./bu.	26	77
Barley at 48 lb./bu.	38.4	53
Shelled corn at 56 lb./bu.	44.8	45
Wheat at 60 lb./bu.	48	42
Corn & cob meal at 70 lb./bu.	28	72
Soybeans at 60 lb./bu.	48	42
Rye at 56 lb./bu.	44.8	45
Soybean oil meal at 54 lb.	—	37
Dairy feed at 35 lb.	—	57

Table A-6 **Measurement Standards, Hay and Straw**

Item	Average Cubic Feet/Ton	Range Cubic Feet/Ton
Hay, baled	275	250–300
Hay, chopped—field cured	425	400–450
Hay, chopped—mow cured	325	300–350
Hay, long	500	475–525
Straw, baled	450	400–500
Straw, chopped	600	575–625
Hay, loose	480	370–390
Straw, loose	800	750–850

Table A-7 **Fahrenheit to Centigrade Temperature Conversions**[1]

°F	°C	°F	°C	°F	°C
100	37.8	77	25.0	54	12.2
99	37.2	76	24.4	53	11.7
98	36.7	75	23.9	52	11.1
97	36.1	74	23.3	51	10.6
96	35.6	73	22.8	50	10.0
95	35.0	72	22.2	49	9.4
94	34.4	71	21.7	48	8.9
93	33.9	70	21.1	47	8.3
92	33.3	69	20.6	46	7.8
91	32.8	68	20.0	45	7.2
90	32.2	67	19.4	44	6.7
89	31.7	66	18.9	43	6.1
88	31.1	65	18.3	42	5.6
87	30.6	64	17.8	41	5.0
86	30.0	63	17.2	40	4.4
85	29.4	62	16.7	39	3.9
84	28.9	61	16.1	38	3.3
83	28.3	60	15.6	37	2.8
82	27.8	59	15.0	36	2.2
81	27.2	58	14.4	35	1.7
80	26.7	57	13.9	34	1.1
79	26.1	56	13.3	33	0.6
78	25.6	55	12.8	32	0.0

[1] Formulas used: $°C = (°F - 32) \times \frac{5}{9}$ *or* $°F = (°C \times \frac{9}{5}) + 32$

Table A-8 Conversion Factors for English and Metric Measurements

To Convert the English	To the Metric Multiply by	To Convert Metric	Multiply by	To Get English
acres	0.4047	hectares	2.47	acres
acres	4047	m.^{2}	0.000247	acres
BTU	1055	joules	0.000948	BTU
BTU	0.0002928	kwh	3415.301	BTU
BTU/hr.	0.2931	watts	3.411805	BTU/hr.
bu.	0.03524	m.^{3}	28.37684	bu.
bu.	35.24	L	0.028377	bu.
ft.^{3}	0.02832	m.^{3}	35.31073	ft.^{3}
ft.^{3}	28.32	L	0.035311	ft.^{3}
in.^{3}	16.39	cm.^{3}	0.061013	in.^{3}
in.^{3}	1.639×10^{-5}	m.^{3}	61012.81	in.^{3}
in.^{3}	0.01639	L	61.01281	in.^{3}
yd.^{3}	0.7646	m.^{3}	1.307873	yd.^{3}
yd.^{3}	764.6	L	0.001308	yd.^{3}
ft.	30.48	cm.	0.032808	ft.
ft.	0.3048	m.	3.28084	ft.
ft./min.	0.508	cm./sec.	1.968504	ft./min.
ft./sec.	30.48	cm./sec.	0.032808	ft./sec.
gal.	3785	cm.^{3}	0.000264	gal.
gal.	0.003785	m.^{3}	264.2008	gal.
gal.	3.785	L	0.264201	gal.
gal./min.	0.06308	L/sec.	15.85289	gal./min.
in.	2.54	cm.	0.393701	in.
in.	0.0254	m.	39.37008	in.
mi.	1.609	km.	0.621504	mi.
mph	26.82	m./min.	0.037286	mph
oz.	28.349	gm.	0.035275	oz.
fl. oz.	0.02947	L	33.93281	fl. oz.
liq. pt.	0.4732	L	2.113271	liq. pt.
lb.	453.59	gm.	0.002205	lb.

Table continues on next page

Table A-8 Conversion Factors for English and Metric Measurements *(continued)*

To Convert the English	To the Metric Multiply by	To Convert Metric	Multiply by	To Get English
qt.	0.9463	L	1.056747	qt.
ft.^{2}	0.0929	m.^{2}	10.76426	ft.^{2}
yd.^{2}	0.8361	m.^{2}	1.196029	yd.^{2}
tons	0.9078	tonnes	1.101564	tons
yd.	0.0009144	km.	1093.613	yd.
yd.	0.9144	m.	1.093613	yd.

Table A-9 Common Measures and Approximate Equivalents

1 liquid teaspoon =	5 milliliters (ml.)
3 liquid teaspoons =	1 liquid tablespoon = 15 ml.
2 liquid tablespoons =	1 liquid ounce = 30 ml.
8 liquid ounces =	1 liquid cup = 0.24 liter
2 liquid cups =	1 liquid pint = 0.47 liter
2 liquid pints =	1 liquid quart = 0.9463 liter
4 liquid quarts =	1 liquid gallon (U.S.) = 3.7854 liter

Table A-10 **More Conversion Factors for Metric and English Units**

Length

1 mile = 1.609 kilometers; 1 kilometer = 0.621 miles

1 yard = 0.914 meters; 1 meter = 1.094 yards

1 inch = 2.54 centimeters; 1 centimeter = 0.394 inches

Area

1 square mile = 2.59 square kilometers; 1 square kilometer = 0.386 square miles

1 acre = 0.00405 square kilometers; 1 square kilometer = 247.1 acres

1 acre = 0.405 hectares; 1 hectare = 2.471 acres

Volume

1 acre/inch = 102.8 cubic meters; 1 cubic meter = 0.00973 acre/inches

1 quart = 0.946 liters; 1 liter = 1.057 quarts

1 bushel = 0.352 hectoliters; 1 hectoliter = 2.838 bushels

Weight

1 pound = 0.454 kilograms; 1 kilogram = 2.205 pounds

1 pound = 0.00454 quintals; 1 quintal = 220.5 pounds

1 ton = 0.9072 metric tons; 1 metric ton = 1.102 tons

Yield or Rate

1 pound/acre = 1.121 kilograms/acre; 1 kilogram/acre = 0.892 pounds/acre

1 ton/acre = 2.242 tons/hectare; 1 ton/hectare = 0.446 tons/acre

1 bushel/acre = 1.121 quintals/hectare; 1 quintal/hectare = 0.892 bushels/acre

1 bushel/acre = (60#) = 0.6726 quintals/hectare; 1 quintal/hectare = 1.487 bushel/acre (60#)

1 bushel/acre = (56#) = 0.6278 quintals/acre; 1 quintal/acre = 1.597 bushels/acre (56#)

Temperature

To convert Fahrenheit (F) to Celsius (C): $0.555 \times (F - 32)$
To convert Celsius (C) to Fahrenheit (F): $1.8 \times (C + 32)$

Table A-11 **Mixing Small Quantities of Liquid Spray**

Concentration per Gallon[1]	Amount to Mix
1 lb.	7 tablespoons or 103 ml.
2 lb.	3.5 tablespoons or 51.5 ml.
3 lb.	2.3 tablespoons or 34.3 ml.
4 lb.	1.7 tablespoons or 25.8 ml.

1. The first column represents concentration of active ingredient per gallon; the second column provides corresponding amount to mix for one thousand square feet to get one pound per acre of active ingredient.

Table A-12 **Checking Planting Rate or Stand per Acre**

Row Spacing (Inches)	Inches of Row per Acre	Row Spacing (Inches)	Inches of Row per Acre
40	155,682	18	348,480
38	165,069	15	418,176
36	174,240	14	448,046
30	209,088	12	522,720
28	224,023	10	627,264
24	261,360	8	784,080
20	313,632	7	896,091

Notes for Table A-12

This table may be useful in checking actual planting rate when planting a crop. It can also be used in obtaining stand counts.

Determine average spacing in inches between seeds (or plants) in the row.

Then divide the appropriate figure in the right-hand column by this figure to determine planting rate (or stand).

Example. Grain sorghum planted in 30-inch rows is found to average 2.5 inches between seeds.

209,088 ÷ 2.5 = 83,635 seeds being planted per acre.

Or the grain sorghum stand averages 1 plant per 3.5 inches of row. Then

209,088 ÷ 3.5 = 59,379 plants per acre.

You can use this method without the table as long as you remember that there are 43,560 square feet per acre and that 144 square inches equal 1 square foot.

Example. 43,560 × 144 = 6,272,640 square inches per acre. Divide 6,272,640 by inches of row width to obtain inches of row per acre (6,272,640 ÷ 30 = 209,088).

Table A-13 **Moisture Conversion for Wheat**

Percent Moisture in Grain	Wheat (lbs.) Equivalent to 60 lbs. of Wheat at 13.5% Moisture
10	57.67
11	58.65
12	59.32
13	59.66
13.5	60.00
14	60.35
15	61.06
16	61.79
17	62.53
18	63.29
19	64.07
20	64.88
21	65.70
22	66.54
23	67.40
24	68.29
25	69.20
26	70.14
27	71.10
28	72.08
29	73.10
30	74.14

Table A-14 **Moisture Conversion for Soybeans**

Percent Moisture in Grain	Soybeans (lbs.) Equivalent to 60 lbs. at 13.0% Moisture
10	58.00
11	58.65
12	59.32
13	60.00
14	60.70
15	61.41
16	62.14
17	62.89
18	63.66
19	64.64
20	65.25
21	66.08
22	66.92
23	67.79
24	68.68
25	69.60
26	70.54
27	71.51
28	72.50
29	73.52
30	74.57

Table A-15 Moisture Conversion for Ear and Shelled Corn

Percent Moisture Shelled Corn in Grain	Harvest Weight (lbs.) of Ear Corn to Yield 56 lbs. at 15.5% Moisture	Shelled Corn (lbs.) Equivalent to 56 lbs. at 15.5% Moisture
10	63.49	52.56
10.5	63.86	52.87
11	64.25	53.16
11.5	64.65	53.46
12	65.06	53.77
12.5	65.60	54.08
13	65.95	54.39
13.5	66.42	54.70
14	66.89	55.02
14.5	67.39	55.34
15	67.89	55.67
15.5	68.40	56.00
16	68.94	56.33
16.5	69.51	56.67
17	70.09	57.01
17.5	70.69	57.35
18	71.31	57.70
18.5	71.95	58.06
19	72.60	58.41
19.5	73.27	58.78
20	73.96	59.15
20.5	74.60	59.52
21	75.36	59.89
21.5	76.07	60.28
22	76.79	60.66
22.5	77.53	61.05

Table continues on next page

Table A-15 **Moisture Conversion for Ear and Shelled Corn** *(continued)*

Percent Moisture Shelled Corn in Grain	Harvest Weight (lbs.) of Ear Corn to Yield 56 lbs. at 15.5% Moisture	Shelled Corn (lbs.) Equivalent to 56 lbs. at 15.5% Moisture
23	78.25	61.45
23.5	79.01	61.85
24	79.76	62.26
24.5	80.50	62.67
25	81.25	63.09
25.5	82.03	63.51
26	82.82	63.94
26.5	83.50	64.38
27	84.19	64.82
27.5	84.90	65.26
28	85.62	65.72
28.5	86.32	66.18
29	87.04	66.64
29.5	87.76	67.12
30	88.50	67.60
30.5	89.22	68.08
31	89.94	68.57
31.5	90.67	69.08
32	91.43	69.58
32.5	92.13	70.10
33	92.85	70.62
33.5	93.55	71.15
34	94.28	71.69
34.5	94.98	72.24
35	95.71	72.80

Note: To obtain bushels of grain, divide total grain weight by appropriate standard bushel weight. To obtain number of hundredweights (cwts.) of grain, divide total grain weight by 100.

Table A-16 How to Create Tables Like A-12 through A-15

Calculating moisture conversion factors covers the most widely grown crops and the most common moisture contents. When you need other conversions, the calculations are relatively simple.

Use percent dry matter in making conversions because the problem is to obtain the same weight of dry matter as is found in a standard bushel. For example, a standard bushel of wheat contains 60 pounds at 13.5 percent moisture. Thus, 86.5% dry matter (100 – 13.5) × 60 lbs. = 51.9 lbs. of dry matter.

Example. How many pounds of 20.5% moisture wheat is equivalent to a standard bushel?

- 13.5% Standard Moisture Content = 100 – 13.5 = 86.5% Dry Matter

- 20.5% Moisture Content = 100 – 20.5 = 79.5% Dry Matter

- 86.5% divided by 79.5 = 108.8%

- (108.8 × Standard Bu. Wt.) (60 for Wheat) = 100

- 65.28 lbs. equivalent to a standard bushel

To check your answer,

$$65.28 \times 79.5\% \text{ dry matter} = 51.9 \text{ lbs. of dry matter.}$$

Table A-17 Relative Effectiveness of Seed Fungicides for Some Diseases of Wheat

Fungicide	Disease				
	Loose Smut	Common Bunt	Septoria	Scab	Pythium
Captan-TBZ	N	P	F	F	F
Thiram-TBZ	N	P	F	G	F
Triadimenol	E	E	E	F	N
Difenconazole	E	E	E	F	N
Captan-TBZ-PCNB	N	N	P	G	F
Carboxin-Thiram	G	G	P/F	P/N	F/N
Thiram	N	P	E	F	F
Carboxin	G	G	P	P/N	N

E = Excellent
F = Fair
G = Good
N = No activity
P = Poor

Table A-18 Suggested Seeding Rates

Crop	Pounds of Seed per Bushel	Rate to Plant (lb./acre)
Alfalfa	60	12–15
Barley, Winter	48	90–120
Bird's-foot Trefoil	60	8
Bromegrass, Smooth[1]	14	10–15
Corn, Dent	56	16–18
Corn, Pop	56	3–6
Clover, Alsike[2]	60	3–4
Clover, Medium[1] or Mammoth Red	60	10
Clover, Ladino[2]	60	1–2
Fescue, Tall	32	10–15
Lespedeza, Korean	40–45	6
Oats, Spring	32	75–100
Orchard grass[1]	14	8–10
Reed Canary Grass	44–48	10
Rye	56	84–112
Ryegrass	—	12–15
Sorghum, Forage	50	8–12
Sorghum, Grain	56	8–12
Sorghum-Sudangrass	40–50	20–25
Soybeans	60	2.7 seeds per foot of 7-inch row
Sudangrass	40	25
Sweetclover	60	10–12
Timothy	45	1–2 (fall) 4 (spring)
Wheat	60	90-120

[1] The rates for these forage crops when seeded alone are higher than those recommended in meadow crop mixtures.
[2] These forages should not be seeded in pure stands; seeding rates are for mixed stands.

Table A-19 Rows per Acre for Common Lengths and Row Spacings

Row Length (feet)	Row Spacing				
	48 inch	**44 inch**	**40 inch**	**36 inch**	**32 inch**
	Rows/acre	Rows/acre	Rows/acre	Rows/acre	Rows/acre
400	27.3	29.7	32.9	36.3	40.8
600	18.2	19.7	21.8	24.2	27.4
800	13.6	14.8	16.4	18.2	20.4
1000	10.8	12.0	13.0	14.5	16.4
1200	9.1	9.6	10.6	12.1	13.2
1400	7.8	8.4	9.2	10.3	11.4
1600	6.8	7.4	8.1	9.1	10.1
1800	6.0	6.5	7.2	8.0	9.0
2000	5.4	5.9	6.5	7.2	8.2

Table A-20 How to Build A Plant Press

The first thing needed for a good plant press is a frame. This can be built from strips of wood cut from an apple box, crate, or other thin strong wood. The strips of wood should be about ¼ inch to ½ inch thick and 1½ inches wide. A local lumber yard should have 48-inch wood lath, which is very good to use. The press should be standard size (13 inches wide and 18 inches long). Slatted construction is necessary so that the plants can dry properly.

Materials and Equipment Needed

- Lath or wood strips about ¼ to ½ inch thick and 1½ inches to 2 inches wide (10 pieces, 13 inches long; 8 pieces, 18 inches long)
- Nails or screws for fastening wood together
- Saw, hammer or screwdriver, square, and pencil
- Two canvas, web, or leather straps at least 48 inches long
- Two sheets of newspaper folded (11 inches × 16 inches) for each plant you plan to collect
- Two pieces of corrugated cardboard or blotting paper (11 inches × 16 inches) for each plant you plan to collect

Construction

1. Cut 8 wood strips 18 inches long, and 10 strips 13 inches long.
2. Use 5 of the 13-inch strips and 4 of the 18-inch strips to make the frame. Two of these frames are needed.
3. Place the folded newspaper and cardboard between the frames in alternating layers.
4. Place the straps around the frame and tighten to hold the frames, newspaper, and cardboard in a tight bundle.

Table A-21 How to Collect, Press, and Mount Plants[1]

Plant mounts make better study material than any manual. A properly dried, pressed, and mounted plant is attractive, easily displayed, and will last a long time. A plant collection makes an interesting conversation piece in the home and can be used as an exhibit at fairs, schools, and other displays.

Equipment

- Digging tool—a shovel, garden digger, or some other digging tool to remove the plant from the soil.

- Trimming tool—a sharp knife or a pair of scissors to cut off woody specimens, to remove excess or old plant material, and to slice thick roots.

- Specimen container—plastic bags are recommended for keeping plants until you can press them.

- Notebook—a field notebook or tablet and a pencil or pen are needed to record all important information about the plant and the location where the plant was found.

- Plant press—a binder-type press, 18 inches long by 12 inches wide with alternating cardboard, blotter, and folded newspaper is recommended to dry and press the plant. Other items such as magazines will work for pressing if enough weight is placed on top. See Appendix Table A-20.

Collection Procedures

1. Because some plants bloom in early spring and others bloom in the late fall, you will not be able to collect all the plants at any one time of year. Plan several collection trips throughout the spring, summer, and fall.

2. Choose plant specimens carefully. Select one, or preferably two, of each plant species to be collected.

3. Avoid plants that are off-color, grazed, overmature, diseased, or otherwise not normal.

4. While at the site, record the plant in your field notebook or tablet by giving it a number. Record the plant name (if it is known) and the information that will be needed when completing the plant labels for your mount. Start a numbering system that will work for you. You may want to include the year, such as 99-1, 99-2, and so on. If you use this format, 99 refers to the year, and each different plant species will be numbered consecutively (1, 2, 3, and so on).

5. When collecting grasses and grass-like plants:
   - Select specimens with seedheads fully emerged from the sheath.
   - Select specimens that are still green including the seedhead.
   - Collect the whole plant, when possible, including a good sample of the roots.
   - Be sure that rhizomes or stolons are attached to the plant if they are typical for that species.

6. When collecting forbs:
   - Select specimens in the flowering stage.
   - Collect the whole plant if possible, including a portion of the root.
   - Some forbs can be collected with both flowers and seeds, or seed pods, on the plant at the same time.
   - Be sure that rhizomes or stolons are attached to the plant if they are typical for that species.
   - Taproots or other thick roots should be sliced away on the underside so that the plant will be fairly flat after pressing.

7. When collecting shrubs and other woody plants:
   - Select a branch about twelve to fourteen inches in length and not over ten inches in width.
   - Collect the plant when it is in bloom.
   - Many shrubs bloom in early spring before they leaf out. In these cases, collect two specimens, one in flower and one after the plant has leafed out.
   - Mount both specimens on the same sheet.
   - It is often useful to include a sample of both the current year and the older bark of woody plants.
   - Roots of large woody plants should not be included on the plant mount.

8. To remove a plant from the soil, dig about six inches straight down around the plant about three inches out from the stem. Carefully lift out the chunk of sod. If the soil is moist, use water to wash away the soil from the roots.

9. Remove all soil particles from the roots. Do not be afraid to wash the roots thoroughly on all the plants collected. In fact, it may take more than one washing. Excess moisture after washing the roots can be removed by firmly pressing the plant between paper towels.

Table continues on next page

Table A-21 **How to Collect, Press, and Mount Plants (*Continued*)**

10. Remove the excess plant material from the roots, stems, leaves, and seedheads. For example, by removing several stems from a large bunchgrass or shrub, it is easier to dry and mount a specimen. If plants are very large and bulky, collect a sample of the stem, leaf arrangement, root, and flower or seedhead.

11. Take several plastic bags with you when collecting plants. Put the plants in the bag with a few drops of water (do not overdo it), then seal the bag, and the specimens will stay fresh for several hours. They should be kept out of direct sunlight. If it isn't possible to press all the plants collected, most plants will stay fresh in the plastic bag if kept cool, such as in a refrigerator, for a day or two. However, put only one kind of plant in a bag and number the bag to match your field notebook.

12. Seeds and/or seed pods are very helpful in identifying many plants. A good way to include seeds is to place several seeds in a small, clear plastic, self-sealing envelope attached to the mount sheet. To prevent new infestations, it is also a good idea to carefully remove and burn all other seeds from any undesirable or weedy plant specimens.

Guidelines for Pressing Plants

The object is to quickly dry the plants under firm pressure to retain plant colors and the plant arrangement.

1. Press the plants as soon as possible after collecting. Once a plant wilts, it will not make an attractive mount.

2. Have your press ready to go before you remove a specimen from the plastic bag. Have plenty of newspaper pages folded lengthwise with about a quarter of the upper and lower edges folded toward the center. This will help keep your specimens from sliding out. A supply of corrugated cardboard sheets (cut to fit your press) are also needed. As you fill your press, alternate the cardboard sheets and folded paper (beginning and ending with a sheet of cardboard) to keep the specimens flat and speed the drying process. Although it is not necessary, blotter sheets can be placed between the newspaper and cardboard to speed the drying process.

3. Remove one plant at a time from the plastic bag. Check the plant closely to make sure that all soil is removed from the roots and remove excess moisture with a paper towel.

4. If the plant is less than 14 inches long, place it between the folded newspaper. Arrange the stems, leaves, roots, and flowers exactly as you want them to appear on the mount. Flowers should be pressed open. Both the upper and lower surfaces of flowers and leaves should be displayed.

5. If the plant is longer than 12 inches, it will be necessary to fold the plant in the shape of a V, N, or W. If the plant is still too large, press a sample of each plant part—stem, leaf, root, and flower or seedhead. For hard-to-handle plants, hold at the stem base firmly and slowly move the plant up and down against the newspaper a few times stopping with an upward stroke. (This will help separate and straighten out the branches and leaves.)

6. Hold the plant firmly in place and fold the upper and lower segments of the newspaper over the plant. While applying pressure to keep the plant in position, write the assigned plant number from your field notebook on the newspaper. Then place the plant into your press (a cardboard sheet should be below and above the folded newspaper).

7. Examine the plant after it has been pressed for 24 hours. This is your last opportunity to do some rearranging while the plant is still flexible. Be sure both upper and lower leaf surfaces show. Change the newspaper or blotter paper every day until the plant is thoroughly dry. Remember that succulent (fleshy) plants will take much longer to press.

8. Plants can be removed from the press in seven to ten days. Keep the plants in folded newspaper until you are ready to mount them.

Mounting Plants

After the plant specimens have been pressed and dried, they are ready to be mounted.

1. Herbarium sheets, standard (white) tag board, or poster board are recommended for mounting sheets. Although herbarium sheets usually have to be ordered through biological supply outlets, poster board can be purchased at most stores selling office and school supplies. If you use tag board, four mount sheets can be cut from one board if each sheet is cut at 11" × 14", or three sheets can be cut if each sheet is cut at 11½" × 16½".

2. Placement of specimen is easy if the plants have been pressed properly. The specimen should be placed upright with roots near the bottom and should provide a pleasing appearance. Leave room in the lower right-hand corner for a 3" × 5" mount label.

3. A transparent glue (Elmer's glue is best) is preferred to spot fasten the specimen to the mountsheet. You can also use small strips of gummed cloth. Scotch tape is not recommended. Small weights, such as lead casts, large nails, heavy washers, or large nuts, will hold the plant to the mount sheet while the glue is drying.

4. Each mount requires a label in the lower right-hand corner. The label must be properly filled out.

Table continues on next page

Table A-21 How to Collect, Press, and Mount Plants *(continued)*

Guidelines for Storing Plants

1. Mounted plants are usually stored in a cabinet or case to protect them from dust and insects. Although protective material is not required, some collectors (especially for 4-H projects) use a protective cover to protect the plant material as it becomes brittle. Use a 4–5 mil clear plastic mylar material and do not use Saran Wrap or 1–2 mil clear plastic. Also, your mounts should not be laminated with a clear seal plastic until a botanist has verified the specimen and signed the label.

2. Your plants should be filed in a logical order that makes it easy to find a specific specimen. By filing all specimens by family, then arranging the family members in alphabetical order by genus and species, it is easy to find a specific specimen.

3. It is usually a good idea to store a few mothballs with your plants to protect them from insects.

Table A-22 Sample Plant Label

Plant common name _____ **Scientific name** _____

Collection Site Information

Date collected _____ **State** _____ **Distance** _____ (miles)

and _____ (direction) from (nearest town/city) nearest landmark _____

Elevation _____ **Slope face** _____

Circle one for each item:

Topography: mountains, foothills, breaks, plains, riparian

Slope: nearly level, rolling, moderate, very steep

Abundance: abundant, occasional, very few

Tree overstory: yes, no

Collector _____

Plant number _____ **Verified by** _____

Table A-23 **Internet Resources for Plant Science**

Topic	URL
Horticulture	http://aggie-horticulture.tamu.edu/vegetable/vegetable.html
Extension Service Index	www.oneglobe.com/agriculture/extnsion.html
Ceres Online	www.ceresgroup.com/col
Commercial Floriculture Production Information	http://www.ces.ncsu.edu/depts/hort/floriculture/cfr/
Florida Electronic Data Information Source (EDIS)	http://edis.ifas.ufl.edu/
Food and Agricultural Organization (FAO) of the United Nations	www.fao.org
Horticulture and Crop Science at Ohio State University	http://hcs.osu.edu/
Horticulture Information Leaflets at North Carolina Extension Service	www.ces.ncsu.edu/depts/hort/hil/index.html
Internet Directory For Botany	http://www.botany.net/IDB/
Biology Lecture Online	http://www.biologie.uni-hamburg.de/b-online/library/botany.htm
Cornell Mann Library Gateway	www.mannlib.cornell.edu
New Crop Resources Online	www.hort.purdue.edu/newcrop/default.html
PLANTS National Database	http://plants.usda.gov/
National Agricultural Library	www.nal.usda.gov
Weekly Weather and Crop Bulletin	http://www.usda.gov/oce/weather/pubs/Weekly/Wwcb/
Economic Research Service (USDA)	http://www.ers.usda.gov/
National Agricultural Statistics Service (USDA)	http://www.nass.usda.gov/
Missouri Extension Publications Library	http://extension.missouri.edu/explore/agguides/
Ag-Links—Links for the Agriculture Industry	www.gennis.com/aglinks.html
Agricultural Research Service (USDA)	www.ars.usda.gov
United States Department of Agriculture (USDA)	www.usda.gov
Animal and Plant Health Inspection Service (APHIS)	www.aphis.usda.gov
Idaho One Plan	www.oneplan.state.id.us
Information Services for Agriculture	www.aginfo.com
Conservation Technology Information Center	www.ctic.purdue.edu/ctic.html
Foreign Agricultural Service	http://ffas.usda.gov

Table A-24 **Sedimentation Test of Soil Texture**

Description

The sedimentation test is an easy way to measure the percentage of sand, silt, and clay in a soil sample. This test is based on the fact that large, heavy particles will settle most rapidly in water, while small, light particles will settle most slowly. Calgon laundry powder is used to "dissolve" the soil aggregates and keep the individual particles separated.

Materials

- Soil sample
- One-quart fruit jar with lid
- Eight percent Calgon solution—mix six tablespoons of Calgon (a laundry powder available in stores) per quart of water
- Metric ruler
- Measuring cup
- Tablespoon

Procedure

1. Place about ½ cup of soil in the jar. Add 3½ cups of water and 5 tablespoons of the Calgon solution.
2. Cap the jar and shake for 5 minutes. Leave the jar on a counter and let settle for 24 hours.
3. After 24 hours, measure the depth of settled soil. All of the soil particles have settled, so this is the TOTAL DEPTH. Write it down and label it.
4. Shake for another 5 minutes. Let stand 40 seconds. This allows sand to settle out. Measure the depth of the settled soil and record as SAND DEPTH.
5. Do not shake again. Let the jar stand for another 30 minutes. Measure the depth, and subtract the sand depth to get the SILT DEPTH.
6. The remaining unsettled particles are clay. Calculate clay by subtracting silt and sand depth from total depth to get CLAY DEPTH.
7. Now calculate the percentage of each soil separate using these formulas:

   - Percent sand = (sand depth/total depth) × 100
   - Percent silt = (silt depth/total depth) × 100
   - Percent clay = (clay depth/total depth) × 100

Table A-25 Soil Horizon Symbol Suffixes *

Symbol	Meaning
a	Highly decomposed (sapric) organic material. This suffix is used with the O horizon.
b	Buried horizon. Such a soil layer is an old mineral horizon buried by sedimentation or other processes.
c	Concretions or hard nodules. A nodule or concretion is a hard "pocket" of soil cemented by iron or other substance.
d	Physical root restriction, either man-made or a naturally dense layer that roots can enter only through fractures.
e	Moderately decomposed (hemic) organic material. Used with the O horizon.
f	Frozen soil. A horizon, usually the C, that contains permanent ice (permafrost).
g	Strong graying. Such a horizon is gray and mottled, the color of reduced (nonoxidized) iron, resulting from saturated conditions.
h	Alluvial accumulation of organic matter. The symbol is used with the B horizon to show that complexes of humus and sesquioxides have washed into the horizon. Includes only small quantities of sesquioxides. May show dark staining.
i	Slightly decomposed (fibric) organic matter. Used with the O horizon.
k	Accumulation of carbonates (CO_3^-). Indicates accumulation of calcium carbonate (lime) or other carbonates in the B or C horizons.
m	Cementation. The symbol indicates a soil horizon that has been cemented hard by carbonates, gypsum, or other material. A second suffix indicates the cementing agent, such as "k" for carbonates. This is a hardpan horizon; roots penetrate only through cracks.
n	Accumulation of sodium. Indicates a high accumulation of exchangeable sodium, as in a sodic soil.
o	Accumulation of sesquioxide clays after intense weathering.
p	Plowing or other human disturbance. Horizon was heavily disturbed by plowing, cultivation, pasturing, or other activity. Applies to O and A horizons.
q	Accumulation of silica (SiO_2).
r	Weathered or soft bedrock. Used with C horizon to indicate bedrock that can be dug with a spade or that roots can enter through cracks.
s	Alluvial accumulation of both sesquioxides and organic matter. Both the organic matter and sesquioxide components of humus-sesquioxide complexes are important.
t	Accumulation of silicate clays. Clay may have formed in horizon or moved into it by alleviations, usually as coatings on ped faces. Mostly used in B horizon, sometimes in C.
v	Plinthite, an iron-rich, humus-poor material common to tropical soils that hardens when exposed to air. B and C horizons.
w	Development of color or structure. The symbol indicates that a B horizon has developed enough to show some color or structure but not enough to show alluvial accumulation of material.
x	Fragipan or other noncemented natural hardpans. These are horizons that are firm, brittle, or have high bulk densities from natural processes.
y	Accumulation of gypsum ($CaSO_4$) in B or C.
z	Accumulation of salts more soluble than gypsum in B or C.

* These lowercase letters are used as suffixes to label certain types of master horizons. One or more suffix may follow a master horizon designation such as Bky, which indicates a B horizon that has accumulated both carbonates (k) and gypsum (y).

Table A-26: Textual Class Groupings

Group	Textural Classes
Fine	sandy clay, silty clay, clay
Moderately fine	sandy clay loam, silty clay loam, clay loam
Medium	silt, silt loam, loam
Moderately coarse	sandy loam
Coarse	sand, loamy sand

Table A-27: Land Evaluation

Land may be evaluated for a number of uses, such as suitability for row crops, home landscapes, or building sites. Such evaluations might be done by soil scientists, land planners, landscape designers, or even students engaged in a soil-judging contest. While land evaluation purposes and methods vary between regions of the country, this table provides some simple and general soil features to use when evaluating a soil.

Definition of Terms	Criteria
Suitability	Land can be said to be good, fair, or poor for any use. The same terms can also be applied to a particular feature of the land, such as slope or soil texture. Looking at the chart that follows, for each land use there is a series of criteria listed on the same line to the right. Each can be rated, and the "total" used to rate the land: • *Good land* is well suited to the use being considered. There are no serious limitations or hazards. Ratings for each criteria are good, or perhaps a couple are fair. • *Fair land* is suitable for the use if a few limitations are corrected. Several of the features are rated as fair. • *Poor land* is unsuitable for the use, or corrections are too expensive to be practical. Even one or two severe limitations may be enough to gain this rating, or a large number of "fair" descriptions.
Slope	Slope is expressed as a percent. For instance, a slope of 2 percent suggests a 2-foot change of elevation over a 100-foot horizontal distance. Changing slope would involve expensive earth moving.
Soil Texture	Table A-26 provides textual class groupings. Except for small areas, such as a garden, texture cannot be changed practically.
Flooding	Flooding concerns how often or how long the land is actually covered with water during the season. Such flooding could be due to stream flooding, heavy rains, or snowmelt. Flooding may be able to be corrected, in some small areas, by earth moving to change drainage patterns. In many cases, only large-scale projects such as the construction of levees can solve these problems.
Internal Drainage	Many cases of poor drainage can be corrected by installing drainage systems, and excessive drainage may be improved by irrigation.
Depth to Restrictive Layer	Restrictive layers can interfere with rooting depth and plant anchorage, roadbeds, foundations, and home drain fields. Such layers could be bedrock, soilpans, water tables, or others. Some can be corrected, like ripping a soilpan, while bedrock cannot be altered easily.
Available Water for Plants	The available water is here rated as the number of inches of water held in the top 5 feet of soil. Since most soils will contain layers, one will have to add the capacity of each layer down to 5 feet. A capacity greater than 9 inches for 5 feet is good, 6 to 9 is fair, and less than 6 is poor. Poor capacity can be alleviated by irrigation, where practical and acceptable.

Table continues on next page

Table A-27: Land Evaluation (*Continued*)

Definition of Terms	Criteria
Erodability	Erodability is related to slope, texture, and other factors. Classes are as follows: • Slight erodability means that under average conditions there is little chance of excessive erosion. Slopes are gentle and short, internal drainage is good. Wind erosion is unlikely. • Moderate erodability could occur on medium- or fine-textured soil on gentle slopes over 300 feet long, or shorter moderate slopes. Wind erosion is possible due to texture or lack of cover. • Severe erodability occurs on slopes over 12 percent, or areas that experience teaches are quite erodable.
Other Features	Other features may also limit a soil for certain uses. Examples include stoniness, soil pH, soil fertility, or salinity. Some of these can be improved easily, like liming agricultural soil to raise pH. On the other hand, improving soil fertility is more difficult for forest uses.
Site Inspection	Much of the information needed for this evaluation can be obtained from published soil surveys. As a lab exercise for students, a soil pit should be dug to allow examination of the soil profile. Select an area of uniform character, preferably a single mapping unit on a soil map. Dig a pit measuring 3 feet by 3 feet at the surface, and about 3½ feet deep. Orient the pit so that the sun shines on the side to be observed. That side should be vertical; the other side can be sloping to save digging. Some of the information needed can be discovered from examining the pit. Other information, like a slope, can be obtained from the surroundings. The instructor will need to provide such information as flooding.

Table A-28: Guidelines for Choosing a Soil-Testing Laboratory*

Test Methods	The use of appropriate test methods is very important in order to accurately determine the concentrations of plant-available nutrients in the soil. Research at many land-grant universities over many decades has resulted in soil-testing methods that are specific for soils in particular regions of the United States. For example, methods developed for testing the predominant soils in the Southern region of the United States may not be applicable for soils in the North Central region or the Northwest region.
Laboratory Proficiency	The proficiency of a laboratory refers to its ability to produce accurate and precise test results—a difficult assessment for laboratories to independently determine. Thus, regional soil-testing research committees and other organizations established the North American Proficiency Testing (NAPT) program in 1998. This program is backed by the professional scientific organization, the Soil Science Society of America. A main purpose of the NAPT is to provide "double-blind" check samples to laboratories who wish to monitor and improve the quality of their soil-testing data. NAPT not only provides the check samples but also collects and statistically analyzes the data from laboratories in the program. Participating laboratories receive a summary of their performance for each soil-test method. Continued self-evaluation and adjustment improves the integrity of the soil-test results.
Laboratory's NAPT Results	A representative of the laboratory should review with the potential client their NAPT quarterly test results with those summarized for all NAPT participating laboratories. Information for each test parameter of interest to the client should be included. Growers should ask for this comparison in order to make a good decision about a laboratory.
Other Customers	A potential client should ask the laboratory to provide the names and telephone numbers of 10 customers. This allows growers to evaluate the laboratory from the perspective of users like themselves.
Units of Results	Potential clients ask a laboratory representative what units are used for each test parameter. Some laboratories use lbs/acre, ppm, or lbs/1,000 square feet. If results from different labs are compared, make sure the units associated with the results are the same. For a valid comparison, a simple conversion may be necessary. For example, to convert ppm to lbs/acre, multiply the ppm value by 2. Certain test parameters may have unfamiliar units, such as meq/100 g for cation exchange capacity. Clients should ask the laboratory representative to explain the meaning of the units if they are unclear.

Table continues on next page

Table A-28: **Guidelines for Choosing a Soil-Testing Laboratory*** *(continued)*

Categories of Quantity	Some laboratories may place test results into categories. Examples are the categories of low, medium, and high. There may be additional categories or different categories than these. These categories usually denote a range of test values. Likely the categories given by one laboratory do not represent the same nutrient concentrations for another laboratory. Clients should ask the laboratory to define each range that is used. In addition, a client should find out if the categories are crop-dependent or calibrated for specific soil conditions (e.g., soil types). That is, results that may be regarded medium for one crop may be considered low for another crop.
Lime and Fertilizer Recommendations	Potential clients need to determine if the soil-testing laboratory provides recommendations for the application of lime and fertilizer for the crops of interest. Also clients ask about the basis for lime and fertilizer recommendations that are used for other crops. Are they calibrated for the client's specific soil types or growing conditions?
Turn-Around Time	How long does it take the laboratory to do the routine soil testing and return the results? In order for the results and recommendations to be useful, the turn-around time must be as short as possible. A good laboratory should be able to provide the results in two to three working days for the routine soil tests of pH, lime requirement, phosphorus, potassium, calcium, and magnesium. The Internet can be a useful system to obtain test results rapidly. Some laboratories can provide the results on the Internet. E-mail between the laboratory and grower allows direct and rapid communication.
Visiting the Laboratory	A visit to the soil-testing laboratory before submitting samples can be helpful. A representative of the laboratory should not hesitate to show a potential client the testing area. During the visit, a potential client can observe the orderliness and cleanliness of the work area.
Reference Check Samples	Laboratories should routinely use internal "blind" and "double-blind" check samples. A "blind" check sample is one that the laboratory technician knows is a check sample and is aware of the range of acceptable values for the parameters being tested. The technician uses this kind of check sample to make sure the method and instrument are performing normally. A "double-blind" check sample is one that the laboratory technician does not know is an internal check sample. In this case, the laboratory manager evaluates the data and determines if the test results produced are in the acceptable range.
Charting Quality Control	A testing laboratory should continuously evaluate its quality by charting its check soil-sample results over time. This allows for measurement and assessment of the variation over time. Ideally, quality-control charts need to be used for each test parameter.
Sample Information and Test Result Forms	Before selecting a laboratory, clients need to ask the laboratory for examples of the information form and the final test result form. An explanation of these forms should include anything that is unclear, how many samples can be represented on each form.
Test Kits	Most soil-testing laboratories supply test kits for their customers. At a minimum, the test kits should contain the sample information form and soil sample container.
Production Professionals	Professionals in the laboratory should be trained in agronomy, horticulture, or soil science.
Laboratory Test Prices	Prices for soil testing often vary greatly from one laboratory to the next. Discounts may be given for large numbers of samples and prices could be negotiable.
Other Testing Services	Other services the laboratory offers could include plant-tissue analysis. This tool can be very useful along with soil testing to monitor the nutrient status of the soil or to isolate problem fertility situations in the field. Some laboratories provide sample collection as an optional service.

* Adapted from an Ohio State University Extension Factsheet

Appendix B

TRACTOR AND LARGE EQUIPMENT STORAGE GUIDELINES

Short-Term Storage

- Do not clean equipment with engine running.
- Keep the equipment and supply of fuel in locked storage and remove the ignition key to prevent children or others from playing or tampering with them.
- To avoid sparks from an accidental short circuit, always disconnect the battery's ground cable (negative terminal) first and reconnect it last.
- Do not store the equipment with fuel in the tank inside a building where fumes may ignite. Allow the engine to cool before storing.
- To avoid the danger of exhaust fume poisoning, do not operate the engine in a closed building without adequate ventilation.
- To reduce fire hazards, clean the equipment thoroughly before storage. Dry grass and leaves around the engine and mufflers may ignite.

Long-Term Storage

- Check for loose bolts and nuts, and tighten if necessary.
- Apply grease to equipment areas where bare metal will rust. Also apply grease to pivot areas.
- Inflate the tires to a pressure a little higher than usual.
- Change the engine oil and run the engine to circulate oil throughout the engine block and internal moving parts.
- Store the equipment indoors in a dry area that is protected from sunlight, rain, and excessive heat. If the equipment must be stored outdoors, cover it with a waterproof tarpaulin.
- Jack the equipment up and place blocks under the front and rear axles so that all four tires are off the ground. Keep the tires out of direct sunlight and extreme heat.

HAND TOOLS AND SMALL POWER EQUIPMENT STORAGE GUIDELINES

- Store equipment and fuel in dry, ventilated area. Storage spaces within a garage area, a storage building, or a good shed are all workable options.
- Remove batteries and fully charge them before storage.
- Clean all dirt, grass, and debris from engines.
- Disconnect spark plug wire(s) or remove spark plugs based on manufacturer's recommendations.
- Inspect power equipment—belts, blades, etc.—and make any needed repairs.
- Drain all gasoline from tanks into a container for disposal.
- Do not blend gasoline, gasohol, or alcohol with diesel fuel, as this creates potential for an explosive hazard.
- Do not store equipment in the same location as fertilizers or swimming pool chemicals. These chemicals are extremely corrosive to metal parts on equipment.
- Start and run engines until they quit to be sure the carburetor is dry to prevent diaphragms from sticking together.
- Before storing hand tools, be sure to clean them and apply a light coat of oil to all metal surfaces to prevent rust from forming.
- For hand tools, check to see if any repairs are needed.

EQUIPMENT MAINTENANCE

All equipment requires preventive maintenance to ensure that it is in proper operating condition. Preventive maintenance lengthens the useful life span of machinery and results in reduced equipment repair and replacement costs. The manager, mechanic, or equipment operator must keep records of when maintenance services are performed on a machine and how many hours the machine is in use. The equipment manual contains a preventive maintenance schedule based on the number of hours of equipment operation. The person responsible for equipment maintenance should rigidly adhere to this schedule.

The list of maintenance services necessary after 50, 100, 300, and 600 hours of equipment use is fairly typical and will help the reader to understand the importance of equipment maintenance records. However, the manager should consult the manual for each piece of equipment under his or her supervision to determine exact maintenance schedule specifications. The oil level must always be checked before a machine is used.

Generalized Preventive Maintenance Schedule

At 50 hours

- Check tire pressure and water level in battery.
- Clean air filter (wash or replace if necessary).
- Remove corrosion from battery terminal connections.
- Check oil level in transmission and hydraulic system.
- Grease lubrication points.

At 100 hours

- Clean engine.
- Drain oil and refill engine crankcase.
- Lubricate clutch and throttle linkage.
- Tighten loose screws and nuts.

At 300 hours (or every six months)

- Inspect and clean spark plugs.
- Check ignition point gap.
- Change hydraulic oil filters.
- Check fan and drive belts.

At 600 hours (or yearly)

- Change air cleaner and spark plugs.
- Touch up with paint.

Hand Tool Maintenance

Remove rust from metal tools and apply rust-inhibiting materials to metal surfaces; repair split or broken wood handles; reshape the heads of driving or driven tools; sharpen blades or cutting tools. Pruners and shears sharpening frequency depends on the usage and the hardness of the steel. Saw sharpening should be done by a professional service and involves the tooth sharpness and tooth set. An indication of the need to have a saw sharpened is being able to run your fingers up and down the saw teeth without getting cut. If the saw binds and becomes stuck in the groove, the teeth may need to be set.

Appendix C

PLANT AND CUT FLOWER STORAGE

Growers produce crops to sell in the market at a price that will give the grower a profit on the investment made. The crops are a very perishable product. They are fragile and easily bruised by mishandling. Plants have a relatively short life compared to vegetable crops such as potatoes. They cannot be canned or frozen and then reprocessed for use later.

The majority of cut flowers and potted plans are offered for sale as fresh products. As such they have a very short useful life. It is estimated that the loss in fresh crops as a result of improper handling is equal to 20 percent of the total farm value of the crop. Any steps that can be taken to reduce this waste will mean a more satisfied consumer and a lower dollar loss to the industry.

PLANT NURSERY PLANTS

Plants set directly from the nursery into the field without wilting are easy to establish but tend to be more vegetative than plants that have been in storage for a few days. Do not set plants directly from the nursery early in the season since growth can be excessive and fruiting may be delayed, especially when fall and early weather is too warm to provide sufficient plant chilling. There is an interaction of photoperiod and temperature in flower initiation. The higher the temperature the shorter the day length required for flower initiation and vice versa. The effect varies with the cultivar. Remove excess dead and diseased foliage before transplanting, and inspect plants for nematodes or diseases. When nematodes are present, plants should not be harvested within 4 to 5 feet of an infected area. If a very light infection of anthracnose is present (a 2-foot diameter area), harvesting of plants 20 feet away from the infection has been satisfactory. However, if anthracnose is found anywhere in a nursery, there is always risk involved in setting plants in the fruiting field. The quality of the transplant determines to a large extent its fruiting capability. Therefore, use transplants of sufficient size and which have not been improperly treated prior to transplanting and are anthracnose- and nematode-free. To reduce the risk of catastrophic loss from disease, weather, or unknown plant defects, growers usually plant more than one cultivar and obtain plants from more than one source. An additional benefit is that the harvest cycles of the two or more cultivars are usually different. This will tend to even out the harvest load.

To obtain maximum plant life, the grower must ensure that certain cultural and storage practices are carried out. These practices include the following:

- Fertilizing for quality yield and growth.
- Supplying adequate water so that the plants are not under stress because of lack of water. Automatic watering systems are an excellent investment in meeting the moisture needs of crops.
- Maintaining accurate temperature control at all stages of growth.
- Keeping the greenhouse glass clean at all times so that the plants can get the maximum light available.
- Controlling the environmental and cultural factors to produce flowers and plants with a high carbohydrate content.
- Harvesting and shipping the flowers and plants at the proper stage of development so that the customer receives the maximum value from them.

Appendix D

Tools of the Trade

Table D-1 Specialized Hand Tools and Their Functions

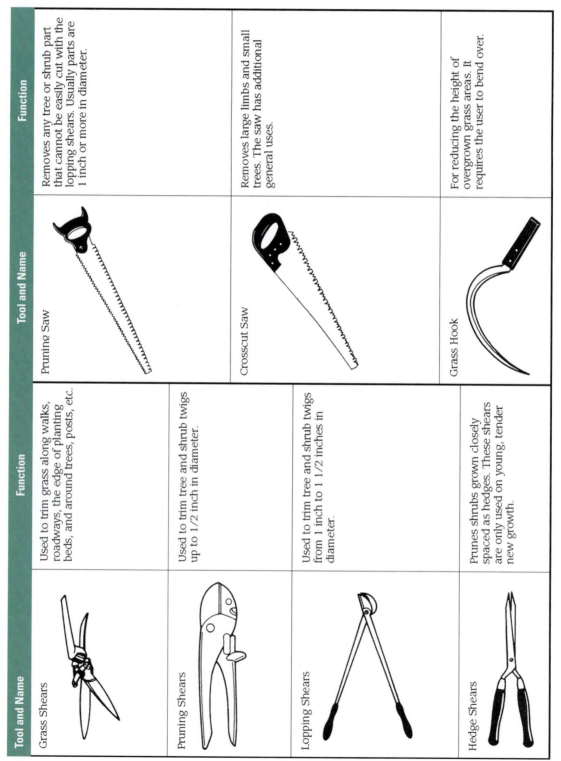

Tool and Name	Function
Grass Shears	Used to trim grass along walks, roadways, the edge of planting beds, and around trees, posts, etc.
Pruning Shears	Used to trim tree and shrub twigs up to 1/2 inch in diameter.
Lopping Shears	Used to trim tree and shrub twigs from 1 inch to 1 1/2 inches in diameter.
Hedge Shears	Prunes shrubs grown closely spaced as hedges. These shears are only used on young, tender new growth.
Pruning Saw	Removes any tree or shrub part that cannot be easily cut with the lopping shears. Usually parts are 1 inch or more in diameter.
Crosscut Saw	Removes large limbs and small trees. The saw has additional general uses.
Grass Hook	For reducing the height of overgrown grass areas. It requires the user to bend over.

Table D-1 Specialized Hand Tools and Their Functions *(continued)*

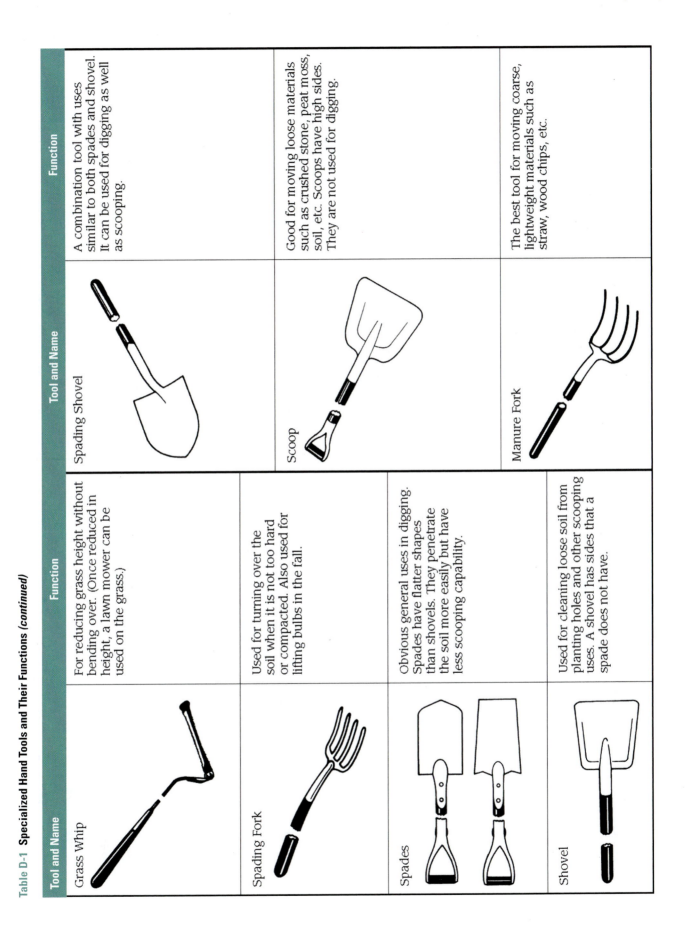

Tool and Name	Function	Tool and Name	Function
Grass Whip	For reducing grass height without bending over. (Once reduced in height, a lawn mower can be used on the grass.)	Spading Shovel	A combination tool with uses similar to both spades and shovel. It can be used for digging as well as scooping.
Spading Fork	Used for turning over the soil when it is not too hard or compacted. Also used for lifting bulbs in the fall.	Scoop	Good for moving loose materials such as crushed stone, peat moss, soil, etc. Scoops have high sides. They are not used for digging.
Spades	Obvious general uses in digging. Spades have flatter shapes than shovels. They penetrate the soil more easily but have less scooping capability.	Manure Fork	The best tool for moving coarse, lightweight materials such as straw, wood chips, etc.
Shovel	Used for cleaning loose soil from planting holes and other scooping uses. A shovel has sides that a spade does not have.		

Table D-1 Specialized Hand Tools and Their Functions *(continued)*

Tool and Name	Function	Tool and Name	Function
Single-Bit and Double-Bit Axes 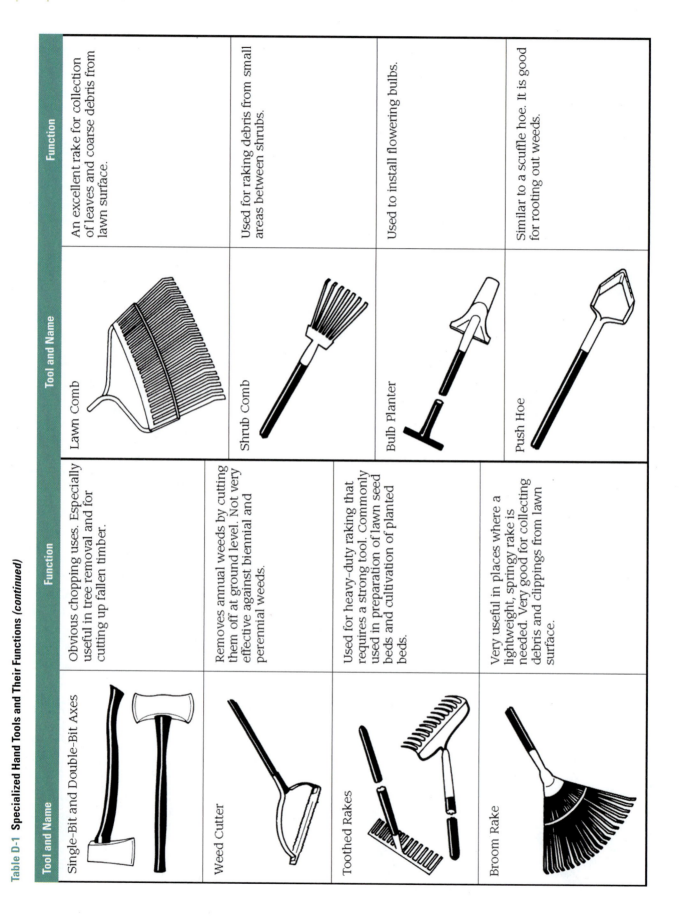	Obvious chopping uses. Especially useful in tree removal and for cutting up fallen timber.	Lawn Comb	An excellent rake for collection of leaves and coarse debris from lawn surface.
Weed Cutter	Removes annual weeds by cutting them off at ground level. Not very effective against biennial and perennial weeds.	Shrub Comb	Used for raking debris from small areas between shrubs.
Toothed Rakes	Used for heavy-duty raking that requires a strong tool. Commonly used in preparation of lawn seed beds and cultivation of planted beds.	Bulb Planter	Used to install flowering bulbs.
Broom Rake	Very useful in places where a lightweight, springy rake is needed. Very good for collecting debris and clippings from lawn surface.	Push Hoe	Similar to a scuffle hoe. It is good for rooting out weeds.

Table D-1 Specialized Hand Tools and Their Functions *(continued)*

Tool and Name	Function	Tool and Name	Function
Pick 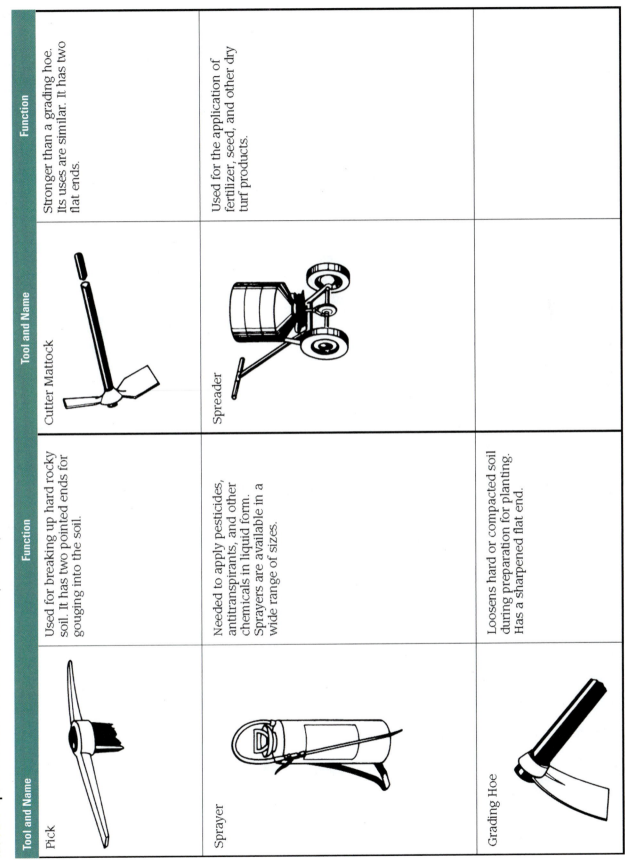	Used for breaking up hard rocky soil. It has two pointed ends for gouging into the soil.	Cutter Mattock	Stronger than a grading hoe. Its uses are similar. It has two flat ends.
Sprayer	Needed to apply pesticides, antitranspirants, and other chemicals in liquid form. Sprayers are available in a wide range of sizes.	Spreader	Used for the application of fertilizer, seed, and other dry turf products.
Grading Hoe	Loosens hard or compacted soil during preparation for planting. Has a sharpened flat end.		

Table D-1 Specialized Hand Tools and Their Functions *(continued)*

Tool and Name	Function	Tool and Name	Function
Walk Behind Mower 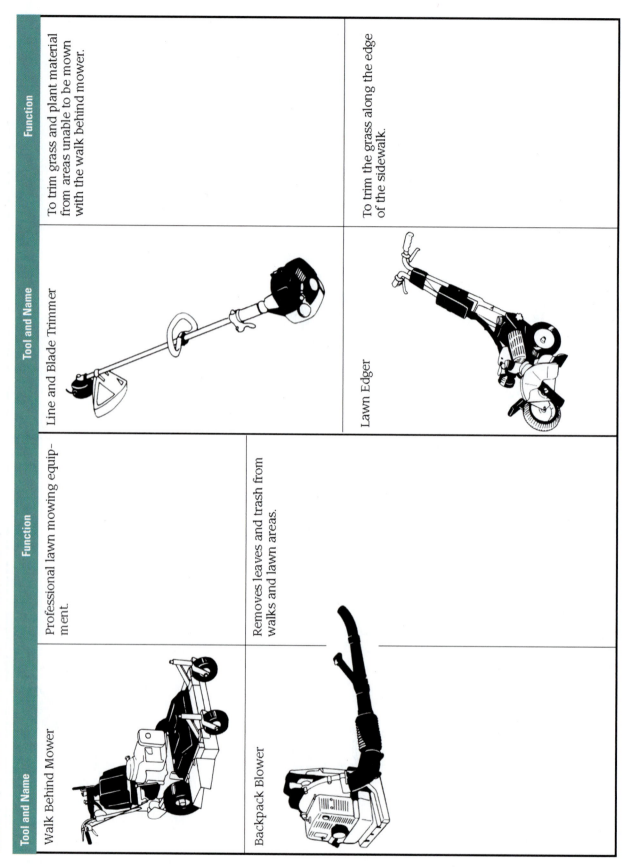	Professional lawn mowing equipment.	Line and Blade Trimmer	To trim grass and plant material from areas unable to be mown with the walk behind mower.
Backpack Blower	Removes leaves and trash from walks and lawn areas.	Lawn Edger	To trim the grass along the edge of the sidewalk.

Index

Page references for illustrations or tables are in **boldface**.